GLENCOE
MATHEMATICS

Mathematics

Applications and Concepts

Course 1

Bailey
Day
Frey
Howard
Hutchens
McClain

Moore-Harris
Ott
Pelfrey
Price
Vielhaber
Willard

McGraw Hill
Glencoe

New York, New York   Columbus, Ohio   Chicago, Illinois   Peoria, Illinois   Woodland Hills, California

The Standardized Test Practice features in this book were aligned and verified
by The Princeton Review, the nation's leader in test preparation. Through its
association with McGraw-Hill, The Princeton Review offers the best way to
help students excel on standardized assessments.

The Princeton Review is not affiliated with Princeton University or Educational Testing Service.

The USA TODAY® service mark, USA TODAY Snapshots® trademark and other
content from USA TODAY® has been licensed by USA TODAY® for use for certain
purposes by Glencoe/McGraw-Hill, a Division of The McGraw-Hill Companies, Inc.
The USA TODAY Snapshots® and the USA TODAY® articles, charts, and photographs
incorporated herein are solely for private, personal, and noncommerical use.

Microsoft® Excel® is a registered trademark of Microsoft Corporation in the United
States and other countries.

Send all inquiries to:
Glencoe/McGraw-Hill
8787 Orion Place
Columbus, OH 43240-4027

ISBN: 0-07-829631-5

7 8 9 10 043/027 10 09 08 07 06 05

Contents in Brief

Authors

Rhonda Bailey
Mathematics Consultant
Mathematics by Design
DeSoto, Texas

Roger Day, Ph.D.
Associate Professor
Illinois State University
Normal, Illinois

Patricia Frey
Director of Staffing and
    Retention
Buffalo City Schools
Buffalo, New York

Arthur C. Howard
Mathematics Teacher
Houston Christian High
    School
Houston, Texas

**Deborah T. Hutchens,
    Ed.D.**
Assistant Principal
Great Bridge Middle School
Chesapeake, Virginia

Kay McClain, Ed.D.
Assistant Professor
Vanderbilt University
Nashville, Tennessee

Beatrice Moore-Harris
Mathematics Consultant
League City, Texas

Jack M. Ott, Ph.D.
Distinguished Professor of
 Secondary Education
 Emeritus
University of South Carolina
Columbia, South Carolina

Ronald Pelfrey, Ed.D.
Mathematics Specialist
Appalachian Rural Systemic
 Initiative
Lexington, Kentucky

Jack Price, Ed.D.
Professor Emeritus
California State Polytechnic
 University
Pomona, California

Kathleen Vielhaber
Mathematics Specialist
Parkway School District
St. Louis, Missouri

Teri Willard, Ed.D.
Assistant Professor of
 Mathematics Education
Central Washington
 University
Ellensburg, Washington

Contributing Authors

USA TODAY
The USA TODAY Snapshots®, created
by USA TODAY®, help students make the
connection between real life and mathematics.

Dinah Zike
Educational Consultant
Dinah-Might Activities, Inc.
San Antonio, Texas

Content Consultants

Each of the Content Consultants reviewed every chapter and gave suggestions for improving the effectiveness of the mathematics instruction.

Mathematics Consultants

L. Harvey Almarode
Curriculum Supervisor, Mathematics K–12
Augusta County Public Schools
Fishersville, VA

Claudia Carter, MA, NBCT
Mathematics Teacher
Mississippi School for Mathematics and Science
Columbus, MS

Carol E. Malloy, Ph.D.
Associate Professor, Curriculum Instruction,
    Secondary Mathematics
The University of North Carolina at Chapel Hill
Chapel Hill, NC

Melissa McClure, Ph.D.
Mathematics Instructor
University of Phoenix On-Line
Fort Worth, TX

Robyn R. Silbey
School-Based Mathematics Specialist
Montgomery County Public Schools
Rockville, MD

Leon L. "Butch" Sloan, Ed.D.
Secondary Mathematics Coordinator
Garland ISD
Garland, TX

Barbara Smith
Mathematics Instructor
Delaware County Community College
Media, PA

Reading Consultant

Lynn T. Havens
Director
Project CRISS
Kalispell, MT

ELL Consultants

Idania Dorta
Mathematics Educational Specialist
Miami–Dade County Public Schools
Miami, FL

Frank de Varona, Ed.S.
Visiting Associate Professor
Florida International University
    College of Education
Miami, FL

Teacher Reviewers

Each Teacher Reviewer reviewed at least two chapters of the Student Edition, giving feedback and suggestions for improving the effectiveness of the mathematics instruction.

Royallee Allen
Teacher, Math Department Head
Eisenhower Middle School
San Antonio, TX

Dennis Baker
Mathematics Department Chair
Desert Shadows Middle School
Scottsdale, AZ

Rosie L. Barnes
Teacher
Fairway Middle School–KISD
Killeen, TX

Charlie Bialowas
Math Curriculum Specialist
Anaheim Union High School District
Anaheim, CA

Stephanie R. Boudreaux
Teacher
Fontainebleau Jr. High School
Mandeville, LA

Dianne G. Bounds
Teacher
Nettleton Junior High School
Jonesboro, AR

Susan Peavy Brooks
Math Teacher
Louis Pizitz Middle School
Vestavia Hills, AL

Karen Sykes Brown
Mathematics Educator
Riverview Middle School
Grundy, VA

Kay E. Brown
Teacher, 7th Grade
North Johnston Middle School
Micro, NC

Renee Burgdorf
Middle Grades Math Teacher
Morgan Co. Middle
Madison, GA

Kelley Summers Calloway
Teacher
Baldwin Middle School
Montgomery, AL

Carolyn M. Catto
Teacher
Harney Middle School
Las Vegas, NV

Claudia M. Cazanas
Math Department Chair
Fairmont Junior High
Pasadena, TX

David J. Chamberlain
Secondary Math Resource Teacher
Capistrano Unified School District
San Juan Capistrano, CA

David M. Chioda
Supervisor Math/Science
Marlboro Township Public Schools
Marlboro, NJ

Carrie Coate
7th Grade Math Teacher
Spanish Fort School
Spanish Fort, AL

Toinette Thomas Coleman
Secondary Mathematics Teacher
Caddo Middle Career & Technology
    School
Shreveport, LA

Linda M. Cordes
Math Department Chairperson
Paul Robeson Middle School
Kansas City, MO

Polly Crabtree
Teacher
Hendersonville Middle School
Hendersonville, NC

Dr. Michael T. Crane
Chairman Mathematics
B.M.C. Durfee High School
Fall River, MA

Tricia Creech, Ph.D.
Curriculum Facilitator
Southeast Guilford Middle School
Greensboro, NC

Lyn Crowell
Math Department Chair
Chisholm Trail Middle School
Round Rock, TX

B. Cummins
Teacher
Crestdale Middle School
Matthews, NC

Debbie Davis
8th Grade Math Teacher
Max Bruner, Jr. Middle School
Ft. Walton Beach, FL

Diane Yendell Day
Math Teacher
Moore Square Museums Magnet
    Middle School
Raleigh, NC

Wendysue Dodrill
Teacher
Barboursville Middle School
Barboursville, WV

Judith F. Duke
Math Teacher
Cranford Burns Middle School
Mobile, AL

Carol Fatta
Math/Computer Instructor
Chester Jr. Sr. M.S.
Chester, NY

Cynthia Fielder
Mathematics Consultant
Atlanta, GA

Georganne Fitzgerald
Mathematics Chair
Crittenden Middle School
Mt. View, CA

Jason M. Fountain
7th Grade Mathematics Teacher
Bay Minette Middle School
Bay Minette, AL

Sandra Gavin
Teacher
Highland Junior High School
Cowiche, WA

Ronald Gohn
8th Grade Mathematics
Dover Intermediate School
Dover, PA

Larry J. Gonzales
Math Department Chairperson
Desert Ridge Middle School
Albuquerque, NM

Shirley Gonzales
Math Teacher
Desert Ridge Middle School
Albuquerque, NM

Paul N. Hartley, Jr.
Mathematics Instructor
Loudoun County Public Schools
Leesburg, VA

Deborah L. Hewitt
Math Teacher
Chester High School
Chester, NY

Steven J. Huesch
Mathematics Teacher/Department
    Chair
Cortney Jr. High
Las Vegas, NV

Sherry Jarvis
8th Grade Math/Algebra 1 Teacher
Flat Rock Middle School
East Flat Rock, NC

Teacher Reviewers *continued*

Mary H. Jones
Math Curriculum Coordinator
Grand Rapids Public Schools
Grand Rapids, MI

Vincent D.R. Kole
Math Teacher
Eisenhower Middle School
Albuquerque, NM

Ladine Kunnanz
Middle School Math Teacher
Sequoyah Middle School
Edmond, OK

Barbara B. Larson
Math Teacher/Department Head
Andersen Middle School
Omaha, NE

Judith Lecocq
7th Grade Teacher
Murphysboro Middle School
Murphysboro, IL

Paula C. Lichiello
7th Grade Math and Pre-Algebra
    Teacher
Forest Middle School
Forest, VA

Michelle Mercier Maher
Teacher
Glasgow Middle School
Baton Rouge, LA

Jeri Manthei
Math Teacher
Millard North Middle School
Omaha, NE

Albert H. Mauthe, Ed.D.
Supervisor of Mathematics (Retired)
Norristown Area School District
Norristown, PA

Karen M. McClellan
Teacher & Math Department Chair
Harper Park Middle
Leesburg, VA

Ken Montgomery
Mathematics Teacher
Tri-Cities High School
East Point, GA

Helen M. O'Connor
Secondary Math Specialist
Harrison School District Two
Colorado Springs, CO

Cindy Ostrander
8th Grade Math Teacher
Edwardsville Middle School
Edwardsville, IL

Michael H. Perlin
8th Grade Mathematics Teacher
John Jay Middle School
Cross River, NY

Denise Pico
Mathematics Teacher
Jack Lund Schofield Middle School
Las Vegas, NV

Ann C. Raymond
Teacher
Oak Ave. Intermediate School
Temple City, CA

M.J. Richards
Middle School Math Teacher
Davis Middle School
Dublin, OH

Linda Lou Rohleder
Math Teacher, Grades 7 & 8
Jasper Middle School
Jasper, IN

Dana Schaefer
Pre-Algebra & Algebra I Teacher
Coachman Fundamental Middle
    School
Clearwater, FL

Donald W. Scheuer, Jr.
Coordinator of Mathematics
Abington School District
Abington, PA

Angela Hardee Slate
Teacher, 7th Grade Math/Algebra
Martin Middle School
Raleigh, NC

Mary Ferrington Soto
7th Grade Math
Calhoun Middle School-Ouachita
    Parish Schools
Calhoun, LA

Diane Stilwell
Mathematics Teacher/Technology
    Coordinator
South Middle School
Morgantown, WV

Pamela Ann Summers
K–12 Mathematics Coordinator
Lubbock ISD–Central Office
Lubbock, TX

Marnita L. Taylor
Mathematics Teacher/Department
    Chairperson
Tolleston Middle School
Gary, IN

Susan Troutman
Teacher
Dulles Middle School
Sugar Land, TX

Barbara C. VanDenBerg
Math Coordinator, K–8
Clifton Board of Education
Clifton, NJ

Mollie VanVeckhoven-Boeving
7th Grade Math and Algebra Teacher
White Hall Jr. High School
White Hall, AR

Mary A. Voss
7th Grade Math Teacher
Andersen Middle School
Omaha, NE

Christine Waddell
Teacher Specialist
Jordan School District
Sandy, UT

E. Jean Ware
Supervisor
Caddo Parish School Board
Shreveport, LA

Karen Y. Watts
9th Grade Math Teacher
Douglas High School
Douglas, AL

Lu Wiggs
Supervisor
I.S. 195
New York, NY

Teacher Advisory Board and Field Test Schools

Teacher Advisory Board

Glencoe/McGraw-Hill wishes to thank the following teachers for their feedback on *Mathematics: Applications and Concepts*. They were instrumental in providing valuable input toward the development of this program.

Katie Davidson
Legg Middle School
Coldwater, MI

Lynanne Gabriel
Bradley Middle School
Huntersville, NC

Kathleen M. Johnson
New Albany-Plain Local Middle School
New Albany, OH

Ronald C. Myer
Indian Springs Middle School
Columbia City, IN

Mike Perlin
John Jay Middle School
Cross River, NY

Reema Rahaman
Brentwood Middle School
Brentwood, MO

Diane T. Scheuber
Elizabeth Middle School
Elizabeth, CO

Deborah Sykora
Hubert H. Humphrey Middle School
Bolingbrook, IL

DeLynn Woodside
Roosevelt Middle School,
Oklahoma City Public Schools
Oklahoma City, OK

Field Test Schools

Glencoe/McGraw-Hill wishes to thank the following schools that field-tested pre-publication manuscript during the 2002–2003 school year. They were instrumental in providing feedback and verifying the effectiveness of this program.

Knox Community Middle School
Knox, IN

Roosevelt Middle School
Oklahoma City, OK

Brentwood Middle School
Brentwood, MO

Elizabeth Middle School
Elizabeth, CO

Legg Middle School
Coldwater, MI

Great Hollow Middle School
Nesconset, NY

Student Advisory Board

The Student Advisory Board gave the authors, editorial staff, and design team feedback on the lesson design, content, and covers of the Student Editions. We thank these students for their hard work and creative suggestions in making *Mathematics: Applications and Concepts* more student friendly.

Front Row: Joey Snyder, Tiffany Pickenpaugh, Craig Hammerstein

Back Row: Brittany Yokum, Alex Johnson, Cimeone Starling, Kristina Smith, Kate Holt, Ben Ball

UNIT 1

Whole Numbers, Algebra, and Statistics 2

UNIT 2

Lesson 3-5, p. 122

Prerequisite Skills
• Getting Started **99**
• Getting Ready for the Next Lesson **105, 110, 113, 119, 124**

 Study Organizer **99**

Study Skills
• Study Tips **108, 109, 112, 116**
• Homework Help **104, 110, 113, 118, 123**

USA TODAY Snapshots **116, 119**

Reading and Writing Mathematics
• Link to Reading **102**
• Reading Math **103**
• Writing Math **101, 104, 106, 107, 109, 112, 118, 123**

Standardized Test Practice
• Multiple Choice **105, 110, 113, 114, 119, 124, 126, 129, 130**
• Short Response/Grid In **105, 113, 131**
• Extended Response **131**
• Worked-Out Example **117**

UNIT 2

Chapter ④ Multiplying and Dividing Decimals 132

Lesson 4-2, p. 141

UNIT 3

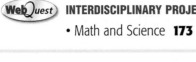 **INTERDISCIPLINARY PROJECT**
 • Math and Science **173**

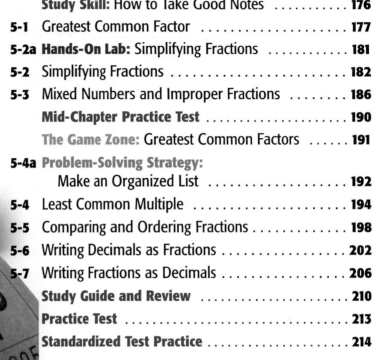

Lesson 5-3, p. 189

Prerequisite Skills
• Getting Started **175**
• Getting Ready for the Next Lesson **180, 185, 189,
                                    197, 201, 205**

FOLDABLES Study Organizer **175**

Study Skills
• Study Tips **178, 202, 203**
• Homework Help **179, 184, 188, 196, 200, 204, 208**

 Snapshots **203**

Reading and Writing Mathematics
• Link to Reading **182, 206**
• Reading Math **187, 207**
• Writing Math **179, 181, 184, 188, 196, 200, 204, 208**

Standardized Test Practice
• Multiple Choice **180, 185, 189, 190, 193, 197, 201,
                   205, 209, 213, 214**
• Short Response/Grid In **180, 185, 190, 201, 205, 215**
• Extended Response **215**
• Worked-Out Example **199**

Chapter 6 Adding and Subtracting Fractions 216

Lesson 6-6, p. 245

UNIT 3

Lesson 7-4, p. 273

Prerequisite Skills
- Getting Started **255**
- Getting Ready for the Next Lesson **258, 264, 267, 275, 279**

FOLDABLES™ Study Organizer **255**

Study Skills
- Study Tips **257, 262, 270, 272, 276**
- Homework Help **258, 263, 267, 274, 278, 284**

Reading and Writing Mathematics
- Writing Math **257, 259, 260, 263, 266, 271, 274, 277, 283**

Standardized Test Practice
- Multiple Choice **258, 264, 267, 268, 275, 279, 281, 284, 288**
- Short Response/Grid In **258, 264, 267, 279, 284, 287, 289**
- Extended Response **289**
- Worked-Out Example **273**

 Snapshots **279**

 INTERDISCIPLINARY PROJECT 284

UNIT 4

Lesson 8-4, p. 311

UNIT 4

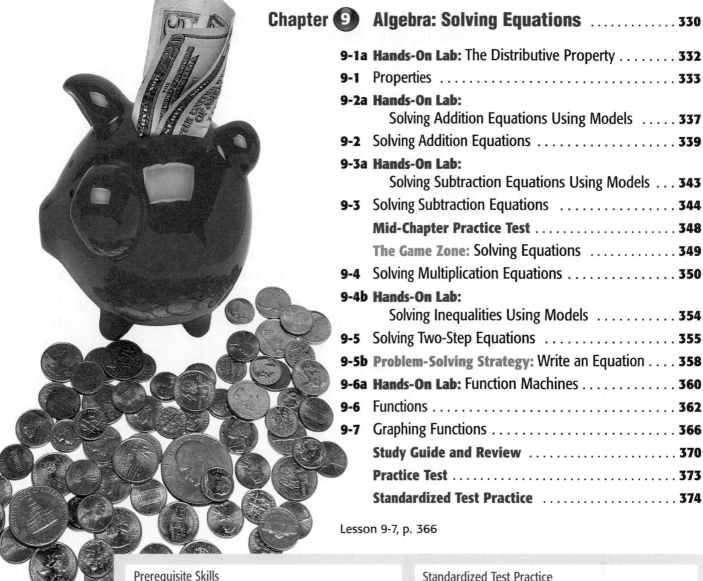

Lesson 9-7, p. 366

Prerequisite Skills
- Getting Started **331**
- Getting Ready for the Next Lesson **336, 342, 347, 353, 357, 365**

FOLDABLES Study Organizer **331**

Study Skills
- Study Tips **338, 340, 344, 351, 354, 366, 367**
- Homework Help **336, 341, 346, 352, 357, 364, 368**

Reading and Writing Mathematics
- Link to Reading **333**
- Writing Math **332, 335, 338, 341, 343, 346, 351, 354, 356, 361, 364, 368**

Standardized Test Practice
- Multiple Choice **336, 342, 347, 348, 353, 357, 359, 365, 369, 373, 374**
- Short Response/Grid In **336, 342, 347, 348, 369, 375**
- Extended Response **375**
- Worked-Out Example **345**

 Snapshots **347, 352**

 INTERDISCIPLINARY PROJECT 369

UNIT 5

Ratio and Proportion 376

Lesson 10-5, p. 401

Prerequisite Skills
• Getting Started **379**
• Getting Ready for the Next Lesson **383, 389, 393,
    397, 403, 406, 412**

FOLDABLES Study Organizer **379**

Study Skills
• Study Tips **380, 387, 395, 400, 401, 404, 405,
    409, 410**
• Homework Help **382, 388, 393, 397, 402, 406,
    411, 417**

Reading and Writing Mathematics
• Writing Math **382, 385, 388, 392, 394, 396, 402,
    405, 408, 411, 416**

Standardized Test Practice
• Multiple Choice **383, 389, 393, 397, 403, 406, 412,
    414, 417, 421, 422**
• Short Response/Grid In **383, 389, 397, 398, 406,
    417, 423**
• Extended Response **423**
• Worked-Out Example **416**

 Snapshots **383, 389, 403, 410**

Prerequisite Skills
- Getting Started **425**
- Getting Ready for the Next Lesson **431, 436, 441, 447**

FOLDABLES Study Organizer **425**

Study Skills
- Study Tips **438, 439, 445, 450**
- Homework Help **430, 435, 440, 446, 452**

Reading and Writing Mathematics
- Reading Math **429, 433**
- Writing Math **427, 430, 432, 435, 437, 440, 446, 451**

Standardized Test Practice
- Multiple Choice **431, 436, 441, 442, 447, 449, 453, 457, 458**
- Short Response/Grid In **431, 436, 442, 459**
- Extended Response **459**
- Worked-Out Example **451**

USA TODAY Snapshots **441**

WebQuest INTERDISCIPLINARY PROJECT **453**

UNIT 6

Lesson 12-4, p. 486

UNIT 6

Lesson 13-6, p. 535

UNIT 6

Lesson 14-5, p. 571

Prerequisite Skills
- Getting Started 545
- Getting Ready for the Next Lesson 549, 554, 559, 566, 573

FOLDABLES Study Organizer 545

Study Skills
- Study Tips 551, 564
- Homework Help 548, 553, 558, 566, 572, 577

Reading and Writing Mathematics
- Reading Math 547, 557, 571
- Writing Math 548, 550, 553, 555, 558, 561, 567, 572, 574, 577

Standardized Test Practice
- Multiple Choice 549, 554, 559, 562, 566, 569, 573, 578, 581, 582
- Short Response/Grid In 549, 559, 562, 573, 583
- Extended Response 583
- Worked-Out Example 552

 Snapshots 559

 INTERDISCIPLINARY PROJECT 578

HOW TO...
Use Your Math Book

Why do I need my math book?

Have you ever been in class and not understood all of what was presented? Or, you understood everything in class, but at home, got stuck on how to solve a couple of problems? Maybe you just wondered when you were ever going to use this stuff?

These next few pages are designed to help you understand everything your math book can be used for . . . besides homework problems!

BEFORE YOU READ

Have a Goal

- What information are you trying to find?
- Why is this information important to you?
- How will you use the information?

Have a Plan

- Read *What You'll Learn* at the beginning of the lesson.
- Look over photos, tables, graphs, and opening activities.
- Locate boldfaced words and read their definitions.
- Find Key Concept and Concept Summary boxes for a preview of what's important.
- Skim the example problems.

Have an Opinion

- Is this information what you were looking for?
  - Do you understand what you have read?
  - How does this information fit with what you already know?

Mathematics
Applications and Concepts
Course 1

IN CLASS

During class is the opportunity to learn as much as possible about that day's lesson. Take advantage of this time! Ask questions about things that you don't understand, and take notes to help you remember important information.

To help keep your notes in order, try making a Foldables Study Organizer. It's as easy as 1-2-3! Here's a Foldable you can use to keep track of the rules for addition, subtraction, multiplication, and division.

FOLDABLES™
Study Organizer

Operations Make this Foldable to help you organize your notes. Begin with a sheet of 11″ × 17″ paper.

STEP 1

Fold
Fold the short sides toward the middle.

STEP 2

Fold Again
Fold the top to the bottom.

STEP 3

Cut
Open. Cut along the second fold to make four tabs.

STEP 4

Label
Label each of the tabs as shown.

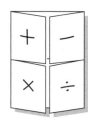

LOOK FOR...

FOLDABLES™

on these pages:

5, 49, 99, 133, 175, 217, 255, 293, 331, 379, 425, 463, 505, and 545.

DOING YOUR HOMEWORK

Regardless of how well you paid attention in class, by the time you arrive at home, your notes may no longer make any sense and your homework seems impossible. It's during these times that your book can be most useful.

- Each lesson has example problems, solved step-by-step, so you can review the day's lesson material.

- A Web site has extra examples to coach you through solving those difficult problems.

- Each exercise set has Homework Help boxes that show you which examples may help with your homework problems.

- Answers to the odd-numbered problems are in the back of the book. Use them to see if you are solving the problems correctly. If you have difficulty on an even problem, do the odd problem next to it. That should give you a hint about how to proceed with the even problem.

LOOK FOR...

The Web site with extra examples on these pages in Chapter 1: 7, 11, 15, 19, 25, 29, 35, and 39.

Homework Help boxes on these pages in Chapter 1: 9, 12, 16, 20, 26, 30, 36, and 41.

Selected Answers starting on page 651.

1. $-2y-7=3$

2. $4+5R=3$

3. $4x+5=13$

4. $6+3.50x=20$

5. $P-H=27$

slope
a $(-2,-4)$ $(1,5)$
b $(0,3)$ $(4,-1)$
c $(2,2)$, $(5,3)$

HOW TO...
Use Your Math Book

BEFORE A TEST

Admit it! You think there is no way to study for a math test! However, there *are* ways to review before a test. Your book offers help with this also.

- Review all of the new vocabulary words and be sure you understand their definitions. These can be found on the first page of each lesson or highlighted in yellow in the text.

- Review the notes you've taken on your Foldable and write down any questions that you still need answered.

- Practice all of the concepts presented in the chapter by using the chapter Study Guide and Review. It has additional problems for you to try as well as more examples to help you understand. You can also take the Chapter Practice Test.

- Take the self-check quizzes from the Web site.

$$3n - 8 = 7$$

$$y = x + 0.5$$

$$\Pi$$

LOOK FOR...

The Web site with self-check quizzes on these pages in Chapter 1: 9, 13, 17, 21, 27, 31, 37, and 41.

The Study Guide and Review for Chapter 1 on page 42.

SCAVENGER HUNT

LET'S GET STARTED

To help you find the information you need quickly, use the Scavenger Hunt below to learn where things are located in each chapter.

CHAPTER
1

1 What is the title of Chapter 1?

2 How can you tell what you'll learn in Lesson 1-1?

3 In Lesson 1-1, there is an exercise dealing with the number of year-round schools in the United States. Where could you find information on the current number of year-round schools?

4 What does the key concept presented in Lesson 1-2 explain?

5 How many Examples are presented in Lesson 1-3?

6 What is the web address where you could find extra examples?

7 In Lesson 1-3, there is a paragraph that reminds you to use divisibility rules when finding factors. What is the main heading above that paragraph?

8 There is a Real-Life Career mentioned in Lesson 1-4. What is it?

9 Sometimes you may ask "When am I ever going to use this?" Name a situation that uses the concepts from Lesson 1-5.

10 List the new vocabulary words that are presented in Lesson 1-6.

11 Suppose you're doing your homework on page 36 and you get stuck on Exercise 14. Where could you find help?

12 What is the web address that would allow you to take a self-check quiz to be sure you understand the lesson?

13 On what pages will you find the Study Guide and Review?

14 Suppose you can't figure out how to do Exercise 10 in the Study Guide on page 43. Where could you find help?

15 You complete the Practice Test on page 45 to study for your chapter test. Where could you find another test for more practice?

UNIT 1

Whole Numbers, Algebra and Statistics

Chapter 1

Number Patterns and Algebra

Chapter 2

Statistics and Graphs

Your understanding of whole numbers forms the foundation of your study of math. In this unit, you will learn how to use a variable and how to represent and interpret real-life data.

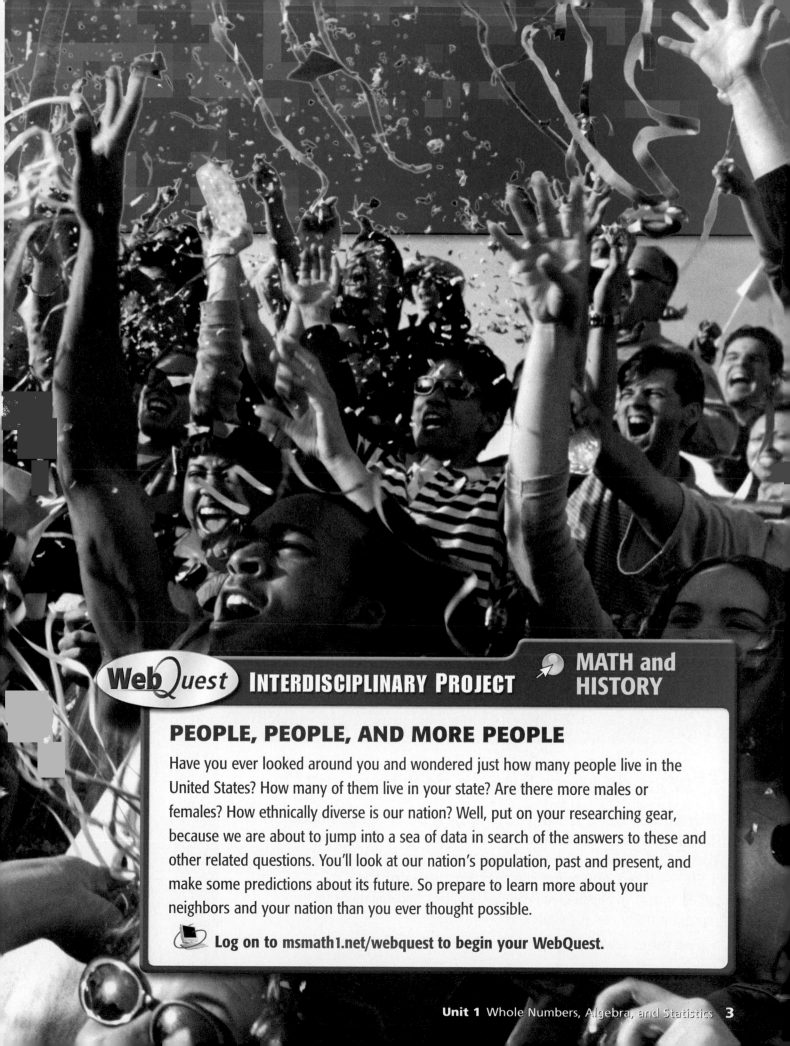

WebQuest INTERDISCIPLINARY PROJECT

MATH and HISTORY

PEOPLE, PEOPLE, AND MORE PEOPLE

Have you ever looked around you and wondered just how many people live in the United States? How many of them live in your state? Are there more males or females? How ethnically diverse is our nation? Well, put on your researching gear, because we are about to jump into a sea of data in search of the answers to these and other related questions. You'll look at our nation's population, past and present, and make some predictions about its future. So prepare to learn more about your neighbors and your nation than you ever thought possible.

Log on to msmath1.net/webquest to begin your WebQuest.

Number Patterns and Algebra

"What do bears have to do with math?"

The Kodiak bear, which is the largest bear, weighs about 1,500 pounds. This is about 1,400 pounds heavier than the smallest bear, the sun bear. You can use the equation **1,500 = x + 1,400** to find the approximate **weight x** of the smallest bear. In algebra, you will use variables and equations to represent many real-life situations.

You will use equations to solve problems about bears and other animals in Lesson 1-7.

GETTING STARTED

Take this quiz to see whether you are ready to begin Chapter 1. Refer to the page number in parentheses if you need more review.

▶ **Vocabulary Review**

Complete each sentence.

1. The smallest place value position in any given whole number is the _<u>?</u>_ place. (Page 586)

2. The result of adding numbers together is called the _<u>?</u>_.

▶ **Prerequisite Skills**

Add. (Page 589)

3. $83 + 29$ 4. $99 + 56$

5. $67 + 42$ 6. $79 + 88$

7. $78 + 97$ 8. $86 + 66$

Subtract. (Page 589)

9. $43 - 7$ 10. $75 - 27$

11. $128 - 34$ 12. $150 - 68$

13. $102 - 76$ 14. $235 - 126$

Multiply. (Page 590)

15. 25×12 16. 18×30

17. 42×15 18. 27×34

19. 50×16 20. 47×22

Divide. (Page 591)

21. $72 \div 9$ 22. $84 \div 6$

23. $126 \div 3$ 24. $146 \div 2$

25. $208 \div 4$ 26. $504 \div 8$

Number Patterns and Algebra Make this Foldable to help you organize your notes. Begin with five sheets of $8\frac{1}{2}$" $\times$ 11" paper.

STEP 1 Stack Pages
Place the sheets of paper $\frac{3}{4}$ inch apart.

STEP 2 Roll Up Bottom Edges
All tabs should be the same size.

STEP 3 Crease and Staple
Staple along the fold.

STEP 4 Label
Label the tabs with the topics from the chapter.

Number Patterns and Algebra

1-1 A Plan for Problem Solving
1-2 Divisibility Patterns
1-3 Prime Factors
1-4 Powers and Exponents
1-5 Order of Operations
1-6 Algebra: Variables and Expressions
1-7 Algebra: Solving Equations
1-8 Geometry: Area of Rectangles

Reading and Writing As you read and study the chapter, take notes and write examples under each tab.

A Plan for Problem Solving

What You'll LEARN

Solve problems using the four-step plan.

WHEN am I ever going to use this?

FUN FACTS If you lined up pennies side by side, how many would be in one mile? Let's find out. Begin by lining up pennies in a row until the row is 1 foot long.

1. How many pennies are in a row that is one mile long? (*Hint:* There are 5,280 feet in one mile.)

2. Explain how to find the value of the pennies in dollars. Then find the value.

3. Explain how you could use the answer to Exercise 1 to estimate the number of quarters in a row one mile long.

When solving math problems, it is often helpful to have an organized problem-solving plan. The four steps listed below can be used to solve any problem.

1. Explore
- Read the problem carefully.
- What facts do you know?
- What do you need to find out?
- Is enough information given?
- Is there extra information?

2. Plan
- How do the facts relate to each other?
- Plan a strategy for solving the problem.
- Estimate the answer.

3. Solve
- Use your plan to solve the problem.
- If your plan does not work, revise it or make a new plan.
- What is the solution?

4. Examine
- Reread the problem.
- Does the answer fit the facts given in the problem?
- Is the answer close to my estimate?
- Does the answer make sense?
- If not, solve the problem another way.

STUDY TIP

Reasonableness In the last step of this plan, you check the reasonableness of the answer by comparing it to the estimate.

Some problems can be easily solved by adding, subtracting, multiplying, or dividing. Key words and phrases play an important role in deciding which operation to use.

Addition	Subtraction	Multiplication	Division
plus	minus	times	divided by
sum	difference	product	quotient
total	less	multiplied by	
in all	subtract	of	

When planning how to solve a problem, it is also important to choose the most appropriate method of computation. You can choose paper and pencil, mental math, a calculator, or estimation.

STUDY TIP

Methods of Computation To determine when an estimate is needed, look for the word "about" in the question portion of the math problem.

EXAMPLE Use the Problem-Solving Plan

1. **BASKETBALL** Refer to the graph below. It shows the all-time three-point field goal leaders in the WNBA. How many more three-point field goals did Katie Smith make than Tina Thompson?

All-Time Three-Point Field Goal Leaders*

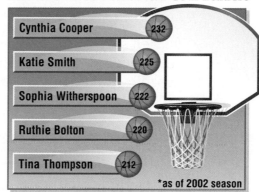

Cynthia Cooper — 232
Katie Smith — 225
Sophia Witherspoon — 222
Ruthie Bolton — 220
Tina Thompson — 212

*as of 2002 season

Source: WNBA

REAL-LIFE MATH

BASKETBALL Katie Smith plays for the Minnesota Lynx. In 2002, she scored 512 points, made 126 out of 153 free throws, made 62 three-point field goals, and played 1,138 minutes.
Source: www.wnba.com

Explore Extra information is given in the graph. You know the number of three-point field goals made by many players. You need to find how many more field goals Katie Smith made than Tina Thompson.

Plan To find the difference, subtract 212 from 225. Since the question asks for an exact answer and there are simple calculations to do, use mental math or paper and pencil. Before you calculate, estimate.

Estimate $220 - 210 = 10$

Solve $225 - 212 = 13$

Katie Smith made 13 more three-point field goals than Tina Thompson.

Examine Compared to the estimate, the answer is reasonable. Since $13 + 212$ is 225, the answer is correct.

The four-step plan can be used to solve problems involving patterns.

EXAMPLE Use the Problem-Solving Plan

2 **SWIMMING** Curtis is on the swim team. The table shows the number of laps he swims in the first four days of practice. If the pattern continues, how many laps will he swim on Friday?

Day	Monday	Tuesday	Wednesday	Thursday	Friday
Laps	5	6	8	11	?

Explore   You know the number of laps he swam daily. You need to find the number of laps for Friday.

Plan   Since an exact answer is needed and the question contains a pattern, use mental math.

Solve   5   6   8   11   ?

              +1   +2   +3

The numbers increase by 1, 2, and 3. The next number should increase by 4. So, if Curtis continues at this rate, he will swim 11 + 4 or 15 laps.

Examine   Since 15 − 4 = 11, 11 − 3 = 8, 8 − 2 = 6, and 6 − 5 = 1, the answer is correct.

Skill and Concept Check

Writing Math
Exercises 1–3

1. **List** and explain each step of the problem-solving plan.

2. **Explain** why you should compare your answer to your estimate.

3. **OPEN ENDED** Write a number pattern containing five numbers. Then write a rule for finding the next number in the pattern.

GUIDED PRACTICE

For Exercises 4 and 5, use the four-step plan to solve each problem.

4. **ANIMALS** An adult male walrus weighs about 2,670 pounds. An adult female walrus weighs about 1,835 pounds. Their tusks measure about 3 feet in length and can weigh over 10 pounds. How much less does an adult female walrus weigh than an adult male walrus?

5. **SCIENCE** The table shows how the number of a certain bacteria increases every 20 minutes. At this rate, how many bacteria will there be after 2 hours?

Time (min)	0	20	40	60	80	100	120
Number of Bacteria	5	10	20	40	80	?	?

Practice and Applications

For Exercises 6–12, use the four-step plan to solve each problem.

HOMEWORK HELP

For Exercises	See Examples
6, 8–10, 12	1
7, 11	2

Extra Practice
See pages 594, 624.

6. **TIME** A bus departed at 11:45 A.M.. It traveled 325 miles at 65 miles per hour. How many hours did it take for the bus to reach its destination?

7. **PATTERNS** Complete the pattern:
6, 11, 16, 21, __?__, __?__, __?__.

8. **SCHOOL** Refer to the graphic. How many more year-round schools were there in 2000–2001 than in 1991–1992?

 Data Update How many year-round schools are there in the U.S. today? Visit msmath1.net/data_update to learn more.

9. **GEOGRAPHY** The land area of the United States is 3,536,278 square miles, and the water area is 181,518 square miles. Find the total area of the United States.

10. **MONEY** The Corbetts want to buy a 36-inch television that costs $788. They plan to pay in four equal payments. What will be the amount of each payment?

USA TODAY Snapshots®

Schools on year-round schedules
Growth of public year-round education in the nation:

Year	Schools
1990–91	859
1991–92	1,646
1992–93	2,017
1993–94	1,941
1994–95	2,214
1995–96	2,368
1996–97	2,400
1997–98	2,681
1998–99	2,856
1999–2000	2,880
2000–01	3,059

Source: National Association for Year-Round Education

By April L. Umminger and Keith Simmons, USA TODAY

11. **SCHOOL** The table at the right lists the times the bell rings every day at Watson Middle School. When do the next three bells ring?

8:50 A.M.
8:54 A.M.
9:34 A.M.
9:38 A.M.
10:18 A.M.

12. **MULTI STEP** Jupiter orbits the Sun at a rate of 8 miles per second. How far does Jupiter travel in one day?

13. **CRITICAL THINKING** Complete the pattern 1, 1, 2, 6, 24, __?__.

Standardized Test Practice and Mixed Review

14. **MULTIPLE CHOICE** Brandon can run one mile in 7 minutes. At this rate, how long will it take him to run 8 miles?

   Ⓐ 58 min   Ⓑ 56 min   Ⓒ 54 min   Ⓓ 15 min

15. **SHORT RESPONSE** Draw the next figure in the pattern.

GETTING READY FOR THE NEXT LESSON

PREREQUISITE SKILL Divide. (Page 591)

16. $42 \div 3$ 17. $126 \div 6$ 18. $49 \div 7$ 19. $118 \div 2$

Divisibility Patterns

What You'll LEARN

Use divisibility rules for 2, 3, 4, 5, 6, 9, and 10.

NEW Vocabulary

divisible
even
odd

Link to READING

Everyday Meaning of Divisible: capable of being divided evenly

HANDS-ON Mini Lab

Materials
- yellow marker
- calculator

Work with a partner.
Copy the numbers shown below.

1	2	3	4	5	6	7	8	9	10
11	12	13	14	15	16	17	18	19	20
21	22	23	24	25	26	27	28	29	30
31	32	33	34	35	36	37	38	39	40
41	42	43	44	45	46	47	48	49	50

- Using a yellow marker, shade all of the numbers that can be divided evenly by 2.
- Underline the numbers that can be divided evenly by 3.
- Circle the numbers that can be divided evenly by 5.
- Place a check mark next to all of the numbers that can be divided evenly by 10.

Describe a pattern in each group of numbers listed.

1. the numbers that can be evenly divided by 2
2. the numbers that can be evenly divided by 5
3. the numbers that can be evenly divided by 10
4. the numbers that can be evenly divided by 3
   (*Hint:* Look at both digits.)

A whole number is **divisible** by another number if the remainder is 0 when the first number is divided by the second. The divisibility rules for 2, 3, 5, and 10 are stated below.

Key Concept
Divisibility Rules

Rule	Examples
A whole number is divisible by:	
• 2 if the ones digit is divisible by 2.	2, 4, 6, 8, 10, 12, …
• 3 if the sum of the digits is divisible by 3.	3, 6, 9, 12, 15, 18, …
• 5 if the ones digit is 0 or 5.	5, 10, 15, 20, 25, …
• 10 if the ones digit is 0.	10, 20, 30, 40, 50, …

A whole number is **even** if it is divisible by 2. A whole number is **odd** if it is not divisible by 2.

1 Tell whether 2,320 is divisible by 2, 3, 5, or 10. Then classify the number as *even* or *odd*.

2: Yes; the ones digit, 0, is divisible by 2.

3: No; the sum of the digits, 7, is not divisible by 3.

5: Yes; the ones digit is 0.

10: Yes; the ones digit is 0.

The number 2,320 is even.

REAL-LIFE MATH

COUNTY FAIRS There were 55 county agricultural fairs in Tennessee in 2002.
Source: State of Tennessee

2 **COUNTY FAIRS** The number of tickets needed to ride certain rides at a county fair is shown. If Alicia has 51 tickets, can she use all of the tickets by riding only the Ferris wheel?

Use divisibility rules to check whether 51 is divisible by 3.

51 → 5 + 1 = 6 and 6 is divisible by 3.

So, 51 is divisible by 3.

Alicia can use all of the tickets by riding only the Ferris wheel.

Canfield County Fair	
Ride	Tickets
roller coaster	5
Ferris wheel	3
bumper cars	2
scrambler	4

The rules for 4, 6, and 9 are related to the rules for 2 and 3.

Key Concept   **Divisibility Rules**

Rule	Examples
A whole number is divisible by:	
• 4 if the number formed by the last two digits is divisible by 4.	4, 8, 12, …, 104, 112, …
• 6 if the number is divisible by both 2 and 3.	6, 12, 18, 24, 30, …
• 9 if the sum of the digits is divisible by 9.	9, 18, 27, 36, 45, …

EXAMPLE Use Divisibility Rules

3 Tell whether 564 is divisible by 4, 6, or 9.

4: Yes; the number formed by the last two digits, 64, is divisible by 4.

6: Yes; the number is divisible by both 2 and 3.

9: No; the sum of the digits, 15, is not divisible by 9.

Your Turn Tell whether each number is divisible by 2, 3, 4, 5, 6, 9, or 10. Then classify the number as *even* or *odd*.

a. 126   b. 684   c. 2,835

Skill and Concept Check

1. **Explain** how you can tell whether a number is even or odd.

2. **OPEN ENDED** Write a number that is divisible by 2, 5 and 10.

3. **Which One Doesn't Belong?** Identify the number that is not divisible by 3. Explain.

| 27 | 96 | 56 | 153 |

GUIDED PRACTICE

Tell whether each number is divisible by 2, 3, 4, 5, 6, 9, or 10. Then classify each number as *even* or *odd*.

4. 40   5. 795   6. 5,067   7. 17,256

8. **GARDENING** Taryn has 2 dozen petunias. She wants to plant the flowers in rows so that each row has the same number of flowers. Can she plant the flowers in 3 equal rows? Explain.

Practice and Applications

Tell whether each number is divisible by 2, 3, 4, 5, 6, 9, or 10. Then classify each number as *even* or *odd*.

9. 60	10. 80	11. 78
12. 45	13. 138	14. 489
15. 605	16. 900	17. 3,135
18. 6,950	19. 8,505	20. 9,948
21. 11,292	22. 15,000	23. 14,980
24. 18,321	25. 151,764	26. 5,203,570

27. Find a number that is divisible by both 3 and 5.

28. Find a number that is divisible by 2, 9, and 10.

29. **FLAGS** Each star on the United States flag represents a state. If another state joins the Union, could the stars be arranged in rows with each row having the same number of stars? Explain.

30. **LEAP YEAR** Any year that is divisible by 4 and does not end in 00 is a leap year. Years ending in 00 are leap years only if they are divisible by 400. Were you born in a leap year? Explain.

Use divisibility rules to find each missing digit. List all possible answers.

31. 8,4■8 is divisible by 3.   32. 12,48■ is divisible by 5.

33. 3,■56 is divisible by 2.   34. 18,45■ is divisible by 10.

35. What is the smallest whole number that is divisible by 2, 3, 5, 6, 9, and 10? Explain how you found this number.

FOOD For Exercises 36 and 37, use the information below.
Rosa is packaging cookies for a bake sale. She has 165 cookies.

36. Can Rosa package the cookies in groups of 5 with none left over?

37. What are the other ways she can package all the cookies in equal-sized packages with 10 or less in a package?

TOYS For Exercises 38 and 39, use the information below.
A toy company sells the sleds and snow tubes listed in the table. One of their shipments weighed 189 pounds and contained only one type of sled or snow tube.

Toy	Weight (lb)
Super Snow Sled	4
Phantom Sled	6
Tundra Snow Tube	3
Snow Tube Deluxe	5

38. Which sled or snow tube does the shipment contain?

39. How many sleds or snow tubes does the shipment contain?

Tell whether each sentence is *sometimes*, *always*, or *never* true.

40. A number that is divisible by 10 is divisible by 2.

41. A number that is divisible by 5 is divisible by 10.

42. **WRITE A PROBLEM** Write about a real-life situation that can be solved using divisibility rules. Then solve the problem.

43. **CRITICAL THINKING** What is true about the product of an odd number and an even number?

44. **CRITICAL THINKING** If all even numbers are divisible by 2, is it true that all odd numbers are divisible by 3? If so, explain your reasoning. If not, give a number for which this is not true.

Standardized Test Practice and Mixed Review

45. **GRID IN** Find the greatest 3-digit number not divisible by 3.

46. **MULTIPLE CHOICE** Which statement *best* describes the numbers shown at the right?

| 1,250 |
| 565 |
| 10,050 |
| 95 |

   Ⓐ divisible by 2

   Ⓑ divisible by 2 and/or 3

   Ⓒ divisible by 4 and/or 5

   Ⓓ divisible by 2 and/or 5

47. **PATTERNS** Complete the pattern: 5, 7, 10, 14, 19, . . . (Lesson 1-1)

48. **MULTI STEP** Find the number of seconds in a day if there are 60 seconds in a minute. (Lesson 1-1)

GETTING READY FOR THE NEXT LESSON

PREREQUISITE SKILL Find each product. (Page 590)

49. $2 \times 2 \times 3$ 50. 3×5 51. $2 \times 3 \times 3$ 52. $3 \times 5 \times 7$

1-3 Prime Factors

What You'll LEARN

Find the prime factorization of a composite number

NEW Vocabulary

factor
prime number
composite number
prime factorization

HANDS-ON Mini Lab

Materials
• square tiles

Work with a partner.

Any given number of squares can be arranged into one or more different rectangles. The table shows the different rectangles that can be made using 2, 3, or 4 squares. A 1×3 rectangle is the same as a 3×1 rectangle.

STEP 1 Copy the table.

Number of Squares	Sketch of Rectangle Formed	Dimensions of Each Rectangle
2		1×2
3		1×3
4		$1 \times 4, 2 \times 2$
5		1×5
6		$1 \times 6, 2 \times 3$
⋮		
20		

STEP 2 Use square tiles to help you complete the table.

1. For what numbers can more than one rectangle be formed?
2. For what numbers can only one rectangle be formed?
3. For the numbers in which only one rectangle is formed, what do you notice about the dimensions of the rectangle?

When two or more numbers are multiplied, each number is called a **factor** of the product.

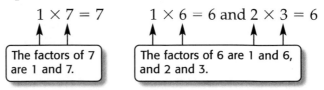

$1 \times 7 = 7$ $1 \times 6 = 6$ and $2 \times 3 = 6$

The factors of 7 are 1 and 7.

The factors of 6 are 1 and 6, and 2 and 3.

A whole number that has exactly two unique factors, 1 and the number itself, is a **prime number**. A number greater than 1 with more than two factors is a **composite number**.

Identify Prime and Composite Numbers

Tell whether each number is *prime*, *composite*, or *neither*.

1 **28**

The factors of 28 are 1 and 28, 2 and 14, and 4 and 7. Since 28 has more than two factors, it is a composite number.

2 **11**

The factors of 11 are 1 and 11. Since there are exactly two factors, 1 and the number itself, 11 is a prime number.

Your Turn Tell whether each number is *prime*, *composite*, or *neither*.

a. 48   b. 7   c. 81

Every composite number can be expressed as a product of prime numbers. This is called a **prime factorization** of the number. A *factor tree* can be used to find the prime factorization of a number.

EXAMPLE **Find Prime Factorization**

3 **Find the prime factorization of 54.**

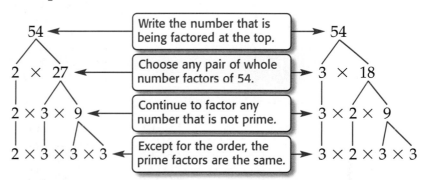

The prime factorization of 54 is 2 × 3 × 3 × 3.

Your Turn Find the prime factorization of each number.

d. 34   e. 72

STUDY TIP

Divisibility Rules
Remember to use divisibility rules when finding factors.

Concept Summary		Prime and Composite
Number	**Definition**	**Examples**
prime	A whole number that has exactly two factors, 1 and the number itself.	11, 13, 23
composite	A number greater than 1 with more than two factors.	6, 10, 18
neither prime nor composite	1 has only one factor. 0 has an infinite number of factors.	0, 1

Skill and Concept Check

Writing Math
Exercise 3

1. **List** the factors of 12.

2. **OPEN ENDED** Give an example of a number that is not composite.

3. **Which One Doesn't Belong?** Identify the number that is not a prime number. Explain your reasoning.

| 57 | 29 | 17 | 83 |

GUIDED PRACTICE

Tell whether each number is *prime, composite,* or *neither*.

4. 10   5. 3   6. 1   7. 61

Find the prime factorization of each number.

8. 36   9. 81   10. 65   11. 19

12. **GEOGRAPHY** The state of Kentucky has 120 counties. Write 120 as a product of primes.

KENTUCKY

Practice and Applications

Tell whether each number is *prime, composite,* or *neither*.

13. 17   14. 0   15. 15   16. 44

17. 23   18. 57   19. 45   20. 29

21. 56   22. 93   23. 53   24. 31

25. 125   26. 114   27. 179   28. 291

HOMEWORK HELP

For Exercises	See Examples
13–28	1–2
31–46	3

Extra Practice
See pages 594, 624.

29. Write 38 as a product of prime numbers.

30. Find the least prime number that is greater than 60.

Find the prime factorization of each number.

31. 24   32. 18   33. 40   34. 75

35. 27   36. 32   37. 49   38. 25

39. 42   40. 104   41. 17   42. 97

43. 102   44. 126   45. 55   46. 77

47. A *counterexample* is an example that shows a statement is not true. Find a counterexample for the statement below.

    All even numbers are composite numbers.

    (*Hint:* You need to find an even number that is *not* a composite number.)

ANIMALS For Exercises 48–50, use the table shown.

Animal	Speed (mph)	Animal	Speed (mph)
cheetah	70	rabbit	35
antelope	60	giraffe	32
lion	50	grizzly bear	30
coyote	43	elephant	25
hyena	40	squirrel	12

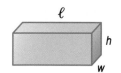

48. Which speed(s) are prime numbers?

49. Which speed(s) have a prime factorization whose factors are all equal?

50. Which speeds have a prime factorization of exactly three factors?

51. **NUMBER SENSE** *Twin primes* are two prime numbers that are consecutive odd integers such as 3 and 5, 5 and 7, and 11 and 13. Find all of the twin primes that are less than 100.

52. **SHOPPING** Amanda bought bags of snacks that each cost the same. She spent a total of $30. Find three possible costs per bag and the number of bags that she could have purchased.

53. **GEOMETRY** To find the volume of a box, you can multiply its height, length, and width. The measure of the volume of a box is 357. Find its possible dimensions.

54. **CRITICAL THINKING** All odd numbers greater than 7 can be expressed as the sum of three prime numbers. Which three prime numbers have a sum of 59?

Standardized Test Practice and Mixed Review

55. **MULTIPLE CHOICE** Which number is *not* prime?
    Ⓐ 7　　　　Ⓑ 31　　　　Ⓒ 39　　　　Ⓓ 47

56. **MULTIPLE CHOICE** What is the prime factorization of 140?
    Ⓕ $2 \times 2 \times 2 \times 5 \times 7$　　　　Ⓖ $2 \times 3 \times 5 \times 7$
    Ⓗ $2 \times 2 \times 5 \times 7$　　　　Ⓘ $3 \times 5 \times 7$

Tell whether each number is divisible by 2, 3, 4, 5, 6, 9, or 10. (Lesson 1-2)

57. 75　　　　　　　58. 462　　　　　　　59. 3,050

60. **SCHOOL** Each class that sells 75 tickets to the school play earns an ice cream party. Kate's class has sold 42 tickets. How many more must they sell to earn an ice cream party? (Lesson 1-1)

GETTING READY FOR THE NEXT LESSON

PREREQUISITE SKILL Multiply. (Page 590)

61. $2 \times 2 \times 2$　　　62. 5×5　　　63. $4 \times 4 \times 4$　　　64. $10 \times 10 \times 10$

Powers and Exponents

What You'll LEARN

Use powers and exponents in expressions

NEW Vocabulary

exponent
base
power
squared
cubed

Materials
• paper
• paper hole punch

Work with a partner.

Any number can be written as a product of prime factors.

STEP 1 Fold a piece of paper in half and make one hole punch. Open the paper and count the number of holes. Copy the table below and record the results.

Number of Folds	Number of Holes	Prime Factorization
1		
⋮		
5		

STEP 2 Find the prime factorization of the number of holes and record the results in the table.

STEP 3 Fold another piece of paper in half twice. Then make one hole punch. Complete the table for two folds.

STEP 4 Complete the table for three, four, and five folds.

1. What prime factors did you record?

2. How does the number of folds relate to the number of factors in the prime factorization of the number of holes?

3. Write the prime factorization of the number of holes made if you folded it eight times.

STUDY TiP

Exponents When no exponent is given, it is understood to be 1. For example, $5 = 5^1$.

A product of prime factors can be written using **exponents** and a **base**.

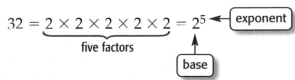

$$32 = \underbrace{2 \times 2 \times 2 \times 2 \times 2}_{\text{five factors}} = 2^5 \leftarrow \boxed{\text{exponent}}$$

$$\boxed{\text{base}}$$

Numbers expressed using exponents are called **powers** .

Powers	Words	Expression	Value
2^5	2 to the fifth power.	$2 \times 2 \times 2 \times 2 \times 2$	32
3^2	3 to the second power or 3 **squared**.	3×3	9
10^3	10 to the third power or 10 **cubed**.	$10 \times 10 \times 10$	1,000

The symbol · means multiplication. For example, 3 · 3 means 3 × 3.

EXAMPLES Write Powers and Products

1 Write 3 · 3 · 3 · 3 using an exponent. Then find the value of the power.

The base is 3. Since 3 is a factor four times, the exponent is 4.
$$3 \cdot 3 \cdot 3 \cdot 3 = 3^4$$
$$= 81$$

2 Write 4^5 as a product. Then find the value of the product.

The base is 4. The exponent is 5. So, 4 is a factor five times.
$$4^5 = 4 \cdot 4 \cdot 4 \cdot 4 \cdot 4$$
$$= 1{,}024$$

Your Turn

a. Write 7 · 7 · 7 using an exponent. Then find the value of the power.

b. Write 2^3 as a product. Then find the value of the product.

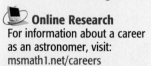

REAL-LIFE MATH

How Does an Astronomer Use Math?
An astronomer uses exponents to represent great distances between galaxies.

Online Research
For information about a career as an astronomer, visit: msmath1.net/careers

EXAMPLE Use Powers to Solve a Problem

3 **SCIENCE** The approximate daytime surface temperature on the moon can be written as 2^8 degrees Fahrenheit. What is this temperature?

Write 2^8 as a product. Then find the value of the product.
$$2^8 = 2 \times 2 \times 2 \times 2 \times 2 \times 2 \times 2 \times 2$$
$$= 256$$

So, the temperature is about 256 degrees Fahrenheit.

Exponents can be used to write the prime factorization of a number.

EXAMPLE Write Prime Factorization Using Exponents

4 Write the prime factorization of 72 using exponents.

The prime factorization of 72 is 2 × 2 × 2 × 3 × 3. This can be written as $2^3 \times 3^2$.

Your Turn Write the prime factorization of each number using exponents.

c. 24   d. 90   e. 120

Skill and Concept Check

1. **Write** each expression in words.

   a. 3^2 b. 2^1 c. 4^5

2. **OPEN ENDED** Write a power whose value is greater than 100.

3. **NUMBER SENSE** List 11^2, 4^4, and 6^3 from least to greatest.

4. **FIND THE ERROR** Anita and Tyree are writing 6^4 as a product. Who is correct? Explain your reasoning.

Anita
$6^4 = 6 \times 6 \times 6 \times 6$

Tyree
$6^4 = 4 \times 4 \times 4 \times 4 \times 4 \times 4$

GUIDED PRACTICE

Write each product using an exponent. Then find the value of the power.

5. $2 \cdot 2 \cdot 2 \cdot 2$ 6. $6 \cdot 6 \cdot 6$

Write each power as a product. Then find the value of the product.

7. 2^6 8. 3^7

Write the prime factorization of each number using exponents.

9. 20   10. 48

Practice and Applications

Write each product using an exponent. Then find the value of the power.

11. $9 \cdot 9$ 12. $8 \cdot 8 \cdot 8 \cdot 8$ 13. $3 \cdot 3 \cdot 3 \cdot 3 \cdot 3 \cdot 3 \cdot 3$

14. $5 \cdot 5 \cdot 5 \cdot 5 \cdot 5$ 15. $11 \cdot 11 \cdot 11$ 16. $7 \cdot 7 \cdot 7 \cdot 7 \cdot 7 \cdot 7$

HOMEWORK HELP

For Exercises	See Examples
11–16	1
19–29	2
38–39	3
17–18, 30–37	4

Extra Practice
See pages 595, 624.

17. **GEOGRAPHY** New Hampshire has a total land area of about 10,000 square miles. Write this number as a power with base of 10.

18. **SCIENCE** The Milky Way galaxy is about 100,000 light years wide. Write 100,000 as a power with 10 as the base.

Write each power as a product. Then find the value of the product.

19. 2^4 20. 3^2 21. 5^3 22. 10^5

23. 9^3 24. 6^5 25. 8^1 26. 4^7

27. seven squared   28. eight cubed   29. four to the fifth power

Write the prime factorization of each number using exponents.

30. 25   31. 56   32. 50   33. 68

34. 88   35. 98   36. 45   37. 189

38. FOOD The number of Calories in one cup of apple juice can be written as the power 3^5. What whole number does 3^5 represent?

39. LANGUAGE An estimated 10^9 people in the world speak Mandarin Chinese. About how many people speak this language?

FISH For Exercises 40 and 41, use the following information and the diagram at the right.
To find how much water the aquarium holds, use the expression $s \times s \times s$, where s is the length of a side.

40. What power represents this expression?

41. The amount of water an aquarium holds is measured in cubic units. How many cubic units of water does it hold?

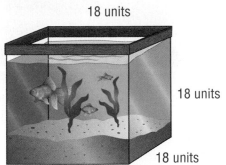

18 units

18 units

18 units

CRITICAL THINKING For Exercises 42–47, refer to the table shown.

Powers of 3	Powers of 5	Powers of 10
$3^4 = 81$	$5^4 = 625$	$10^4 = 10{,}000$
$3^3 = 27$	$5^3 = 125$	$10^3 = 1{,}000$
$3^2 = 9$	$5^2 = 25$	$10^2 = 100$
$3^1 = 3$	$5^1 = 5$	$10^1 = ?$
$3^0 = ?$	$5^0 = ?$	$10^0 = ?$

42. Copy and complete the table.

43. Describe the pattern for the powers of 3. What is the value of 3^0?

44. Describe the pattern for the powers of 5. What is the value of 5^0?

45. Describe the pattern for the powers of 10. Find the values of 10^1 and 10^0.

46. Extend the pattern for the powers of 10 to find 10^5 and 10^6.

47. How can you easily write the value of any power of 10?

Standardized Test Practice and Mixed Review

48. MULTIPLE CHOICE Rewrite $2 \times 3 \times 3 \times 11$ using exponents.

   A $2 \times 2 \times 3 \times 11$ **B** $3^2 \times 11$

   C $2 \times 3^2 \times 11$ **D** $2 \times 2^3 \times 11$

49. SHORT RESPONSE Write the power that represents 9^4.

Tell whether each number is *prime*, *composite*, or *neither*. (Lesson 1-3)

50. 63   **51.** 0   **52.** 29   **53.** 71

Tell whether each number is divisible by 2, 3, 4, 5, 6, 9, or 10. (Lesson 1-2)

54. 360   **55.** 2,022   **56.** 7,525

GETTING READY FOR THE NEXT LESSON

PREREQUISITE SKILL Divide. (Page 591)

57. $36 \div 3$ **58.** $45 \div 5$ **59.** $104 \div 8$ **60.** $120 \div 6$

 msmath1.net/self_check_quiz

Vocabulary and Concepts

1. **Write** two words or phrases that represent multiplication. (Lesson 1-1)

2. **Explain** why 51 is a composite number. (Lesson 1-3)

Skills and Applications

3. **TECHNOLOGY** A recordable CD for a computer holds 700 megabytes of data. Pedro records the songs and music videos listed at the right. What is the amount of storage space left on the CD? (Lesson 1-1)

song or video	size (MB)
song #1	35
song #2	40
song #3	37
video #1	125
video #2	140

Tell whether each number is divisible by 2, 3, 4, 5, 6, 9, or 10. Then classify each number as *even* or *odd*. (Lesson 1-2)

4. 42
5. 135
6. 3,600

Tell whether each number is *prime, composite,* or *neither.* (Lesson 1-3)

7. 57
8. 97
9. 0

Find the prime factorization of each number. (Lesson 1-3)

10. 45
11. 72

Write each power as a product. Then find the value of the product. (Lesson 1-4)

12. 3^4
13. 6^3

14. Write *eight squared* as a power. (Lesson 1-4)

Standardized Test Practice

15. **MULTIPLE CHOICE** Which expression shows 75 as a product of prime factors? (Lesson 1-3)

   Ⓐ $2 \times 3 \times 5$ Ⓑ 3×25

   Ⓒ $3 \times 5 \times 5$ Ⓓ $3 \times 3 \times 25$

16. **GRID IN** A principal has 144 computers for 24 classrooms. How many computers will each classroom have if each classroom has the same number of computers? (Lesson 1-1)

The GameZone

A Place To Practice Your Math Skills

Facto Bingo

- **GET READY!**

  Players: three or four
  Materials: calculator

- **GET SET!**

  - Each player copies the Facto Bingo card shown.

Facto Bingo				
		Free Space		

  - Each player selects 24 different numbers from the list below and writes them in the upper right hand corner boxes.

Facto Bingo Numbers									
6	9	10	13	15	18	20	21	24	25
28	32	37	40	43	45	48	50	52	55
59	60	62	64	66	67	69	70	72	74
75	76	79	80	85	88	89	90	96	98

- **GO!**

  - The caller reads one number at a time at random from the list above.
  - If a player has the number, the player marks the space by writing the prime factorization in the larger box.
  - **Who Wins?** The first player with bingo wins.

Order of Operations

What You'll LEARN

Evaluate expressions using the order of operations.

NEW Vocabulary

numerical expression
order of operations

WHEN am I ever going to use this?

HEALTH When you exercise, you burn Calories. The table shows the number of Calories burned in one hour for different types of activities.

Activity	Calories Burned per Hour
jogging (5.5 mph)	490
walking (4.5 mph)	290
bike riding (6 mph)	160

1. How many Calories would you burn by walking for 2 hours?

2. Find the number of Calories a person would burn by walking for 2 hours and bike riding for 3 hours.

3. Explain how you found the total number of Calories.

A **numerical expression** like $4 + 3 \times 5$ or $2^2 + 6 \div 2$ is a combination of numbers and operations. The **order of operations** tells you which operation to perform first so that everyone gets the same final answer.

Key Concept   Order of Operations

1. Simplify the expressions inside grouping symbols, like parentheses.
2. Find the value of all powers.
3. Multiply and divide in order from left to right.
4. Add and subtract in order from left to right.

EXAMPLES   Use Order of Operations

Find the value of each expression.

1 $4 + 3 \times 5$

$$4 + 3 \times 5 = 4 + 15 \quad \text{Multiply 3 and 5.}$$
$$= 19 \quad \text{Add 4 and 15.}$$

2 $10 - 2 + 8$

$$10 - 2 + 8 = 8 + 8 \quad \text{Subtract 2 from 10 first.}$$
$$= 16 \quad \text{Add 8 and 8.}$$

3 $20 \div 4 + 17 \times (9 - 6)$

$$20 \div 4 + 17 \times (9 - 6) = 20 \div 4 + 17 \times 3 \quad \text{Subtract 6 from 9.}$$
$$= 5 + 17 \times 3 \quad \text{Divide 20 by 4.}$$
$$= 5 + 51 \quad \text{Multiply 17 by 3.}$$
$$= 56 \quad \text{Add 5 and 51.}$$

Use Order of Operations with Powers

Find the value of each expression.

4 $3 \cdot 6^2 + 4$

$$3 \cdot 6^2 + 4 = 3 \cdot 36 + 4 \quad \text{Find } 6^2.$$
$$= 108 + 4 \quad \text{Multiply 3 and 36.}$$
$$= 112 \quad \text{Add 108 and 4.}$$

5 $60 \div (12 + 2^3) \times 9$

$$60 \div (12 + 2^3) \times 9 = 60 \div (12 + 8) \times 9 \quad \text{Find } 2^3.$$
$$= 60 \div 20 \times 9 \quad \text{Add 12 and 8.}$$
$$= 3 \times 9 \quad \text{Divide 60 by 20.}$$
$$= 27 \quad \text{Multiply 3 and 9.}$$

Your Turn Find the value of each expression.

a. $10 + 2 \times 15$ **b.** $10 + (6 - 5)$ **c.** $25 \times (5 - 2) \div 5$

d. $24 \div 2^3 + 6$ **e.** $(27 + 2) \times 4^2$ **f.** $50 - (16 \div 2^2) + 3$

You can use the order of operations to solve real-life problems involving more than one operation.

EXAMPLE **Apply Order of Operations**

6 **MOVIES** Javier and four friends go to the movies. Each person buys a movie ticket, snack, and a soda. Find the total cost of the trip to the movies.

Cost of Going to the Movies			
Item	ticket	snack	soda
Cost ($)	6	3	2

To find the total cost, write an expression and then evaluate it using the order of operations.

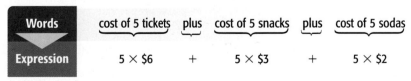

Words	cost of 5 tickets	plus	cost of 5 snacks	plus	cost of 5 sodas
Expression	5 × $6	+	5 × $3	+	5 × $2

$$5 \times \$6 + 5 \times \$3 + 5 \times \$2 = \$30 + 5 \times \$3 + 5 \times \$2$$
$$= \$30 + \$15 + 5 \times \$2$$
$$= \$30 + \$15 + \$10$$
$$= \$55$$

The total cost of the trip to the movies is $55.

Skill and Concept Check

1. **OPEN ENDED** Write an expression that contains three different operations. Then find the value of the expression.

2. **FIND THE ERROR** Haley and Ryan are finding $7 - 3 + 2$. Who is correct? Explain.

Haley
$7 - 3 + 2 = 7 - 5$
$= 2$

Ryan
$7 - 3 + 2 = 4 + 2$
$= 6$

GUIDED PRACTICE

Find the value of each expression.

3. $9 + 3 - 5$

4. $10 - 3 + 9$

5. $(26 + 5) \times 2 - 15$

6. $5 + (21 - 3)$

7. $18 \div (2 + 7) \times 2 + 1$

8. $21 \div (3 + 4) \times 3 - 8$

9. $5^2 + 8 \div 2$

10. $19 - (3^2 + 4) + 6$

11. $7^3 - (5 + 9) \times 2$

For Exercises 12 and 13, use the following information.
MONEY Tickets to the museum cost $4 for children and $8 for adults.

12. Write an expression to find the total cost of 4 adult tickets and 2 children's tickets.

13. Find the value of the expression.

Practice and Applications

Find the value of each expression.

14. $8 + 4 - 3$

15. $9 + 12 - 15$

16. $38 - 19 + 12$

17. $22 - 17 + 8$

18. $7 + 9 \times (3 + 8)$

19. $(9 + 2) \times 6 - 5$

20. $63 \div (10 - 3) \times 3$

21. $66 \times (6 \div 2) + 1$

22. $27 \div (3 + 6) \times 5 - 12$

23. $55 \div 11 + 7 \times (2 + 14)$

24. $5^3 - 12 \div 3$

25. $26 + 6^2 \div 4$

26. $15 - 2^3 \div 4$

27. $22 \div 2 \times 3^2$

28. $8 \times (2^4 - 3) + 8$

29. $12 \div 4 + (5^2 - 6)$

30. $9 + 4^3 \times (20 - 8) \div 2 + 6$

31. $96 \div 4^2 + (25 \times 2) - 15 - 3$

32. What is the value of 100 divided by 5 times 4 minus 79?

33. Using symbols, write the product of 7 and 6 minus 2.

MUSIC For Exercises 34 and 35, use the following information.
A store sells DVDs for $20 each and CDs for $12.

34. Write an expression for the total cost of 4 DVDs and 2 CDs.

35. What is the total cost of the items?

HOMEWORK HELP

For Exercises	See Examples
14–32	1–5
34–37	6

Extra Practice
See pages 595, 624.

ICE CREAM For Exercises 36 and 37, use the table shown and the following information.
Suppose you and your friends place an order for 2 scoops of cookie dough ice cream, 1 scoop of chocolate ice cream, and 3 scoops of strawberry ice cream.

Fat Grams per Ice Cream Scoop	
Flavor	Fat (g)
cookie dough	17
vanilla	7
chocolate	5
strawberry	7

36. Write an expression to find the total number of fat grams in the order.

37. Find the total number of fat grams in the order.

38. **CRITICAL THINKING** Write an expression whose value is 10. It should contain four numbers and two different operations.

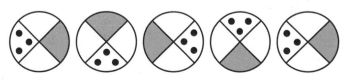
Standardized Test Practice and Mixed Review

39. **SHORT RESPONSE** Find the value of $6 \times 5 - 9 \div 3$.

40. **MULTIPLE CHOICE** Which expression could be used to find the total cost of 2 adult and 4 student admissions to the football game?

Football Game Ticket
Adult Admission: $5
Student Admission: $2

  Ⓐ $2 + \$5 \times 4 + \$2$ Ⓑ $\$5 + \$2 + 6$

  Ⓒ $\$2 + 4 \times \$5 + 2$ Ⓓ $2 \times \$5 + 4 \times \$2$

41. Write 7^4 as a product. What is the value of 7^4? (Lesson 1-4)

Find the prime factorization of each number. (Lesson 1-3)

42. 42   43. 75   44. 110   45. 130

46. Tell whether 15,080 is divisible by 2, 3, 4, 5, 6, 9, or 10. Then classify the number as *even* or *odd*. (Lesson 1-2)

47. **PATTERNS** Draw the next three figures in the pattern. (Lesson 1-1)

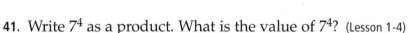

PREREQUISITE SKILL Add. (Page 589)

48. $26 + 98$ 49. $23 + 16$ 50. $61 + 19$ 51. $54 + 6$

Algebra: Variables and Expressions

What You'll LEARN

Evaluate algebraic expressions.

NEW Vocabulary

algebra
variable
algebraic expression
evaluate

HANDS-ON Mini Lab

Materials
• cups
• counters

Work with a partner.

You can use cups and counters to model and find the value of an expression. Follow these steps to model the expression *the sum of two and some number*.

STEP 1 Let two counters represent the known value, 2. Let one cup represent the unknown value.

STEP 2 Have your partner assign a value to the unknown number by placing any number of counters into the cup.

STEP 3 Empty the cup and count the number of counters. Add this amount to the 2 counters. What is the value of the expression?

1. Model *the sum of five and some number*.

2. Find the value of the expression if the unknown value is 4.

3. Write a sentence explaining how to evaluate an expression like *the sum of some number and seven* when the unknown value is given.

Algebra is a language of symbols. One symbol that is often used is a variable. A **variable** is a symbol, usually a letter, used to represent a number. The expression $2 + n$ represents *the sum of two and some number*.

The expression $2 + n$ is an algebraic expression. **Algebraic expressions** are combinations of variables, numbers, and at least one operation.

STUDY TIP

Expression
When a plus sign or minus sign separates an algebraic expression into parts, each part is called a *term*.

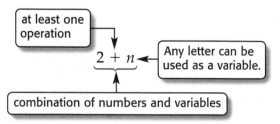

The letter x is often used as a variable. It is also common to use the first letter of the value you are representing.

The variables in an expression can be replaced with any number. Once the variables have been replaced, you can **evaluate**, or find the value of, the algebraic expression.

EXAMPLES Evaluate Algebraic Expressions

1 Evaluate $16 + b$ if $b = 25$.

$$16 + b = 16 + 25 \quad \text{Replace } b \text{ with 25.}$$
$$ = 41 \quad \text{Add 16 and 25.}$$

2 Evaluate $x - y$ if $x = 64$ and $y = 27$.

$$x - y = 64 - 27 \quad \text{Replace } x \text{ with 64 and } y \text{ with 27.}$$
$$ = 37 \quad \text{Subtract 27 from 64.}$$

In addition to the symbol ×, there are several other ways to show multiplication.

$2 \cdot 3$	means	2×3
$5t$	means	$5 \times t$
st	means	$s \times t$

EXAMPLE Evaluate an Algebraic Expression

3 Evaluate $5t + 4$ if $t = 3$.

$$5t + 4 = 5 \times 3 + 4 \quad \text{Replace } t \text{ with 3.}$$
$$ = 15 + 4 \quad \text{Multiply 5 and 3.}$$
$$ = 19 \quad \text{Add 15 and 4.}$$

Your Turn Evaluate each expression if $a = 6$ and $b = 4$.

a. $a + 8$ b. $a - b$ c. $a \cdot b$ d. $2a - 5$

EXAMPLE Evaluate an Algebraic Expression

4 **MULTIPLE-CHOICE TEST ITEM** Find the value of $2x^2 - 6 \cdot 2$ if $x = 7$.

Ⓐ 86   Ⓑ 90   Ⓒ 95   Ⓓ 98

Read the Test Item

You need to find the value of the expression.

Solve the Test Item

$$2x^2 - 6 \cdot 2 = 2 \cdot 7^2 - 6 \cdot 2 \quad \text{Replace } x \text{ with 7.}$$
$$ = 2 \cdot 49 - 6 \cdot 2 \quad \text{Evaluate } 7^2.$$
$$ = 98 - 6 \cdot 2 \quad \text{Multiply 2 and 49.}$$
$$ = 98 - 12 \quad \text{Multiply 6 and 2.}$$
$$ = 86 \quad \text{Subtract 12 from 98.}$$

The answer is A.

Writing
Math
Exercise 3

1. **Complete** the table.

Algebraic Expressions	Variables	Numbers	Operations
$7a - 3b$	?	?	?
$2w - 3x + 4y$	?	?	?

2. **OPEN ENDED** Write an expression that means the same as $3t$.

3. **Which One Doesn't Belong?** Identify the expression that is not an algebraic expression. Explain your reasoning.

| $5x$ | $3 + 4$ | ab | $7x + 1$ |

GUIDED PRACTICE

Evaluate each expression if $m = 4$ and $z = 9$.

4. $3 + m$ 5. $z - m$ 6. $4m - 2$

Evaluate each expression if $a = 2$, $b = 5$, and $c = 7$.

7. $4c - 5 + 3$ 8. $6a \div 3 + 4$ 9. $10 + 6b \times 3$

10. Find the value of $5a \div 2 + b^2 - c$ if $a = 2$, $b = 5$, and $c = 7$.

Practice and Applications

Evaluate each expression if $m = 2$ and $n = 16$.

11. $m + 10$ 12. $n + 8$ 13. $9 - m$

14. $22 - n$ 15. $n \div 4$ 16. $12 \div m$

17. $n \times 3$ 18. $6 \cdot m$ 19. $m + n$

HOMEWORK HELP

For Exercises	See Examples
11–45	1–4

Extra Practice
See pages 595, 624.

Evaluate each expression if $a = 4$, $b = 15$, and $c = 9$.

20. $b - a$ 21. $5c + 6$ 22. $2 \cdot b + 7$ 23. $c^2 + a$

24. $b^2 - 5c$ 25. $a^2 \times b$ 26. $3c - 4$ 27. $4b - 10$

28. $3a \div 4$ 29. $4b \div 5$ 30. $5b \times 2$ 31. $2ac$

Evaluate each expression if $x = 3$, $y = 12$, and $z = 8$.

32. $4z + 8 - 6$ 33. $6x - 9 \div 3$ 34. $5y \div 4 - 7$ 35. $15 + 9x \div 3$

36. $19 - 2z + 1$ 37. $7z \div 4 + 5x$ 38. $x^2 + 5y - z^2$ 39. $y^2 - 3z + 7$

40. $y^2 \div (3 \cdot z)$ 41. $z^2 - (5x)$ 42. $(2z)^2 + 3x^2 - y$ 43. $2z^2 + 3x - z$

44. Find the value of $2t + 2w$ if $t = 350$ and $w = 190$.

45. What is the value of $st \div 6r$ if $r = 5$, $s = 32$, and $t = 45$?

46. **BALLOONING** Distance traveled can be found using the expression rt, where r represents rate and t represents time. How far did a hot air balloon travel at a rate of 15 miles per hour for 6 hours?

47. **CIRCLES** The expression $2r$ can be used to find the diameter of a circle, where r is the length of the radius. Find the diameter of the compact disc shown.

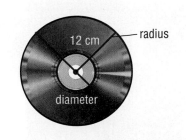

12 cm — radius
diameter

48. **PLANES** To find the speed of an airplane, use the expression $d \div t$ where d represents distance and t represents time. Find the speed s of a plane that travels 3,636 miles in 9 hours.

WEATHER For Exercises 49–51, use the following information and the table shown.
To change a temperature given in degrees Celsius to degrees Fahrenheit, first multiply the Celsius temperature by 9. Next, divide the answer by 5. Then add 32 to the result.

49. Write an expression that can be used to change a temperature from degrees Celsius to degrees Fahrenheit.

50. What is the average high February temperature in degrees Fahrenheit for Cabo San Lucas, Mexico?

51. Find the average high April temperature in degrees Fahrenheit for Cabo San Lucas, Mexico.

Average Monthly High Temperatures for Cabo San Lucas, Mexico	
Month	Temp. (°C)
February	25
April	30
July	35

Source: *The World Almanac*

52. **CRITICAL THINKING** Elan and Robin each have a calculator. Elan starts at zero and adds 3 each time. Robin starts at 100 and subtracts 7 each time. Will their displays ever show the same number if they press the keys at the same time? If so, what is the number?

Standardized Test Practice and Mixed Review

53. **MULTIPLE CHOICE** If $m = 6$ and $n = 4$, find the value of $2m - n$.
   Ⓐ 2 Ⓑ 4 Ⓒ 6 Ⓓ 8

54. **MULTIPLE CHOICE** The table shows the results of a survey in which people were asked to name the place they use technology when away from home. Which expression represents the total number of people who participated in the survey?

   Ⓕ $x - 183$ Ⓖ $x + 173$
   Ⓗ $x + 183$ Ⓘ $2x + 183$

Where People Use Technology	
Place	Number of Responses
on vacation	57
while driving	57
outdoors	42
mall	x
gym	17
public restrooms	10

Source: Market Facts for Best Buy

Find the value of each expression. (Lesson 1-5)

55. $12 - 8 \div 2 + 1$

56. $5^2 + (20 \div 2) - 7$

57. **SCIENCE** The distance from Earth to the Sun is close to 10^8 miles. How many miles is this? (Lesson 1-4)

GETTING READY FOR THE NEXT LESSON

PREREQUISITE SKILL Subtract. (Page 589)

58. $18 - 9$ 59. $25 - 18$ 60. $104 - 39$ 61. $211 - 105$

What You'll Learn
Solve problems by using the guess and check strategy.

Guess and Check

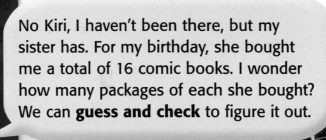

Hey Trent, have you been to that new comic book store? They sell used comic books in packages of 5 and the new comic books in packages of 3.

No Kiri, I haven't been there, but my sister has. For my birthday, she bought me a total of 16 comic books. I wonder how many packages of each she bought? We can **guess and check** to figure it out.

Explore	We know that the store sells 3-book packages and 5-book packages. We need to find the number of packages of each that were bought.
Plan	Make a guess until you find an answer that makes sense for your problem.
Solve	

Number of 3-book packages	Number of 5-book packages	Total Number of Comic Books
1	1	1(3) + 1(5) = 8
1	2	1(3) + 2(5) = 13
2	1	2(3) + 1(5) = 11
2	2	2(3) + 2(5) = 16 ✔

So, Trent's sister bought two 3-book packages and two 5-book packages.

Examine	Two 3-book packages result in 6 books. Two 5-book packages result in 10 books. Since 6 + 10 is 16, the answer is correct.

Analyze the Strategy

1. **Explain** how you can use the guess and check strategy to solve the following problem.

   A wallet contains 14 bills worth $200. If all of the money was in $5 bills, $10 bills, and $20 bills, how many of each bill was in the wallet?

2. **Explain** when to use the guess and check strategy to solve a problem.

3. **Write** a problem that can be solved using guess and check. Then tell the steps you would take to find the solution of the problem.

Solve. Use the guess and check strategy.

4. **MONEY** Mateo has seven coins that total $1.50. What are the coins?

5. **NUMBER THEORY** What prime number between 60 and 80 is one more than the product of two factors that are consecutive numbers?

Mixed Problem Solving

Solve. Use any strategy.

6. **GEOGRAPHY** Refer to the table. How much higher is Mount Hood than Mount Marcy?

Mountain	Elevation (ft)
Mount Marcy	5,344
Mount Kea	13,796
Mount Mitchell	6,684
Mount Hood	11,239

7. **SCIENCE** Each hand in the human body has 27 bones. There are 6 more bones in your fingers than in your wrist. There are 3 fewer bones in your palm than in your wrist. How many bones are in each part of your hand?

8. **SCHOOL** Paige studied 115 spelling words in five days. How many words did she study each day if she studied the same amount of words each day?

9. **PATTERNS** Draw the next figure in the pattern.

10. **JOBS** Felisa works after school at a bicycle store. Her hourly wage is $6. If Felisa works 32 hours, how much does she make?

11. **SCHOOL** Use the graph below. How many more 6th graders made the honor roll than 5th graders in the first trimester?

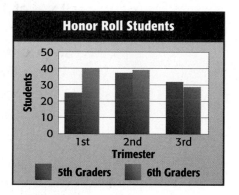

12. **NUMBERS** Maxine is thinking of four numbers from 1 through 9 whose sum is 23. Find the numbers.

13. **MULTI STEP** James is collecting money for a jump-a-thon. His goal is to collect $85. So far he has collected $10 each from two people and $5 each from six people. How far away is he from his goal?

14. **STANDARDIZED TEST PRACTICE** Willow purchased a new car. Her loan, including interest, is $12,720. How much are her monthly payments if she has 60 monthly payments to make?

  Ⓐ $250   Ⓑ $212

  Ⓒ $225   Ⓓ $242

You will use the guess and check strategy in the next lesson.

Algebra: Solving Equations

What You'll LEARN

Solve equations by using mental math and the guess and check strategy.

NEW Vocabulary

equation
equals sign
solve
solution

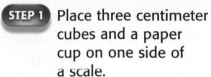 **HANDS-ON** **Mini Lab**

Materials

- balance scale
- paper cup
- centimeter cubes

Work with a partner.

When the amounts on each side of a scale are equal, the scale is balanced.

STEP 1 Place three centimeter cubes and a paper cup on one side of a scale.

STEP 2 Place eight centimeter cubes on the other side of the scale.

1. Suppose the variable x represents the number of cubes in the cup. What equation represents this situation?

2. Replace the cup with centimeter cubes until the scale balances. How many centimeter cubes did you need to balance the scale?

Let x represent the cup. Model each sentence on a scale. Find the number of centimeter cubes needed to balance the scale.

3. $x + 1 = 4$ 4. $x + 3 = 5$

5. $x + 7 = 8$ 6. $x + 2 = 2$

An **equation** is a sentence that contains an **equals sign**, $=$. A few examples are shown below.

$$2 + 7 = 9 \qquad 10 - 6 = 4 \qquad 5 - 1 = 4$$

Some equations contain variables.

$$2 + x = 9 \qquad k - 6 = 4 \qquad 5 - m = 4$$

When you replace a variable with a value that results in a true sentence, you **solve** the equation. The value for the variable is the **solution** of the equation.

The equation is $2 + x = 9$. → $2 + x = 9$

$2 + 7 = 9$

The value for the variable that results in a true sentence is 7. So, 7 is the solution. → $9 = 9$ ← This sentence is true.

Find the Solution of an Equation

1 Which of the numbers 3, 4, or 5 is the solution of the equation $a + 7 = 11$?

Value of a	$a + 7 \stackrel{?}{=} 11$	Are Both Sides Equal?
3	$3 + 7 = 11$ $10 \neq 11$	no
4	$4 + 7 = 11$ $11 = 11$	yes ✔
5	$5 + 7 = 11$ $12 \neq 11$	no

The solution of $a + 7 = 11$ is 4 because replacing a with 4 results in a true sentence.

Your Turn Identify the solution of each equation from the list given.

a. $15 - k = 6$; 7, 8, 9   b. $n + 12 = 15$; 2, 3, 4

Some equations can be solved mentally.

EXAMPLE **Solve an Equation Mentally**

2 Solve $5 + h = 12$ mentally.

$5 + h = 12$ THINK   What number plus 5 equals 12?

$5 + 7 = 12$ You know that $5 + 7 = 12$.

$h = 7$ The solution is 7.

Your Turn Solve each equation mentally.

c. $18 - d = 10$ d. $w + 7 = 16$

REAL-LIFE MATH

ANIMALS The combined head and body length of a ground squirrel can range from 4 inches to 14 inches.

Source: www.infoplease.com

You can also solve equations by using guess and check.

EXAMPLE **Use Guess and Check**

3 **ANIMALS** An antelope can run 49 miles per hour faster than a squirrel. Solve the equation $s + 49 = 61$ to find the speed of a squirrel.

Estimate Since $10 + 50 = 60$, the solution should be about 10.
Use guess and check.

Try 10.

$s + 49 = 61$
$10 + 49 \stackrel{?}{=} 61$
$59 \neq 61$

Try 11.

$s + 49 = 61$
$11 + 49 \stackrel{?}{=} 61$
$60 \neq 61$

Try 12.

$s + 49 = 61$
$12 + 49 \stackrel{?}{=} 61$
$61 = 61$ ✔

The solution is 12. So, the speed of the squirrel is 12 miles per hour.

1. **Write** the meaning of the term *solution*.

2. **OPEN ENDED** Give an example of an equation that has a solution of 5.

GUIDED PRACTICE

Identify the solution of each equation from the list given.

3. $9 + w = 17; 7, 8, 9$

4. $d - 11 = 5; 14, 15, 16$

5. $4 = y + 2; 2, 3, 4$

6. $8 - c = 3; 4, 5, 6$

Solve each equation mentally.

7. $x + 6 = 18$

8. $n - 10 = 30$

9. $15 + k = 30$

10. **MONEY** Lexie bought a pair of in-line skates for $45 and a set of kneepads. The total cost of the items before tax was $63. The equation $45 + k = 63$ represents this situation. Find the cost of the kneepads.

Practice and Applications

Identify the solution of each equation from the list given.

11. $a + 15 = 23; 6, 7, 8$

12. $29 + d = 54; 24, 25, 26$

13. $35 = 45 - n; 10, 11, 12$

14. $p - 12 = 19; 29, 30, 31$

15. $59 + w = 82; 23, 24, 25$

16. $95 = 18 + m; 77, 78, 79$

17. $45 - h = 38; 6, 7, 8$

18. $66 = z - 23; 88, 89, 90$

HOMEWORK HELP	
For Exercises	See Examples
11–18	1
23–34	2
21, 35, 37	3

Extra Practice
See pages 596, 624.

19. *True* or *False*? If $27 = x + 11$, then $x = 16$.

20. *True* or *False*? If $14 + b = 86$, then $b = 78$.

21. **ANIMALS** The equation $b - 4 = 5$ describes the approximate difference in length between the largest bear and smallest bear. If b is the length of the largest bear in feet, what is the approximate length of the largest bear?

22. **RESEARCH** Use the Internet or another reference source to find the only bear species that is found in the Southern Hemisphere. What is the approximate length and weight of this bear?

Solve each equation mentally.

23. $j + 7 = 13$

24. $m + 4 = 17$

25. $16 + h = 24$

26. $14 + t = 30$

27. $15 - b = 12$

28. $25 - k = 20$

29. $22 = 30 - m$

30. $12 = 24 - y$

31. $3.5 + k = 8.0$

32. $24 + d = 40$

33. $100 - c = 25$

34. $15 = 50 - m$

35. **FOOD** The equation $b + 7 = 12$ describes the number of boxes of cereal and the number of breakfast bars in a kitchen cabinet. If b is the number of breakfast bars, how many breakfast bars are there?

36. Tell whether the statement below is *sometimes*, *always*, or *never* true. *Equations like $a + 4 = 8$ and $4 - m = 2$ have one solution.*

37. SCHOOL Last year, the number of students attending Glenwood Middle School was 575. There are 650 students this year. The equation $575 + n = 650$ shows the increase in the number of students from one year to the next. Find the number of new students n.

CRITICAL THINKING For Exercises 38 and 39, tell whether each statement is *true* or *false*. Then explain your reasoning.

38. In $m + 8$, the variable m can have any value.

39. In $m + 8 = 12$, the variable m can have any value for a solution.

EXTENDING THE LESSON A sentence that contains a greater than sign, $>$, or a less than sign, $<$, is an *inequality*. To *solve an inequality*, find the values of the variable that makes the inequality true.

Example
Which of the numbers 2, 3, or 4 is a solution of $x + 3 > 5$?

The solutions are 3 and 4.

Identify the solution(s) of each inequality from the list given.

Value for x	$x + 3 > 5$	Is $x + 3 > 5$?
2	$2 + 3 \overset{?}{>} 5$ $5 \not> 5$	no
3	$3 + 3 \overset{?}{>} 5$ $6 > 5$	yes
4	$4 + 3 \overset{?}{>} 5$ $7 > 5$	yes

40. $x + 2 > 6$; 4, 5, 6

41. $m + 5 > 7$; 3, 4, 5

42. $10 < a + 3$; 6, 7, 8

43. $12 < 8 + h$; 3, 4, 5

Standardized Test Practice and Mixed Review

44. MULTIPLE CHOICE The equation $16 + g = 64$ represents the total cost of a softball and a softball glove. The variable g represents the cost of the softball glove. What is the cost of the softball glove?

   A $8   **B** $48   **C** $64   **D** $80

45. MULTIPLE CHOICE Which equation is true when $m = 8$?

   F $24 = 15 + m$ **G** $m - 4 = 12$

   H $26 - m = 18$ **I** $16 = 21 - m$

Evaluate the expression if $r = 2$, $s = 4$, and $t = 6$. (Lesson 1-6)

46. $3rst + 14$ **47.** $9 \div 3 \cdot s + t$ **48.** $4 + t \div r \cdot 4s$

49. Find the value of $12 \div 3 \times 3 - 8$. (Lesson 1-5)

GETTING READY FOR THE NEXT LESSON

PREREQUISITE SKILL Multiply. (Page 590)

50. 8×12 **51.** 6×15 **52.** 4×18 **53.** 5×17

Study Skill

HOW TO...

Read Math Problems

BEST WAY TO SOLVE

When you have a math problem to solve, don't automatically reach for your paper and pencil. There is more than one way to solve a problem.

Before you start to solve a problem, read it carefully. Look for key words and numbers that will give you hints about how to solve it.

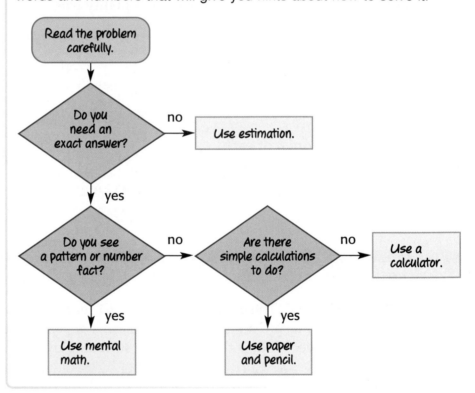

SKILL PRACTICE

Choose the best method of computation. Then solve each problem.

1. **SHOPPING** The Mall of America in Minnesota contains 4,200,000 square feet of shopping. The Austin Mall in Texas contains 1,100,000 square feet. How many more square feet are there in the Mall of America than the Austin Mall?

2. **MONEY** One kind of money used in the country of Armenia is called a *dram*. In 2001, one U.S. dollar was equal to 552 drams. If you exchanged $25 for drams, how many drams would you have received?

3. **MUSIC** Suppose you want to buy a drum set for $359 and drum cases for $49. You've already saved $259 toward buying these items. About how much more do you need to have?

Geometry: Area of Rectangles

What You'll LEARN

Find the areas of rectangles.

NEW Vocabulary

area
formula

WHEN am I ever going to use this?

PATTERNS Checkered patterns can often be found on game boards and flags.

1. Complete the table below.

Object	Squares Along the Length	Squares Along the Width	Squares Needed to Cover the Surface
flag			
game board			

2. What relationship exists between the length and the width, and the number of squares needed to cover the surface?

The **area** of a figure is the number of square units needed to cover a surface. The rectangle shown has an area of 24 square units.

4 units | 24 square units | 6 units

You can also use a formula to find the area of a rectangle. A **formula** is an equation that shows a relationship among certain quantities.

Key Concept — Area of a Rectangle

Words	The area A of a rectangle is the product of the length ℓ and width w.	Model	ℓ w
Formula	$A = \ell \times w$		

EXAMPLE Find the Area of a Rectangle

1 Find the area of a rectangle with length 12 feet and width 7 feet.

$A = \ell \times w$ Area of a rectangle

$A = 12 \times 7$ Replace ℓ with 12 and w with 7.

$A = 84$

The area is 84 square feet.

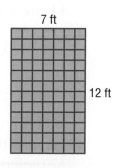

7 ft

12 ft

SPORTS A high school volleyball court is 60 feet long and 30 feet wide.

Source: NFHS

EXAMPLE Use Area to Solve a Problem

2 **SPORTS** Use the information at the left. A high school basketball court measures 84 feet long and 50 feet wide. What is the difference between the area of a basketball court and a volleyball court?

Area of a Basketball Court

$A = \ell \times w$ Area of a rectangle

$A = 84 \times 50$ Replace ℓ with 84 and w with 50.

$A = 4{,}200$ Multiply.

Area of a Volleyball Court

$A = \ell \times w$ Area of a rectangle

$A = 60 \times 30$ Replace ℓ with 60 and w with 30.

$A = 1{,}800$ Multiply.

To find the difference, subtract.
$4{,}200 - 1{,}800 = 2{,}400$

The area of a high school basketball court is 2,400 square feet greater than the area of a high school volleyball court.

Your Turn Find the area of each rectangle.

a.
12 in.

6 in.

b.
10 m
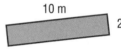
2 m

Skill and Concept Check

1. **Explain** how to find the area of a rectangle.

2. **OPEN ENDED** Draw and label a rectangle that has an area of 48 square units.

3. **NUMBER SENSE** Give the dimensions of two different rectangles that have the same area.

GUIDED PRACTICE

Find the area of each rectangle.

4.
5 cm
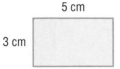
3 cm

5.
8 ft

15 ft

6.
40 cm
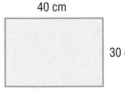
30 cm

7. What is the area of a rectangle with a length of 9 meters and a width of 17 meters?

Practice and Applications

Find the area of each rectangle.

8.
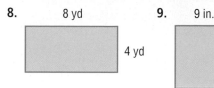
8 yd
4 yd

9.
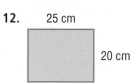
9 in.
10 in.

10.
14 ft
6 ft

HOMEWORK HELP

For Exercises	See Examples
8–15	1
19–20	2

Extra Practice
See pages 596, 624.

11. 16 m
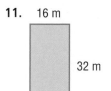
32 m

12. 25 cm
20 cm

13. 48 ft
17 ft

14. Find the area of a rectangle with a length of 26 inches and a width of 12 inches.

15. What is the area of a rectangle having a width of 22 feet and a length of 19 feet?

Estimate the area of each rectangle.

16.
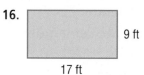
9 ft
17 ft

17.
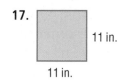
11 in.
11 in.

18. 55 yd
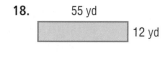
12 yd

19. **TECHNOLOGY** A television screen measures 9 inches by 12 inches. What is the area of the viewing screen?

20. **SCHOOL** A 3-ring binder measures 11 inches by 10 inches. What is the area of the front cover of the binder?

21. **CRITICAL THINKING** Suppose opposite sides of a rectangle are increased by 3 units. Would the area of the rectangle increase by 6 square units? Use a model in your explanation.

Standardized Test Practice and Mixed Review

22. **MULTIPLE CHOICE** Which rectangle has an area of 54 square units?
    Ⓐ length: 8 units; width: 6 units Ⓑ length: 9 units; width: 4 units
    Ⓒ length: 8 units; width: 8 units Ⓓ length: 9 units; width: 6 units

23. **SHORT RESPONSE** What is the area of a rectangle with length of 50 feet and width of 55 feet?

Solve each equation mentally. (Lesson 1-7)

24. $x + 4 = 12$ 25. $9 - m = 5$ 26. $k - 8 = 20$

27. **ALGEBRA** Evaluate $ab - c$ if $a = 3$, $b = 16$, and $c = 5$. (Lesson 1-6)

Study Guide and Review

Vocabulary and Concept Check

algebra (p. 28)	equation (p. 34)	order of operations (p. 24)
algebraic expression (p. 28)	evaluate (p. 29)	power (p. 18)
area (p. 39)	even (p. 10)	prime factorization (p. 15)
base (p. 18)	exponent (p. 18)	prime number (p. 14)
composite number (p. 14)	factor (p. 14)	solution (p. 34)
cubed (p. 18)	formula (p. 39)	solve (p. 34)
divisible (p. 10)	numerical expression (p. 24)	squared (p. 18)
equals sign (p. 34)	odd (p. 10)	variable (p. 28)

Choose the correct term or number to complete each sentence.

1. When two or more numbers are multiplied, each number is called a (base, factor) of the product.

2. A(n) (variable, algebraic expression) represents an unknown value.

3. A whole number is (squared, divisible) by another number if the remainder is 0 when the first is divided by the second.

4. Numbers expressed using exponents are (powers, prime numbers).

5. A whole number that has exactly two factors, 1 and the number itself, is a (prime number, composite number).

6. An example of an algebraic expression is $(3w - 8, 2)$.

7. The value of the variable that makes the two sides of an equation equal is the (solution, factor) of the equation.

Lesson-by-Lesson Exercises and Examples

1-1 **A Plan for Problem Solving** (pp. 6–9)

Use the four-step plan to solve each problem.

8. **SCHOOL** Tickets to a school dance cost $4 each. If $352 was collected, how many people bought tickets?

9. **HISTORY** In 1932, there were two presidential candidates who won electoral votes. Franklin Roosevelt won 472 electoral votes while Herbert Hoover had 59. How many electoral votes were cast?

Example 1 Meg studied 2 hours each day for 10 days. How many hours has she studied?

Explore You need to find the total number of hours Meg studied.

Plan Multiply 2 by 10.

Solve $2 \times 10 = 20$
So, Meg studied 20 hours.

Examine Since $20 \div 2 = 10$, the answer makes sense.

 msmath1.net/vocabulary_review

1-2 Divisibility Patterns (pp. 10–13)

Tell whether each number is divisible by 2, 3, 4, 5, 6, 9, or 10. Then classify each number as *even* or *odd*.

10. 85   11. 90
12. 60   13. 75
14. 72   15. 128
16. 554   17. 980
18. 3,065   19. 44,000

20. **SCHOOL** Can 157 calculators be divided evenly among 6 classrooms? Explain.

21. **FOOD** List all of the ways 248 fruit bars can be packaged so that each package has the same number, but less than 10 fruit bars.

Example 2 Tell whether 726 is divisible by 2, 3, 4, 5, 6, 9, or 10. Then classify 726 as *even* or *odd*.

2: Yes; the ones digit is divisible by 2.

3: Yes; the sum of the digits, 15, is divisible by 3.

4: No, the number formed by the last two digits, 26, is not divisible by 4.

5: No, the ones digit is not a 0 or a 5.

6: Yes; the number is divisible by both 2 and 3.

9: No; the sum of the digits is not divisible by 9.

10: No, the ones digit is not 0.

So, 726 is divisible by 2, 3, and 6. The number 726 is even.

1-3 Prime Factors (pp. 14–17)

Tell whether each number is *prime*, *composite*, or *neither*.

22. 44   23. 67

Find the prime factorization of each number.

24. 42   25. 75
26. 96   27. 88

Example 3 Find the prime factorization of 18.

Make a factor tree.

18	Write the number to be factored.
2 × 9	Choose any two factors of 18.
2 × 3 × 3	Continue to factor any number that is not prime until you have a row of prime numbers.

The prime factorization of 18 is 2 × 3 × 3.

1-4 Powers and Exponents (pp. 18–21)

Write each product using an exponent. Then find the value of the power.

28. $5 \cdot 5 \cdot 5 \cdot 5$

29. $1 \cdot 1 \cdot 1 \cdot 1 \cdot 1$

30. $12 \cdot 12 \cdot 12$

31. $3 \cdot 3 \cdot 3 \cdot 3 \cdot 3 \cdot 3 \cdot 3 \cdot 3$

Example 4 Write $4 \cdot 4 \cdot 4 \cdot 4 \cdot 4 \cdot 4$ using an exponent. Then find the value of the power.

The base is 4. Since 4 is a factor 6 times, the exponent is 6.

$4 \cdot 4 \cdot 4 \cdot 4 \cdot 4 \cdot 4 = 4^6$ or 4,096

1-5 **Order of Operations** (pp. 24–27)

Find the value of each expression.
32. $16 - 9 + 3^2$
33. $24 \div 8 + 4 - 7$
34. $4 \times 6 + 2 \times 3$
35. $8 + 3^3 \times 4$
36. $10 + 15 \div 5 - 6$
37. $11^2 - 6 + 3 \times 15$

Example 5 Find the value of
$28 \div 2 - 1 \times 5$.
$28 \div 2 - 1 \times 5$
$\quad = 14 - 1 \times 5$ Divide 28 by 2.
$\quad = 14 - 5$ Multiply 1 and 5.
$\quad = 9$ Subtract 5 from 14.
The value of $28 \div 2 - 1 \times 5$ is 9.

1-6 **Algebra: Variables and Expressions** (pp. 28–31)

Evaluate each expression if $a = 18$ and $b = 6$.
38. $a - 7$
39. $b + a$
40. $a \times b$
41. $a^2 \div b$
42. $3b^2 + a$
43. $2a - 10$

Evaluate each expression if $x = 6$, $y = 8$, and $z = 12$.
44. $8y - 9$
45. $2x + 4y$
46. $3z^2 - 4x$
47. $z \div 3 + x \times y$

Example 6 Evaluate $9 - k^3$ if $k = 2$.
$9 - k^3 = 9 - 2^3$ Replace k with 2.
$\quad = 9 - 8$ $2^3 = 8$
$\quad = 1$ Subtract 8 from 9.

Example 7 Evaluate $10 + m \times n$ if $m = 3$ and $n = 5$.
$10 + m \times n = 10 + 3 \times 5$ $m = 3$ and $n = 5$
$\quad = 10 + 15$ Multiply 3 and 5.
$\quad = 25$ Add 10 and 15.

1-7 **Algebra: Solving Equations** (pp. 34–37)

Solve each equation mentally.
48. $p + 2 = 9$
49. $c + 7 = 14$
50. $20 + y = 25$
51. $40 = 15 + m$
52. $10 - w = 7$
53. $16 - n = 10$
54. $27 = x - 3$
55. $17 = 25 - h$

Example 8 Solve $x + 9 = 13$ mentally.
$x + 9 = 13$ What number plus 9 is 13?
$4 + 9 = 13$ You know that $4 + 9$ is 13.
$\quad x = 4$ The solution is 4.

1-8 **Geometry: Area of Rectangles** (pp. 39–41)

56. Find the area of the rectangle below.

13 cm

2 cm

57. **SCHOOL** Find the area of a chalkboard that measures 4 feet by 12 feet.

Example 9
Find the area of the rectangle at the right.

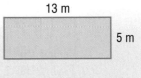
13 m

5 m

$A = \ell \times w$ Area of a rectangle
$\quad = 13 \times 5$ Replace ℓ with 13 and w with 5.
$\quad = 65$
The area is 65 square meters.

Practice Test

Vocabulary and Concepts

1. **OPEN ENDED** Give an example of a prime number.

2. **Define** *solution* of an equation.

Skills and Applications

Tell whether each number is divisible by 2, 3, 4, 5, 6, 9, or 10. Then classify each number as *even* or *odd*.

3. 95

4. 106

5. 3,300

Write each product using an exponent. Then find the value of the power.

6. $8 \cdot 8 \cdot 8$

7. $2 \cdot 2 \cdot 2 \cdot 2 \cdot 2 \cdot 2$

8. $9 \cdot 9 \cdot 9 \cdot 9$

9. **SOCCER** The number of goals Jacob scored in the first four years of playing soccer is shown. At this rate, how many goals will he score at the end of the fifth year?

Year	Goals
1	4
2	5
3	7
4	10
5	?

Find the prime factorization of each number.

10. 42

11. 68

12. 105

13. Write 3^5 as a product. Then find the value of the product.

Find the value of each expression.

14. $12 - 3 \times 2 + 15$

15. $24 \div 8 + 4 \times 9$

16. $72 \div 2^3 - 4 \times 2$

Find the area of each rectangle.

17.
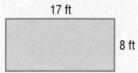
17 ft
8 ft

18.
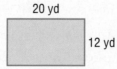
20 yd
12 yd

Evaluate each expression if $a = 4$ and $b = 3$.

19. $a + 12$

20. $27 \div b$

21. $a^3 - 2b$

Solve each equation mentally.

22. $d + 9 = 14$

23. $18 = 22 - m$

24. $27 - h = 18$

Standardized Test Practice

Ⓐ Ⓑ Ⓒ Ⓓ

25. **SHORT RESPONSE** Refer to the figure shown. How many times greater is the area of the large rectangle than the area of the small rectangle?

Standardized Test Practice

Record your answers on the answer sheet provided by your teacher or on a sheet of paper.

1. The total cost of 4 bikes was $1,600. Which expression can be used to find the cost of one bike if each bike cost the same amount? (Lesson 1-1)

   (A) $1,600 × 4   (B) $1,600 ÷ 4
   (C) $1,600 + 4   (D) 4 ÷ $1,600

2. Which figure is next in the pattern? (Lesson 1-1)

   WM∑WM

   (F) M   (G) ∑

   (H) ∑   (I) W

3. Which number divides into 120, 240, and 360 with no remainder, but will not divide evenly into 160 and 320? (Lesson 1-2)

   (A) 2   (B) 3   (C) 4   (D) 5

4. A number divides evenly into 27, 30, 33, and 36. Which of the following will the number also divide evenly into? (Lesson 1-2)

   (F) 47   (G) 48   (H) 49   (I) 50

5. Which of the following *best* describes the numbers 6, 15, 21, 28, and 32? (Lesson 1-3)

   (A) composite   (B) even
   (C) odd   (D) prime

Question 3 By eliminating the choices that divide evenly into the second set of numbers, you will then be left with the correct answer.

6. Which figure represents a prime number? (Lesson 1-3)

7. How is $3 × 3 × 3 × 3$ expressed using exponents? (Lesson 1-4)

   (A) 3^3 (B) 4^3
   (C) $3 × 3^3$ (D) 3^4

8. Which operation should be done first in $20 - 3 × 6 ÷ 2 + 3$? (Lesson 1-5)

   (F) $20 - 3$ (G) $3 × 6$
   (H) $6 ÷ 2$ (I) $2 + 3$

9. Find $10 + 50 ÷ 5 × 2 - 6$. (Lesson 1-5)

   (A) 0   (B) 9   (C) 18   (D) 24

10. What is the value of $(8 × r) - 4$ if $r = 4$? (Lesson 1-6)

    (F) 0   (G) 8   (H) 12   (I) 28

11. What is the solution of $12 = k + 5$? (Lesson 1-7)

    (A) 4   (B) 5   (C) 6   (D) 7

12. How many square units cover the rectangle shown? (Lesson 1-8)

    (F) 12 square units
    (G) 20 square units
    (H) 24 square units
    (I) 28 square units

PART 2 Short Response/Grid In

Record your answers on the answer sheet provided by your teacher or on a sheet of paper.

13. Order 769, 452, and 515 from least to greatest. (Prerequisite Skill, p. 588)

14. The table shows the distance Trevor rode his bike each day for a week. At this rate, how many miles will he ride on Saturday? (Lesson 1-1)

Day	Distance (mi)
Sunday	2
Monday	5
Tuesday	8
Wednesday	11
Thursday	?
Friday	?
Saturday	?

15. Sandra bought 3 pounds of grapes and 2 pounds of apples. What is the total amount she spent on fruit? (Lesson 1-1)

$2 per pound $3 per pound

16. Tell which number greater than one divides evenly into all of the numbers in the list. (Lesson 1-2)

| 36 |
| 51 |
| 81 |
| 99 |

17. Write 38 as a product of prime factors. (Lesson 1-3)

18. How is 4^6 written as a product? (Lesson 1-4)

19. Evaluate $12 - 10 \div 2 + 3 \times 2$. (Lesson 1-5)

20. What is the value of $42 - 3^2 + 7 \times 5$? (Lesson 1-5)

21. What is the value of $4b \div 3$, if $b = 3$? (Lesson 1-6)

22. Find the solution of $d + 32 = 45$. (Lesson 1-7)

23. Find the area of the rectangle. (Lesson 1-8)

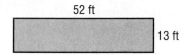

52 ft

13 ft

24. What is the width of the rectangle? (Lesson 1-8)

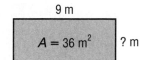

9 m

$A = 36$ m^2 ? m

PART 3 Extended Response

Record your answers on a sheet of paper. Show your work.

25. The table below shows how much Kyle has earned in each of the last four weeks doing chores. (Lesson 1-1)

Week	Amount ($)
1	2
2	3
3	5
4	9

a. How much can he expect to earn the fifth week if this pattern continues?

b. Explain the pattern.

26. List the numbers from 1 to 100. (Lesson 1-3)

a. Explain how you would find all of the prime numbers between 1 and 100.

b. What are the prime numbers between 1 and 100?

c. List three prime numbers between 200 and 250.

Statistics and Graphs

❝ What do butterflies have to do with statistics? ❞

Monarch butterflies are commonly found in eastern North America. Each fall, they migrate up to 3,000 miles to warmer climates. Scientists monitor the butterflies as they pass certain watch points along the migration route. By recording and analyzing the data, scientists can determine the distance and direction butterflies travel when migrating.

You will analyze data about butterflies in Lesson 2-5.

GETTING STARTED

Take this quiz to see whether you are ready to begin Chapter 2. Refer to the lesson or page number in parentheses if you need more review.

▶ Vocabulary Review

State whether each sentence is *true* or *false*. If *false*, replace the underlined word or number to make a true sentence.

1. According to the <u>rules</u> of operations, multiply or divide first. Then add or subtract. (Lesson 1-5)

2. An <u>expression</u> must include at least one operation. (Lesson 1-6)

▶ Prerequisite Skills

Add. (Page 589)

3. $16 + 28$

4. $39 + 25 + 11$

5. $63 + 9 + 37$

6. $74 + 14$

7. $8 + 56 + 10 + 7$

8. $44 + 18 + 5$

Divide. (Page 591)

9. $72 \div 9$

10. $96 \div 8$

11. $84 \div 2$

12. $102 \div 6$

13. $125 \div 5$

14. $212 \div 4$

Find the value of each expression.
(Lesson 1-5)

15. $15 - 4 + 2$

16. $6 + 35 \div 7$

17. $30 \div (8 - 3)$

18. $(2^5 \div 4) - 5$

19. $12 \times (4 \div 2) + 3^3$

20. $5^2 \times 2 - (5 \times 4)$

Statistics and Graphs
Make this Foldable to help you organize information about statistics and graphs. Begin with four sheets of graph paper.

STEP 1 **Fold Pages**
Fold each sheet of graph paper in half along the width.

STEP 2 **Unfold and Tape**
Unfold each sheet and tape to form one long piece.

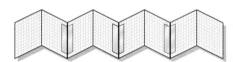

STEP 3 **Label**
Label the pages with the lesson numbers as shown.

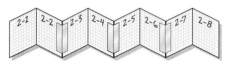

STEP 4 **Refold**
Refold the pages to form a journal.

Reading and Writing As you read and study the chapter, write notes and examples from each lesson.

Frequency Tables

What You'll LEARN

Make and interpret frequency tables.

NEW Vocabulary

statistics
data
frequency table
scale
interval
tally mark

WHEN am I ever going to use this?

TREES A state champion tree is the tallest tree of each species found in the state. The heights of New York's champion trees are listed in the table.

1. What is the height of the tallest tree?

2. How many trees are between 41 and 80 feet tall?

3. Tell how you might organize the heights of the trees so that the information is easier to find and read.

Heights (ft) of New York's State Champion Trees					
76	88	91	90	99	72
70	98	110	135	114	120
83	108	95	110	104	112
58	100	75	105	56	63
61	112	72	68	93	85
77	95	102	75	96	95
96	114	82	91	72	83

Source: NYS Big Tree Register, 2001

Statistics involves collecting, organizing, analyzing, and presenting data. **Data** are pieces of information that are often numerical. The data above can be organized in a frequency table. A **frequency table** shows the number of pieces of data that fall within given intervals.

The **scale** allows you to record all of the data. It includes the least number, 56, and the greatest, 135. The scale is 51 to 140.

Heights (ft) of New York's State Champion Trees		
Height	Tally	Frequency
51–80	̶H̶T̶ ̶H̶T̶ III	13
81–110	̶H̶T̶ ̶H̶T̶ ̶H̶T̶ ̶H̶T̶ III	23
111–140	̶H̶T̶ I	6

The scale is separated into equal parts called **intervals**. The interval is 30.

Tally marks are counters used to record items in a group.

EXAMPLE Make a Frequency Table

1. **SOCCER** The number of points scored by major league soccer teams in the 2001 season is shown. Make a frequency table of the data.

Major League Soccer Points Scored, 2001 Season			
53	26	35	45
42	53	14	36
27	45	47	23

Source: Major League Soccer

Step 1 Choose an appropriate scale and interval for the data.

scale: 1 to 60 The scale includes all of the data, and the
interval: 15   interval separates the scale into equal parts.

Step 2 Draw a table with three columns and label the columns *Points*, *Tally*, and *Frequency*.

Step 3 In the first column, list the intervals. In the second column, tally the data. In the third column, add the tallies.

Major League Soccer Points Scored, 2001 Season		
Points	**Tally**	**Frequency**
1–15	I	1
16–30	III	3
31–45	IIII	5
46–60	III	3

Some frequency tables may not have scales and intervals.

EXAMPLE Make a Frequency Table

2 **FOOD** Mr. Thompson asked his students to name their favorite food. Make a frequency table of the data that resulted from the survey.

C T H P
P C D H
D P T P
H P P P
P P D H
H C T D

P = pizza
T = taco
H = hamburger
D = hot dog
C = chicken

Draw a table with three columns. In the first column, list each food. Then complete the table.

Favorite Food		
Food	**Tally**	**Frequency**
pizza	IIII IIII	9
taco	III	3
hamburger	IIII	5
hot dog	IIII	4
chicken	III	3

You can analyze and interpret the data in a frequency table.

REAL-LIFE MATH

ENGLISH Some of the most common words in the English language are: *the, of, to, and, a, in, is, it, you,* and *that*.

Source: www.about.com

EXAMPLE Interpret a Frequency Table

3 **ENGLISH** The frequency table shows how often the five most common words in English appeared in a magazine article. What do you think is the most common word in English? How did you reach your conclusion?

Most Common Words		
Word	**Tally**	**Frequency**
to	IIII IIII IIII IIII I	21
of	IIII IIII III	13
the	IIII IIII IIII IIII IIII IIII IIII IIII IIII II	47
and	IIII IIII IIII IIII IIII I	26
a	IIII IIII IIII	14

According to the frequency table, the word "the" was used most often. This and other such articles may suggest that "the" is the most common word in English.

Skill and Concept Check

1. **Describe** how to find an appropriate scale for a set of data.

2. **Tell** an advantage of organizing data in a frequency table.

3. **OPEN ENDED** Write a data set containing 12 pieces of data whose frequency table will have a scale from 26 to 50.

GUIDED PRACTICE

Make a frequency table for each set of data.

4.

Tallest Buildings (ft) in Miami, Florida			
400	450	625	420
794	480	484	510
425	456	405	520
764	400	487	487

Source: *The World Almanac*

5.

Pets Owned by Various Students							
F	D	T	G	D	C	C	G
C	D	C	F	C	D	D	C
D	D	C	D	H	T	F	D

F = fish   D = dog   C = cat
T = turtle G = gerbil H = hamster

BASEBALL For Exercises 6 and 7, use the frequency table shown at the right.

6. Describe the scale and the interval.

7. How many players hit more than 64 home runs?

Most Home Runs in a Single Season		
Home Runs	Tally	Frequency
50–54	IIII IIII IIII III	18
55–59	IIII II	7
60–64	IIII	4
65–69	II	2
70–74	II	2

Source: Major League Baseball

Practice and Applications

Make a frequency table for each set of data.

8.

Cost ($) of Various Skateboards				
99	67	139	63	75
59	89	59	70	78
99	55	125	64	110

9.

Students' Monthly Trips to the Mall					
5	10	0	1	11	4
12	4	3	6	8	5
8	9	6	2	13	2

10.

Favorite Color					
R	G	B	R	R	K
B	P	P	Y	R	B
P	R	Y	K	B	Y
B	B	P	B	P	R

R = red   B = blue   Y = yellow
G = green   P = purple K = pink

11.

Favorite Type of Movie							
A	H	R	A	A	C	C	D
C	C	C	A	C	H	A	C
H	S	D	C	D	S	S	C
C	H	A	R	S	A	A	C

C = comedy   D = drama   A = action
R = romantic H = horror S = science fiction

HOMEWORK HELP

For Exercises	See Examples
8–11	1, 2
14–15	3

Extra Practice
See pages 596, 625.

12. Which scale is more appropriate for the data set 3, 17, 9, 6, 2, 5, and 12: 0 to 10 or 0 to 20? Explain your reasoning.

13. What is the best interval for the data set 245, 144, 489, 348, 36, 284, 150, 94, and 220: 10, 100, or 1,000? Explain your reasoning.

SCIENCE For Exercises 14 and 15, use the frequency table that lists the years of the total solar eclipses.

Total Solar Eclipses, 1960–2019		
Year	Tally	Frequency
1960–1969	ЖHТ I	6
1970–1979	ЖHТ II	7
1980–1989	ЖHТ III	8
1990–1999	ЖHТ III	8
2000–2009	ЖHТ II	7
2010–2019	ЖHТ II	7

Source: *The World Almanac*

14. In what decade did the least number of total solar eclipses occur?

15. Which decades had more than seven total solar eclipses?

16. **CRITICAL THINKING** Tell why a frequency table cannot contain the intervals 0–25, 25–50, 50–100, and 100–150.

EXTENDING THE LESSON In a frequency table, *cumulative frequencies* are the sums of all preceding frequencies. Refer to the table in Exercises 14 and 15. The cumulative frequency for the first three frequencies is 6 + 7 + 8 or 21. There were 21 solar eclipses between 1960 and 1989.

Find the cumulative frequencies for each frequency in the data set.

17.

Touchdowns Scored per Season		
Touchdowns	Tally	Frequency
2	ЖHТ	5
4	II	2
6	III	3

18.

Monthly Sales of Tennis Shoes		
Size	Tally	Frequency
7	ЖHТ ЖHТ II	12
8	ЖHТ ЖHТ ЖHТ I	16
9	ЖHТ ЖHТ ЖHТ ЖHТ III	23
10	ЖHТ ЖHТ	10

Standardized Test Practice and Mixed Review

19. **SHORT RESPONSE** Write an appropriate scale for the data set $12, $20, $15, $10, $11, $13, $9, $7, $17, $6, $13, and $8.

20. **MULTIPLE CHOICE** Refer to the frequency table at the right. Twenty-two students went bowling. How many students scored 130–139 points?

    (A) 2   (B) 3   (C) 4   (D) 5

Bowling Scores		
Score	Tally	Frequency
100–109	ЖHТ	5
110–119	ЖHТ I	6
120–129	ЖHТ III	8
130–139		?

21. **GEOMETRY** What is the area of a rectangle with sides 23 feet and 14 feet? (Lesson 1-8)

22. Solve $m + 7 = 16$ mentally. (Lesson 1-7)

GETTING READY FOR THE NEXT LESSON

BASIC SKILL For Exercises 23 and 24, use the pictograph.

23. How much Brand A ice cream was sold?

24. Which brand sold less than 5 gallons?

Gallons of Ice Cream Sold

Brand	Amount
A	
B	
C	

Key: = 2 gallons

2-2a Problem-Solving Strategy
A Preview of Lesson 2-2

What You'll Learn
Solve problems by using a graph.

Use a Graph

Hey, Aisha! did you know that cheetahs can run 70 miles per hour? That's fast!

I did know that, Emily. Cheetahs are the fastest land animals. We can **use the graph** to compare the speed of a cheetah to the speed of a giraffe.

Explore	The graph shows the speeds of various animals. We need to compare the speed of a cheetah to the speed of a giraffe.	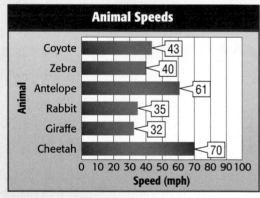

Source: www.infoplease.com

Plan	Use the information given in the graph. Subtract 32 from 70 to find the difference in the speeds.
Solve	$70 - 32 = 38$ The cheetah can run 38 miles per hour faster than the giraffe.
Examine	On the graph, the bar that shows a cheetah's speed is about twice as long as the bar that shows a giraffe's speed. Since 70 is about two times 32, the answer makes sense.

Analyze the Strategy

1. **Explain** how a graph is used to solve a problem involving a graph.
2. **Describe** three examples of data that can be displayed on a graph.
3. **Write** a problem that can be solved by using the information given in the graph above.

Solve. Use a graph.

MAGAZINES For Exercises 4 and 5, use the graph at the right.

4. During which week were 12 subscriptions sold? Explain your reasoning.

5. How many more subscriptions were sold during the fifth week than during the third week?

Subscriptions Sold Per Week

= 4 subscriptions

Mixed Problem Solving

Solve. Use any strategy.

6. **MULTI STEP** How much money will Liseta save if she saves $2 a day for 25 weeks?

7. **WEATHER** Which city has the lower normal temperature for July?

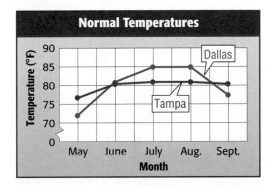

8. **SALES** Refer to the graph below. On how many of the days were more smoothies sold than snow cones?

9. **MULTI STEP** Jorge has $125 in his savings account. He deposits $20 every week and withdraws $25 every four weeks. What will his balance be in 8 weeks?

10. **STANDARDIZED TEST PRACTICE** Which sentence about the graphed data is *not* true?

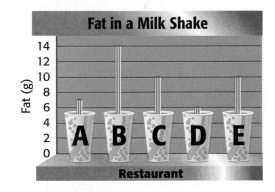

Ⓐ Restaurant B's milkshake contains the most fat.

Ⓑ Restaurant E's milkshake contains 10 grams of fat.

Ⓒ Restaurant A's milkshake contains 8 grams of fat.

Ⓓ Restaurant B's milkshake has twice as many fat grams as Restaurant A's.

You will use graphs in the next lesson.

Bar Graphs and Line Graphs

What You'll LEARN

Make and interpret bar graphs and line graphs.

NEW Vocabulary

graph
bar graph
vertical axis
horizontal axis
line graph

WHEN am I ever going to use this?

ROLLER COASTERS The types of roller coasters found in the United States are listed in the table.

1. What type of roller coaster is most common?

2. What might be an advantage of organizing data in a table? Are there any disadvantages of organizing data in this way?

Types of Roller Coasters in the United States	
Type	Frequency
inverted	37
stand up	9
steel	457
suspended	12
wild mouse	32
wood	116

Source: Roller Coaster Database

Data are often shown in a table. However, a **graph** is a more visual way to display data. One type of graph is a bar graph. A **bar graph** is used to compare data.

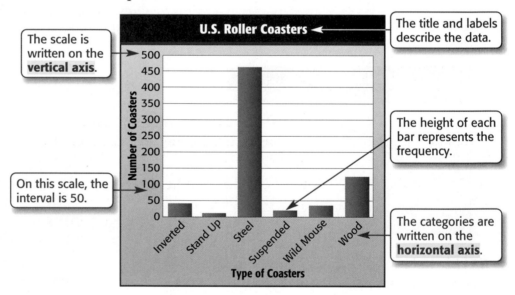

The title and labels describe the data.

The scale is written on the **vertical axis**.

On this scale, the interval is 50.

The height of each bar represents the frequency.

The categories are written on the **horizontal axis**.

Another type of graph is a line graph. A **line graph** is used to show how a set of data changes over a period of time.

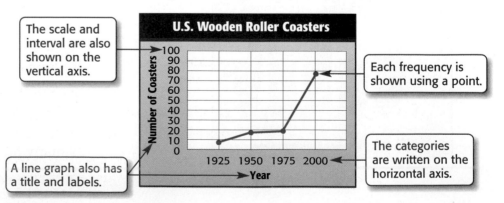

The scale and interval are also shown on the vertical axis.

Each frequency is shown using a point.

A line graph also has a title and labels.

The categories are written on the horizontal axis.

Make and Interpret a Bar Graph

1 SCHOOL Make a vertical bar graph of the data. Compare the number of students who scored a B to the number who scored a C.

Step 1 Decide on the scale and interval. The data includes numbers from 2 to 13. So, a scale from 0 to 14 and an interval of 2 is reasonable.

Step 2 Label the horizontal and vertical axes.

Step 3 Draw bars for each grade. The height of each bar shows the number of students earning that grade.

Step 4 Label the graph with a title.

Math Scores	
Grade	Frequency
A	10
B	13
C	7
D	2

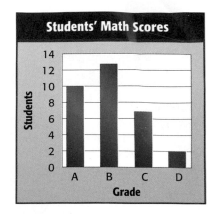

About twice as many students scored a B than a C.

REAL-LIFE MATH

WEATHER The number of tornadoes in the United States from 1996 to 2001 is given in the table.

Year	Tornadoes
1996	1,173
1997	1,148
1998	1,424
1999	1,343
2000	1,071
2001	1,213

Source: National Weather Service

EXAMPLE **Make and Interpret a Line Graph**

2 WEATHER Make a line graph of the data at the left. Then describe the change in the number of tornadoes from 1997 to 1999.

Step 1 Decide on the scale and the interval. The data includes numbers from 1,071 to 1,424. The scale is 1,000 to 1,500, and the interval is 100. Use a break to show that numbers are left out.

Step 2 Label the horizontal and vertical axes.

Step 3 Draw and connect the points for each year. Each point shows the number of tornadoes in that year.

Step 4 Label the graph with a title.

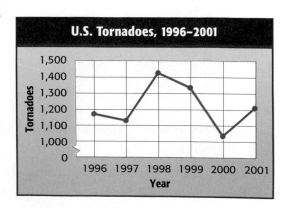

The number of tornadoes greatly increased from 1997 to 1998 and then decreased from 1998 to 1999.

...nd Concept Check

Writing Math
Exercises 1 & 2

Compare and contrast a bar graph and a line graph.

2. **OPEN ENDED** Describe a situation that would be best represented by a line graph.

GUIDED PRACTICE

Make the graph listed for each set of data.

3. bar graph

U.S. Endangered Species	
Species	**Frequency**
Fish	70
Reptiles	14
Amphibians	10
Birds	78
Mammals	63

Source: Fish and Wildlife Service

4. line graph

Basic Cable TV Subscribers (nearest million)	
Year	**Frequency**
1980	18
1985	40
1990	55
1995	63
2000	69

Source: *The World Almanac*

5. **FISH** Refer to the bar graph you made in Exercise 3. How does the number of endangered fish compare to the number of endangered reptiles?

Practice and Applications

Make a bar graph for each set of data.

6.

Heisman Trophy Winners	
Position	**Frequency**
Cornerback	1
End	2
Fullback	2
Halfback	19
Quarterback	20
Running Back	19
Wide Receiver	3

Source: *The World Almanac*

7.

U.S. Gulf Coast Shoreline	
State	**Amount (mi)**
Alabama	607
Florida	5,095
Louisiana	7,721
Mississippi	359
Texas	3,359

Source: NOAA

HOMEWORK HELP

For Exercises	See Examples
6–9	1
11–14	2

Extra Practice
See pages 597, 625.

GEOGRAPHY For Exercises 8 and 9, refer to the bar graph you made in Exercise 7.

8. Which state has about half the amount of shoreline as Louisiana?

9. Name the state that has about twice as much shoreline as Mississippi.

10. **RESEARCH** Use the Internet or another source to find the number of different species of animals in the zoos in Denver, Cleveland, Detroit, and Dallas. Make a bar graph of the data set. Then compare the data.

Make a line graph for each set of data.

11.

Australia's Population	
Year	Population (millions)
1960	10
1970	13
1980	15
1990	17
2000	19

Source: U.S. Census Bureau

12.

Public School Enrollment, K-12	
Year	Students (thousands)
1997	46,127
1998	46,535
1999	46,857
2000	47,223
2001*	47,213

Source: National Center for Education Statistics
*Estimated

 Data Update What was Australia's population in 2001? How does the change in data affect the line graph in Exercise 11? Visit **msmath1.net/data_update** to learn more.

13. POPULATION Refer to the line graph made in Exercise 11. Describe the change in Australia's population from 1960 to 2000.

14. SCHOOL Refer to the line graph made in Exercise 12. What two years showed the greatest increase in enrollment?

WEATHER For Exercises 15 and 16, refer to the table.

15. Write an appropriate scale and interval for the data set.

16. Would this data be best represented by a bar graph or line graph? Explain.

Average Temperatures, Louisville, Kentucky					
Month	Temp.	Month	Temp.	Month	Temp.
Jan.	32	May	65	Sept.	70
Feb.	36	June	73	Oct.	58
Mar.	46	July	77	Nov.	47
Apr.	56	Aug.	76	Dec.	37

Source: *The World Almanac*

17. CRITICAL THINKING Explain how the vertical scale and interval affect the look of a bar graph or line graph.

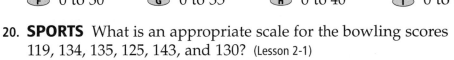
Standardized Test Practice and Mixed Review

For Exercises 18 and 19, refer to the graph at the right.

18. MULTIPLE CHOICE What is the interval for the data?

     Ⓐ 35   Ⓑ 10   Ⓒ 5   Ⓓ 1

19. MULTIPLE CHOICE What is the scale of the graph?

     Ⓕ 0 to 30   Ⓖ 0 to 35   Ⓗ 0 to 40   Ⓘ 0 to 45

20. SPORTS What is an appropriate scale for the bowling scores 119, 134, 135, 125, 143, and 130? (Lesson 2-1)

Find the area of each rectangle described. (Lesson 1-8)

21. length: 4 feet, width: 6 feet

22. length: 12 yards, width: 7 yards

GETTING READY FOR THE NEXT LESSON

PREREQUISITE SKILL Add. (Page 589)

23. 13 + 41   **24.** 57 + 31   **25.** 5 + 18 + 32   **26.** 14 + 45 + 27

What You'll LEARN

Use a spreadsheet to make a line graph and a bar graph.

Making Line and Bar Graphs

You can use a Microsoft Excel® spreadsheet to make graphs.

ACTIVITY 1

For a science project, Coty and Ciera tracked the number of robins they saw each day for one week. The calendar below shows their results.

SUN.	MON.	TUE.	WED.	THUR.	FRI.	SAT.
1 4	**2** 7	**3** 3	**4** 5	**5** 10	**6** 4	**7** 12

Set up a spreadsheet like the one shown below.

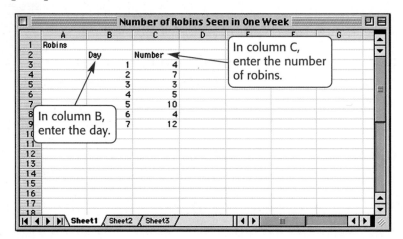

The next step is to "tell" the spreadsheet to make a line graph for the data. In a spreadsheet, the Chart Wizard icon is used to make a graph.

1. Highlight the data in column C. ◄──── This tells the spreadsheet to read the data in column C.

2. Click on the Chart Wizard icon.

3. Click on Custom.

4. Highlight "Two Axes" or "Lines on 2 Axes".

5. Click Next.

6. Click Next.

7. Enter a title. Then enter "Day" in the Category (X) axis box. Enter "Number of Robins" in the Category (Y) axis box.

8. Click Next.

9. Click Finish.

The line graph is shown below.

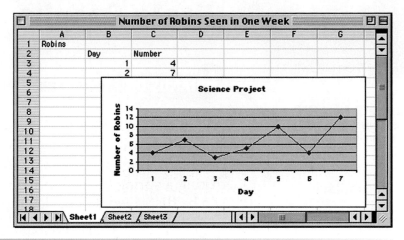

ACTIVITY 2

Use the same data to make a bar graph. Set up the spreadsheet shown in Activity 1. The next step is to "tell" the spreadsheet to make a bar graph for the data.

- Highlight the data in column C.
- Click on the Chart Wizard icon.
- Make sure that Standard Types and Column are selected.
- Complete steps 5–9 from Activity 1.

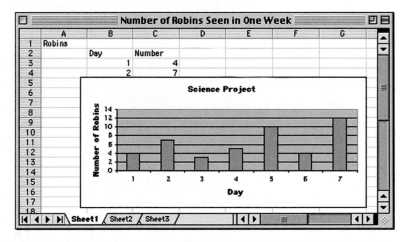

EXERCISES

1. Explain the steps you would take to make a line graph of the number of robins Coty and Ciera saw in the first four days.

2. **OPEN ENDED** Collect some data that can be recorded in a spreadsheet. Then make a graph to display the data.

3. Discuss the advantages and disadvantages of using a spreadsheet to make a graph instead of using paper and pencil.

2-3

Circle Graphs

What You'll LEARN

Interpret circle graphs.

NEW Vocabulary

circle graph

HANDS-ON Mini Lab

Materials
- tape measure
- tape
- adding machine tape
- string

Work with a partner.

The table below shows the number of people driving together in one vehicle during a spring break trip. This data can be displayed in a circle graph.

STEP 1 For each category, let 1 centimeter equal 1%. Mark the length, in centimeters, that represents each percent on a piece of adding machine tape. Label each section.

People in Vehicle	Percent
1–3	11%
4–5	35%
6–7	29%
8–9	11%
10 or more	14%

Source: Carmax.com

1–3 people 11%	4–5 people 35%	6–7 people 29%	8–9 people 11%	10 or more 14%

STEP 2 Tape the ends together to form a circle.

STEP 3 Tape one end of a piece of string to the center of the circle. Tape the other end to the point where two sections meet. Repeat with four more pieces of string.

1. Make a bar graph of the data.
2. Which graph represents the data better, a circle graph or a bar graph? Explain.

A **circle graph** is used to compare parts of a whole.

Driving Together in One Vehicle to Spring Break

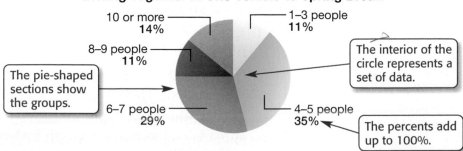

The interior of the circle represents a set of data.

The pie-shaped sections show the groups.

The percents add up to 100%.

You can analyze data displayed in a circle graph.

EXAMPLES Interpret Circle Graphs

CHORES Lisa surveyed her class to see which rooms in their homes were the messiest. The circle graph at the right shows the results of her survey.

Messiest Rooms in Home

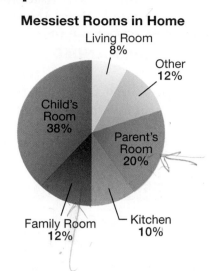

① **Which room did students say is the messiest?**

The largest section of the graph is the section that represents a child's room. So, students said that a child's room is the messiest room in the home.

② **How does the number of students that say their parent's room is the messiest compare to the number of students that say the family room is the messiest?**

The section representing parent's room is about twice the size of the section representing family room. So, twice as many student's say that the parent's room is the messiest room.

Many graphs found in newspapers contain circle graphs.

EXAMPLE Interpret Circle Graphs

③ **FAMILY TIME** The graph shows the average amount of time families spend together during the school year. In which two intervals do families spend about the same amount of time together?

By comparing the percents, you find that the sections labeled 12–24 hours and more than 48 hours are about the same. So, about the same number of families spend these amounts of time together during the school year.

USA TODAY Snapshots®

Family time

The amount of time adults with children ages 6-17 say they spend together as a family in an average week during the school year:

28% 12-24 hours
33% 25-48 hours
27% More than 48 hours
12% Less than 12 hours

Source: Market Facts for Siemens Communcation Devices

By Cindy Hall and Quin Tian, USA TODAY

Skill and Concept Check

Writing
Math
Exercises 1 & 3

1. **Explain** how to identify the greatest and least values of a set of data when looking at a circle graph.

2. **OPEN ENDED** List three characteristics of circle graphs.

3. **Which One Doesn't Belong?** Identify the display that is not the same as the other three. Explain your reasoning.

circle graph	line graph	bar graph	frequency table

GUIDED PRACTICE

FOOD For Exercises 4 and 5, use the graph.

4. What is the most popular reason for choosing a fast food restaurant?

5. How do reasonable prices compare to quality of food as a reason for choosing a fast food restaurant?

Reasons for Choosing a Fast Food Restaurant

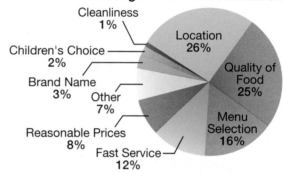

Source: Maritz Marketing Ameripol

INTERNET For Exercises 6 and 7, use the graph that shows the ages at which libraries in the U.S. stop requiring parental permission for children to use the Internet.

6. At what age do all libraries stop requiring parental permission?

7. How does the number of libraries that stop requiring parental permission at ages 14–15 compare to the number of libraries that stop requiring parental permission at ages 16–17?

Ages at Which Libraries Stop Requiring Permission to Use the Internet

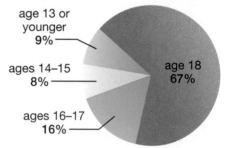

Source: University of Illinois

Practice and Applications

For Exercises 8–10, use the graph.

8. Who most influences kids to read?

9. Which two groups are least influential in getting kids to read?

10. About how much more do parents influence kids to read than teachers?

Who Influences Kids to Read

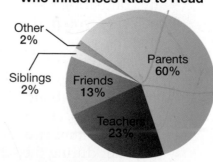

Source: SWR Worldwide for shopforschool.com

HOMEWORK HELP

For Exercises	See Examples
8–14	1, 2, 3

Extra Practice
See pages 597, 625.

SUMMER For Exercises 11–14, use the graph at the right.

11. Name the two least summer nuisances.

12. Name the two biggest summer nuisances.

13. Name two nuisances that together make up about half or 50% of the summer nuisances.

14. How does humidity compare to yard work as a summer nuisance?

15. **CRITICAL THINKING** Which is the least appropriate data to display in a circle graph? Explain.

    a. data that shows what age groups buy athletic shoes

    b. data that shows what most motivates teens to volunteer

    c. data that shows average price of a hamburger every five years

    d. data that shows how many hours adults help their children study

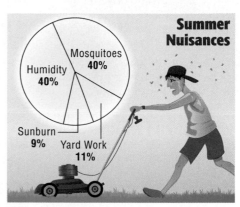
Summer Nuisances

Mosquitoes 40%
Humidity 40%
Sunburn 9%
Yard Work 11%

Source: Impulse Research Corporation

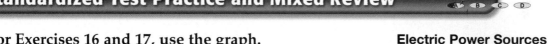
Standardized Test Practice and Mixed Review

For Exercises 16 and 17, use the graph.

16. **SHORT RESPONSE** Which fuel is used to generate more electricity than any of the others?

17. **SHORT RESPONSE** Which two fuels together generate about the same amount of electricity as nuclear fuel?

Electric Power Sources

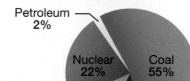
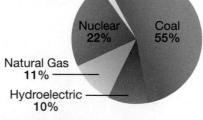

Petroleum 2%
Nuclear 22%
Coal 55%
Natural Gas 11%
Hydroelectric 10%

Source: Energy Information Administration

18. Make a bar graph for the data shown at the right. (Lesson 2-2)

19. Write an appropriate scale for the set of data 32, 45, 32, 27, 28, 45, 23, and 34. (Lesson 2-1)

Favorite Type of Book	
Type	**Frequency**
Fiction	12
Nonfiction	8
Mystery	15
Romance	6
Adventure	13

GETTING READY FOR THE NEXT LESSON

PREREQUISITE SKILL Make a line graph for each set of data. (Lesson 2-2)

20.

Cell Phones Sold			
Day	**Amount**	**Day**	**Amount**
1	7	4	5
2	10	5	14
3	8	6	12

21.

Distance (mi) of Daily Walks			
Day	**Distance**	**Day**	**Distance**
M	2	Th	2
T	3	F	5
W	1	S	4

Making Predictions

WHEN am I ever going to use this?

AUTO RACING The table shows the amount of money won by each winner of the Daytona 500 from 1985 to 2002.

Year	Amount ($)	Year	Amount ($)	Year	Amount ($)
\multicolumn{6}{c}{**Money Won by Daytona 500 Winners, 1985–2002**}
1985	185,500	1991	233,000	1997	377,410
1986	192,715	1992	244,050	1998	1,059,805
1987	204,150	1993	238,200	1999	1,172,246
1988	202,940	1994	258,275	2000	1,277,975
1989	184,900	1995	300,460	2001	1,331,185
1990	188,150	1996	360,775	2002	1,409,017

Source: www.news-journalonline.com

1. Describe the trends in the winning amounts.

2. Make a prediction as to the amount of money the winner of the 2005 Daytona 500 will receive.

Line graphs are often used to predict future events because they show trends over time.

EXAMPLES Make Predictions

1 **AUTO RACING** The data given in the table above is shown in the line graph below. Describe the trend. Then predict how much the 2005 Daytona 500 winner will receive.

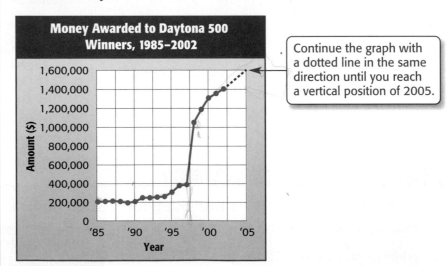

Continue the graph with a dotted line in the same direction until you reach a vertical position of 2005.

Notice that the increase since 1998 has been steady. By extending the graph, you can predict that the winner of the 2005 Daytona 500 will receive about $1,600,000.

2 SNOWBOARDING What does the graph below tell you about the popularity of snowboarding?

Snowboard Sales at SportsCo

The graph shows that snowboard sales have been increasing each year. You can assume that the popularity of the sport is increasing.

Skill and Concept Check

Writing Math Exercise 1

1. **Explain** why line graphs are often used to make predictions.

2. **OPEN ENDED** Give an example of a situation when being able to make a prediction based on a line graph would be useful.

GUIDED PRACTICE

POPULATION For Exercises 3–6, use the graph that shows the change in the world population from 1950 to 2000.

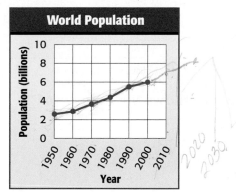

World Population

Source: U.S. Census Bureau

3. Describe the pattern or trend in the world population.

4. Predict the population in 2010.

5. What do you think will be the world population in 2030?

6. Make a prediction for the world population in 2050.

FOOD For Exercises 7 and 8, use the graph that shows the change in the amount of chicken people ate from 1990 to 2000.

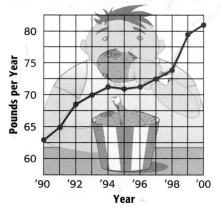

How Much Chicken Do We Eat?

Source: Agriculture Department, Economic Research Service

7. Describe the change in the amount of chicken eaten from 1990 to 2000.

8. Based on the graph, do you think the amount of chicken consumed in 2001 was over 100 pounds per person per year? Explain.

Practice and Applications

HOMEWORK HELP

For Exercises	See Examples
9–18, 20–21, 23–24	1, 2

Extra Practice
See pages 597, 625.

SPORTS For Exercises 9 and 10, refer to Example 2 on page 67.

9. Predict the number of snowboard sales in 2005. Explain your reasoning.

10. About how many snowboards were sold in 1998? How did you reach this conclusion?

FITNESS For Exercises 11–13, use the graph at the right.

11. Describe the change in the amount of time it takes Taryn to run 1,500 meters.

12. Predict when Taryn will run 1,500 meters in less than 6 minutes.

13. Predict the number of weeks it will take until Taryn runs 1,500 meters in 5 minutes.

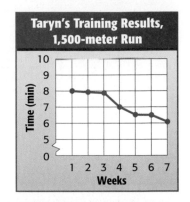

WEATHER For Exercises 14–18, use the graph at the right.

14. Predict the average temperature for Miami in February.

15. Predict the average temperature for Albany in October.

16. What do you think is the average temperature for Anchorage in October?

17. How much warmer would you expect it to be in Miami than in Albany in February?

18. How much colder would you expect it to be in Anchorage than in Miami in October?

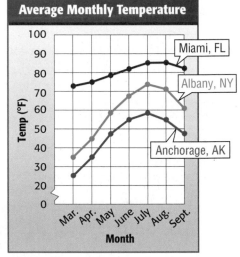

Source: *The World Almanac*

BASEBALL For Exercises 19–22, use the table that shows the number of games won by the Detroit Tigers from 1995 to 2001.

Tigers' Statistics							
Year	1995	1996	1997	1998	1999	2000	2001
Games Won	60	53	79	65	69	79	66

Source: DetroitTigers.com

19. Make a line graph of the data.

20. In what year did the team have the greatest increase in the number of games won?

21. In what year did the team have the greatest decrease in the number of games won?

22. Explain the disadvantages of using this line graph to make a prediction about the number of games that the team will win in 2002, 2003, and 2004.

SCIENCE For Exercises 23 and 24, use the table at the right that shows the results of a science experiment.

Distance Ball is Dropped (cm)	25	32	45	57	68	80	98
Height of Ball on First Bounce (cm)	15	27	39	48	57	63	70

23. What would you predict about the ball if the distance the ball is dropped continues to increase?

24. What would happen if the distance the ball is dropped is decreased?

25. **CRITICAL THINKING** Give an example of a set of sports data that can be graphed and used to make predictions.

Standardized Test Practice and Mixed Review

For Exercises 26 and 27, refer to the graph.

26. **MULTIPLE CHOICE** What is the best prediction for the number of cars sold in 2005?

   Ⓐ 3,000 Ⓑ 380 Ⓒ 3,500 Ⓓ 325

27. **MULTIPLE CHOICE** Which prediction for the number of cars sold from one year to the next appears to have been the most difficult to predict?

   Ⓕ 2002 to 2003 Ⓖ 2001 to 2002
   Ⓗ 2003 to 2004 Ⓘ 2000 to 2001

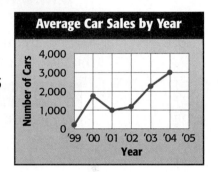

For Exercises 28 and 29, use the circle graph.
(Lesson 2-3)

28. Which season is least preferred to travel?

29. Which two seasons are preferred equally?

30. Make a line graph of the data shown below. (Lesson 2-2)

Gasoline Price per Gallon, 1994–2001			
Year	Cost (¢)	Year	Cost (¢)
1994	111	1998	106
1995	115	1999	117
1996	123	2000	151
1997	123	2001	156

Source: *The World Almanac*

GETTING READY FOR THE NEXT LESSON

PREREQUISITE SKILL Order each set of data from least to greatest. (Page 588)

31.

Candy Costs (¢)				
50	45	75	55	25
60	40	65	80	30

32.

Roller Coaster Speeds (mph)				
40	42	48	60	72
93	57	72	85	70

Vocabulary and Concepts

1. **Describe** a frequency table. (Lesson 2-1)
2. **Define** *data*. (Lesson 2-1)

Skills and Applications

3. Make a frequency table for the data below. (Lesson 2-1)

Speed (mph) of Cars					
65	62	81	72	69	75
66	73	58	79	70	72

4. Make a bar graph for the data shown at the right. (Lesson 2-2)

Favorite Car Color	
Color	Frequency
blue	12
red	19
silver	9
green	11
black	9

5. Refer to Exercise 3. Describe an appropriate scale and interval.

FOOD For Exercises 6 and 7, use the graph below. (Lesson 2-3)

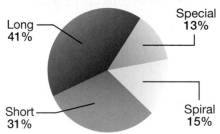

Pasta Sales by Pasta Shape

6. What percent of the sales came from spiral pasta?

7. Which two types of pasta have about the same amount of sales?

MONEY For Exercises 8 and 9, use the graph below. (Lessons 2-2 and 2-4)

Keisha's Savings

8. Describe a trend in Keisha's savings.

9. Predict the total amount saved in 8 weeks.

Standardized Test Practice

10. **MULTIPLE CHOICE** Which display uses a scale, intervals, and tally marks? (Lesson 2-1)

    Ⓐ frequency table   Ⓑ line graph

    Ⓒ circle graph   Ⓓ bar graph

The GameZone

A Place To Practice Your Math Skills

Great Graph Race

● **GET READY!**

Players: two or more
Materials: two spinners

● **GET SET!**

- Label the vertical and horizontal axes of a bar graph as shown at the right.

- Each player chooses one of the products 1, 2, 3, 4, 5, 6, 8, 10, 12, or 20.

Great Graph Race

Frequency: 10, 8, 6, 4, 2, 0

Product: 1 2 3 4 6 12

● **GO!**

- Each player takes a turn spinning each spinner once. Then the player finds the product of the numbers on the spinners and either creates a bar or extends the bar on the graph for that product.

These spinners represent the product 6.

- **Who Wins?** The player whose bar reaches a frequency of 10 first wins.

Stem-and-Leaf Plots

What You'll LEARN

Construct and interpret stem-and-leaf plots.

NEW Vocabulary

stem-and-leaf plot
stems
leaves
key

WHEN am I ever going to use this?

SPORTS The number of points scored by the winning team in each NCAA women's basketball championship game from 1982 to 2002 is shown.

NCAA Division I Women's Basketball Points Scored by Winning Teams, 1982–2002						
76	70	56	70	60	68	71
69	97	76	78	70	93	68
72	67	88	84	83	62	82

Source: *The World Almanac*

1. What were the least and greatest number of points scored?

2. Which number of points occurred most often?

Sometimes it is hard to read data in a table. You can use a stem-and-leaf plot to organize large data sets so that they can be read and interpreted easily. In a **stem-and-leaf plot**, the data is ordered from least to greatest and is organized by place value.

EXAMPLE Construct a Stem-and-Leaf Plot

1 **SPORTS** Make a stem-and-leaf plot of the data above.

Step 1 Order the data from least to greatest.

56 60 62 67 68 68 69 70 70 70 71
72 76 76 78 82 83 84 88 93 97

Step 2 Draw a vertical line and write the tens digits from least to greatest to the left of the line. These digits form the **stems**. Since the least value is 56 and the greatest value is 97, the stems are 5, 6, 7, 8, and 9.

Step 3 Write the units digits in order to the right of the line with the corresponding stem. The units digits form the **leaves**.

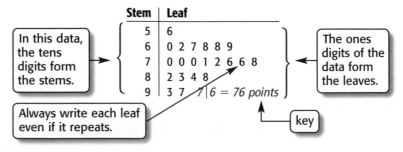

Step 4 Include a **key** that explains the stems and leaves.

You can interpret data that are displayed in a stem-and-leaf plot.

EXAMPLE **Interpret Stem-and-Leaf Plots**

2 **NATURE** Suppose you are doing a science project for which you counted the number of Monarch butterflies you saw each day for one month. The stem-and leaf plot below shows your results.

Stem	Leaf
0	5 5 6 7 8 9 9 9 9 9
1	0 1 1 3 4 6 7 8 8 9 9
2	2 2 3 6 7 7 9
3	7
4	1 $1|3 = 13$ butterflies

Write a few sentences that analyze the data you collected.

By looking at the plot, it is easy to see that the least number of butterflies was 5 and the greatest number was 41. You can also see that most of the data falls between 0 and 29.

Skill and Concept Check

Writing Math
Exercises 1 & 2

1. **Describe** an advantage of displaying a set of data in a stem-and-leaf plot.

2. **FIND THE ERROR** Eduardo and Tenisha are writing the stems for the data set 5, 45, 76, 34, 56, 2, 11, and 20. Who is correct? Explain.

> Eduardo
> 0, 1, 2, 4, 5, 6

> Tenisha
> 0, 1, 2, 3, 4, 5, 6,

3. **OPEN ENDED** Make a stem-and-leaf plot that displays a set of data containing fifteen numbers from 25 to 64.

GUIDED PRACTICE

Make a stem-and-leaf plot for each set of data.

4. 37, 28, 25, 29, 31, 45, 32, 31, 46, 39, 27, 21, 20, 21, 31, 31

5. 81, 76, 55, 90, 71, 80, 83, 85, 79, 99, 70, 75, 70, 92, 93, 93, 82, 94

SPORTS For Exercises 6–8, use the stem-and-leaf plot that shows the ages of the drivers in the 2002 Daytona 500 auto race.

Stem	Leaf
2	1 3 4 6 6 6 7 9
3	0 0 1 2 3 4 4 5 7 7 8 8 8 8 8 9 9 9
4	0 1 2 2 2 3 4 4 4 5 5 5 5 6 6
5	2
6	0 $3|0 = 30$ years

6. What was the age of the youngest driver?

7. What was the age of the oldest driver?

8. Write two additional sentences that analyze the data.

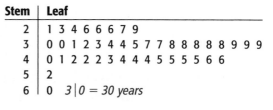

Practice and Applications

Make a stem-and-leaf plot for each set of data.

9. 24, 7, 14, 25, 28, 47, 2, 13, 8, 9, 17, 30, 35, 1, 16, 39

10. 53, 64, 15, 22, 16, 42, 12, 38, 68, 63, 23, 35, 30, 33, 34, 35

11. 62, 65, 67, 67, 62, 67, 51, 73, 72, 70, 65, 63, 72, 78, 60, 61, 54

12. 76, 82, 70, 93, 71, 80, 63, 73, 90, 92, 74, 79, 82, 91, 95, 93, 75

13. Construct a stem-and-leaf plot for the temperature data set 67°, 85°, 73°, 65°, 68°, 79°, 60°, 82°, 68°, 72°, 79°, 66°, 70°, 71°, 69°, 55°.

14. Display the amounts $116, $128, $129, $100, $98, $135, $101, $95, $132, $136, $99, $137, $154, $138, and $126 in a stem-and-leaf plot. (*Hint*: Use the hundreds and tens digits to form the stems.)

HOMEWORK HELP

For Exercises	See Examples
9–15	1
16, 18–20, 24–26	2

Extra Practice See pages 598, 625.

RACING For Exercises 15–17, use the data at the right.

15. Make a stem-and-leaf plot for the data.

16. Write a sentence that analyzes the data.

17. **WRITE A PROBLEM** Write a problem you can solve using the plot.

Distance Between Checkpoints for the 2002 Iditarod Dog Sled Race (mi)

20	29	14	52	34	45	30
48	93	48	23	38	60	112
52	52	42	90	40	58	48
28	18	55	22			

Source: www.oregontrail.net

SHOPPING For Exercises 18–20, use the stem-and-leaf plot at the right that shows costs for various bicycle helmets.

18. How much is the least expensive helmet?

19. How many helmets cost less than $30?

20. Write a sentence that analyzes the data.

Stem	Leaf
1	7 8 8 8 8 9 9 9
2	2 4 5 5 5 9
3	0 0 0 5 5 9
4	9 9 9 2|5 = $25

Tell whether a *frequency table, bar graph, line graph, circle graph,* or *stem-and-leaf plot* is most appropriate to display each set of data. Explain your reasoning.

21. the number of people who subscribe to digital cable television from 2000 to today

22. the amount of the total monthly income spent on expenses

23. survey results listing the number of pets owned by each student in a classroom

DAMS For Exercises 24–26, use the data at the right that shows the heights of the world's twenty highest dams.

24. The height of the Hoover dam is 233 meters. How is 233 shown on the plot?

25. What is the height of the highest dam?

26. The Grand Dixence dam in Switzerland is the second highest dam. Find the height of this dam.

Stem	Leaf
22	0 0 3 5 6
23	3 4 5 7
24	0 2 3 3
25	0
26	0 1 2
27	2
28	5
29	
30	0 24|3 = 243 meters

Source: *The World Almanac*

27. CRITICAL THINKING Explain how you could change a stem-and-leaf plot into a bar graph.

28. MULTIPLE CHOICE The table shows the number of years that the leading NFL passers played in the NFL. Which list shows the stems for the data?

Ⓐ 0, 1

Ⓒ 1, 2

Ⓑ 0, 1, 2

Ⓓ 1, 2, 3, 4, 5, 6, 7, 8, 9

Number of Years the Leading Lifetime NFL Passers Played in NFL				
15	15	9	17	7
11	11	8	11	18
19	10	16	12	13
19	14	12	11	14

Source: *The World Almanac*

29. MULTIPLE CHOICE The stem-and-leaf plot shows the number of hours the students in Mrs. Cretella's class watch television each week. Which interval contains most of the students?

Ⓕ 0–9

Ⓗ 20–29

Ⓖ 10–19

Ⓘ 30–39

Stem	Leaf
0	2 3 3 3 4 4 5 8 9 9
1	0 0 1 2 3 3 4 5 6 6 8 9 9
2	1 3
3	0 1 2 1|4 = 14 hours

SPORTS For Exercises 30–32, use the graph. (Lesson 2-4)

30. What does the graph tell you about the winning times?

31. Between what years was the greatest decrease in the winning time?

32. Make a prediction of the winning time for the 2004 Olympics in the men's 100-meter freestyle.

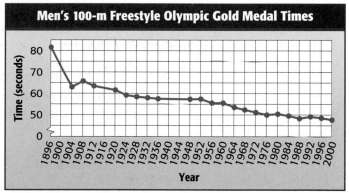

Source: *The World Almanac*

33. FOOD The circle graph shows pie sales at a local bakery. What part of the total sales is peanut butter and strawberry? (Lesson 2-3)

34. MULTI STEP Each day Monday through Saturday, about 2,300 pieces of mail are delivered by each residential mail carrier. About how many pieces of mail are delivered by a mail carrier in five years? (Lesson 1-1)

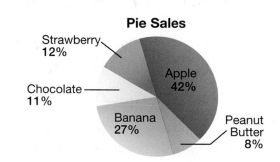

Pie Sales

Strawberry 12%
Chocolate 11%
Apple 42%
Banana 27%
Peanut Butter 8%

GETTING READY FOR THE NEXT LESSON

PREREQUISITE SKILL Find the value of each expression. (Lesson 1-5)

35. $(15 + 17) \div 2$

36. $(4 + 8 + 3) \div 3$

37. $(10 + 23 + 5 + 18) \div 4$

Mean

What You'll LEARN

Find the mean of a set of data.

NEW Vocabulary

average
measure of central tendency
mean
outlier

HANDS-ON Mini Lab

Materials
• 40 pennies
• 5 plastic cups

Work with a partner.

Suppose the table at the right shows your scores for five quizzes.

• Place pennies in each cup to represent each score.

8	7	9	6	10

Quiz	Score
1	8
2	7
3	9
4	6
5	10

• Move the pennies from one cup to another cup so that each cup has the same number of pennies.

1. How many pennies are in each cup?

2. For the five quizzes, your average score was ___?___ points.

3. Suppose your teacher gave you another quiz and you scored 14 points. How many pennies would be in each cup?

A number that helps describe all of the data in a data set is an **average**, or a **measure of central tendency**. One of the most common measures of central tendency is the mean.

Key Concept — Mean

Words The **mean** of a set of data is the sum of the data divided by the number of pieces of data.

Example data set: 8, 7, 9, 6, 10 → mean: $\dfrac{8 + 7 + 9 + 6 + 10}{5} = \dfrac{40}{5}$ or 8

EXAMPLE Find Mean

1. **MONEY** The cost of fifteen different backpacks is shown. Find the mean.

Backpack Costs ($)				
19	25	30	30	27
45	40	50	46	25
22	35	45	49	22

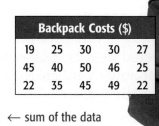

$\text{mean} = \dfrac{19 + 25 + 30 + ... + 22}{15}$ ← sum of the data
                     ← number of data items

$= \dfrac{510}{15}$ or 34

The mean cost of the backpacks is $34.

In statistics, a set of data may contain a value much higher or lower than the other values. This value is called an **outlier**. Outliers can significantly affect the mean.

EXAMPLE Determine How Outliers Affect Mean

2 **WEATHER** Identify the outlier in the temperature data. Then find the mean with and without the outlier. Describe how the outlier affects the mean of the data.

Day	Temp. (°F)
Monday	80
Tuesday	81
Wednesday	40
Thursday	77
Friday	82

Compared to the other values, 40°F is extremely low. So, it is an outlier.

mean with outlier	mean without outlier
$\text{mean} = \dfrac{80 + 81 + 40 + 77 + 82}{5}$	$\text{mean} = \dfrac{80 + 81 + 77 + 82}{4}$
$= \dfrac{360}{5} \text{ or } 72$	$= \dfrac{320}{4} \text{ or } 80$

With the outlier, the mean is less than all but one of the data values. Without the outlier, the mean better represents the values in the data set.

Skill and Concept Check

Writing Math
Exercise 1

1. **Explain** how to find the mean of a set of data.

2. **OPEN ENDED** Write a set of data that has an outlier.

3. **DATA SENSE** Choose the correct value for n in the data set 40, 45, 48, n, 42, 41 that makes each sentence true.

   a. The mean is 44.   b. The mean is 45.

GUIDED PRACTICE

Find the mean for each set of data.

4. 25, 30, 33, 23, 27, 31, 27   5. 38, 52, 54, 48, 40, 32

GEOGRAPHY For Exercises 6–8, use the table at the right. It lists the average depths of the oceans.

Depths of World's Oceans	
Ocean	**Depth (ft)**
Pacific	15,215
Atlantic	12,881
Indian	13,002
Arctic	3,953
Southern	14,749

Source: www.enchantedlearning.com

6. What is the mean of the data?

7. Which depth is an outlier? Explain.

8. How does this outlier affect the mean of the data?

 msmath1.net/extra_examples

Practice and Applications

Find the mean for each set of data.

9. 13, 15, 17, 12, 13

10. 28, 30, 32, 21, 29, 28, 28

11. 76, 82, 75, 87

12. 13, 17, 14, 16, 16, 14, 16, 14

13.

Price	Tally	Frequency
$25	II	2
$50	IIII	4
$60	I	1
$70	III	3

14.

Stem	Leaf
7	0 2 5 6
8	0 0 0 0 2
9	1 3 6
10	3 4 8 $9|3 = 93$

HOMEWORK HELP

For Exercises	See Examples
9–14, 15, 17–18	1
16, 19	2

Extra Practice
See pages 598, 625.

NATURE For Exercises 15–17, use the table that shows the approximate heights of some of the tallest U.S. trees.

15. Find the mean of the data.

16. Identify the outlier.

17. Find the mean if the Coast Redwood is not included in the data set.

18. **BABY-SITTING** Danielle earned $15, $20, $10, $12, $20, $16, $18, and $25 baby-sitting. What is the mean of the amounts she earned?

19. **CRITICAL THINKING** Write a set of data in which the mean is affected by an outlier.

Largest Trees in U.S.	
Tree	**Height (ft)**
Western Red Cedar	160
Coast Redwood	320
Monterey Cypress	100
California Laurel	110
Sitka Spruce	200
Port-Orford-Cedar	220

Source: *The World Almanac*

Standardized Test Practice and Mixed Review

20. **MULTIPLE CHOICE** Which piece of data in the data set 98, 103, 96, 147, 100, 85, 546, 120, 98 is an outlier?

   Ⓐ 103   Ⓑ 120   Ⓒ 147   Ⓓ 546

21. **MULTIPLE CHOICE** Benito scored 72 points in 6 games. What was the mean number of points he scored per game?

   Ⓕ 6   Ⓖ 8   Ⓗ 12   Ⓘ 13

22. **SCHOOL** The ages of the teachers at Fairview Middle School are shown in the stem-and-leaf plot. Into what intervals do most of the data fall? (Lesson 2-5)

Stem	Leaf
2	3 8
3	8 9
4	1 4 5 6 7 9
5	0 0 5 8
6	3 $4|1 = 41$ years

23. Which type of graph is best used to make predictions over time? (Lesson 2-4)

GETTING READY FOR THE NEXT LESSON

PREREQUISITE SKILL Subtract. (Page 589)

24. $125 - 76$ **25.** $236 - 89$ **26.** $175 - 106$ **27.** $224 - 156$

msmath1.net/self_check_quiz

What You'll LEARN

Use a spreadsheet to find the mean.

Spreadsheets and Mean

Spreadsheets can be used to find the average of your test scores in school.

ACTIVITY

The table at the right shows Tyrone's test scores in math, language arts, science, and social studies.

Subject	Test 1	Test 2	Test 3	Test 4
Math	85	93	84	89
Language Arts	78	72	86	90
Science	76	83	82	92
Social Studies	83	85	88	91

Set up a spreadsheet like the one shown.

Column A lists the subjects.

Column C lists the scores on the second tests.

Column E lists the scores on the fourth tests.

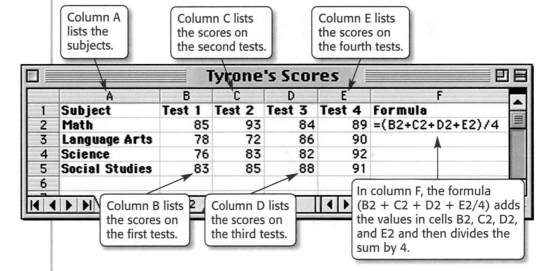

Tyrone's Scores

	A	B	C	D	E	F
1	Subject	Test 1	Test 2	Test 3	Test 4	Formula
2	Math	85	93	84	89	=(B2+C2+D2+E2)/4
3	Language Arts	78	72	86	90	
4	Science	76	83	82	92	
5	Social Studies	83	85	88	91	
6						

Column B lists the scores on the first tests.

Column D lists the scores on the third tests.

In column F, the formula (B2 + C2 + D2 + E2/4) adds the values in cells B2, C2, D2, and E2 and then divides the sum by 4.

The next step is to find the mean. To do this, place the cursor in cell F2 and press the Enter key.

EXERCISES

1. What formulas should you enter to find Tyrone's average score in language arts, science, and social studies?

2. What is Tyrone's average score in each subject? Round to the nearest whole number.

3. Tyrone had a test in each of these subjects last week. His scores are shown below. Use the spreadsheet to calculate Tyrone's new averages in each subject.

   math: 89   language arts: 86   science: 74   social studies: 94

Median, Mode, and Range

What You'll LEARN

Find the median, mode, and range of a set of data.

NEW Vocabulary

median
mode
range

WHEN am I ever going to use this?

BIRDS The table shows the approximate wingspans of the birds having the widest wingspans.

Widest Wingspans (ft)			
12	6	10	12
11	9	10	

Source: www.swishweb.com

1. Find the mean wingspan.
2. List the data in order from least to greatest.
3. Which data is in the middle of the arranged data?
4. Compare the number that is in the middle of the data set to the mean of the data.

You have already learned that the mean is one type of measure of central tendency. Two other types are the median and the mode.

Key Concept — Median

Words   The **median** of a set of data is the middle number of the ordered data, or the mean of the middle two numbers.

Examples   data set: 3, 4, ⑧ 10, 12 → median: 8

data set: 2, 4, ⑥, 8, 11, 12 → median: $\dfrac{6+8}{2}$ or 7

Mode

Words   The **mode** of a set of data is the number or numbers that occur most often.

Example   data set: 12, 23, ㉘, 28, 32, ㊻, 46 → modes: 28 and 46

EXAMPLE   Find Median and Mode

1 MONEY The table shows the cost of nine books. Find the median and mode of the data.

Book Costs ($)		
20	7	10
15	11	20
25	15	8

To find the median, order the data from least to greatest.

median: 7, 8, 10, 11, ⑮ 15, 20, 20, 25

To find the mode, find the number or numbers that occur most often.

mode: 7, 8, 10, 11, ⑮, 15, ⑳, 20, 25

The median is $15. There are two modes, $15 and $20.

Standardized tests often contain mean, median, and mode questions.

Standardized Test Practice

EXAMPLE Find Mean, Median, and Mode

2 **MULTIPLE-CHOICE TEST ITEM** What are the mean, median, and mode of the temperature data 64°, 70°, 56°, 58°, 60°, 70°, respectively?

    Ⓐ 64°, 60°, 70°   Ⓑ 63°, 62°, 70°

    Ⓒ 64°, 62°, 70°   Ⓓ 63°, 61°, 70°

Read the Test Item

You need to find the mean, median, and mode of the data.

Solve the Test Item

mean: $\dfrac{64 + 70 + 56 + 58 + 60 + 70}{6} = \dfrac{378}{6}$ or 63

median: 56, 58, <u>60, 64</u>, 70, 70

$$\dfrac{60 + 64}{2} = \dfrac{124}{2} \text{ or } 62$$

> There is an even number of data values. So, to find the median, find the mean of the two middle numbers.

mode: 70

The mean is 63°, the median is 62°, and the mode is 70°. So, the answer is B.

Some averages may describe a set of data better than other averages.

EXAMPLE Use Mean, Median, and Mode

3 **ELECTIONS** The number of electors for the states in two U.S. regions is shown. Which measure of central tendency best describes the data?

mean: $\dfrac{3 + 3 + 4 + 4 + 5 + 7 + 8 + 11 + 54}{9}$

           $= \dfrac{99}{9}$ or 11

median: 3, 3, 4, 4, <u>5</u>, 7, 8, 11, 54

                median

mode: 3 and 4

The modes are closer to the lower end of the data, and the mean is closer to the higher end. The median is closer in value to most of the data. It tells you that one-half of the states have more than 5 electors. So, the median best describes these data.

State Electors Far West and Mountain Regions		
11	7	54
3	4	4
3	5	8

Source: *The World Almanac*

The guidelines on the next page will help you to choose which measure of central tendency best describes a set of data.

Concept Summary	Mean, Median, and Mode
Measure	**Best Used to Describe the Data When...**
Mean	• the data set has no very high or low numbers.
Median	• the data set has some high or low numbers and most of the data in the middle are close in value.
Mode	• the data set has many identical numbers.

The mean, median, and mode of a data set describe the center of a set of data. The range of a set of data describes how much the data vary. The **range** of a set of data is the difference between the greatest and the least values of the set.

EXAMPLE Find Range

4 Find the range of the data set {125, 45, 67, 150, 32, 12}. Then write a sentence describing how the data vary.

The greatest value is 150. The least value is 12. So, the range is 150 − 12 or 138.

The range tells us that the data are spread out.

Skill and Concept Check

Writing Math

Exercise 1

1. **Explain** how you would find the median and mode of a data set.

2. **OPEN ENDED** Write a set of data in which the mode best describes the data. Then write a set of data in which the mode is not the best choice to describe the data.

GUIDED PRACTICE

Find the mean, median, mode, and range for each set of data.

3. 15, 20, 23, 13, 17, 21, 17

4. 46, 62, 63, 57, 50, 42, 56, 40

TUNNELS For Exercises 5–8, use the table below.

Five Longest Underwater Car Tunnels in U.S.					
State	NY	NY	MA	NY	VA
Length (ft)	8,220	8,560	8,450	9,120	8,190

Source: *The World Almanac*

5. What are the mean, median, and mode of the data?

6. Which measure of central tendency best describes the data? Explain.

7. What is the range of the data?

8. Write a sentence describing how the lengths of the tunnels vary.

HOMEWORK HELP

For Exercises	See Examples
9–12, 15, 18	1, 2, 4
14, 16	3

Extra Practice
See pages 598, 625.

Find the mean, median, mode, and range for each set of data.

9. 23, 22, 15, 36, 44

10. 18, 20, 22, 11, 19, 18, 18

11. 97, 85, 92, 86

12. 23, 27, 24, 26, 26, 24, 26, 24

13. Write a sentence that describes how the data in Exercise 11 vary.

14. Which measure best describes the data in Exercise 9? Explain.

MUSIC For Exercises 15 and 16, use the following information.
Jessica's friends bought CDs for $12, $14, $18, $10, $14, $12, $12, and $12.

15. Find the mean, median, and mode of the data set.

16. Which measure best describes the cost of the CDs? Explain.

WEATHER For Exercises 17–19, refer to the table at the right.

17. Compare the median high temperatures.

18. Find the range for each data set.

19. Write a statement that compares the daily high temperatures for the two cities.

Daily High Temperatures (°F)	
Cleveland	Cincinnati
75 50 80 75	80 72 75 74
70 84 70	72 76 76

20. **CRITICAL THINKING** Write a set of data in which the mean is 15, the median is 15, and the modes are 13 and 15.

21. **MULTIPLE CHOICE** Which measure of central tendency is always found in the data set itself?

    Ⓐ median Ⓑ mean Ⓒ mode Ⓓ all of the above

22. **MULTIPLE CHOICE** Find the median of 25, 30, 25, 15, 27, and 28.

    Ⓕ 24 Ⓖ 25 Ⓗ 26 Ⓘ 30

23. Identify any outliers in the data set 50, 42, 56, 50, 48, 18, 45, 46.
    (Lesson 2-6)

24. Display 27, 31, 25, 19, 31, 32, 24, 26, 33, and 31 in a stem-and-leaf plot.
    (Lesson 2-5)

GETTING READY FOR THE NEXT LESSON

PREREQUISITE SKILL For Exercises 25–27, use the graph.
(Lesson 2-2)

25. Who spent the most on lunch?

26. How much more did Kiyo spend than Hugo?

27. How does the amount Matt spent compare to the amount Jacob spent?

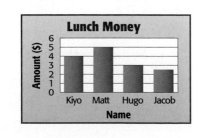

Graphing Calculator Investigation

A Follow-Up of Lesson 2-7

Box-and-Whisker Plots

The monthly mean temperatures for Burlington, Vermont, are shown. You can display the data in a **box-and-whisker plot**.

Monthly Normal Temperatures (°F)												
	J	**F**	**M**	**A**	**M**	**J**	**J**	**A**	**S**	**O**	**N**	**D**
Burlington, VT	16	18	31	44	56	65	71	68	59	48	37	23

ACTIVITY 1

STEP 1 Write the data shown in the table from least to greatest.

16   18   23   31   37   44   48   56   59   65   68   71

STEP 2 Draw a number line that includes all of the data.

STEP 3 Mark the least and greatest number as the *lower extreme* and *upper extreme*. Find and label the median.

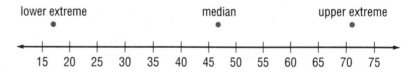

STEP 4 The median of a data set separates the set in half. Find the medians of the lower and upper halves.

$$16 \quad 18 \quad \underbrace{23 \quad 31}_{\frac{23 + 31}{2} = 27} \quad 37 \quad \underset{\text{median}}{44 \uparrow 48} \quad 56 \quad \underbrace{59 \quad 65}_{\frac{59 + 65}{2} = 62} \quad 68 \quad 71$$

Label these values as *lower quartile* and *upper quartile*. Draw a *box* around the quartile values, and *whiskers* that extend from each quartile to the extreme data points.

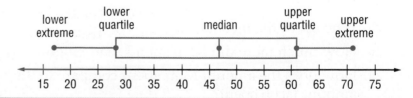

You can also use a TI-83 Plus graphing calculator to make a box-and-whisker plot.

ACTIVITY 2

Use the temperature data from Activity 1.

STEP 1 Enter the data into the calculator's memory. Press STAT ENTER to see the lists. Then enter the data by entering each number and pressing ENTER .

STEP 2 Choose the graph. Press 2nd [STAT PLOT] to display the menu. Choose the first plot by pressing ENTER . Use the arrow and ENTER keys to highlight the modified box-and-whisker plot, **L1** for the **Xlist** and 1 as the frequency.

STEP 3 Press WINDOW to choose the display window. Choose appropriate range settings for the *x* values. The window 0 to 75 with a scale of 5 includes all of this data.

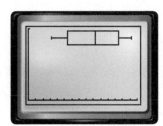

STEP 4 Display the graph by pressing GRAPH . In order to see the important parts of the graph, press TRACE and ENTER .

Your Turn

a. Use the Internet or another source to find the monthly normal temperatures for a city in your state. Draw a box-and-whisker plot of the data. Then, use a graphing calculator to display the data.

EXERCISES

1. **Draw** a box-and-whisker plot for the set of data below.

Baseball Games Won by Teams in National League, 2002															
95	94	79	67	65	95	85	73	72	69	65	97	86	85	82	76

Source: *The World Almanac*

2. **Describe** the data values that are located in the box of a box-and-whisker plot.

3. **MAKE A CONJECTURE** Write a sentence describing what the length of the box of the box-and-whisker plot tells you about the data set.

STUDY TIP

Clear Memory
Before entering data in a table, be sure to clear any existing data. To clear the calculator's memory, position the cursor at the top of each table and press CLEAR ENTER .

2-8 Analyzing Graphs

What You'll LEARN

Recognize when statistics and graphs are misleading.

WHEN am I ever going to use this?

FOOTBALL The graph at the right shows the number of touchdown passes thrown by some of the NFL's leading quarterbacks.

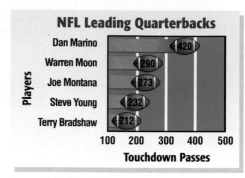

NFL Leading Quarterbacks

Source: National Football League

1. Suppose you look at the lengths of the bars that represent Dan Marino and Terry Bradshaw. You might conclude that Dan Marino threw three times as many touchdown passes as Terry Bradshaw. Why is this conclusion incorrect?

Graphs let readers analyze and interpret data easily. However, graphs are sometimes drawn to influence conclusions by misrepresenting the data.

EXAMPLE Drawing Conclusions from Graphs

1 **ANIMALS** Refer to the graphs below. Which graph suggests that an African elephant drinks twice as much water a day as an Asian elephant? Is this a valid conclusion? Explain.

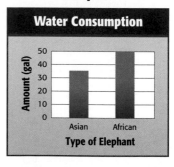

In graph A, the lengths of the bars indicate that an African elephant drinks twice as much water a day as an Asian elephant. However, the amount of water an African elephant drinks, 50 gallons, is not more than twice the amount for the Asian elephant, 36 gallons.

A break in the scale in Graph A is a reminder that the scale has been compressed.

The use of a different scale on a graph can influence conclusions drawn from the graph.

EXAMPLE **Changing the Interval of Graphs**

2 **SCHOOL DANCES** The graphs show how the price of spring dance tickets increased. Which graph makes it appear that the cost increased more rapidly? Explain.

Graph A

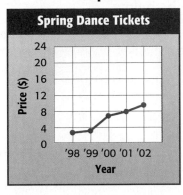

Graph B

The graphs show the same data. However, Graph A uses an interval of 4, and Graph B uses an interval of 2. Graph B makes it appear that the cost increased more rapidly even though the price increase is the same in each graph.

Statistics can also be used to influence conclusions.

EXAMPLE **Misleading Statistics**

3 **MARKETING** An amusement park boasts that the average height of their roller coasters is 170 feet. Explain how using this average to attract visitors is misleading.

Coaster	Height (ft)
Viper	109
Monster	135
Red Zip	115
Tornado	365
Riptide	126

mean: 170 ft   median: 126 ft   mode: none

The average used by the park was the mean. This measure is much greater than most of the heights listed because of the outlier. So, it is misleading to use this measure to attract visitors.

A more appropriate measure to describe the data would be the median, which is closer to the height of most of the coasters.

Writing Math
Exercise 1

1. **Explain** how to redraw the graph on page 86 so that it shows that Dan Marino had twice as many touchdown passes as Terry Bradshaw.

2. **OPEN ENDED** Find a set of data and display it in two separate graphs using different intervals.

GUIDED PRACTICE

For Exercises 3–4, use the graphs below.

Graph A — Career Home Run Leaders* — Barry Bonds 611, Willie Mays 660, Babe Ruth 714, Hank Aaron 755. Home Runs axis: 0, 600, 700, 800. *through 2002 regular season

Graph B — Career Home Run Leaders* — Barry Bonds 611, Willie Mays 660, Babe Ruth 714, Hank Aaron 755. Home Runs axis: 0, 100, 200, 300, 400, 500, 600, 700, 800. *through 2002 regular season

Source: Major League Baseball

3. In Graph A, how many more home runs does Willie Mays appear to have hit than Barry Bonds? Explain.

4. Suppose you want to show that Willie Mays hit almost as many home runs as Babe Ruth. Which graph would you use? Explain.

Practice and Applications

MONEY For Exercises 5–7, use the graph shown.

Cost of Cameras — Price ($): 0, 100, 120, 140, 160, 180, 200, 220 — Brand A, Brand B

5. Based on the size of the bars, compare the costs of Brand A and Brand B.

6. Explain how this graph may be misleading.

7. Redraw the graph to show that Brand A and Brand B cost about the same.

HOMEWORK HELP

For Exercises	See Examples
5–7	1
11	2
8–10	3

Extra Practice
See pages 599, 625.

TRAVEL For Exercises 8–10, use the table at the right.

8. Find the mean, median, and mode of the data.

9. Which measure might be misleading in describing the average number of yearly visitors that visit these sights? Explain.

10. Which measure would be best if you wanted a value close to the most number of visitors? Explain.

Annual Sight-Seeing Visitors	
Sight	**Visitors***
Cape Cod	4,600,000
Grand Canyon	4,500,000
Lincoln Memorial	4,000,000
Castle Clinton	4,600,000
Smoky Mountains	10,200,000

Source: *The World Almanac*
*Approximation

11. **STOCKS** The graphs below show the increases and decreases in the monthly closing prices of Skateboard Depot's stock.

Graph A

Graph B

Suppose you are a stockbroker and want to show a customer that the price of the stock is consistent. Which graph would you show the customer? Explain your reasoning.

12. **CRITICAL THINKING** Find a graph in a newspaper or another source. Then redraw the graph so that the data appear to show different results. Which graph describes the data better? Explain.

Standardized Test Practice and Mixed Review

13. **MULTIPLE CHOICE** The table lists recent top five American Kennel Club registrations. Which measure would be most misleading as to the average number of dogs in any given breed?

    A mean   **B** median

    C mode   **D** none of the averages

Breed	Registered
Labrador Retriever	172,841
Golden Retriever	66,300
German Shepherd	57,660
Dachshund	54,773
Beagle	52,026

Source: *The World Almanac*

14. **MULTIPLE CHOICE** Which measure would most accurately describe the data in Exercise 13?

    F mean   **G** median   **H** mode   **I** none of the averages

15. Find the median, mode, and range for the set of data 68°, 70°, 73°, 75°, 76°, 76°, and 82°. (Lesson 2-7)

Find the mean for each set of data. (Lesson 2-6)

16. 25, 20, 19, 22, 24, 28   17. 25°, 14°, 21°, 16°, 19°   18. 95, 97, 98, 90

WebQuest **Interdisciplinary Project**

People, People, and More People

It's time to complete your project. Use the information and data you have gathered about the U.S. population, past and present, to prepare a booklet or poster. Be sure to include all the required graphs with your project.

msmath1.net/webquest

Vocabulary and Concept Check

average (p. 76)	key (p. 72)	outlier (p. 77)
bar graph (p. 56)	leaves (p. 72)	range (p. 82)
circle graph (p. 62)	line graph (p. 56)	scale (p. 50)
data (p. 50)	mean (p. 76)	statistics (p. 50)
frequency table (p. 50)	measure of central	stem-and-leaf plot (p. 72)
graph (p. 56)	tendency (p. 76)	stems (p. 72)
horizontal axis (p. 56)	median (p. 80)	tally mark (p. 50)
interval (p. 50)	mode (p. 80)	vertical axis (p. 56)

Choose the letter of the term that best matches each phrase.

1. separates the scale into equal parts
2. the sum of a set of data divided by the number of pieces of data
3. a display that uses place value to display a large set of data
4. the middle number when a set of data is arranged in numerical order
5. shows how data change over time
6. pieces of numerical information
7. explains the information given in a stem-and-leaf plot

a. stem-and-leaf plot
b. mean
c. median
d. interval
e. key
f. line graph
g. data

Lesson-by-Lesson Exercises and Examples

2-1 Frequency Tables (pp. 50–53)

Make a frequency table for each set of data.

8.
Number of Siblings					
3	1	2	1	3	1
0	4	1	0	2	1
2	1	0	3	1	0

9.
Favorite Color							
R	B	R	B	Y	G	B	B
G	R	Y	Y	B	B	R	B

R = red   G = green
B = blue   Y = yellow

Example 1 Make a frequency table for the data in the table.

Favorite Fruit	
banana	grapes
apple	apple
banana	grapes
apple	apple

Draw a table with three columns. Complete the table.

Favorite Fruit		
Fruit	Tally	Frequency
banana	II	2
apple	IIII	4
grapes	II	2

 msmath1.net/vocabulary_review

2-2 Bar Graphs and Line Graphs (pp. 56–59)

10. Make a line graph for the set of data shown.

Zoo Visitors	
Year	Visitors
2000	12,300
2001	13,400
2002	15,900
2003	15,100
2004	16,200

Example 2 Make a line graph for the set of data that shows the attendance at a spring dance.

Year	Attendance
2001	350
2002	410
2003	425
2004	450

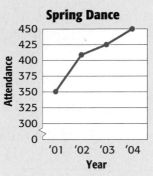

2-3 Circle Graphs (pp. 62–65)

11. **FOOD** Name two muffins that together are preferred by half the people surveyed.

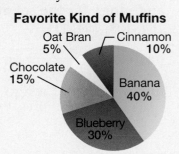

Favorite Kind of Muffins
Oat Bran 5% — Cinnamon 10%
Chocolate 15%
Banana 40%
Blueberry 30%

Example 3 **FOOD** Refer to the circle graph at the left. Which kind of muffin do twice as many people prefer than those who prefer chocolate?

By comparing the percents, you find that the section for blueberry is twice the size of the section for chocolate. So, twice as many people prefer blueberry muffins than chocolate muffins.

2-4 Making Predictions (pp. 66–69)

SPORTS For Exercises 12 and 13, refer to the graph.

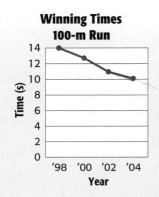

12. Describe the trend in the winning times.

13. Predict the winning time for 2006.

Example 4 **SALES** Refer to the graph. Predict the number of CDs that will be sold in 2005.

By extending the graph, it appears that about 700 CDs will be sold in 2005.

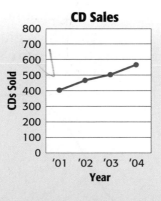

2-5 Stem-and-Leaf Plots (pp. 72–75)

Make a stem-and-leaf plot for each set of data.

14. 83, 72, 95, 64, 90, 88, 78, 84, 61, 73
15. 18, 35, 27, 56, 19, 22, 41, 28, 31, 29
16. 20, 8, 43, 39, 10, 47, 2, 27, 27, 39, 40

Example 5 Make a stem-and-leaf plot for the set of data 65, 72, 68, 60, 75, 78, 69, 70, and 64.

Stem	Leaf
6	0 4 5 8 9
7	0 2 5 8 $7\mid2 = 72$

2-6 Mean (pp. 76–78)

Find the mean for each set of data.

17. 23, 34, 29, 36, 18, 22, 27
18. 81, 72, 84, 72, 72, 81
19. 103, 110, 98, 104, 110

Example 6 Find the mean for the set of data 117, 98, 104, 108, 104, 111.

$$\text{mean} = \frac{117 + 98 + 104 + 108 + 104 + 111}{6}$$

$$= 107$$

2-7 Median, Mode, and Range (pp. 80–83)

Find the median, mode, and range for each set of data.

20. 21, 23, 27, 30
21. 36, 42, 48, 36, 82

36, 36, 42, 48, 82

Example 7 Find the median, mode, and range for the set of data 117, 98, 104, 108, 104, 111.

$$\text{median} = \frac{104 + 108}{2} \text{ or } 106$$

$$\text{mode} = 104 \qquad \text{range} = 117 - 98 \text{ or } 19$$

2-8 Analyzing Graphs (pp. 86–89)

MONEY For Exercises 22 and 23, use the graphs at the right.

22. Which graph would you show to a telephone customer? Explain.
23. Which graph would you show to a company stockholder? Explain.

Example 8
MONEY Which graph makes a call to Port Arthur look more economical?

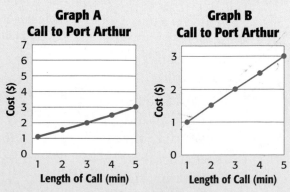

Graph A appears more economical.

Practice Test

Vocabulary and Concepts

1. **Define** *statistics*.
2. **Write** an interval and scale for the data set 55, 30, 78, 98, 7, and 45.

Skills and Applications

Find the mean, median, mode, and range for each set of data.

3. 67, 68, 103, 65, 80, 54, 53

4. 232, 200, 242, 242

LAKES For Exercises 5 and 6, refer to the table. It shows maximum depths of the Great Lakes.

5. Are there any outliers? Explain.

6. Write a sentence explaining how the outlier would affect the mean of the data.

Lake	Max. Depth (ft)
Superior	1,333
Michigan	923
Huron	750
Erie	210
Ontario	802

Source: *The World Almanac*

For Exercises 7 and 8, use the data set 86, 85, 92, 73, 75, 96, 84, 92, 74, 87.

7. Make a stem-and-leaf plot for the set of data.

8. Which might be misleading in describing the set of data, the mean, median, or mode?

VEGETABLES For Exercises 9 and 10, refer to the circle graph.

9. Which vegetable is most favored?

10. How do peas compare to corn as a favorite vegetable?

Favorite Vegetables

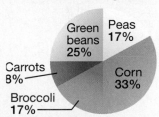

Green beans 25%
Peas 17%
Carrots 8%
Corn 33%
Broccoli 17%

Standardized Test Practice

Ⓐ Ⓑ Ⓒ Ⓓ

For Exercises 11 and 12, refer to the bar graph at the right.

11. **MULTIPLE CHOICE** Mark is in the 6th grade. How many students in his class bought their lunch?

   Ⓐ 60   Ⓑ 80

   Ⓒ 100   Ⓓ 120

12. **GRID IN** How many students in these four grades bought their lunch?

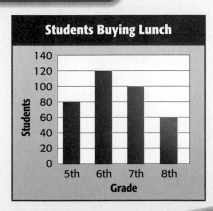

Students Buying Lunch

PART 1 Multiple Choice

Record your answers on the answer sheet provided by your teacher or on a sheet of paper.

1. Which number divides into 324, 630, and 228 with no remainder, but does not divide evenly into 369? (Lesson 1-2)

   (A) 5   (B) 6

   (C) 9   (D) 10

2. Which number sentence represents the cost of three paintbrushes, two pencils, and five tubes of paint? (Lesson 1-5)

Item	Cost
paintbrush	$2
art pencil	$1
tube of paint	$4

   (F) $(3 \times 2) + (2 \times 1) + (5 \times 4) = 28$

   (G) $(3 \times 2) + (2 \times 1) + (5 \times 4) = 32$

   (H) $(3 \times 2) + (2 \times 1) + (5 \times 4) = 52$

   (I) $(3 \times 2) + (2 \times 1) + (5 \times 4) = 240$

3. The frequency table shows the results of a survey of favorite ice cream flavors. Which flavor is the least favorite? (Lesson 2-1)

Flavor	Tally	Frequency
chocolate	JHT I	6
vanilla	JHT III	8
strawberry	IIII	4
cookies 'n cream	JHT JHT I	11

   (A) chocolate

   (B) vanilla

   (C) strawberry

   (D) cookies 'n cream

4. The graph shows the temperature at different times of the day on a Monday. What was the temperature at 11:00 A.M.? (Lesson 2-2)

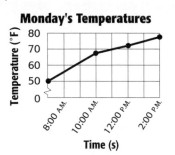

Monday's Temperatures

   (F) about 65°F   (G) about 68°F

   (H) about 70°F   (I) about 75°F

5. Which stem-and-leaf plot shows the data 75, 93, 66, 72, 80, 84, 72, 87? (Lesson 2-5)

   (A)
Stem	Leaf
6	1
7	3
8	3
9	1

   (B)
Stem	Leaf
6	6
7	2 5
8	0 4 7
9	3

   (C)
Stem	Leaf
6	6
7	2 2 5
8	4 7
9	1

   (D)
Stem	Leaf
6	6
7	2 2 5
8	0 4 7
9	3

6. Eneas wanted to buy a computer. The range in prices for the models he looked at was $2,700. If the most expensive model was $3,350, how much was the least expensive? (Lesson 2-7)

   (F) $450   (G) $650

   (H) $1,700   (I) $6,050

TEST-TAKING TIP

Question 6 Since you know the range and the highest value, work backward to find the lowest value.

PART 2 Short Response/Grid In

Record your answers on the answer sheet provided by your teacher or on a sheet of paper.

7. Martina collected $38 for a holiday children's fund. Kenji collected half as much. How much money did Kenji collect? (Lesson 1-1)

8. Write the prime factorization of 120. (Lesson 1-3)

9. What is the area of the rectangle shown below? (Lesson 1-8)

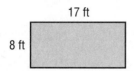

17 ft

8 ft

10. The circle graph shows the results of a survey of Mr. Yan's class. What percent of students enjoy reading fashion or sports magazines? (Lesson 2-3)

Magazine Preferences

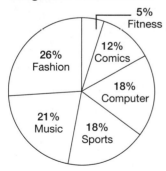

5% Fitness
12% Comics
26% Fashion
18% Computer
21% Music
18% Sports

11. Vanessa counted the number of blooms on her sweet pea plants every week. They are listed below.

44, 63, 66, 63, 60, 45, 55, 59, 42, 71

Construct a stem-and-leaf plot of this data. (Lesson 2-5)

12. What is the mean of the following set of data? (Lesson 2-6)

9, 16, 9, 12, 16, 14, 13, 12, 16, 9, 13, 16

13. Jennifer's test scores for five history tests were 66, 73, 92, 90, and 73. How do the median and mode of her scores compare? (Lesson 2-7)

PART 3 Extended Response

Record your answers on a sheet of paper. Show your work.

14. The line graph shows Josh's times for 4 races in the 100-meter event. (Lesson 2-4)

   a. What is the best prediction for the time he will run in his next 100-meter race?

   b. Explain how you reached this conclusion.

Josh's 100-m Race Times

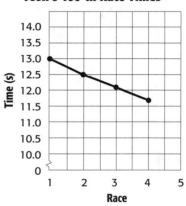

Time (s)
14.0
13.5
13.0
12.5
12.0
11.5
11.0
10.5
10.0
0
1 2 3 4 5
Race

15. Brian drew a graph showing the heights in centimeters of his three younger siblings. (Lesson 2-8)

Height of Brian's Siblings

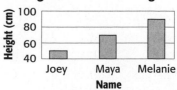

Height (cm)
100
80
60
40
Joey Maya Melanie
Name

   a. The bar for Maya's height is three times the bar for Joey's height. How is this a misrepresentation?

   b. How could the graph be changed to be less misleading?

UNIT 2
Decimals

Your study of math includes more than just whole numbers. In this unit, you will use decimals to describe many real-life situations and learn how to add, subtract, multiply, and divide with them in order to solve problems.

INTERDISCIPLINARY PROJECT 🔍 **MATH and FINANCE**

DOWN TO THE LAST PENNY!

On your mark, get set, SHOP! Being the cost-conscious shopper that you are, you have been asked to help a family make and maintain a grocery budget that will meet their needs. On this shopping adventure, you'll gather data about the cost of common grocery items, find their total cost, and compare this cost to the amount a family can spend on groceries. You'll also compare costs by calculating the unit cost of items. This family really needs your help, so put on your thinking cap and let's get shopping!

📖 **Log on to msmath1.net/webquest to begin your WebQuest.**

CHAPTER 3

Adding and Subtracting Decimals

❝What does money have to do with math?❞

Any time you spend money, you use decimals. If you need to find out how much money you earn over a period of time, you add decimals. If you need to know whether you have enough money when you reach the checkout counter, you round decimals. You use decimals almost every day of your life.

You will solve problems about money in Lessons 3-3, 3-4, and 3-5.

GETTING STARTED

Take this quiz to see whether you are ready to begin Chapter 3. Refer to the lesson or page number in parentheses if you need more review.

▶ ## Vocabulary Review

Complete each sentence.

1. The four steps of the problem-solving plan are: Explore, ___?___, Solve, and Examine. (Lesson 1-1)

2. To find the value of an algebraic expression, you must ___?___ it for given values of the variables. (Lesson 1-6)

3. The ___?___ of a set of data is the sum of the data divided by the number of pieces of data. (Lesson 2-6)

▶ ## Prerequisite Skills

Evaluate each expression if $a = 3$ and $b = 4$. (Lesson 1-6)

4. $3a - 2b$ 5. $5 + 2a$

6. $b - 1 + a$ 7. $16 - b$

8. $3b + a$ 9. $b + 3a$

Add or subtract. (Page 589)

10. $82 - 67$ 11. $29 + 54$

12. $48 - 33$ 13. $61 + 19$

Multiply or divide. (Pages 590, 591)

14. $36 \div 4$ 15. 9×3

16. 6×5 17. $56 \div 8$

Round each number to the nearest tens place. (Page 592)

18. 5   19. 75

20. 148   21. 156

Adding and Subtracting Decimals Make this Foldable to help you organize your notes. Begin with two sheets of notebook paper.

STEP 1 **Fold and Cut One Sheet**
Fold in half. Cut along fold from edges to margin.

STEP 2 **Fold and Cut the Other Sheet**
Fold in half. Cut along fold between margins.

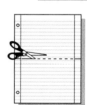

STEP 3 **Fold**
Insert first sheet through second sheet and along folds.

STEP 4 **Label**
Label each side of each page with a lesson number and title.

Chapter 3: Adding and Subtracting Decimals

Reading and Writing As you read and study the chapter, fill the journal with notes, diagrams, and examples for each lesson.

Modeling Decimals

What You'll LEARN

Use models to represent, compare, order, add, and subtract decimals.

Materials

• centimeter grid paper

Decimal models can be used to represent decimals.

Ones (1)	Tenths (0.1)	Hundredths (0.01)
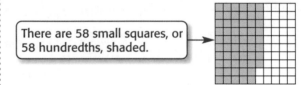		
One whole 10-by-10 grid represents 1 or 1.0.	One whole grid is made up of 10 rows or columns. Each row or column represents one tenth or 0.1.	One whole grid has 100 small squares. Each square represents one hundredth or 0.01.

ACTIVITIES *Work with a partner.*

1 **Write the decimal shown by the model.**

There are 58 small squares, or 58 hundredths, shaded. →

The model represents 58 hundredths or 0.58.

2 **Compare 0.35 and 0.32 by using models.**

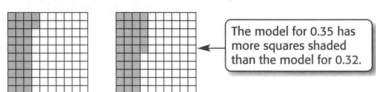

The model for 0.35 has more squares shaded than the model for 0.32.

So, 0.35 is greater than 0.32. That is, 0.35 > 0.32.

Your Turn Write the decimal shown by each model.

a.   b.   c.

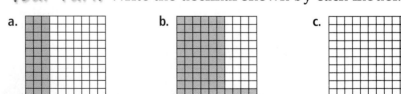

Compare each pair of decimals using models.

    d. 0.68 and 0.65   **e.** 0.2 and 0.28   **f.** 0.35 and 0.4

You can also use models to add and subtract decimals.

ACTIVITIES *Work with a partner.*

3 Find 0.16 + 0.77 using decimal models.

STEP 1 Shade 0.16 green.

STEP 2 Shade 0.77 blue.

The sum is represented by the total shaded area.

So, 0.16 + 0.77 = 0.93.

4 Find 0.52 − 0.08 using decimal models.

STEP 1 Shade 0.52 green.

STEP 2 Use x's to cross out 0.08 from the shaded area.

The difference is represented by the amount of shaded area that does not have an x in it.

So, 0.52 − 0.08 = 0.44.

Your Turn Find each sum using decimal models.

g. 0.14 + 0.67   **h.** 0.35 + 0.42   **i.** 0.03 + 0.07

Find each difference using decimal models.

j. 0.75 − 0.36   **k.** 0.68 − 0.27   **l.** 0.88 − 0.49

Writing Math

1. **Explain** why 0.3 is equal to 0.30. Use a model in your explanation.

2. **MAKE A CONJECTURE** Explain how you could compare decimals without using models.

3. **Explain** how you can use grid paper to model the following.
   a. 0.25 + 0.3   **b.** 0.8 − 0.37

4. **MAKE A CONJECTURE** Write a rule you can use to add or subtract decimals without using models.

Representing Decimals

What You'll LEARN

Represent decimals in word form, standard form, and expanded form.

NEW Vocabulary

standard form
expanded form

Link to READING

Deci-: a prefix meaning tenth part

HANDS-ON Mini Lab

Materials
- base-ten blocks
- decimal models
- play money

Work with a partner.

The models below show some ways to represent the decimal 1.34.

Place-Value Chart

1,000	100	10	1	0.1	0.01	0.001
thousands	hundreds	tens	ones	tenths	hundredths	thousandths
O	O	O	1.	3	4	O

Money

1 dollar   3 dimes   4 pennies

Decimal Model

one   34 hundredths

Base-Ten Blocks

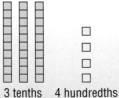

1 one   3 tenths   4 hundredths

Model each decimal using a place-value chart, money, a decimal model, and base-ten blocks.

1. 1.56   **2.** 0.85   **3.** 0.08   **4.** $2.25

Decimals, like whole numbers, are based on the number ten. The digits and the position of each digit determine the value of a decimal. The decimal point separates the whole number part of the decimal from the part that is less than one.

Place-Value Chart

1,000	100	10	1	0.1	0.01	0.001	0.0001
thousands	hundreds	tens	ones	tenths	hundredths	thousandths	ten-thousandths
O	O	O	1.	3	4	O	O

whole number   less than one

You can use place value to write decimals in word form.

READING Math

Decimal Point Use the word *and* only to read the decimal point. For example, read 0.235 as *two hundred thirty-five thousandths.* Read 235.035 as *two hundred thirty-five and thirty-five thousandths.*

16

EXAMPLE **Write a Decimal in Word Form**

1 Write 35.376 in word form.

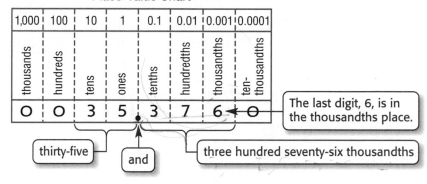

Place-Value Chart

1,000	100	10	1	0.1	0.01	0.001	0.0001
thousands	hundreds	tens	ones	tenths	hundredths	thousandths	ten-thousandths
O	O	3	5 .	3	7	6	O

The last digit, 6, is in the thousandths place.

thirty-five   and   three hundred seventy-six thousandths

35.376 is thirty-five and three hundred seventy-six thousandths.

Decimals can be written in standard form and expanded form. **Standard form** is the usual way to write a number. **Expanded form** is a sum of the products of each digit and its place value.

word form	standard form	expanded form
twelve hundredths	0.12	$(1 \times 0.1) + (2 \times 0.01)$

EXAMPLE **Standard Form and Expanded Form**

2 Write *fifty-four and seven ten-thousandths* in standard form and in expanded form.

Place-Value Chart

1,000	100	10	1	0.1	0.01	0.001	0.0001
thousands	hundreds	tens	ones	tenths	hundredths	thousandths	ten-thousandths
O	O	5	4 .	O	O	O	7

Standard form: 54.0007

Expanded form: $(5 \times 10) + (4 \times 1) + (0 \times 0.1) + (0 \times 0.01) +$
$(0 \times 0.001) + (7 \times 0.0001)$

Your Turn Write each decimal in word form.

a. 0.825   **b.** 16.08   **c.** 142.67

d. Write *twelve and four tenths* in standard form and in expanded form.

Skill and Concept Check

Writing Math
Exercises 1 & 3

1. **Explain** the difference between word form, standard form, and expanded form.

2. **OPEN ENDED** Draw a model that represents 2.75.

3. **Which One Doesn't Belong?** Identify the number that does not have the same value as the other three. Explain your reasoning.

0.75	seven and five hundredths	(7 × 0.1) + (5 × 0.01)	seventy-five hundredths

GUIDED PRACTICE

Write each decimal in word form.

4. 0.7
5. 0.08
6. 5.32
7. 0.022
8. 34.542
9. 8.6284

Write each decimal in standard form and in expanded form.

10. nine tenths
11. twelve thousandths
12. three and twenty-two hundredths
13. forty-nine and thirty-six ten-thousandths

14. **FOOD** A bottle of soda contains 1.25 pints. Write this number in two other forms.

Practice and Applications

Write each decimal in word form.

15. 0.4
16. 0.9
17. 3.56
18. 1.03
19. 7.17
20. 4.94
21. 0.068
22. 0.387
23. 78.023
24. 20.054
25. 0.0036
26. 9.0769

27. How is 301.0019 written in word form?
28. How is 284.1243 written in word form?

Write each decimal in standard form and in expanded form.

29. five tenths
30. eleven and three tenths
31. two and five hundredths
32. thirty-four and sixteen hundredths
33. forty-one and sixty-two ten-thousandths
34. one hundred two ten-thousandths
35. eighty-three ten-thousandths
36. fifty-two and one hundredths

37. Write (5 × 0.1) + (2 × 0.01) in word form.
38. Write (4 × 1) + (2 × 0.1) in word form.
39. How is (3 × 10) + (3 × 1) + (4 × 0.1) written in standard form?
40. Write (4 × 0.001) + (8 × 0.0001) in standard form.

HOMEWORK HELP

For Exercises	See Examples
15–28, 41	1
29–40, 44	2

Extra Practice
See pages 599, 626.

41. WRITING CHECKS To safeguard against errors, the dollar amount on a check is written in both standard form and word form. Write $23.79 in words.

FOOD For Exercises 42–44, use the information at the right.

42. Which numbers have their last digit in the hundredths place?

43. How did you identify the hundredths place?

44. Write each of these numbers in expanded form.

Unpopped Popcorn Kernel	
Ingredient	Grams
water	0.125
fat	0.03
protein	0.105
carbohydrates	0.71
mineral water	0.02

Source: www.popcornpopper.com

45. RESEARCH Use the Internet or another source to find the definition of decimal. Then write two ways that decimals are used in everyday life.

CRITICAL THINKING For Exercises 46 and 47, use the following information.

A decimal is made using each digit 5, 8, and 2 once.

46. What is the greatest possible decimal greater than 5 but less than 8?

47. Find the least possible decimal greater than 0 but less than one.

Standardized Test Practice and Mixed Review

48. **MULTIPLE CHOICE** Choose the decimal that represents *twelve and sixty-three thousandths.*

   Ⓐ 1,206.3   Ⓑ 120.63   Ⓒ 12.063   Ⓓ 0.12063

49. **SHORT RESPONSE** Write 34.056 in words.

50. **FOOTBALL** Would the mean be misleading in describing the average football scores in the table? Explain. (Lesson 2-8)

51. **SCHOOL** Find the median for the set of Tim's history test scores: 88, 90, 87, 91, 49. (Lesson 2-7)

Hayes Middle School Football Scores	
Game 1	20
Game 2	27
Game 3	22
Game 4	13
Game 5	30

Find the value of each expression. (Lesson 1-5)

52. $45 \div 3 \times 3 - 7 + 12$ **53.** $5 \times 6 + 6 - 12 \div 2$

GETTING READY FOR THE NEXT LESSON

BASIC SKILL Choose the letter of the point that represents each decimal.

54. 6.3   **55.** 6.7

56. 6.2   **57.** 6.5

58. 7.2   **59.** 6.9

A F   D   B   C   E

6.0   6.8   7.4

3-1b HANDS-ON LAB

A Follow-Up of Lesson 3-1

Other Number Systems

What You'll LEARN

Write numbers using Roman and Egyptian numerals.

INVESTIGATE *Work with a partner.*

At the very end of a movie, you'll find the year the movie was made. However, instead of seeing the year 2005, you'll usually see it written using Roman numerals as MMV.

The Roman numeral system uses combinations of seven letters to represent numbers. These letters are shown in the table at the right.

All other numbers are combinations of these seven letters. Several numbers and their Roman numerals are shown below.

Roman Numeral	Number
I	1
V	5
X	10
L	50
C	100
D	500
M	1,000

Number	2	3	4	6	7	8	9
Roman Numeral	II	III	IV	VI	VII	VIII	IX
Number	14	16	20	40	42	49	90
Roman Numeral	XIV	XVI	XX	XL	XLII	XLIX	XC

Writing Math

1. **MAKE A CONJECTURE** Study the patterns in the table. **Write** a sentence or two explaining the rule for forming Roman numerals.

Write each number using Roman numerals.

2. 6   3. 40   4. 23   5. 15   6. 55

Write the number for each Roman numeral.

7. XLIX   8. C   9. XCVIII   10. XXIV   11. XVIII

12. The page numbers at the front of your math book are written using Roman numerals. Write the number for the greatest Roman numeral you find there.

13. **Describe** a disadvantage of using Roman numerals.

14. *True* or *False*? The Roman numeral system is a place-value system. Explain.

INVESTIGATE *Work with a partner.*

The ancient Egyptian numbering system was very straightforward. A unique symbol represented each decimal place value in the whole number system.

The ancient Egyptian numeral system uses combinations of seven symbols. These symbols are shown in the table at the right.

All other numbers are combinations of these seven symbols. Several numbers and their Ancient Egyptian numerals are shown below.

Ancient Egyptian Numerals	Number
I	1
∩	10
ℓ	100
⚓	1,000
𝕊	10,000
𝕁	100,000
𝕏	1,000,000

Number	2	4	12	110
Egyptian Numeral	II	IIII	∩II	ℓ∩
Number	**1,200**	**11,110**	**221,100**	**1,111,000**
Egyptian Numeral	⚓ℓℓ	𝕊⚓∩	𝕁𝕁ℓℓ⚓	𝕏𝕁ℓ⚓

Writing Math

15. **Compare and contrast** our decimal number system with the ancient Egyptian numbering system.

Write each number using Egyptian numerals.

16. 4   17. 20   18. 112   19. 1,203

Write the number for each Egyptian numeral.

20. IIIII   21. ∩

22. ℓ∩∩∩   23. ⚓ℓℓℓ

24. **Describe** a disadvantage of using Egyptian numerals.

25. **Identify** any similarities between the Roman numeral system and the ancient Egyptian numbering system.

26. **MAKE A CONJECTURE** How do you think you would add numbers written with Egyptian numerals? How is it similar to adding in a place-value system?

Comparing and Ordering Decimals

What You'll LEARN

Compare and order decimals.

NEW Vocabulary

equivalent decimals

MATH Symbols

< less than
> greater than

WHEN am I ever going to use this?

SNOWBOARDING The table lists the top five finishers at the 2002 Olympic Games Men's Halfpipe.

Men's Halfpipe Results		
Snowboarders	Country	Score
Danny Kass	USA	42.5
Giacomo Kratter	Italy	42.0
Takaharu Nakai	Japan	40.7
Ross Powers	USA	46.1
Jarret Thomas	USA	42.1

Source: www.mountainzone.com

1. Which player had the highest score? Explain.

Comparing decimals is similar to comparing whole numbers. You can use place value or a number line to compare decimals.

EXAMPLE Compare Decimals

1 **SNOWBOARDING** Refer to the table above. Use > or < to compare Danny Kass' score with Jarret Thomas' score.

Method 1 Use place value.

> First, line up the decimal points.

Danny Kass: 42.5

Jarret Thomas: 42.1

> Then, starting at the left, find the first place the digits differ. Compare the digits.

Since 5 > 1, 42.5 > 42.1. So, Danny Kass's score was higher than Jarret Thomas's score.

Method 2 Use a number line.

Numbers to the right are greater than numbers to the left. Since 42.5 is to the right of 42.1, 42.5 > 42.1.

Your Turn Use >, <, or = to compare each pair of decimals.

a. 5.67 ● 5.72

b. 0.293 ● 0.253

STUDY TIP

< **and** > Recall that the symbol always points toward the lesser number.

Decimals that name the same number are called **equivalent decimals**. Examples are 0.6 and 0.60.

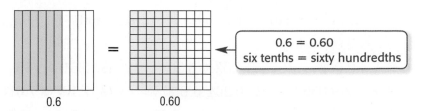

0.6   0.60

0.6 = 0.60
six tenths = sixty hundredths

When you *annex*, or place zeros to the right of the last digit in a decimal, the value of the decimal does not change. Annexing zeros is useful when ordering a group of decimals.

EXAMPLE **Order Decimals**

2 Order 15, 14.95, 15.8, and 15.01 from least to greatest.

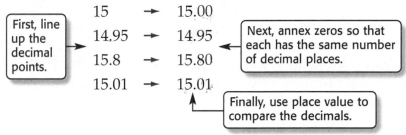

15	→	15.00
14.95	→	14.95
15.8	→	15.80
15.01	→	15.01

First, line up the decimal points.

Next, annex zeros so that each has the same number of decimal places.

Finally, use place value to compare the decimals.

STUDY TIP

Checking Reasonableness
You can check the reasonableness of the order by using a number line.

The order from least to greatest is 14.95, 15, 15.01, and 15.8.

Your Turn

c. Order 35.06, 35.7, 35.5, and 35.849 from greatest to least.

Skill and Concept Check

Writing Math
Exercises 2 & 3

1. **Draw** a number line to show the order of 2.5, 2.05, 2.55, and 2.35 from least to greatest.

2. **OPEN ENDED** Write a decimal that is equivalent to 0.4. Then draw a model to show that your answer is correct.

3. **FIND THE ERROR** Mark and Carlos are ordering 0.4, 0.5, and 0.49 from least to greatest. Who is correct? Explain.

Mark
0.4, 0.5, 0.49

Carlos
0.4, 0.49,

GUIDED PRACTICE

Use >, <, or = to compare each pair of decimals.

4. 2.7 ● 2.07   **5.** 0.4 ● 0.5   **6.** 25.5 ● 25.50

7. Order 0.002, 0.09, 0.2, 0.21, and 0.19 from least to greatest.

Practice and Applications

Use >, <, or = to compare each pair of decimals.

8. 0.2 ● 2.0

9. 3.3 ● 3.30

10. 0.08 ● 0.8

11. 0.4 ● 0.004

12. 6.02 ● 6.20

13. 5.51 ● 5.15

14. 9.003 ● 9.030

15. 0.204 ● 0.214

16. 7.107 ● 7.011

17. 23.88 ● 23.880

18. 0.0624 ● 0.0264

19. 2.5634 ● 2.5364

HOMEWORK HELP

For Exercises	See Examples
8–19, 24	1
20–23, 25	2

Extra Practice
See pages 599, 626.

Order each set of decimals from least to greatest.

20. 16, 16.2, 16.02, 15.99

21. 5.545, 4.45, 4.9945, 5.6

Order each set of decimals from greatest to least.

22. 2.1, 2.01, 2.11, 2.111

23. 32.32, 32.032, 32.302, 3.99

24. AUTO RACING In 1999, Jeff Gordon drove 161.551 miles per hour to win the Daytona 500. In 2001, Michael Waltrip won, driving 161.794 miles per hour. Who was faster?

25. BOOKS Most library books are placed on shelves so that their call numbers are ordered from least to greatest. Use the information at the right to find the order the books should be placed on the shelf.

Book	Number
Baleen Whales	599.52
The Blue Whale	599.5248
The Whale	599.5

26. CRITICAL THINKING Della has more money than Sara but less money than Eric. Halley has 10¢ more than Hector. The amounts are $0.89, $1.70, $1.18, $0.79, and $1.07. How much does each person have?

Standardized Test Practice and Mixed Review

27. MULTIPLE CHOICE Which number is between 3.18 and 4.03?

Ⓐ 3.082   Ⓑ 3.205   Ⓒ 4.052   Ⓓ 4.352

28. MULTIPLE CHOICE Which of these decimals is least?

Ⓕ 94.7   Ⓖ 98.5   Ⓗ 99.7   Ⓘ 101.1

Write each decimal in standard form. (Lesson 3-1)

29. thirty-seven thousandths

30. nine and sixteen thousandths

31. STATISTICS Is the mode a misleading measure of central tendency for the set of data 21, 20, 19, 13, 21, 18, 12, and 21? Explain. (Lesson 2-8)

32. Determine whether 315 is divisible by 2, 3, 5, or 10. (Lesson 1-2)

GETTING READY FOR THE NEXT LESSON

PREREQUISITE SKILL Identify each underlined place-value position. (Lesson 3-1)

33. 14.0<u>6</u>

34. 3.<u>0</u>54

35. 0.42<u>78</u>

36. 2.960<u>0</u>

Rounding Decimals

What You'll LEARN

Round decimals.

WHEN **am I ever going to use this?**

SCHOOL The Jackson Middle School lunch menu is shown at the right.

1. Round each cost to the nearest dollar.

2. How did you decide how to round each number?

3. **MAKE A CONJECTURE** about how to round each cost to the nearest dime.

Lunch Menu	
Item	**Cost ($)**
Pizza	$1.20
Salad	$2.65
Taco	$1.30
Soda	$0.85
Milk	$0.75
Fruit	$1.25

You can round decimals just as you round whole numbers.

Key Concept **Rounding Decimals**

To round a decimal, first underline the digit to be rounded. Then look at the digit to the right of the place being rounded.

- If the digit is 4 or less, the underlined digit remains the same.
- If the digit is 5 or greater, add 1 to the underlined digit.

EXAMPLES **Round Decimals**

① **Round 1.324 to the nearest whole number.**

> Underline the digit to be rounded. In this case, the digit is in the ones place.

1.324

> Then look at the digit to the right. Since 3 is less than 5, the digit 1 remains the same.

On the number line, 1.3 is closer to 1.0 than to 2.0. To the nearest whole number, 1.324 rounds to 1.0.

1.3

1.0 ——————— 2.0

② **Round 99.96 to the nearest tenth.**

> Underline the digit to be rounded. In this case, the digit is in the tenths place.

99.96

> Then look at the digit to the right. Since the digit is 6, add one to the underlined digit.

On the number line, 99.96 is closer to 100.0 than to 99.9. To the nearest tenth, 99.96 rounds to 100.0.

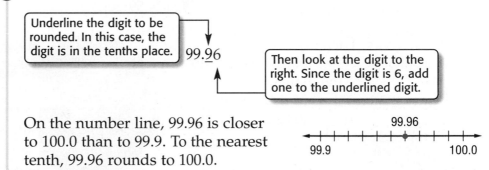

99.96

99.9 ——————— 100.0

Rounding is often used in real-life problems involving money.

Exercises 1 & 3

EXAMPLE **Use Rounding to Solve a Problem**

STUDY TIP

Rounding There are 100 cents in a dollar. So, rounding to the nearest cent means to round to the nearest hundredth.

3 **FOOD** A bag of potato chips costs $0.2572 per ounce. How much is this to the nearest cent?

To round to the nearest cent, round to the nearest hundredths place.

Underline the digit in the hundredths place.

$0.2572

Then look at the digit to the right. The digit is greater than 5. So, add one to the underlined digit.

To the nearest cent, the cost is $0.26 per ounce.

Your Turn Round each decimal to the indicated place-value position.

a. 0.27853; ten-thousandths   b. $5.8962; cent

Skill and Concept Check

Writing Math

1. **Draw** a number line to show why 3.47 rounded to the nearest tenth is 3.5. Write a sentence explaining the number line.

2. **OPEN ENDED** Give an example of a number that when rounded to the nearest hundredth is 45.39.

3. **Which One Doesn't Belong?** Identify the decimal that is not the same as the others when rounded to the nearest tenth. Explain.

| 34.62 | 34.59 | 34.49 | 34.56 |

GUIDED PRACTICE

Round each decimal to the indicated place-value position.

4. 0.329; tenths
5. 1.75; ones
6. 45.522; hundredths
7. 0.5888; thousandths
8. 7.67597; ten-thousandths
9. $34.59; tens

10. **SCIENCE** The table shows the rate of acceleration due to gravity for a few of the planets. To the nearest tenth, what is the rate of acceleration due to gravity for each planet?

Planet	Acceleration (meters per second)
Jupiter	23.12
Saturn	8.96
Uranus	8.69
Mars	3.69

Source: *Science Scope*

Practice and Applications

Round each decimal to the indicated place-value position.

11. 7.445; tenths

12. 7.999; tenths

13. $5.68; ones

14. 10.49; ones

15. 2.499; hundredths

16. 40.458; hundredths

17. 5.4572; thousandths

18. 45.0099; thousandths

19. 9.56303; ten-thousandths

20. 988.08055; ten-thousandths

21. $87.09; tens

22. 1,567.893; tens

23. Round $67.37 to the nearest dollar.

24. What is 67,234.63992 rounded to the nearest thousandth?

25. **FOOD** The United States is considered the "Ice Cream Capital of the World." Each person eats an average of nearly 5.75 gallons per year. Round 5.75 gallons to the nearest gallon.

26. **MEDIA** Disc jockeys often refer to a radio station by rounding its call number to the nearest whole number. What number would DJs use to refer to a radio station whose call number is 102.9?

27. **CRITICAL THINKING** Write three different decimals that round to 10.0 when rounded to the nearest tenth.

HOMEWORK HELP	
For Exercises	See Examples
11–24	1, 2
25–26	3

Extra Practice
See pages 600, 626.

Standardized Test Practice and Mixed Review

28. **MULTIPLE CHOICE** The atomic weights of certain elements are given in the table. What is the atomic weight of sodium to the nearest tenth?

   Ⓐ 22.98   Ⓑ 22.99

   Ⓒ 23.0   Ⓓ 23.1

Element	Atomic Weight
Sodium	22.9898
Neon	20.180
Magnesium	24.305

Source: www.webelements.com

29. **SHORT RESPONSE** Round 1,789.8379 to the nearest thousandth.

Use >, <, or = to compare each pair of decimals. (Lesson 3-2)

30. 8.64 ● 8.065

31. 2.5038 ● 25.083

32. 12.004 ● 12.042

33. Write *thirty-two and five hundredths* in standard form. (Lesson 3-1)

34. Find the prime factorization of 40. (Lesson 1-3)

GETTING READY FOR THE NEXT LESSON

PREREQUISITE SKILL Add or subtract. (Page 589)

35. 43 + 15

36. 68 + 37

37. 85 − 23

38. 52 − 29

Vocabulary and Concepts

1. **Explain** how to write a decimal in word form. (Lesson 3-1)

2. **State** the rule used for rounding decimals. (Lesson 3-3)

Skills and Applications

3. Write 12.65 in word form. (Lesson 3-1)

4. Write *four and two hundred thirty-two thousandths* in standard form and in expanded form. (Lesson 3-1)

Use >, <, or = to compare each pair of decimals. (Lesson 3-2)

5. 0.06 ● 0.6

6. 6.3232 ● 6.3202

7. 2.15 ● 2.150

Order each set from least to greatest. (Lesson 3-2)

8. 8.2, 8.02, 8.025, 8.225

9. 0.001, 0.101, 0.0101, 0.011

Round each decimal to the indicated place-value position. (Lesson 3-3)

10. 8.236; tenths

11. 10.0879; thousandths

12. 7.84; ones

13. 431; hundreds

Standardized Test Practice

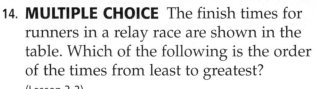

14. **MULTIPLE CHOICE** The finish times for runners in a relay race are shown in the table. Which of the following is the order of the times from least to greatest?
(Lesson 3-2)

Runner	Finish Time (s)
1	32.02
2	31.95
3	32.2004
4	32.0029

Ⓐ 32.2004, 32.02, 32.0029, 31.95

Ⓑ 32.02, 32.0029, 32.2004, 31.95

Ⓒ 31.95, 32.0029, 32.02, 32.2004

Ⓓ 31.95, 32.2004, 32.0029, 32.02

15. **MULTIPLE CHOICE** The cost per gallon of gasoline is often listed as a decimal in thousandths. To the nearest cent, what would you pay for a gallon of gasoline that costs $1.239? (Lesson 3-3)

Ⓔ $1.25   Ⓕ $1.24   Ⓖ $1.23   Ⓟ $1.22

The Game Zone

A Place To Practice Your Math Skills

Decimal War

● **GET READY!**

Players: two
Materials: spinner with digits 0 through 9, paper

● **GET SET!**

- Each player creates ten game sheets like the one shown at the right, one for each of ten rounds.

- Make a spinner as shown.

● **GO!**

- One player spins the spinner.

- Each player writes the number in one of the blanks on his or her game sheet.

- The other player spins the spinner, and each player writes the number in a blank.

- Play continues once more so that all blanks are filled.

- The person with the greater decimal scores 1 point.

- Repeat for ten rounds.

- **Who Wins?** The person with the greater number of points after ten rounds is the winner.

Estimating Sums and Differences

What You'll LEARN

Estimate sums and differences of decimals.

NEW Vocabulary

front-end estimation
clustering

WHEN am I ever going to use this?

TRAVEL The graph shows about how many passengers travel through the busiest United States airports.

1. Round each number to the nearest million.

2. About how many more people travel through Hartsfield Atlanta than San Francisco?

USA TODAY Snapshots®

Atlanta busiest U.S. airport
Passengers (in millions of per year):

Hartsfield Atlanta International **80.2**

Chicago-O'Hare International **72.1**

Los Angeles International **68.5**

Dallas-Fort Worth International **60.7**

San Francisco International **41.2**

Source: Department of Transportation, Federal Aviation Administration, Airports Council International

By Lori Joseph and Dave Merrill, USA TODAY

To estimate sums and differences of decimals, you can use the same methods you used for whole numbers.

EXAMPLES Use Estimation to Solve Problems

1 Estimate the total amount of passengers that travel through Dallas-Fort Worth and Los Angeles.

Round each number to the nearest ten for easier adding.

$$
\begin{array}{rcl}
60.7 & \to & 60 \quad \text{60.7 rounds to 60.} \\
+\ 68.5 & \to & +\ 70 \quad \text{68.5 rounds to 70.} \\
\hline
& & 130
\end{array}
$$

There are about 130 million passengers.

2 Estimate how many more passengers travel through Hartsfield Atlanta than through Chicago-O'Hare.

$$
\begin{array}{rcl}
80.2 & \to & 80 \quad \text{80.2 rounds to 80.} \\
-\ 72.1 & \to & -\ 70 \quad \text{72.1 rounds to 70.} \\
\hline
& & 10
\end{array}
$$

There are about 10 million more passengers.

STUDY TIP

Using estimation
There is no one correct answer when estimating. To estimate means to find an approximate value. However, *reasonableness* is important.

Another type of estimation is front-end estimation. When you use **front-end estimation**, add or subtract the front digits. Then add or subtract the digits in the next place value position.

 EXAMPLE **Use Front-End Estimation**

3 Estimate 34.6 + 55.3 using front-end estimation.

Add the front digits.   Then add the next digits.

$$
\begin{array}{r}
34.6 \\
+\ 55.3 \\
\hline
8
\end{array}
\qquad \rightarrow \qquad
\begin{array}{r}
34.6 \\
+\ 55.3 \\
\hline
89
\end{array}
$$

Using front-end estimation, 34.6 + 55.3 is about 89.

 Your Turn Estimate using front-end estimation.

a. $\begin{array}{r} 22.35 \\ -\ 11.14 \\ \hline \end{array}$ b. $\begin{array}{r} 5.45 \\ +\ 0.57 \\ \hline \end{array}$ c. $37.92 − $21.62

When estimating a sum in which all of the addends are close to the same number, you can use **clustering**.

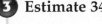
Standardized Test Practice

EXAMPLE   **Use Clustering**

4 **MULTIPLE-CHOICE TEST ITEM** Use the information in the table to estimate the total number of hours worked in the four months.

Month	Hours Worked
May	72.50
June	68.50
July	69.75
August	71.75

(A) 210   (B) 280

(C) 350   (D) 420

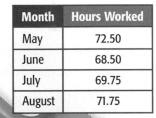

Test-Taking Tip
The Princeton Review

Clustering Clustering is good for problems in which the addends are close together.

Read the Test Item The addends are clustered around 70.

$$
\begin{array}{rcr}
72.50 & \rightarrow & 70 \\
68.50 & \rightarrow & 70 \\
69.75 & \rightarrow & 70 \\
+\ 71.75 & \rightarrow & +\ 70 \\
\hline
 & & 280
\end{array}
$$

Solve the Test Item Multiplication is repeated addition. So, a good estimate is 4 × 70, or 280. The answer is B.

Concept Summary   **Estimation Methods**

Rounding	Estimate by rounding each decimal to the nearest whole number that is easy for you to add or subtract mentally.
Front-End Estimation	Estimate by first adding or subtracting the front digits. Then add or subtract the next digits.
Clustering	Estimate by rounding a group of close numbers to the same number.

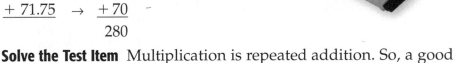

1. **Explain** how you would estimate $1.843 - 0.328$.

2. **OPEN ENDED** Describe a situation where it makes sense to use the clustering method to estimate a sum.

3. **NUMBER SENSE** How do you know that the sum of 5.4, 6.3, and 9.6 is greater than 20?

GUIDED PRACTICE

Estimate using rounding.

4. $0.36 + 0.83$

5. $4.44 - 2.79$

Estimate using front-end estimation.

6. $179 + 188 + 213$

7. $\$442 - \$126$

Estimate using clustering.

8. $5.32 + 4.78 + 5.42$

9. $0.95 + 0.79 + 1.02$

10. **PHONE COSTS** Use the information in the table. Estimate the total cost of the phone calls using clustering.

Phone Calls				
Minutes	8.7	9.1	9.0	8.9
Amount ($)	1.04	1.09	1.08	1.07

Practice and Applications

Estimate using rounding.

11. $49.59 + 16.22$

12. $\$41.59 - \$19.72$

13. $2.33 + 4.88 + 5.5$

14. $\$102.55 - \$52.77$

15. $0.36 + 0.83$

16. $0.7 - 0.6363$

17. About how much more is $74.50 than $29.95?

18. Estimate the sum of $2.456 + 1.925 + 2.395 + 1.695$.

HOMEWORK HELP

For Exercises	See Examples
11–18, 34	1, 2
19–27	3
28–33, 35	4

Extra Practice
See pages 600, 626.

Estimate using front-end estimation.

19.
$$75.45 - 5.23$$

20.
$$27.09 - 12.05$$

21.
$$28.65 + 71.53$$

22.
$$124.82 + 64.98$$

23.
$$\$315.65 + 30.42$$

24.
$$186.25 - 86.49$$

25.
$$116.22 - 14.67$$

26.
$$50.96 + 19.28$$

27. **RECYCLING** Two classes recycled paper. One class earned $16.52. The other class earned $28.80. About how much more did the second class earn?

Estimate using clustering.

28. $6.99 + 6.59 + 7.02 + 7.44$

29. $\$3.33 + \$3.45 + \$2.78 + \$2.99$

30. $5.45 + 5.3948 + 4.7999$

31. $\$55.49 + \$54.99 + \$55.33$

32. $10.33 + 10.45 + 10.89 + 9.79$

33. $99.8 + 100.2 + 99.5 + 100.4$

SPORTS For Exercises 34–37, use the graph.

34. About how much more would you expect to pay for purchases at a National Football League game than at a Major League Baseball game?

35. Use clustering to estimate the total cost for purchases at a National Basketball Association, National Hockey League, and a National Football League game.

36. **MULTI STEP** Suppose the average price of one ticket for a Major League Baseball game is $35.00. About how much would a family of four pay for four tickets and optional purchases?

37. **WRITE A PROBLEM** Write and solve a problem using the information in the graph. Then solve your problem using estimation.

USA TODAY Snapshots®

Optional purchases add up at games
Average amount spent by one person at a professional game besides the price of a ticket:

Major League Baseball	$15.40
National Basketball Association	$18.20
National Hockey League	$18.25
National Football League	$19.00

Source: *American Demographics; 2000 Inside the Ownership of Professional Sports Teams*

By Ellen J. Horrow and Sam Ward, USA TODAY

38. **CRITICAL THINKING** Five same-priced items are purchased. Based on rounding, the estimate of the total was $15. What is the maximum and minimum price the item could be?

Standardized Test Practice and Mixed Review

39. **MULTIPLE CHOICE** Zack plans on buying 4 shirts. The cost of each shirt ranges from $19.99 to $35.99. What would be a reasonable total cost for the shirts?

Ⓐ $60 Ⓑ $70 Ⓒ $120 Ⓓ $160

40. **MULTIPLE CHOICE** Refer to the table. Which is the best estimate for the total number of acres of land burned?

Ⓕ 25 million Ⓖ 30 million

Ⓗ 35 million Ⓘ 40 million

41. **WEATHER** Washington, D.C., has an average annual precipitation of 35.86 inches. Round this amount to the nearest tenth. (Lesson 3-3)

42. Order the decimals 27.025, 26.98, 27.13, 27.9, and 27.131 from least to greatest. (Lesson 3-2)

Land Burned in Wildfires	
Year	Acres Burned (millions)
2000	8.4
1996	6.7
1988	7.4
1969	6.7
1963	7.1

Source: www.nife.gov

GETTING READY FOR THE NEXT LESSON

PREREQUISITE SKILL Add or subtract. (Page 589)

43. 278
  + 199

44. 1,297
  + 86

45. 700
  − 235

46. 1,252
  − 79

47. 2,378
  − 195

 msmath1.net/self_check_quiz

Take Good Notes

NUMBER MAP

Have you heard the expression *a picture is worth a thousand words*? Sometimes the best notes you take in math class might be in the form of a drawing.

Just as a road map shows how cities are related to each other, a number map can show how numbers are related to each other. Start by placing a number in the center of the map.

Below is a number map that shows various meanings of the decimal 0.5. Notice that you can add both mathematical meanings and everyday meanings to the number map.

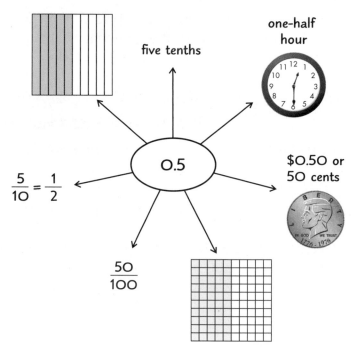

SKILL PRACTICE

Make a number map for each number. (*Hint*: For whole numbers, think of factors, prime factors, divisibility, place value, and so on.)

1. 0.75
2. 0.1
3. 0.01
4. 1.25
5. 2.5
6. 25
7. 45
8. 60
9. 100

10. Refer to Exercise 1. Explain how each mathematical or everyday meaning on the number map relates to the decimal 0.75.

3-5

Adding and Subtracting Decimals

What You'll LEARN

Add and subtract decimals.

REVIEW Vocabulary

evaluate: find the value of an expression by replacing variables with numerals **(Lesson 1-6)**

WHEN am I ever going to use this?

MOVIES The table shows the top five movies based on gross amount earned in a weekend.

Movie	Money Earned (millions)
#1	$25.0
#2	$23.1
#3	$13.1
#4	$10.0
#5	$5.5

1. Estimate the total amount of money earned by the top five movies.

2. Add the digits in the same place-value position. Use estimation to place the decimal point in the sum.

To add or subtract decimals, add or subtract digits in the same place-value position. Be sure to line up the decimal points before you add or subtract.

EXAMPLE   Add Decimals

1 Find the sum of 23.1 and 5.8.

Estimate $23.1 + 5.8 \rightarrow 23 + 6 = 29$

$\begin{array}{r} 23.1 \\ + 5.8 \\ \hline 28.9 \end{array}$ Line up the decimal points.

Add as with whole numbers.

> Compare the answer to the estimate. Since 28.9 is close to 29, the answer is reasonable.

The sum of 23.1 and 5.8 is 28.9.

EXAMPLE   Subtract Decimals

2 Find $5.774 - 2.371$.

Estimate $5.774 - 2.371 \rightarrow 6 - 2 = 4$

$\begin{array}{r} 5.774 \\ - 2.371 \\ \hline 3.403 \end{array}$ Line up the decimal points.

Subtract as with whole numbers.

So, $5.774 - 2.371 = 3.403$.   Compare to the estimate.

Your Turn Add or subtract.

a. $54.7 + 21.4$ b. $9.543 - 3.67$ c. $72.4 + 125.82$

Sometimes it is necessary to annex zeros before you subtract.

EXAMPLE **Annex Zeros**

3 Find 6 − 2.38.

Estimate $6 - 2.38 \rightarrow 6 - 2 = 4$

$$\begin{array}{r} 6.00 \quad \text{Annex zeros.} \\ -\ 2.38 \\ \hline 3.62 \end{array}$$

So, 6 − 2.38 = 3.62.   Compare to the estimate.

Your Turn Subtract.

d. 2 − 1.78   e. 14 − 9.09   f. 23 − 4.216

EXAMPLE **Use Decimals to Solve a Problem**

4 **SPEED SKATING** The table shows the top three times for the speed skating event in the 2002 Winter Olympics. What is the time difference between first place and third place?

1,000-Meter Women's Speed Skating	
Skater	Time (s)
Chris Witty	73.83
Sabine Volker	73.96
Jennifer Rodriguez	74.24

Source: www.weasel.student

Estimate $74.24 - 73.83 \rightarrow 74 - 74 = 0$

$$\begin{array}{r} 74.24 \\ -\ 73.83 \\ \hline 0.41 \end{array}$$

So, the difference between first place and third place is 0.41 second.

You can also use decimals to evaluate algebraic expressions.

EXAMPLE **Evaluate an Expression**

5 **ALGEBRA** Evaluate $x + y$ if $x = 2.85$ and $y = 17.975$.

$x + y = 2.85 + 17.975$ Replace x with 2.85 and y with 17.975.

Estimate $2.85 + 17.975 \rightarrow 3 + 18 = 21$

$$\begin{array}{r} 2.850 \quad \text{Line up the decimal points. Annex a zero.} \\ +17.975 \\ \hline 20.825 \quad \text{Add as with whole numbers.} \end{array}$$

The value is 20.825. This value is close to the estimate. So, the answer is reasonable.

Your Turn Evaluate each expression if $a = 2.56$ and $b = 28.96$.

g. $3.23 + a$ h. $68.96 - b$ i. $b - a$

Skill and Concept Check

Writing Math
Exercises 1 & 3

1. **Explain** how you would find the sum of 3.3 and 2.89.

2. **OPEN ENDED** Write a subtraction problem in which it is helpful to annex a zero.

3. **FIND THE ERROR** Ryan and Akiko are finding $8.9 - 3.72$. Who is correct? Explain.

> Ryan
> $8.9 - 3.72 = 5.22$

> Akiko
> $8.9 - 3.72 =$

4. **NUMBER SENSE** Pick five numbers from the list below whose sum is 10.2. Use each number only once.

   1.9   3   2.7   3.9   2.4   0.6   1.1   3.1   0.15

GUIDED PRACTICE

Add or subtract.

5. $\begin{aligned} 5.5 \\ + 3.2 \end{aligned}$

6. $\begin{aligned} 5.78 \\ - 5 \end{aligned}$

7. $\begin{aligned} 9.67 \\ + 2.35 \end{aligned}$

8. $\begin{aligned} 0.40 \\ - 0.20 \end{aligned}$

9. $5.5 - 1.24$

10. $1.254 + 0.3 + 4.15$

11. **ALGEBRA** Evaluate $s - t$ if $s = 8$ and $t = 4.25$.

Practice and Applications

Add or subtract.

12. $\begin{aligned} 7.2 \\ + 9.5 \end{aligned}$

13. $\begin{aligned} 4.9 \\ + 3.0 \end{aligned}$

14. $\begin{aligned} 1.34 \\ + 2 \end{aligned}$

15. $\begin{aligned} 0.796 \\ + 13 \end{aligned}$

16. $\begin{aligned} 5.6 \\ - 3.5 \end{aligned}$

17. $\begin{aligned} 19.86 \\ - 4.94 \end{aligned}$

18. $\begin{aligned} 97 \\ - 16.98 \end{aligned}$

19. $\begin{aligned} 82 \\ - 67.18 \end{aligned}$

20. $58.67 - 28.72$

21. $14.39 - 12.16$

22. $2.649 + 0.75 + 1.784$

ALGEBRA Evaluate each expression if $a = 128.9$ and $d = 22.035$.

23. $a - d$

24. $d + a$

25. $a - 11.25 - d$

26. $75 + d + a$

Find the value of each expression.

27. $2.3 + 6 \times 2$

28. $15.3 - 3^2$

29. $3 + 6.5 - 2.8$

30. $2^2 - 1.58 + 6.5$

31. **STATISTICS** Use the table to find out how many more students per teacher there are in California than in Nevada.

32. **MONEY** How much change would you receive if you gave a cashier $20 for a purchase that costs $18.74?

HOMEWORK HELP	
For Exercises	See Examples
12–22	1, 2, 3
31–32, 36–38	4
23–26	5
Extra Practice See pages 600, 626.	

Student-per-Teacher Ratio	
State	Ratio
Washington	19.9
Oregon	19.6
Nevada	18.7
California	21

Source: National Center for Education Statistics

 msmath1.net/self_check_quiz

For Exercises 33 and 34, find a counterexample for each statement.

33. The sum of two decimals having their last nonzero digit in the hundredths place also has its last nonzero digit in the hundredths place.

34. The difference of two decimals having their last nonzero digits in the tenths place also has its last nonzero digit in the tenths place.

35. **WRITE A PROBLEM** Write about a real-life situation that can be solved using addition or subtraction of decimals.

CARS For Exercises 36–39, use the information and the table. The top five choices for every 100 people are listed in the table.

36. Find the total number of people per 100 who chose the top five most popular colors.

37. How many more people per 100 chose black over white?

38. **MULTI STEP** How many more people per 100 chose the top three colors than the last two?

39. Do you believe that the colors chosen from year to year would be the same? Explain your reasoning.

Favorite Color for a Sport Compact Car–2001	
Silver	25.4
Black	14.5
Med./Dk. Blue	11.3
White	9.8
Med. Red	7.4

Source: infoplease.com

40. **ALGEBRA** What is the value of $a + b$ if $a = 126.68$ and $b = 1,987.9$?

41. **CRITICAL THINKING** Arrange the digits 1, 2, 3, 4, 5, 6, 7, and 8 into two decimals so that their difference is as close to 0 as possible. Use each digit only once.

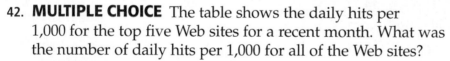

Standardized Test Practice and Mixed Review

42. **MULTIPLE CHOICE** The table shows the daily hits per 1,000 for the top five Web sites for a recent month. What was the number of daily hits per 1,000 for all of the Web sites?

    (A) 48.9622   (B) 50.484

    (C) 53.654   (D) 56.484

Web Sites	Hits per 1,000
A	14.699
B	14.295
C	7.790
D	7.563
E	6.137

Source: U.S. News and World Report

43. **MULTIPLE CHOICE** Dasan purchased $13.72 worth of gasoline. He received $10 back from the attendant. How much did Dasan give the attendant?

    (F) $15.72   (G) $20.00   (H) $25.00   (I) none of these

Estimate. Use an appropriate method. (Lesson 3-4)

44. $4.231 + 3.98$ 45. $3.945 + 1.92 + 3.55$ 46. $9.345 - 6.625$ 47. $\$11.11 - \$6.45$

48. Round 28.561 to the nearest tenth. (Lesson 3-3)

GETTING READY FOR THE NEXT LESSON

PREREQUISITE SKILL Add, subtract, multiply, or divide. (Pages 589–591)

49. $25 + 16$ 50. $96 - 25$ 51. 2×8 52. $24 \div 8$

Problem-Solving Strategy
A Follow-Up of Lesson 3-5

What You'll Learn

Solve problems by choosing an appropriate method of computation.

Choose the Method of Computation

> I heard your family drove to North Carolina for a vacation. About how far did you travel?

> We drove 356.2 miles the first day, 304.8 miles the second day, and 283.1 miles the third day. Then we drove the same route back home. Let's estimate to figure it out.

Explore	We don't need an exact answer, and it's too hard to compute mentally. Since we need to find *about* how far, we can estimate.
Plan	Let's start by estimating the number of miles traveled each day. Add the total for the three days and double that for the trip back home.
Solve	Day One -----------→ 356.2 **400** Day Two -----------→ 304.8 **300** Day Three -----------→ 283.1 **+300** **1,000** The trip was about 1,000 miles one way. The return trip was approximately another 1,000 miles for a total of 2,000 miles. Draw a sketch of the distance traveled and the return trip.
Examine	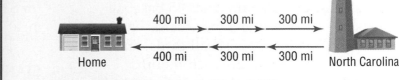

400 + 300 + 300 + 300 + 300 + 400 = 2,000
So, our answer of about 2,000 miles is correct. |

Analyze the Strategy

1. **Explain** when you would use estimation as the method of computation.

2. **Describe** how to mentally find the product of 40 and 3.

3. **Write** a problem in which you would use a calculator as the method of computation. Explain.

Apply the Strategy

Choose the best method of computation to solve each problem. Explain why you chose your method.

4. **TREES** It costs $283 to plant an acre of trees in a national forest. About how much will it cost to plant 640 acres?

5. **MEASUREMENT** How many seconds are in one week?

6. **SCHOOL** Each of 10 teachers donated $25 to the school scholarship fund. How much money was donated in all?

7. **MONEY** Ruben's mother gave him $20 to go to the grocery store. If the groceries cost $15.38, how much change will he receive?

Mixed Problem Solving

Solve. Use any strategy.

8. **GEOGRAPHY** The area of Rhode Island is 1,212 square miles. The area of Alaska is 591,004 square miles. About how many times larger is Alaska than Rhode Island?

9. **PATTERNS** How many triangles are in the bottom row of the fifth figure of this pattern?

  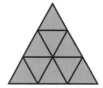

10. **MONEY** You have $100.75 in your checking account. You write checks for $21.78, $43, and $7.08. What is your new balance?

11. **BASEBALL CARDS** Jamal has 45 baseball cards. He is collecting 5 more cards each month. Alicia has 30 baseball cards, and she is collecting 10 more each month. How many months will it be before Alicia has more cards than Jamal?

12. **MEASUREMENT** If there are 8 fluid ounces in 1 cup, 2 cups in 1 pint, 2 pints in 1 quart, and 4 quarts in 1 gallon, how many fluid ounces are in 1 gallon?

13. **MONEY** Jane's lunch cost $3.64. She gives the cashier a $10 bill. How much change should Jane receive?

14. **HURRICANES** Refer to the table below.

Hurricanes	
Category	**Wind Speed (miles per hour)**
one	74–95
two	96–110
three	111–130
four	131–155
five	above 155

Source: www.carteretnewtimes.com

Hurricanes can be classified according to their wind speeds. What is the average of the minimum and maximum speeds for a category four hurricane?

15. **FOOD** Is $7 enough money to buy a loaf of bread for $0.98, one pound of cheese for $2.29, and one pound of luncheon meat for $3.29? Explain.

16. **STANDARDIZED TEST PRACTICE**
Alita, Alisa, and Alano are sharing the cost of their mother's birthday gift, which costs $147. About how much will each child need to contribute?

Ⓐ between $30 and $35

Ⓑ between $35 and $40

Ⓒ between $40 and $45

Ⓓ between $45 and $50

CHAPTER 3 Study Guide and Review

Vocabulary and Concept Check

clustering (p. 117)   expanded form (p. 103)   standard form (p. 103)
equivalent decimals (p. 109)   front-end estimation (p. 117)

State whether each sentence is *true* or *false*. If *false*, replace the underlined word or number to make a true sentence.

1. The number 0.07 is <u>greater</u> than 0.071.
2. When rounding decimals, the digit in the place being rounded should be rounded up if the digit to its right is <u>6</u>.
3. In 643.082, the digit 2 names the number two <u>hundredths</u>.
4. *Six hundred and twelve thousandths* written as a decimal is <u>0.612</u>.
5. Decimals that name the same number are called <u>equivalent</u> decimals.

Lesson-by-Lesson Exercises and Examples

3-1 Representing Decimals (pp. 102–105)

Write each decimal in standard form and in expanded form.

6. thirteen hundredths

7. six and five tenths

8. eighty-three and five thousandths

9. **GARDENING** A giant pumpkin weighed fifty-three and one hundred seventy-five thousandths pounds. Write this weight in standard form.

Example 1 Write 21.62 in word form.
21.62 is twenty-one and sixty-two hundredths

Example 2 Write three hundred forty-six thousandths in standard form and in expanded form.
Standard form: 0.346
Expanded form:
$(3 \times 0.1) + (4 \times 0.01) + (6 \times 0.001)$

3-2 Comparing and Ordering Decimals (pp. 108–110)

Use >, <, or = to compare each pair of decimals.

10. 0.35 ● 0.3   11. 6.024 ● 6.204

12. 0.10 ● 0.1   13. 8.34 ● 9.3

Example 3 Use <, >, or = to compare 4.153 and 4.159.

4.15<u>3</u>   Line up the decimal points.

4.15<u>9</u>   Starting at the left, find the first place the digits differ.

Since $3 < 9$, $4.153 < 4.159$.

3-3 Rounding Decimals (pp. 111–113)

Round each decimal to the indicated place-value position.

14. 5.031; hundredths

15. 0.00042; ten-thousandths

16. **FOOD COSTS** A box of cereal costs $0.216 per ounce. Round this price to the nearest cent.

Example 4 Round 8.0314 to the hundredths place.

8.0314   Underline the digit to be rounded.

8.0314   Then look at the digit to the right.
↑   Since 1 is less than 5, the digit 3 stays the same.

So, 8.0314 rounds to 8.03.

3-4 Estimating Sums and Differences (pp. 116–119)

Estimate using rounding.

17. 37.82 + 14.24   18. $72.18 − $29.93

19. 6.8 + 4.2 + 3.5   20. 129.6 − 9.7

Estimate using front-end estimation.

21.   31.29
     + 58.07

22.   93.65
     − 62.13

23.   145.91
    + 31.65

24.   87.25
     − 63.97

Estimate using clustering.

25. 12.045 + 11.81 + 12.3 + 11.56

26. $6.45 + $5.88 + $5.61 + $6.03

27. 1.15 + 0.74 + 0.99 + 1.06

Example 5 Estimate 38.61 − 14.25 using rounding.

$$
\begin{array}{rcl}
38.61 & \to & 39 \\
-\,14.25 & \to & -\,14 \\
\hline
 & & 25
\end{array}
$$
Round each number to the nearest whole number.

Example 6 Estimate 24.6 + 35.1 using front-end estimation.

24.6 + 35.1   Add the front digits to get 5. Then add the next digits. An estimate is 59.

Example 7 Estimate 8.12 + 7.65 + 8.31 + 8.08 using clustering.

All addends of the sum are close to 8. So, an estimate is 4 × 8 or 32.

3-5 Adding and Subtracting Decimals (pp. 121–124)

Add or subtract.

28.   18.35
     + 23.61

29.   148.93
    − 121.36

30.   1.325
     + 0.081

31. 248 − 131.28

32. **TRAVEL** Mr. Becker drove 11.3 miles to the dentist, 7.5 miles to the library, and 5.8 miles back home. How far did he travel?

Example 8 Find the sum of 48.23 and 11.65.

Estimate 48.23 + 11.65 → 48 + 12 = 60

$$
\begin{array}{r}
48.23 \\
+\,11.65 \\
\hline
59.88
\end{array}
$$
Line up the decimals.
Add as with whole numbers.

The sum is 59.88.

Practice Test

Vocabulary and Concepts

1. **Define** *expanded form* and give an example.
2. **Describe** how place value is used to compare decimals.

Skills and Applications

Write each decimal in word form.

3. 0.07
4. 8.051
5. 43.43

6. **SCIENCE** The weight of a particular molecule is given as 0.0003 ounce. Write the weight in word form.

Write each decimal in standard form and in expanded form.

7. six tenths
8. two and twenty-one thousandths
9. one and nine hundredths

Use >, <, or = to compare each pair of decimals.

10. 0.06 ● 0.60
11. 4.888 ● 4.880
12. 2.03 ● 2.030

Order each set of decimals from greatest to least.

13. 5.222, 5.202, 5.022, 5.2222
14. 0.04, 0.0404, 0.404, 0.0444

Round each decimal to the indicated place-value position.

15. 2.059; hundredths
16. 27.35; tens
17. 4.86273; ten-thousandths
18. 3.4553; thousandths

19. **ARCHITECTURE** Round each ceiling height in the table to the nearest tenth of a foot.

20. Estimate 38.23 + 11.84 using rounding.

21. Estimate 75.38 + 22.04 using front-end estimation.

Room	Porch	Living Room	Bedroom
Ceiling Height (ft)	8.12	12.35	8.59

Add or subtract.

22. 43.28
   + 31.45

23. 392.802
   − 173.521

24. 0.724
   + 6.458

Standardized Test Practice

25. **MULTIPLE CHOICE** Matthew ordered juice for $0.89, scrambled eggs for $3.69, and milk for $0.59. About how much did he spend?

   Ⓐ $10
   Ⓑ $8
   Ⓒ $6
   Ⓓ $4

Standardized Test Practice

Record your answers on the answer sheet provided by your teacher or on a sheet of paper.

1. There are ten million people living in a city. How is ten million written in standard form? (Prerequisite Skill, p. 586)

    Ⓐ 10,000
    Ⓑ 1,000,000
    Ⓒ 10,000,000
    Ⓓ 10,000,000,000

2. Xavier, Justin, Leslie, and Cree are playing a game of prime and composites. Points are given when a card with a prime number is selected. Who did *not* score any points on this turn? (Lesson 1-3)

| Xavier | Justin | Leslie | Cree |

    Ⓕ Cree   Ⓖ Justin
    Ⓗ Leslie   Ⓘ Xavier

3. Mrs. Sabina graded the spelling test for her sixth-grade class. The range in scores is 56. If the lowest grade is 42, what is the highest grade? (Lesson 2-7)

    Ⓐ 14   Ⓑ 56   Ⓒ 96   Ⓓ 98

4. How is 2.6251 read? (Lesson 3-1)

    Ⓕ two and six thousand two hundred fifty-one thousandths

    Ⓖ two and six thousand two hundred fifty-one hundred thousandths

    Ⓗ two and six thousand two hundred fifty-one ten thousandths

    Ⓘ two and six thousand two hundred fifty-one hundredths

5. Which decimal number can replace *P* on the number line? (Lesson 3-1)

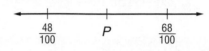

    Ⓐ 0.32   Ⓑ 0.48   Ⓒ 0.56   Ⓓ 0.69

6. Michael practiced his long jump after school. He recorded each jump. Which jump is the longest? (Lesson 3-2)

    Ⓕ 7.008 m   Ⓖ 7.049 m
    Ⓗ 7.073 m   Ⓘ 7.080 m

7. The store calculates sales tax and rounds it to the nearest cent. The tax on a coat totaled $3.02. Which of the following could be the actual amount of the tax before it was rounded? (Lesson 3-3)

    Ⓐ $3.000   Ⓑ $3.024
    Ⓒ $3.036   Ⓓ $3.030

8. The table shows the distance Jillian walked each day. What is the *best* estimate of the distance she walked over the five days? (Lesson 3-4)

Jillian's Walks	
Day	**Distance (km)**
Monday	2.4
Tuesday	5.2
Wednesday	3.6
Thursday	7.9
Friday	4.1

    Ⓕ 21 km   Ⓖ 23 km
    Ⓗ 25 km   Ⓘ 26 km

9. How much greater is 11.2 than 10.8? (Lesson 3-5)

    Ⓐ 0.4   Ⓑ 1.4   Ⓒ 1.6   Ⓓ 22.0

PART 2 Short Response/Grid In

Record your answers on the answer sheet provided by your teacher or on a sheet of paper.

10. Write $5 \times 5 \times 5 \times 5$ using an exponent. Then find the value of the power. (Lesson 1-3)

11. The graph shows the approximate top speeds of the world's fastest athletes. How much faster is a speed skater than a swimmer? (Lesson 2-2)

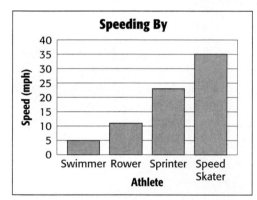

Speeding By

Source: *Chicago Tribune*

12. The table shows the prices of various recycled products. What is the median price per ton? (Lesson 2-7)

Price Per Ton ($)			
62	60	53	97
88	42	69	119
84	132	153	165
153	121	30	17

13. Sasha checked her reaction time using a stopwatch. During which trial was Sasha's reaction time the slowest? (*Hint*: The slowest time is the longest). (Lesson 3-2)

Reaction Time (s)	
Trial 1	1.031
Trial 2	1.016
Trial 3	1.050
Trial 4	1.007

14. Kelly scored an average of 12.16 points per game. What is her average score rounded to the nearest tenth? (Lesson 3-3)

15. Miranda rode her bike to the park and then to Dave's house. By the end of the day, she had biked a total of about 13 kilometers. From Dave's house, did Miranda bike back to the park, and then home? Explain. (Lesson 3-4)

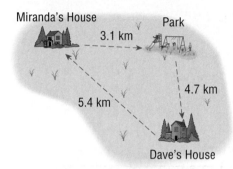

16. Is 6.14, 6.2, or 1.74 the solution of $x - 3.94 = 2.2$? (Lesson 3-5)

PART 3 Extended Response

Record your answers on a sheet of paper. Show your work.

17. Mr. Evans had a yard sale. He wrote both the original cost and the yard sale price on each item. (Lesson 3-5)

Item	Original Price ($)	Selling Price ($)
Table	95.15	12
Mirror	42.14	8
Picture Frame	17.53	2
Television	324.99	52

a. If Mr. Evans sold all four items, how much did he make at the yard sale?

b. What was his loss?

c. Explain how you calculated his loss.

TEST-TAKING TIP

Question 16 To check the solution of an equation, replace the variable in the equation with your solution.

 msmath1.net/standardized_test

CHAPTER 4

Multiplying and Dividing Decimals

"How do you use decimals on vacation?"

If you traveled to Australia, you would need to exchange U.S. dollars for Australian dollars. **In a recent month, every U.S. dollar could be exchanged for 1.79662 Australian dollars.** To find how many Australian dollars you would receive, you multiply by a decimal.

You will solve a problem about exchanging U.S. currency in Lesson 4-2.

GETTING STARTED

Take this quiz to see whether you are ready to begin Chapter 4. Refer to the lesson or page number in parentheses if you need more review.

▶ ## Vocabulary Review

Complete each sentence.

1. To find the closest value of a number based on a given place, you must __?__ the number. (Page 592)

2. In 4^3, 4 is raised to the third __?__. (Lesson 1-4)

▶ ## Prerequisite Skills

Write each power as a product. Then find the value of the power. (Lesson 1-4)

3. 10^2　　　4. 10^3　　　5. 10^5

Evaluate each expression. (Lesson 1-5)

6. $2 \times 14 + 2 \times 6$　　　7. $2 \times 1 + 2 \times 1$

8. $2 \times 2 + 2 \times 5$　　　9. $2 \times 5 + 2 \times 9$

10. $2 \times 7 + 2 \times 3$　　　11. $2 \times 8 + 2 \times 11$

12. Find the area of the rectangle. (Lesson 1-8)

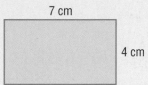

7 cm

4 cm

Add. (Lesson 3-5)

13. $6.8 + 6.8 + 10.2 + 10.2$

14. $7.1 + 7.1 + 13.3 + 13.3$

15. $4.6 + 4.6 + 2.25 + 2.25$

16. $11 + 11 + 9.9 + 9.9$

17. $8 + 8 + 3.7 + 3.7$

18. $12.4 + 12.4 + 5.5 + 5.5$

Decimals Make this Foldable to help you organize your notes. Begin with one sheet of construction paper.

STEP 1 Fold
Fold widthwise to within 1 inch of the bottom edge.

STEP 2 Fold again
Fold in half.

STEP 3 Cut
Open and cut along fold line, forming two tabs.

STEP 4 Label
Label as shown.

Reading and Writing As you read and study the chapter, write examples under each tab.

Multiplying Decimals by Whole Numbers

You can use decimal models to multiply a decimal by a whole number. Recall that a 10-by-10 grid represents the number one.

ACTIVITY *Work with a partner.*

1. Model 0.5 × 3 using decimal models.

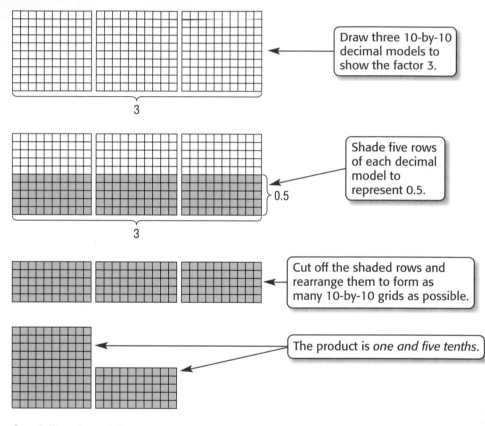

Draw three 10-by-10 decimal models to show the factor 3.

Shade five rows of each decimal model to represent 0.5.

Cut off the shaded rows and rearrange them to form as many 10-by-10 grids as possible.

The product is *one and five tenths.*

So, 0.5 × 3 = 1.5.

Your Turn Use decimal models to show each product.

a. 3 × 0.5   b. 2 × 0.7   c. 0.8 × 4

Writing Math

1. **MAKE A CONJECTURE** Is the product of a whole number and a decimal greater than the whole number or less than the whole number? Explain your reasoning.

2. Test your conjecture on 7 × 0.3. Check your answer by making a model or with a calculator.

Multiplying Decimals by Whole Numbers

What You'll LEARN

Estimate and find the product of decimals and whole numbers.

NEW Vocabulary

scientific notation

Link to READING

Everyday Meaning of Annex: to add something

WHEN am I ever going to use this?

SHOPPING CDs are on sale for $7.99. Diana wants to buy two. The table shows different ways to find the total cost.

Cost of Two CDs	
Add.	$7.99 + $7.99 = $15.98
Estimate.	$7.99 rounds to $8. 2 × $8 = $16
Multiply.	2 × $7.99 = ■

1. Use the addition problem and the estimate to find 2 × $7.99.

2. Write an addition problem, an estimate, and a multiplication problem to find the total cost of 3 CDs, 4 CDs, and 5 CDs.

3. **Make a conjecture** about how to find the product of $0.35 and 3.

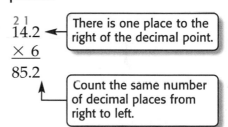

When multiplying a decimal by a whole number, multiply as with whole numbers. Then use estimation to place the decimal point in the product. You can also count the number of decimal places.

EXAMPLES Multiply Decimals

1 Find 14.2 × 6.

Method 1 Use estimation.

Round 14.2 to 14.

$14.2 \times 6 \rightarrow 14 \times 6$ or 84

$$\begin{array}{r} {\overset{2\ 1}{14.2}} \\ \times\ 6 \\ \hline 85.2 \end{array}$$
Since the estimate is 84, place the decimal point after the 5.

Method 2 Count decimal places.

$$\begin{array}{r} {\overset{2\ 1}{14.2}} \\ \times\ 6 \\ \hline 85.2 \end{array}$$

There is one place to the right of the decimal point.

Count the same number of decimal places from right to left.

2 Find 9 × 0.83.

Method 1 Use estimation.

Round 0.83 to 1.

$9 \times 0.83 \rightarrow 9 \times 1$ or 9

$$\begin{array}{r} {\overset{2}{0.83}} \\ \times\ 9 \\ \hline 7.47 \end{array}$$
Since the estimate is 9, place the decimal point after the 7.

Method 2 Count decimal places.

$$\begin{array}{r} {\overset{2}{0.83}} \\ \times\ 9 \\ \hline 7.47 \end{array}$$

There are two places to the right of the decimal point.

Count the same number of decimal places from right to left.

Your Turn Multiply.

a. 3.4 × 5   b. 11.4 × 8   c. 7 × 2.04

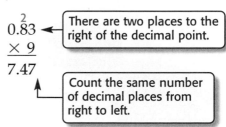

If there are not enough decimal places in the product, you need to annex zeros to the left.

EXAMPLES **Annex Zeros in the Product**

3 Find 2×0.018.

Estimate $2 \times 0.018 \rightarrow 2 \times 0$ or 0. The product is close to zero.

$$\begin{array}{r} 0.01\overset{1}{8} \\ \times2 \\ \hline 0.036 \end{array}$$

There are three decimal places.

> Annex a zero on the left of 36 to make three decimal places.

Check $0.018 + 0.018 = 0.036$ ✔

4 **ALGEBRA** Evaluate $4c$ if $c = 0.0027$.

$4c = 4 \times 0.0027$ Replace c with 0.0027.

$$\begin{array}{r} 0.00\overset{2}{2}7 \\ \times4 \\ \hline 0.0108 \end{array}$$

There are four decimal places.

Annex a zero to make four decimal places.

Your Turn Multiply.

d. 3×0.02 e. 8×0.12 f. 11×0.045

When the number 450 is expressed as the product of 4.5 and 10^2 (a power of ten), the number is written in **scientific notation**. You can use the order of operations or mental math to write numbers like 4.5×10^2 in standard form.

EXAMPLE **Scientific Notation**

5 **DINOSAURS** Write 6.5×10^7 in standard form.

Method 1 Use order of operations.

Evaluate 10^7 first. Then multiply.

$6.5 \times 10^7 = 6.5 \times 10,000,000$

$ = 65,000,000$

So, $6.5 \times 10^7 = 65,000,000$.

Method 2 Use mental math.

Move the decimal point 7 places.

$6.5 \times 10^7 = 6.5000000$

$ = 65,000,000$

Your Turn Write each number in standard form.

g. 7.9×10^3 h. 4.13×10^4 i. 2.3×10^6

Skill and Concept Check

Writing
Math
Exercises 1–4

1. **Explain** two methods of placing the decimal point in the product.

2. **OPEN ENDED** Write a multiplication problem where one factor is a decimal and the other is a whole number. The product should be between 2 and 3.

3. **FIND THE ERROR** Amanda and Kelly are finding the product of 0.52 and 2. Who is correct? Explain.

Amanda	Kelly
0.52	0.52
× 2	× 2
0.104	1.04

4. **NUMBER SENSE** Is the product of 0.81 and 15 greater than 15 or less than 15? How do you know?

GUIDED PRACTICE

Multiply.

5. 0.7 × 6

6. 0.3 × 2

7. 0.52 × 3

8. 2.13 × 6

9. 4×0.9

10. 5×0.8

11. 9×0.008

12. 3×0.015

13. **ALGEBRA** Evaluate $129t$ if $t = 2.9$.

14. Write 2.5×10^3 in standard form.

Practice and Applications

Multiply.

15. 1.2 × 7

16. 0.9 × 4

17. 0.65 × 6

18. 6.32 × 8

19. 0.7 × 9

20. 1.7 × 5

21. 3.62 × 4

22. 0.97 × 2

23. 2×1.3

24. 3×0.5

25. 1.8×9

26. 2.4×8

27. 4×0.02

28. 7×0.012

29. 9×0.0036

30. 0.0198×2

HOMEWORK HELP

For Exercises	See Examples
15–26, 42–45	1, 2
27–30	3
34–35	4
36–41	5

Extra Practice
See pages 601, 627.

GEOMETRY Find the area of each rectangle.

31.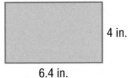
4 in.
6.4 in.

32.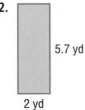
5.7 yd
2 yd

33.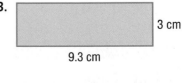
3 cm
9.3 cm

34. ALGEBRA Evaluate $3.05n$ if $n = 27$.

35. ALGEBRA Evaluate $80.05w$ if $w = 2$.

Write each number in standard form.

36. 5×10^4 **37.** 4×10^6 **38.** 1.5×10^3

39. 9.3×10^5 **40.** 3.45×10^3 **41.** 2.17×10^6

42. MULTI STEP Laura is trying to eat less than 750 Calories at dinner. A 4-serving, thin crust cheese pizza has 272.8 Calories per serving. A dinner salad has 150 Calories. Will Laura be able to eat the salad and two pieces of pizza for under 750 Calories? Explain.

SOCCER For Exercises 43–45, use the table.

The table shows soccer ball prices that Nick found online. He decided to buy one dozen Type 3 soccer balls.

Soccer Ball	Type 1	Type 2	Type 3	Type 4	Type 5
Price	6.99	14.99	19.99	34.99	99.99

43. What is the total cost?

44. What is the cost for one dozen of the highest price soccer balls?

45. How much would one dozen of the lowest priced soccer balls cost?

46. WRITE A PROBLEM Write a problem about a real-life situation that can be solved using multiplication. One factor should be a decimal. Then solve the problem.

47. Which of the numbers 4, 5, or 6 is the solution of $3.67a = 18.35$?

48. CRITICAL THINKING Write an equation with one factor containing a decimal where it is necessary to annex zeros in the product.

Standardized Test Practice and Mixed Review

49. MULTIPLE CHOICE Ernesto bought 7 spiral notebooks. Each notebook cost $2.29, including tax. What was the total cost of the notebooks?

    A $8.93   **B** $16.03   **C** $16.93   **D** $17.03

50. MULTIPLE CHOICE Before sales tax, what is the total cost of three CDs selling for $13.98 each?

    F $13.98   **G** $20.97   **H** $27.96   **I** $41.94

51. Add 15.783 and 390.81. (Lesson 3-5)

Estimate using rounding. (Lesson 3-4)

52. $29.34 - 9.0$ **53.** $42.28 - 1.52$ **54.** $26.48 + 3.95$

GETTING READY FOR THE NEXT LESSON

PREREQUISITE SKILL Find the value of each expression. (Page 590)

55. 43×25 **56.** 126×13 **57.** 18×165

Multiplying Decimals

What You'll LEARN

Use decimal models to multiply decimals.

In the Hands-On Lab on page 134, you used decimal models to multiply a decimal by a whole number. You can use similar models to multiply a decimal by a decimal.

Materials

• grid paper
• colored pencils
• scissors

ACTIVITY *Work with a partner.*

1 Model 0.8 × 0.4 using decimal models.

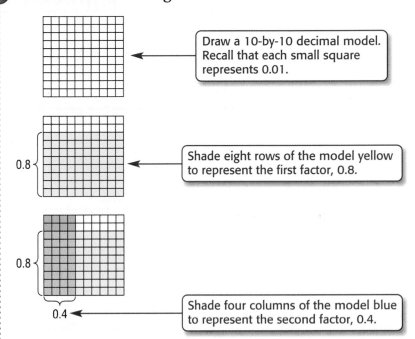

Draw a 10-by-10 decimal model. Recall that each small square represents 0.01.

0.8 — Shade eight rows of the model yellow to represent the first factor, 0.8.

0.8

0.4 — Shade four columns of the model blue to represent the second factor, 0.4.

There are 32 hundredths in the region that is shaded green. So, 0.8 × 0.4 = 0.32.

Your Turn Use decimal models to show each product.

a. 0.3 × 0.3   **b.** 0.4 × 0.9   **c.** 0.9 × 0.5

Writing Math

1. Tell how many decimal places are in each factor and in each product of Exercises a–c above.

2. **MAKE A CONJECTURE** Use the pattern you discovered in Exercise 1 to find 0.6 × 0.2. Check your conjecture with a model or a calculator.

3. Find two decimals whose product is 0.24.

ACTIVITY *Work with a partner.*

2 Model 0.7 × 2.5 using decimal models.

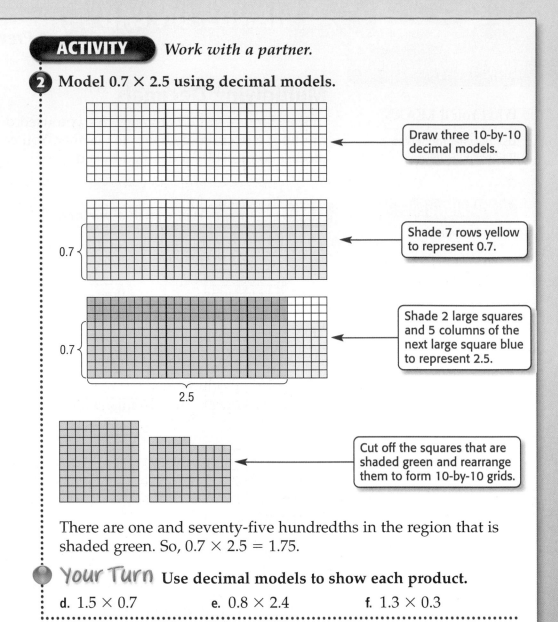

Draw three 10-by-10 decimal models.

Shade 7 rows yellow to represent 0.7.

Shade 2 large squares and 5 columns of the next large square blue to represent 2.5.

Cut off the squares that are shaded green and rearrange them to form 10-by-10 grids.

There are one and seventy-five hundredths in the region that is shaded green. So, 0.7 × 2.5 = 1.75.

Your Turn Use decimal models to show each product.

d. 1.5 × 0.7   **e.** 0.8 × 2.4   **f.** 1.3 × 0.3

Writing Math

4. **MAKE A CONJECTURE** How does the number of decimal places in the product relate to the number of decimal places in the factors?

5. Analyze each product.

   a. Explain why the first product is less than 0.6.

   b. Explain why the second product is equal to 0.6.

   c. Explain why the third product is greater than 0.6.

First Factor		Second Factor		Product
0.9	×	0.6	=	0.54
1.0	×	0.6	=	0.6
1.5	×	0.6	=	0.90

Multiplying Decimals

What You'll LEARN

Multiply decimals by decimals.

SHOPPING A candy store is having a sale. The sale prices are shown in the table.

1. Suppose you fill a bag with 1.3 pounds of jellybeans. The product 1.3×2 can be used to estimate the total cost. Estimate the total cost.

2. Multiply 13 by 200.

Candy Store (Cost per lb)	
jellybeans	$2.07
gummy worms	$2.21
snow caps	$2.79

3. How are the answers to Exercises 1 and 2 related?

Repeat Exercises 1–3 for each amount of candy.

4. 1.7 pounds of gummy worms   5. 2.28 pounds of snow caps

6. **Make a conjecture** about how to place the decimal point in the product of two decimals.

When multiplying a decimal by a decimal, multiply as with whole numbers. To place the decimal point, find the sum of the number of decimal places in each factor. The product has the same number of decimal places.

EXAMPLES   Multiply Decimals

1 Find 4.2×6.7.   **Estimate** $4.2 \times 6.7 \rightarrow 4 \times 7$ or 28

$$
\begin{array}{r}
4.2 \quad \leftarrow \text{one decimal place} \\
\times\, 6.7 \quad \leftarrow \text{one decimal place} \\
\hline
294 \\
252 \quad\;\; \\
\hline
28.14 \quad \leftarrow \text{two decimal places}
\end{array}
$$

The product is 28.14.   Compared to the estimate, the product is reasonable.

2 Find 1.6×0.09.   **Estimate** $1.6 \times 0.09 \rightarrow 2 \times 0$ or 0

$$
\begin{array}{r}
1.6 \quad \leftarrow \text{one decimal place} \\
\times\, 0.09 \quad \leftarrow \text{two decimal places} \\
\hline
0.144 \quad \leftarrow \text{three decimal places}
\end{array}
$$

The product is 0.144.   Compared to the estimate, the product is reasonable.

Your Turn Multiply.

a. 5.7×2.8 b. 4.12×0.07 c. 0.014×3.7

EXAMPLE Evaluate an Expression

3 **ALGEBRA** Evaluate $1.4x$ if $x = 0.067$.

$1.4x = 1.4 \times 0.067$ Replace x with 0.067.

$$
\begin{array}{r}
0.067 \quad \leftarrow \text{three decimal places}\\
\times \ 1.4 \quad \leftarrow \text{one decimal place}\\
\hline
268\\
67\\
\hline
0.0938 \quad \leftarrow \text{Annex a zero to make four decimal places.}
\end{array}
$$

Your Turn Evaluate each expression.

d. $0.04t$, if $t = 3.2$ **e.** $2.6b$, if $b = 2.05$ **f.** $1.33c$, if $c = 0.06$

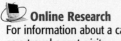

There are many real-life situations when you need to multiply two decimals.

EXAMPLE Multiply Decimals to Solve a Problem

4 **TRAVEL** Ryan and his family are traveling to Mexico. One U.S. dollar is worth 8.9 pesos. How many pesos would he receive for $75.50?

Estimate $8.9 \times 75.50 \to 9 \times 80$ or 720

$$
\begin{array}{r}
75.50 \quad \leftarrow \text{two decimal places}\\
\times \ 8.9 \quad \leftarrow \text{one decimal place}\\
\hline
67950\\
60400\\
\hline
671.950
\end{array}
$$

The product has three decimal places. You can drop the zero at the end because $671.950 = 671.95$.

Ryan would receive 671 pesos.

Skill and Concept Check

Writing Math
Exercise 2

1. **OPEN ENDED** Write a multiplication problem in which the product has three decimal places.

2. **NUMBER SENSE** Place the decimal point in the answer to make it correct. Explain your reasoning. $3.9853 \times 8.032856 = 32013341\ldots$

GUIDED PRACTICE

Multiply.

3. 0.6×0.5 **4.** 1.4×2.56 **5.** 27.43×1.089

6. 0.3×2.4 **7.** 0.52×2.1 **8.** 0.45×0.053

9. **MONEY** Juan is buying a video game that costs $32.99. The sales tax is found by multiplying the cost of the video game by 0.06. What is the cost of the sales tax for the video game?

Practice and Applications

Multiply.

10. 0.7×0.4 **11.** 1.5×2.7 **12.** 0.4×3.7

13. 1.7×0.4 **14.** 0.98×7.3 **15.** 2.4×3.48

16. 6.2×0.03 **17.** 14.7×11.36 **18.** 0.28×0.08

19. 0.45×0.05 **20.** 25.24×6.487 **21.** 9.63×2.045

HOMEWORK HELP

For Exercises	See Examples
10–21	1, 2
22–25	3
26–28	4

Extra Practice
See pages 601, 627.

ALGEBRA Evaluate each expression if $a = 1.3$, $b = 0.042$, and $c = 2.01$.

22. $ab + c$ **23.** $a \times 6.023 - c$ **24.** $3.25c + b$ **25.** abc

26. TRAVEL A steamboat travels 36.5 miles each day. How far will it travel in 6.5 days?

27. ALGEBRA Which of the numbers 9.2, 9.5, or 9.7 is the solution of $2.65t = 25.705$?

28. GEOMETRY To the nearest tenth, find the area of the figure at the right.

6.9 in.

3 in.

6 in.

3 in.

Tell whether each sentence is *sometimes*, *always*, or *never* true. Explain.

29. The product of two decimals less than one is less than one.

30. The product of a decimal greater than one and a decimal less than one is greater than one.

CRITICAL THINKING Evaluate each expression.

31. $0.3(3 - 0.5)$ **32.** $0.16(7 - 2.8)$ **33.** $1.06(2 + 0.58)$

Standardized Test Practice and Mixed Review

34. MULTIPLE CHOICE A U.S. dollar equals 0.623 English pound. About how many pounds will Dom receive in exchange for $126?

Ⓐ 86 pounds   Ⓑ 79 pounds   Ⓒ 75 pounds   Ⓓ 57 pounds

35. MULTIPLE CHOICE Katelyn makes $5.60 an hour. If she works 16.75 hours in a week, how much will she earn for the week?

Ⓕ $9.38   Ⓖ $93.80   Ⓗ $938.00   Ⓘ $9380

Multiply. (Lesson 4-1)

36. 45×0.27 **37.** 3.2×109 **38.** 27×0.45 **39.** 2.94×16

40. What is the sum of 14.26 and 12.43? (Lesson 3-5)

GETTING READY FOR THE NEXT LESSON

PREREQUISITE SKILL Divide. (Page 591)

41. $21 \div 3$ **42.** $81 \div 9$ **43.** $56 \div 8$ **44.** $63 \div 7$

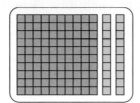 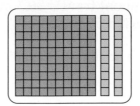

4-3

Dividing Decimals by Whole Numbers

What You'll LEARN

Divide decimals by whole numbers.

REVIEW Vocabulary

quotient: the solution in division

HANDS-ON Mini Lab

Materials
- base-ten blocks
- markers

Work with a partner.

To find 2.4 ÷ 2 using base-ten blocks, model 2.4 as 2 wholes and 4 tenths. Then separate into two equal groups.

There is one whole and two tenths in each group.

So, 2.4 ÷ 2 = 1.2.

Use base-ten blocks to show each quotient.

1. 3.4 ÷ 2   **2.** 4.2 ÷ 3   **3.** 5.6 ÷ 4

Find each whole number quotient.

4. 34 ÷ 2   **5.** 42 ÷ 3   **6.** 56 ÷ 4

7. Compare and contrast the quotients in Exercises 1–3 with the quotients in Exercises 4–6.

8. MAKE A CONJECTURE Write a rule how to divide a decimal by a whole number.

Dividing a decimal by a whole number is similar to dividing whole numbers.

EXAMPLE   Divide a Decimal by a 1-Digit Number

1 Find 6.8 ÷ 2.   **Estimate** 6 ÷ 2 = 3

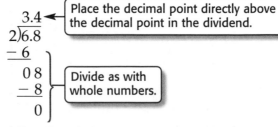

$$
\begin{array}{r}
3.4 \\
2\overline{)6.8} \\
-6 \\
\hline
0\,8 \\
-8 \\
\hline
0
\end{array}
$$

Place the decimal point directly above the decimal point in the dividend.

Divide as with whole numbers.

6.8 ÷ 2 = 3.4   Compared to the estimate, the quotient is reasonable.

EXAMPLE **Divide a Decimal by a 2-Digit Number**

2 Find $7.49 \div 14$.

Estimate $10 \div 10 = 1$

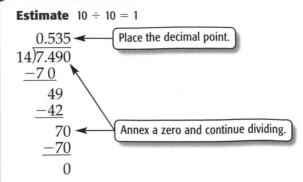

```
      0.535  ← Place the decimal point.
14)7.490
   −7 0
      49
     −42
      70  ← Annex a zero and continue dividing.
     −70
       0
```

$7.392 \div 14 = 0.535$ Compared to the estimate, the quotient is reasonable.

 **Your Turn** Divide.

a. $3)\overline{7.5}$ b. $7)\overline{3.5}$ c. $3.49 \div 4$

STUDY TIP

Checking your answer To check that the answer is correct, multiply the quotient by the divisor. In Example 2, $0.535 \times 14 = 7.49$.

Usually, when you divide decimals the answer does not come out evenly. You need to round the quotient to a specified place-value position. Always divide to one more place-value position than the place to which you are rounding.

EXAMPLE **Round a Quotient**

Standardized
Test Practice

3 **GRID-IN TEST ITEM** Seth purchased 3 video games for $51.79, including tax. If each game costs the same amount, what was the price of each game in dollars?

Read the Test Item To find the price of one game, divide the total cost by the number of games. Round to the nearest cent, or hundredths place, if necessary.

Solve the Test Item

```
     17.263    Place the decimal point.
3)51.790
  −3           Divide as with whole numbers.
   21
  −21
   07
  −06
    19
   −18
    10     Divide until you place a digit
    −9     in the thousandths place.
     1
```

To the nearest cent, the cost in dollars is 17.26.

Fill in the Grid

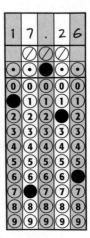

Test-Taking Tip
The Princeton Review

Grid In
Write the answer in the answer boxes on the top line. Then grid in 17, the decimal point, and 26.

1. **...xplain** how you can use estimation to place the decimal point in the quotient $42.56 \div 22$.

2. **OPEN ENDED** Write a real-life problem that involves dividing a decimal by a whole number.

3. **NUMBER SENSE** Is the quotient $8.3 \div 10$ greater than one or less than one? Explain.

4. **FIND THE ERROR** Toru and Amber are finding $11.2 \div 14$. Who is correct? Explain.

Toru
$$\begin{array}{r} 8. \\ 14\overline{)11.2} \\ -112 \\ \hline 0 \end{array}$$

Amber
$$\begin{array}{r} 0.8 \\ 14\overline{)11.2} \\ -112 \\ \hline 0 \end{array}$$

GUIDED PRACTICE

Divide. Round to the nearest tenth if necessary.

5. $3\overline{)39.39}$
6. $2\overline{)9.6}$
7. $6\overline{)8.53}$
8. $46\overline{)1087.9}$
9. $12.32 \div 22$
10. $69.904 \div 34$

11. **MONEY** Brianna and 5 of her friends bought a six-pack of fruit juice after their lacrosse game. If the six-pack costs $3.29, how much does each person owe to the nearest cent if the cost is divided equally?

Practice and Applications

Divide. Round to the nearest tenth if necessary.

12. $2\overline{)36.8}$
13. $4\overline{)3.6}$
14. $5\overline{)118.5}$
15. $19\overline{)11.4}$
16. $10.22 \div 14$
17. $55.2 \div 46$
18. $7\overline{)7.24}$
19. $4\overline{)6.27}$
20. $6\overline{)232.22}$
21. $31\overline{)336.75}$
22. $751.2 \div 25$
23. $48.68 \div 7$

HOMEWORK HELP

For Exercises	See Examples
12–14, 24–26	1
15–17	2
18–23, 27–29	3

Extra Practice
See pages 601, 627.

24. **SPORTS** Four girls of a track team ran the 4-by-100 meter relay in a total of 46.8 seconds. What was the average time for each runner?

25. **MUSIC** Find the average time of a track on a CD from the times in the table.

Time of Track (minutes)				
4.73	3.97	2.93	2.83	3.44

 Data Update What is the average time of all the tracks on your favorite CD? Visit msmath1.net/data_update to learn more.

26. **MONEY** Tyler's father has budgeted $64.50 for his three children's monthly allowance. Assuming they each earn the same amount, how much allowance will Tyler receive?

27. LANDMARKS Each story in an office building is about 4 meters tall. The Eiffel Tower in Paris, France, is 300.51 meters tall. To the nearest whole number, about how many stories tall is the Eiffel Tower?

28. MULTI STEP A class set of 30 calculators would have cost $4,498.50 in the early 1970s. However, in 2002, 30 calculators could be purchased for $352.50. How much less was the average price of one calculator in 2002 than in 1970?

29. FOOD The spreadsheet shows the unit price for a jar of peanut butter. To find the unit price, divide the cost of the item by its size. Find the unit price for the next three items. Round to the nearest cent.

	A	B	C	D
	Item	Cost	Size	Unit Price
1	Peanut			
2	Butter	$2.99	12 oz	0.25
3	Bread	$1.19	16 oz	
4	Orange Juice	$0.89	8 oz	
5	Cereal	$3.35	18 oz	

Cost of Grocery Items — Sheet1 / Sheet2

30. SHARING If 8 people are going to share a 2-liter bottle of soda equally, how much will each person get?

Find the mean for each set of data.

31. 22.6, 24.8, 25.4, 26.9

32. 1.43, 1.78, 2.45, 2.78, 3.25

33. CRITICAL THINKING Create a division problem that meets all of the following conditions.

- The divisor is a whole number, and the dividend is a decimal.
- The quotient is 1.265 when rounded to the nearest thousandth.
- The quotient is 1.26 when rounded to the nearest hundredth.

Standardized Test Practice and Mixed Review

34. MULTIPLE CHOICE Three people bought pens for a total of $11.55. How much did each person pay if they shared the cost equally?

 Ⓐ $3.25 Ⓑ $3.45 Ⓒ $3.65 Ⓓ $3.85

35. SHORT RESPONSE The table shows how much money Halley made in one week for a variety of jobs. To the nearest cent, what was her average pay for these three jobs?

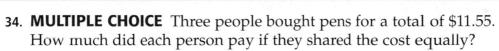

Jobs	Pay in a week
baby-sitting	$50.00
pet sitting	$10.50
lawn work	$22.50

Multiply. (Lesson 4-2)

36. 2.4×5.7 **37.** 1.6×2.3 **38.** $0.32(8.1)$ **39.** $2.68(0.84)$

40. What is the product of 4.156 and 12? (Lesson 4-1)

41. Find the least prime number that is greater than 25. (Lesson 1-3)

GETTING READY FOR THE NEXT LESSON

PREREQUISITE SKILL Divide. (Page 591 and Lesson 4-3)

42. $5\overline{)25}$ **43.** $81 \div 9$ **44.** $14\overline{)114.8}$ **45.** $516.06 \div 18$

Vocabulary and Concepts

1. **OPEN ENDED** Write a multiplication problem in which one factor is a decimal and the other is a whole number. The product should be less than 5. (Lesson 4-1)

2. **Explain** how to place the decimal point in the quotient when dividing a decimal by a whole number. (Lesson 4-3)

Skills and Applications

Multiply. (Lesson 4-1)

3. 4.3×5

4. 0.78×9

5. 1.4×3

6. 5.34×3

7. 0.09×8

8. 4.6×5

9. **MONEY EXCHANGE** If the Japanese yen is worth 0.0078 of one U.S. dollar, what is the value of 3,750 yen in U.S. dollars? (Lesson 4-1)

10. **CAR PAYMENTS** Mr. Dillon will pay a total of $9,100.08 for his car lease over a period of 36 months. How much are his payments each month? (Lesson 4-1)

11. **ALGEBRA** Evaluate $4.2y$ if $y = 0.98$. (Lesson 4-2)

12. **GEOMETRY** Find the area of the rectangle. (Lesson 4-2)

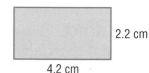

2.2 cm

4.2 cm

Divide. Round to the nearest tenth if necessary. (Lesson 4-3)

13. $4\overline{)24.8}$

14. $9\overline{)34.2}$

15. $24\overline{)19.752}$

16. $48.6 \div 6$

17. $54.45 \div 55$

18. $2.08 \div 5$

Standardized Test Practice

19. **MULTIPLE CHOICE** Yoko wants to buy 3 necklaces that cost $12.99 each. How much money will she need? (Lesson 4-1)

   Ⓐ $29.67
   Ⓑ $31.52
   Ⓒ $38.97
   Ⓓ $42.27

20. **SHORT RESPONSE** T-shirts are on sale at 3 for $29.97. How much will Jessica pay for one T-shirt? (Lesson 4-3)

The Game Zone

A Place To Practice Your Math Skills

Decimos

- **GET READY!**

    Players: two, three, or four
    Materials: spinner, index cards

- **GET SET!**

    - Each player makes game sheets like the one shown at the right.

    - Make a spinner as shown.

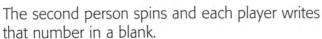

- **GO!**

    - The first person spins the spinner. Each player writes the number in one of the blanks on his or her game sheet.

        The second person spins and each player writes that number in a blank.

        The next person spins and players fill in their game sheets. A zero cannot be placed as the divisor.

    - All players find their quotients. The player with the greatest quotient earns one point. In case of a tie, those players each earn one point.

    - **Who Wins?** The first person to earn 5 points wins.

Dividing by Decimals

The model below shows $15 \div 3$.

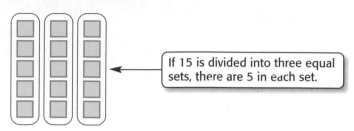

If 15 is divided into three equal sets, there are 5 in each set.

Dividing decimals is similar to dividing whole numbers. In the Activity below, 1.5 is the *dividend* and 0.3 is the *divisor*.

• Use base-ten blocks to model the dividend.
• Replace any ones block with tenths.
• Separate the tenths into groups represented by the divisor.
• The quotient is the number of groups.

ACTIVITY   *Work with a partner.*

1 Model $1.5 \div 0.3$.

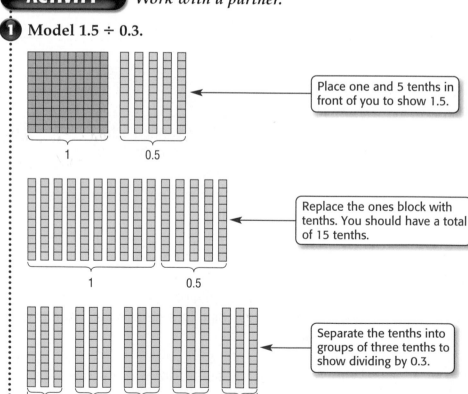

Place one and 5 tenths in front of you to show 1.5.

Replace the ones block with tenths. You should have a total of 15 tenths.

Separate the tenths into groups of three tenths to show dividing by 0.3.

There are five groups of three tenths in 1.5. So, $1.5 \div 0.3 = 5$.

You can use a similar model to divide by hundredths.

ACTIVITY   *Work with a partner.*

2 **Model 0.4 ÷ 0.05.**

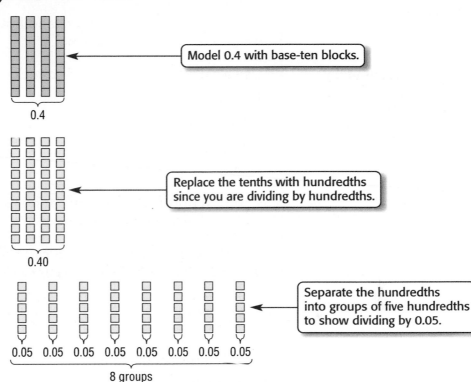

Model 0.4 with base-ten blocks.

0.4

Replace the tenths with hundredths since you are dividing by hundredths.

0.40

Separate the hundredths into groups of five hundredths to show dividing by 0.05.

0.05 0.05 0.05 0.05 0.05 0.05 0.05 0.05

8 groups

There are eight groups of five hundredths in 0.4.

So, 0.4 ÷ 0.05 = 8.

Your Turn **Use base-ten blocks to find each quotient.**

a. 2.4 ÷ 0.6   **b.** 1.2 ÷ 0.4   **c.** 1.8 ÷ 0.6

d. 0.9 ÷ 0.09   **e.** 0.8 ÷ 0.04   **f.** 0.6 ÷ 0.05

Writing Math

1. **Explain** why the base-ten blocks representing the dividend must be replaced or separated into the smallest place value of the divisor.

2. **Tell** why the quotient 0.4 ÷ 0.05 is a whole number. What does the quotient represent?

3. **Determine** the missing divisor in the sentence 0.8 ÷ __?__ = 20. Explain.

4. **Tell** whether 1.2 ÷ 0.03 is *less than*, *equal to*, or *greater than* 1.2. Explain your reasoning.

Dividing by Decimals

What You'll LEARN

Divide decimals by decimals.

REVIEW Vocabulary

power: numbers expressed using exponents **(Lesson 1-4)**

HANDS-ON Mini Lab

Materials
• calculator

Work with a partner.

Patterns can help you understand how to divide a decimal by a decimal.

Use a calculator to find each quotient.

1. $0.048 \div 0.06$

    $0.48 \div 0.6$

    $4.8 \div 6$

    $48 \div 60$

2. $0.0182 \div 0.13$

    $0.182 \div 1.3$

    $1.82 \div 13$

    $18.2 \div 130$

3. Which of the quotients in Exercises 1 and 2 would be easier to find *without* a calculator? Explain your reasoning.

Rewrite each problem so you can find the quotient without using a calculator. Then find the quotient.

4. $0.42 \div 0.7$ **5.** $1.26 \div 0.3$ **6.** $1.55 \div 0.5$

When dividing by decimals, change the divisor into a whole number. To do this, multiply both the divisor and the dividend by the same power of 10. Then divide as with whole numbers.

EXAMPLE   Divide by Decimals

1 Find $14.19 \div 2.2$.   **Estimate** $14 \div 2 = 7$

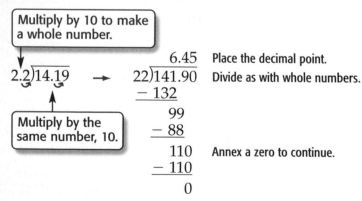

Multiply by 10 to make a whole number.

$2.2\overline{)14.19}$ → $22\overline{)141.90}$ Place the decimal point.

Multiply by the same number, 10.

        6.45   Divide as with whole numbers.

    $- 132$

       99

    $- 88$

     110   Annex a zero to continue.

   $- 110$

       0

14.19 divided by 2.2 is 6.45.   Compare to the estimate.

Check $6.45 \times 2.2 = 14.19$ ✔

Your Turn Divide.

a. $1.7\overline{)54.4}$ **b.** $0.36\overline{)8.424}$ **c.** $0.0063 \div 0.007$

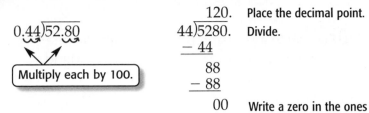

EXAMPLES **Zeros in the Quotient and Dividend**

2 **Find 52.8 ÷ 0.44.**

0.44)52.80

Multiply each by 100.

```
         120.    Place the decimal point.
  44)5280.       Divide.
   − 44
     88
   − 88
     00         Write a zero in the ones
                place of the quotient
                because 0 ÷ 44 = 0.
```

So, 52.8 ÷ 0.44 = 120.

Check 120 × 0.44 = 52.8 ✔

3 **Find 0.09 ÷ 1.8.**

1.8)0.09

Multiply each by 10.

```
      0.05    Place the decimal point.
  18)0.90     18 does not go into 9, so
   − 0        write a 0 in the tenths place.
     09
   − 00
     90       Annex a 0 in the dividend
   − 90       and continue to divide.
      0
```

So, 0.09 ÷ 1.8 is 0.05.

Check 0.05 × 1.8 = 0.09 ✔

Your Turn Divide.

d. 0.014)5.6   e. 0.002)62.4   f. 0.4 ÷ 0.0025

There are times when it is necessary to round the quotient.

EXAMPLE **Round Quotients**

4 **INTERNET** How many times more homes in the U.S. have Internet access than in Japan? Round to the nearest tenth.

Find 104.8 ÷ 21.6.

21.6)104.8 →

```
          4.85
  216)1048.00
   − 864
    1840
  − 1728
    1120
  − 1080
      40
```

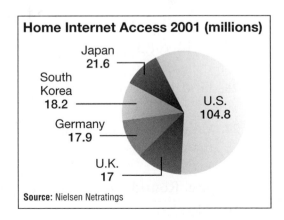

Home Internet Access 2001 (millions)

Japan
21.6

South
Korea
18.2

Germany
17.9

U.K.
17

U.S.
104.8

Source: Nielsen Netratings

STUDY TIP

Rounding You can stop dividing when there is a digit in the hundredths place.

To the nearest tenth, 104.8 ÷ 21.6 = 4.9. So there are about 4.9 times more homes in the U.S. with Internet access.

1. **Explain** why $1.92 \div 0.51$ should be about 4.

2. **OPEN ENDED** Write a division problem with decimals in which it is necessary to annex one or more zeros to the dividend.

3. **Which One Doesn't Belong?** Identify the problem that does not have the same quotient as the other three. Explain your reasoning.

$0.5\overline{)0.35}$ $5\overline{)3.5}$ $0.05\overline{)0.035}$ $5\overline{)35}$

GUIDED PRACTICE

Divide. Round to the nearest hundredth if necessary.

4. $0.3\overline{)3.69}$ 5. $0.8\overline{)9.92}$ 6. $0.3\overline{)0.45}$

7. $3.4\overline{)0.68}$ 8. $0.0025\overline{)0.4}$ 9. $4.27 \div 0.35$

10. $0.464 \div 0.06$ 11. $0.321 \div 0.4$ 12. $8.4 \div 2.03$

13. **GARDENING** A flower garden is 11.25 meters long. Mrs. Owens wants to make a border along one side using bricks that are 0.25 meter long. How many bricks does she need?

Practice and Applications

Divide.

14. $0.5\overline{)4.55}$ 15. $0.9\overline{)2.07}$ 16. $0.14\overline{)16.24}$

17. $2.7\overline{)1.08}$ 18. $0.42\overline{)96.6}$ 19. $0.03\overline{)13.5}$

20. $1.3\overline{)0.0338}$ 21. $3.4\overline{)0.16728}$ 22. $1.44 \div 0.4$

23. $29.12 \div 1.3$ 24. $0.12 \div 0.15$ 25. $0.242 \div 0.4$

26. Find 10.272 divided by 2.4.   27. What is $6.24 \div 0.00012$?

HOMEWORK HELP	
For Exercises	See Examples
14–17, 22–28	1
18–19	2
20–21	3
29–40	4
Extra Practice See pages 602, 627.	

28. **CARPENTRY** If a board 7.5 feet long is cut into 2.5 foot-pieces, how many pieces will there be?

Divide. Round each quotient to the nearest hundredth.

29. $0.4\overline{)0.231}$ 30. $0.7\overline{)1.32}$ 31. $0.26\overline{)0.249}$ 32. $0.71\overline{)0.24495}$

33. $0.07625 \div 2.5$ 34. $2.582 \div 34.2$ 35. $6.453 \div 12.8$ 36. $3.792 \div 4.25$

37. **TRAVEL** The Vielhaber family drove 315.5 miles for a soccer tournament and used 11.4 gallons of gas. How many miles did they get per gallon of gas to the nearest hundredth? Estimate the answer before calculating.

TECHNOLOGY For Exercises 38 and 39, use the information in the graphic. Estimate your answers first.

38. The 2001 sales are how many times as great as the 2000 sales? Round to the nearest tenth.

39. If the sales were to increase the same amount in 2002, what would be the predicted amount for 2002?

40. **RESEARCH** Use the Internet or another source to find the average speed of Ward Burton's car in the 2002 Daytona 500 race. If a passenger car averages 55.5 mph, how many times as fast was Ward's car, to the nearest hundredth?

41. If a decimal greater than 0 and less than 1 is divided by a lesser decimal, would the quotient be *always, sometimes,* or *never* less than 1? Explain.

42. **SCIENCE** Sound travels through air at 330 meters per second. How long will it take a bat's cry to reach its prey and echo back if the prey is 1 meter away?

43. **CRITICAL THINKING**
Replace each ■ with digits to make a true sentence.
■.8■3 ÷ 0.82 = 4.6■

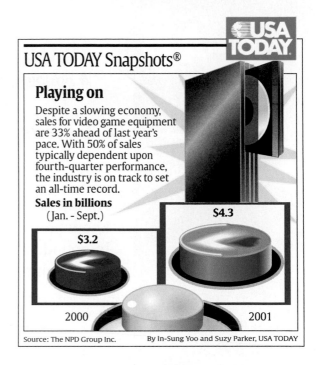

USA TODAY Snapshots®

Playing on
Despite a slowing economy, sales for video game equipment are 33% ahead of last year's pace. With 50% of sales typically dependent upon fourth-quarter performance, the industry is on track to set an all-time record.

Sales in billions
(Jan. - Sept.)

$3.2 $4.3

2000   2001

Source: The NPD Group Inc.   By In-Sung Yoo and Suzy Parker, USA TODAY

Standardized Test Practice and Mixed Review

44. **MULTIPLE CHOICE** To the nearest tenth, how many times greater was the average gasoline price on May 14 than on August 20?

  Ⓐ 0.8   Ⓑ 1.2   Ⓒ 1.3   Ⓓ 1.4

45. **GRID IN** Solve $z = 20.57 \div 3.4$.

46. Find the quotient when 68.52 is divided by 12. (Lesson 4-3)

Multiply. (Lesson 4-2)

47. 19.2×2.45 48. 7.3×9.367 49. 8.25×12.42 50. 9.016×51.9

Average U.S. Gasoline Price (per gallon)	Date
$1.71	May 14, 2001
$1.42	August 20, 2001

Source: Energy Information Administration

GETTING READY FOR THE NEXT LESSON

PREREQUISITE SKILL Evaluate each expression. (Lesson 1-5)

51. $2(1) + 2(3)$ 52. $2(18) + 2(9)$ 53. $2(3) \times 2(5)$ 54. $2(36) + 2(20)$

Problem-Solving Strategy
A Follow-Up of Lesson 4-4

What You'll Learn
Determine if an answer is reasonable.

Determine Reasonable Answers

For our science project we need to know how much a gray whale weighs in pounds. I found a table that shows the weights of whales in tons.

Well, I know there are 2,000 pounds in one ton. Let's use this to **find a reasonable answer**.

Explore	We know the weight in tons. We need to find a reasonable weight in pounds.
Plan	One ton equals 2,000 pounds. So, estimate the product of 38.5 and 2,000 to find a reasonable weight.
Solve	2,000 × 38.5 → 2,000 × 40 or 80,000 A reasonable weight is 80,000 pounds.
Examine	Since 2,000 × 38.5 = 77,000, 80,000 pounds is a reasonable answer.

Whale	Weight (ton)
Blue	151.0
Bowhead	95.0
Fin	69.9
Gray	38.5
Humpback	38.1

Source: *Top 10 of Everything*

Analyze the Strategy

1. **Explain** when you would use the strategy of determining reasonable answers to solve a problem.

2. **Describe** a situation where determining a reasonable answer would help you solve a problem.

3. **Write** a problem using the table above that can be solved by determining a reasonable answer. Then tell the steps you would take to find the solution of the problem.

Solve. Use the determine reasonable answers strategy.

4. **BASEBALL** In 2002, 820,590 people attended 25 of the Atlanta Braves home games. Which is a more reasonable estimate for the number of people that attended each game: 30,000 or 40,000? Explain.

5. **MONEY MATTERS** Courtney wants to buy 2 science fiction books for $3.95 each, 3 magazines for $2.95 each, and 1 bookmark for $0.39 at the school book fair. Does she need to bring $20 or will $15 be enough? Explain.

Mixed Problem Solving

Solve. Use any strategy.

6. **PATTERNS** What are the next two figures in the pattern?

ᒉᔭᒍᐸᒉᒉ

7. **ENTERTAINMENT** In music, a gold album award is presented to an artist who has sold at least 500,000 units of a single album or CD. If an artist has 16 gold albums, what is the minimum number of albums that have been sold?

8. **AGES** Erin's mother is 4 times as old as Erin. Her grandmother is twice as old as Erin's mother. The sum of their three ages is 104. How old is Erin, her mother, and her grandmother?

9. **GEOGRAPHY** The graphic below shows the lengths in miles of the longest rivers in the world. About how many total miles long are the three rivers?

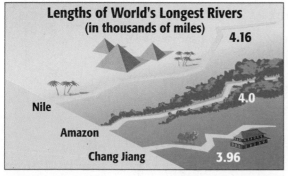

Lengths of World's Longest Rivers
(in thousands of miles)
4.16
Nile
4.0
Amazon
Chang Jiang   3.96

Source: *The World Almanac*

10. **EDUCATION** Use the graph below to predict the population of Dorsey Intermediate School in 2006.

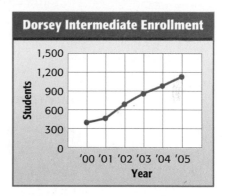

Dorsey Intermediate Enrollment
Students: 1,500 / 1,200 / 900 / 600 / 300 / 0
Year: '00 '01 '02 '03 '04 '05

11. Estimate the product of 56.2 and 312.

12. **EDUCATION** The high school gym will hold 2,800 people and the 721 seniors who are graduating. Is it reasonable to offer each graduate four tickets for family and friends? Explain.

13. **BIRTHDAYS** Suppose a relative matches your age with dollars each birthday. You are 13 years old. How much money have you been given over the years by this relative?

14. **STANDARDIZED TEST PRACTICE**
The median price of five gifts was $17. The least amount spent was $11, and the most was $22.50. Which amount is a reasonable total for what was spent?

Ⓐ $65.80   Ⓑ $77.25

Ⓒ $88.50   Ⓓ $98.70

4-5 Perimeter

HANDS-ON Mini Lab

Materials
- ruler
- grid paper

Work with a partner.

What is the distance around the front cover of your textbook?

STEP 1 Use the ruler to measure each side of the front cover. Round to the nearest inch.

STEP 2 Draw the length and width of the book on the grid paper. Label the length ℓ and the width w.

1. Find the distance around your textbook by adding the measures of each side.

2. Can you think of more than one way to find the distance around your book? If so, describe it.

The distance around any closed figure is called its **perimeter**.

Key Concept — Perimeter of a Rectangle

Words The perimeter P of a rectangle is the sum of the lengths and widths. It is also two times the length ℓ plus two times the width w.

Symbols $P = \ell + w + \ell + w$
$P = 2\ell + 2w$

Model

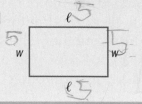

EXAMPLE Find the Perimeter

1 Find the perimeter of the rectangle.

Estimate $10 + 4 + 10 + 4 = 28$

$P = 2\ell + 2w$ Write the formula.

$P = 2(10.2) + 2(3.9)$ Replace ℓ with 10.2 and w with 3.9.

$P = 20.4 + 7.8$ Multiply.

$P = 28.2$ Add.

The perimeter is 28.2 inches.   Compare to the estimate.

3.9 in.

10.2 in.   10.2 in.

3.9 in.

Your Turn Find the perimeter of each rectangle.

a. 2 ft by 3 ft   b. 6 in. by 10 in.   c. 15 mm by 12 mm

Since each side of a square has the same length, you can multiply the measure of any of its sides *s* by 4 to find its perimeter.

Key Concept   **Perimeter of a Square**

Words	The perimeter *P* of a square is four times the measure of any of its sides *s*.	**Model**	
Symbols	$P = 4s$		

EXAMPLE   **Find the Perimeter of a Square**

2 **ANIMALS** The sleeping quarters for a bear at the zoo is a square that measures 4 yards on each side. What is the perimeter of the sleeping area?

Words	Perimeter of a square is equal to four times the measure of any side.
Variables	$P = 4s$
Equation	$P = 4(4)$

$P = 4(4)$ Write the equation.

$P = 16$ Multiply.

The perimeter of the bear's sleeping area is 16 yards.

Skill and Concept Check

Writing
Math

Exercises 2 & 3

1. **OPEN ENDED** Draw a rectangle that has a perimeter of 14 inches.

2. **NUMBER SENSE** What happens to the perimeter of a rectangle if you double its length and width?

3. **FIND THE ERROR** Crystal and Luanda are finding the perimeter of a rectangle that is 6.3 inches by 2.8 inches. Who is correct? Explain.

Crystal
6.3 × 2.8 = 17.64 in.

Luanda
6.3 + 6.3 + 2.8 + 2.8 = 18.2 in.

GUIDED PRACTICE

Find the perimeter of each figure.

4.
5.1 in.
3.5 in.
3.5 in.
5.1 in.

5.
17 cm
22 cm
22 cm
17 cm

6.
12.5 m
9.2 m
9.2 m
12.5 m

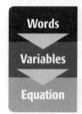

Practice and Applications

HOMEWORK HELP

For Exercises	See Examples
7–15	1, 2

Extra Practice
See pages 602, 627.

Find the perimeter of each figure.

7.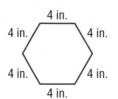
89 yd / 43 yd / 43 yd / 89 yd

8.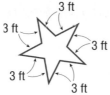
96 mm / 104 mm / 104 mm / 96 mm

9.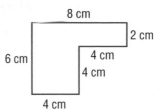
32 ft / 12 ft / 12 ft / 32 ft

10. 12.4 cm by 21.6 cm   **11.** 11.4 m by 12.9 m   **12.** 9.5 mi by 11.9 mi

13.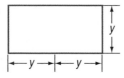
4 in. / 4 in. / 4 in. / 4 in. / 4 in. / 4 in.

14.
3 ft / 3 ft / 3 ft / 3 ft / 3 ft

15.
8 cm / 2 cm / 6 cm / 4 cm / 4 cm / 4 cm

How many segments y units long are needed for the perimeter of each figure?

16.
y / y / y

17.
y / y

18. **BASKETBALL** A basketball court measures 26 meters by 14 meters. Find the perimeter of the court.

19. **CRITICAL THINKING** Refer to Exercise 18. Suppose 10 meters of seating is added to each side of the basketball court. Find the perimeter of the seating area.

Standardized Test Practice and Mixed Review

20. **MULTIPLE CHOICE** The perimeter of a rectangular playground is 121.2 feet. What is the length if the width is 25.4 feet?

   Ⓐ 41.7 ft   Ⓑ 38.6 ft   Ⓒ 35.2 ft   Ⓓ 30.6 ft

21. **SHORT RESPONSE** Find the distance around the football field.

120 yd / 53 yd

Divide. (Lesson 4-4)

22. $16.4\overline{)94.3}$

23. $4.9\overline{)14.798}$

24. $95.5 \div 0.05$

25. $21.112 \div 5.2$

26. Five people share 8.65 ounces of juice equally. How much does each receive? (Lesson 4-3)

GETTING READY FOR THE NEXT LESSON

PREREQUISITE SKILL Multiply. (Lesson 4-2)

27. 17×23 **28.** 28×42 **29.** 6.4×5.8 **30.** 3.22×6.7

4-6 Circumference

What You'll LEARN

Find the circumference of circles.

NEW Vocabulary

circle
center
diameter
circumference
radius

MATH Symbols

π (pi) ≈ 3.14

HANDS-ON Mini Lab

Work with a partner.

The Olympic rings are made from circles. In this Mini Lab, you'll look for a relationship between the distance around a circle (circumference) and the distance across the circle (diameter).

Materials
- string
- ruler
- calculator
- jar lid
- other circular objects

STEP 1 Cut a piece of string the length of the distance around a jar lid C. Measure the string. Copy the table and record the measurement.

STEP 2 Measure the distance across the lid d. Record the measurement in the table.

Object	C	d	$\frac{C}{d}$

STEP 3 Repeat steps 1 and 2 for several circular objects.

STEP 4 Use a calculator to divide the distance around each circle by the distance across the circle. Record the quotient in the table

1. What do you notice about each quotient?
2. What conclusion can you make about the circumference and diameter of a circle?
3. Predict the distance around a circle that is 4 inches across.

A **circle** is the set of all points in a plane that are the same distance from a point called the **center**.

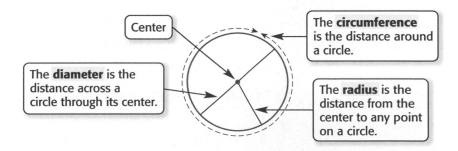

Center

The **circumference** is the distance around a circle.

The **diameter** is the distance across a circle through its center.

The **radius** is the distance from the center to any point on a circle.

In the Mini Lab, you discovered that the circumference of a circle is a little more than three times its diameter. The exact number of times is represented by the Greek letter π (pi).

Key Concept — Circumference

Words	The circumference of a circle is equal to π times its diameter or π times twice its radius.
Symbols	$C = \pi d$ or $C = 2\pi r$

Model

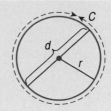

READING Math

The symbol ≈ means *approximately equal to*.

The real value of π is 3.1415926… . It never ends. We use 3.14 as an approximation. So, π ≈ 3.14.

EXAMPLE Find the Circumference of a Circle

1 Find the circumference of a circle whose diameter is 4.5 inches. Round to the nearest tenth.

You know the diameter. Use $C = \pi d$.

$C = \pi d$ Write the formula.

$\approx 3.14 \times 4.5$ Replace π with 3.14 and *d* with 4.5.

≈ 14.13 Multiply.

The circumference is about 14.1 inches.

4.5 in.

Your Turn

a. Find the circumference of a circle whose diameter is 15 meters. Round to the nearest tenth.

EXAMPLE Use Circumference to Solve a Problem

2 **HOBBIES** Ashlee likes to fly model airplanes. The plane flies in circles at the end of a 38-foot line. What is the circumference of the largest circle in which the plane can fly?

You know the radius of the circle.

$C = 2\pi r$ Write the formula.

$\approx 2 \cdot 3.14 \cdot 38$ π ≈ 3.14, *r* = 38

≈ 238.64 Multiply.

To the nearest tenth, the circumference is 238.6 feet.

38 ft

Your Turn Find the circumference of each circle. Round to the nearest tenth.

b. *r* = 23 in.   c. *r* = 4.5 cm   d. *r* = 6.5 ft

Skill and Concept Check

Writing Math
Exercises 3 & 4

1. **Draw** a circle and label the center, a radius, and a diameter.

2. **OPEN ENDED** Draw and label a circle whose circumference is more than 5 inches, but less than 10 inches.

3. **FIND THE ERROR** Alvin and Jerome are finding the circumference of a circle whose radius is 2.5 feet. Who is correct? Explain.

> Alvin
> $C \approx 2 \times 3.14 \times 2.5$

> Jerome
> $C \approx 3.14 \times 2.5$

4. **NUMBER SENSE** Without calculating, will the circumference of a circle with a radius of 4 feet be greater or less than 24 feet? Explain your answer.

GUIDED PRACTICE

Find the circumference of each circle shown or described. Round to the nearest tenth.

5.

21 ft

6.

4 in.

7. $d = 0.875$ yard

8. Find the circumference of a circle with a radius of 0.75 meter.

Practice and Applications

Find the circumference of each circle shown or described. Round to the nearest tenth.

9.
3.5 in.

10.
5.25 yd

11.

6.2 m

12.

10.7 km

HOMEWORK HELP	
For Exercises	See Examples
9–10, 13–14 19–22	1
11–12, 15–16	2
Extra Practice See pages 602, 627.	

13. $d = 6$ ft   14. $d = 28$ cm   15. $r = 21$ mm   16. $r = 2.25$ in.

17. Find the circumference of a circle whose diameter is 4.8 inches.

18. The radius of a circle measures 3.5 kilometers. What is the measure of its circumference?

ENTERTAINMENT For Exercises 19–21, refer to the table. How far do passengers travel on each revolution? Round to the nearest tenth.

19. The Big Ferris Wheel

20. London Eye

21. Texas Star

Ferris Wheel	Diameter (feet)
The Big Ferris Wheel	250
London Eye	442.9
Texas Star	213.3

22. **MULTI STEP** The largest tree in the world has a diameter of about 26.5 feet at 4.5 feet above the ground. If a person with outstretched arms can reach 6 feet, how many people would it take to reach around the tree?

23. **GEOMETRY** You can find the diameter of a circle if you know its circumference. To find the circumference, you *multiply* π times the diameter. So, to find the diameter, *divide* the circumference by π.

    a. Find the diameter of a circle with circumference of 3.14 miles.

    b. Find the diameter of a circle with circumference of 15.7 meters.

24. **CRITICAL THINKING** How would the circumference of a circle change if you doubled its diameter?

25. **CRITICAL THINKING** Suppose you measure the diameter of a circle to be about 12 centimeters and use 3.14 for π. Is it reasonable to give 37.68 as the exact circumference? Why or why not?

EXTENDING THE LESSON A *chord* is a segment whose endpoints are on a circle. A diameter is one example of a chord.

26. Draw a circle. Draw an example of a chord that is *not* a diameter.

Standardized Test Practice and Mixed Review

27. **MULTIPLE CHOICE** Find the circumference of the circle to the nearest tenth.

    A 26.7 cm   B 53.4 cm
    C 78.1 cm   D 106.8 cm

28. **MULTIPLE CHOICE** Awan rode his mountain bike in a straight line for a total of 565.2 inches. If his tires have a diameter of 12 inches, about how many times did the tires revolve?

    F 180   G 15   H 13   I 12

Find the perimeter of each rectangle with the dimensions given. (Lesson 4-5)

29. 3.8 inches by 4.9 inches

30. 15 feet by 17.5 feet

31. 17 yards by 24 yards

32. 1.25 miles by 4.56 miles

33. Find the quotient if 160.896 is divided by 12.57. (Lesson 4-4)

Web Quest Interdisciplinary Project

Down to the Last Penny!

It's time to complete your project. Use the information and data you have gathered about grocery costs for your family to prepare a spreadsheet. Be sure to include all the required calculations with your project.

msmath1.net/webquest

What You'll LEARN

Use a spreadsheet to plan a budget.

Spreadsheet Basics

Spreadsheets allow users to perform many calculations quickly and easily. They can be used to create a budget.

 **ACTIVITY**

The Hoffman children are planning budgets for their allowances. Megan receives $30 per week, Alex, $25, and Kevin, $20. This money is to be used for snacks, entertainment, and savings. Each child has decided what part of his or her allowance will be placed in each category. This information is summarized below.

	Snacks	Entertainment	Savings
Megan	25% or 0.25	60% or 0.60	15% or 0.15
Alex	30% or 0.30	60% or 0.60	10% or 0.10
Kevin	15% or 0.15	55% or 0.55	30% or 0.30

A spreadsheet can be used to find how much money the children have for snacks, entertainment, and savings each week. Each child's allowance and the decimal part for each category are entered into the spreadsheet. Copy the information below into your spreadsheet.

	A	B	C	D	E	F	G	H	I
1	Child	Allowance	Snacks	Categories Entertainment	Savings		Snacks	$ Allotted Entertainment	Savings
2	Megan	30	0.25	0.6	0.15		=B2*C2		
3	Alex	25	0.3	0.6	0.1		=B3*C3		
4	Kevin	20	0.15	0.55	0.3		=B4*C4		
5									

Budgeting Allowances — Sheet1 / Sheet2 / Sheet3

 EXERCISES

1. Explain each of the formulas in column G.

2. Complete the formulas for columns H and I. Place these formulas in your spreadsheet.

3. How much money will Alex put into savings? How long will it take Alex to save $50.00?

4. Add an extra row into the spreadsheet and insert your name. Enter a reasonable allowance. Then select the portion of the allowance you would put in each category. Find how much money you would actually have for each category by adding formulas for each category.

Vocabulary and Concept Check

center (p. 161) diameter (p. 161) radius (p. 161)
circle (p. 161) perimeter (p.158) scientific notation (p. 136)
circumference (p. 161)

Choose the correct term or number to complete each sentence.

1. To find the circumference of a circle, you must know its (radius, center).

2. When (multiplying, dividing) two decimals, count the number of decimal places in each factor to determine the number of decimal places in the answer.

3. To check your answer for a division problem, you can multiply the quotient by the (dividend, divisor).

4. The number of decimal places in the product of 6.03 and 0.4 is (5, 3).

5. The (perimeter, area) is the distance around any closed figure.

6. The (radius, diameter) of a circle is the distance across its center.

7. To change the divisor into a whole number, multiply both the divisor and the dividend by the same power of (10, 100).

8. When dividing a decimal by a whole number, place the decimal point in the quotient directly (below, above) the decimal point found in the dividend.

Lesson-by-Lesson Exercises and Examples

4-1 **Multiplying Decimals by Whole Numbers** (pp. 135–138)

Multiply.

9. 1.4×6 10. 3×9.95
11. 0.82×4 12. 12.9×7
13. 5×0.48 14. 24.7×3
15. 6×6.65 16. 2.6×8

17. **SHOPPING** Three pairs of shoes are priced at $39.95 each. Find the total cost for the shoes.

18. **MONEY** If you work 6 hours at $6.35 an hour, how much would you make?

Example 1 Find 6.45×7.

Method 1 Use estimation.

Round 6.45 to 6.

$6.45 \times 7 \rightarrow 6 \times 7$ or 42

$$\begin{array}{r} 3\ 3 \\ 6.45 \\ \times\ 7 \\ \hline 45.15 \end{array}$$ Since the estimate is 42, place the decimal point after the 5.

Method 2 Count decimal places.

$$\begin{array}{r} 3\ 3 \\ 6.45 \\ \times\ 7 \\ \hline 45.15 \end{array}$$ There are two decimal places to the right of the decimal in 6.45.

Count the same number of places from right to left in the product.

4-2 **Multiplying Decimals** (pp. 141–143)

Multiply.

19. 0.6×1.3
20. 8.74×2.23
21. 0.04×5.1
22. 2.6×3.9
23. 4.15×3.8
24. 0.002×50.5

25. Find the product of 0.04 and 0.0063.

26. **GEOMETRY** Find the area of the rectangle.

5.4 in.

1.3 in.

Example 2 Find 38.76×4.2.

$$
\begin{array}{r}
38.76 \\
\times\ 4.2 \\
\hline
7752 \\
15504\ \\
\hline
162.792
\end{array}
$$

← two decimal places
← one decimal place

← three decimal places

4-3 **Dividing Decimals by Whole Numbers** (pp. 144–147)

Divide.

27. $12.24 \div 36$
28. $32)\overline{203.84}$
29. $35)\overline{136.5}$
30. $14)\overline{37.1}$
31. $4.41 \div 5$
32. $8)\overline{26.96}$

33. **SPORTS BANQUET** The cost of the Spring Sports Banquet is to be divided equally among the 62 people attending. If the cost is $542.50, find the cost per person.

Example 3 Find the quotient $16.1 \div 7$.

$$
\begin{array}{r}
2.3 \\
7)\overline{16.1} \\
-14\ \ \\
\hline
2\ 1 \\
-2\ 1 \\
\hline
0
\end{array}
$$

Place the decimal point.
Divide as with whole numbers.

4-4 **Dividing by Decimals** (pp. 152–155)

Divide.

34. $0.96 \div 0.6$
35. $11.16 \div 6.2$
36. $0.276 \div 0.6$
37. $5.88 \div 0.4$
38. $0.5)\overline{18.45}$
39. $0.08)\overline{5.2}$
40. $2.6)\overline{0.65}$
41. $0.25)\overline{0.155}$

42. **SPACE** The Aero Spacelines Super Guppy, a converted Boeing C-97, can carry 87.5 tons. Tanks that weigh 4.5 tons each are to be loaded onto the Super Guppy. What is the most number of tanks it can transport?

Example 4 Find $11.48 \div 8.2$.

$$8.2)\overline{11.48}$$

Multiply the divisor and the dividend by 10 to move the decimal point one place to the right so that the divisor is a whole number.

$$
\begin{array}{r}
1.4 \\
82)\overline{114.8} \\
-82\ \ \\
\hline
32\ 8 \\
-32\ 8 \\
\hline
0
\end{array}
$$

Place the decimal point.
Divide as with whole numbers.

4-5 Perimeter (pp. 158–160)

Find the perimeter of each rectangle.

43.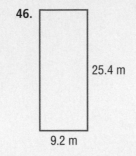
5 in.
8 in.

44.
9 cm
12.8 cm

45.
34.5 ft
18.6 ft

46.
25.4 m
9.2 m

47. Find the perimeter of a rectangle that measures 10.4 inches wide and 6.4 inches long.

Example 5 Find the perimeter of the rectangle.

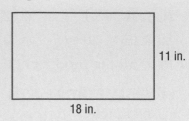
11 in.
18 in.

$P = 2\ell + 2w$ Write the formula.
$P = 2(18) + 2(11)$ $\ell = 18; w = 11$
$P = 36 + 22$ Multiply.
$P = 58$ Simplify.

The perimeter is 58 inches.

4-6 Circumference (pp. 161–164)

Find the circumference of each circle. Round to the nearest tenth.

48.
5 yd

49.
16 m

50.
13.2 cm

51.
124.6 ft

52. **SWIMMING** The radius of a circular pool is 10 feet. Find the circumference of the pool. Round to the nearest tenth.

53. **SCIENCE** A radio telescope has a circular dish with a diameter of 112 feet. What is the circumference of the circular dish? Round to the nearest tenth.

Example 6 Find the circumference of the circle. Round to the nearest tenth.

7 ft

$C = 2\pi r$ Write the formula.
$\approx 2(3.14)(7)$ $\pi \approx 3.14; r = 7$
≈ 43.96 Multiply.
≈ 44.0 Round to the nearest tenth.

The circumference is 44.0 feet.

Example 7 Find the circumference of the circle whose diameter is 26 meters. Round to the nearest tenth.

$C = \pi d$ Write the formula.
$\approx (3.14)(26)$ $\pi \approx 3.14; d = 26$
≈ 81.64 Multiply.
≈ 81.6 Round to the nearest tenth.

The circumference is 81.6 meters.

Practice Test

Vocabulary and Concepts

1. **Explain** the counting method for determining where to place the decimal when multiplying two decimals.

2. **Define** *perimeter*.

Skills and Applications

Multiply.

3. 2.3×9 4. 4×0.61 5. 5.22×12

6. 0.6×2.3 7. 3.05×2.4 8. 2.9×0.16

9. **MONEY MATTERS** David wants to purchase a new baseball glove that costs $49.95. The sales tax is found by multiplying the price of the glove by 0.075. How much sales tax will David pay? Round to the nearest cent.

Divide.

10. $19.36 \div 44$ 11. $9\overline{)37.8}$ 12. $60.34 \div 7$

13. $1.4\overline{)3.29}$ 14. $93.912 \div 4.3$ 15. $0.02\overline{)0.015}$

16. **SPORTS** At the 1996 Olympics, American sprinter Michael Johnson set a world record of 19.32 seconds for the 200-meter dash. A honeybee can fly the same distance in 40.572 seconds. About how many times faster than a honeybee was Michael Johnson?

Find the circumference of each circle. Round to the nearest hundredth.

17.

8.25 cm

18.

4 in.

19. Find the perimeter of the rectangle.

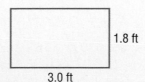
1.8 ft
3.0 ft

Standardized Test Practice

20. Tony ordered a pizza with a circumference of 44 inches. To the nearest whole number, what is the radius of the pizza?

   (A) 7 in.   (B) 7.1 in.   (C) 14 in.   (D) 41 in.

PART 1 Multiple Choice

Record your answers on the answer sheet provided by your teacher or on a sheet of paper.

1. What is 3,254 × 6? (Prerequisite Skill, p. 590)

   Ⓐ 18,524   Ⓑ 19,524

   Ⓒ 19,536   Ⓓ 24,524

2. For their vacation, the Borecki family drove from their house to the beach in 4 hours. Driving at the same rate, the Boreckis drove from the beach to a historical site.

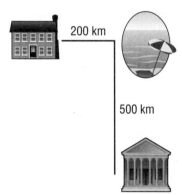

200 km

500 km

   Which expression finds the total amount of time it took them to drive from the beach to the historical site? (Lesson 1-1)

   Ⓕ 500 − 200

   Ⓖ 500 ÷ 4

   Ⓗ (500 + 200) ÷ 4

   Ⓘ 500 ÷ (200 ÷ 4)

3. Which of the following is the greatest? (Lesson 3-1)

   Ⓐ four thousand

   Ⓑ four hundred

   Ⓒ four-thousandths

   Ⓓ four and one-thousandth

4. What is 12 × 0.4? (Lesson 4-1)

   Ⓕ 0.0048   Ⓖ 0.048

   Ⓗ 0.48   Ⓘ 4.8

5. You can drive your car 19.56 miles with one gallon of gasoline. How many miles can you drive with 11.86 gallons of gasoline? (Lesson 4-2)

   Ⓐ 210.45 mi   Ⓑ 231.98 mi

   Ⓒ 280.55 mi   Ⓓ 310.26 mi

6. Ron paid $6.72 for 40 sheets of stickers. What was the average price of each sheet of stickers rounded to the nearest cent? (Lesson 4-3)

   Ⓕ $0.17   Ⓖ $0.28

   Ⓗ $0.39   Ⓘ $0.59

7. What is the value of 8.7 ÷ 0.6? (Lesson 4-4)

   Ⓐ 0.00145   Ⓑ 0.145

   Ⓒ 1.45   Ⓓ 14.5

8. Which of the following is the perimeter of the rectangle? (Lesson 4-5)

   3.7 yd

   6.2 yd

   Ⓕ 6.5 yd   Ⓖ 9.4 yd

   Ⓗ 12.2 yd   Ⓘ 19.8 yd

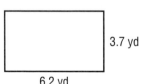

TEST-TAKING TIP

Question 8 Use estimation to eliminate any unreasonable answers. For example, eliminate answer F because one of the sides by itself is almost 6.5 yards.

PART 2 Short Response/Grid In

Record your answers on the answer sheet provided by your teacher or on a sheet of paper.

9. Jillian was planning a party and told 2 friends. The next day, each of those friends told 2 more friends. Then those friends each told 2 more friends.

Day 1	3
Day 2	7
Day 3	15
Day 4	31
Day 5	?

If the pattern continues, how many people will know about the party by Day 5? (Lesson 1-1)

10. What is the value of $2^4 + 3^2$? (Lesson 1-5)

11. The height of each student in a class was measured and recorded. The range in heights was 13 inches. The tallest and shortest students are shown below.

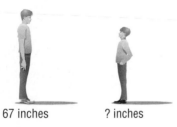

67 inches   ? inches

What is the height of the shortest student? (Lesson 2-7)

12. Yvette is training for a local run. Her goal is to run 30 miles each week. So far this week, she has run 6.5 miles, 5.2 miles, 7.8 miles, 3 miles, and 6.9 miles. How many more miles does Yvette need to run this week to reach her 30-mile goal? (Lesson 3-5)

13. Florida's population in 2025 is projected to be about 2.08×10^7. Write the number in standard form. (Lesson 4-1)

14. Impulses in the human nervous system travel at a rate of 188 miles per hour. Find the speed in miles per minute. Round to the nearest hundredth. (Lesson 4-3)

15. The streets on Trevor's block form a large square with each side measuring 0.3 mile. If he walks around the block twice, how far does he go? (Lesson 4-5)

PART 3 Extended Response

Record your answers on a sheet of paper. Show your work.

16. The dimensions of a rectangle are shown below. (Lesson 4-1)

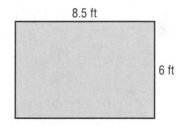

8.5 ft

6 ft

a. What is the area of the rectangle?

b. What is the perimeter of the rectangle?

c. How does the perimeter and area change if each dimension is doubled? Explain.

17. Use the circle graph to find how many times more CD albums were sold than cassette singles. Round to the nearest tenth. (Lesson 4-4)

Music Sales at Music Hut (percent of total)

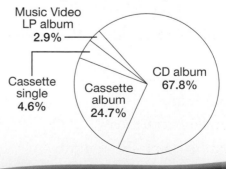

Music Video LP album 2.9%

Cassette single 4.6%

Cassette album 24.7%

CD album 67.8%

UNIT 3
Fractions

In addition to decimals, many of the numbers that you encounter in your daily life are fractions. In this unit, you will add, subtract, multiply, and divide fractions in order to solve problems.

 **INTERDISCIPLINARY PROJECT**   **MATH and SCIENCE**

COOKING UP A MYSTERY!

Have you ever wanted to be a scientist so you could mix substances together to create chemical reactions? Come join us on an explosive scientific adventure. Along the way, you'll discover the recipe for making a homemade volcano erupt. You'll also gather data about real volcanoes. So pack your math tools and don't forget a fireproof suit. This adventure is hot, hot, hot!

 Log on to msmath1.net/webquest to begin your WebQuest.

CHAPTER 5

Fractions and Decimals

""How do seashells relate to math?""

Many different seashells are found along the seashores. **The measurements of seashells often contain fractions and decimals.** The Florida Fighting Conch shell commonly found in Florida, North Carolina, Texas, and Mexico has a height of $2\frac{3}{4}$ to $4\frac{1}{4}$ inches. This is much larger than the California Cone shell, which has a height of $\frac{3}{4}$ to $1\frac{5}{8}$ inches.

You will solve problems about seashells in Lesson 5-7.

GETTING STARTED

Take this quiz to see whether you are ready to begin Chapter 5. Refer to the lesson or page number in parentheses if you need more review.

▶ Vocabulary Review

Choose the correct term to complete each sentence.

1. 1, 2, and 9 are all (multiples, factors) of 18. (Lesson 1-3)

2. Because 13 is only divisible by 1 and 13, it is a (prime, composite) number. (Lesson 1-3)

▶ Prerequisite Skills

Tell whether each number is divisible by 2, 3, 4, 5, 6, 9, or 10. (Lesson 1-2)

3. 67

4. 891

5. 145

6. 202

Find the prime factorization of each number. (Lesson 1-3)

7. 63

8. 264

9. 120

10. 28

Write each decimal in standard form. (Lesson 3-1)

11. five and three tenths

12. seventy-four hundredths

13. two tenths

14. sixteen thousandths

Divide. (Lesson 4-3)

15. $1.2 \div 3$

16. $3.5 \div 4$

17. $4.0 \div 5$

18. $6.3 \div 2$

FOLDABLES™
Study Organizer

Fractions and Decimals
Make this Foldable to help you understand fractions and decimals. Begin with one sheet of $8\frac{1}{2}" \times 11"$ paper.

STEP 1 **Fold Paper**
Fold top of paper down and bottom of paper up as shown.

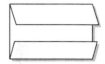

STEP 2 **Label**
Label the top fold Fractions and the bottom fold Decimals.

STEP 3 **Unfold and Draw**
Unfold the paper and draw a number line in the middle of the paper.

STEP 4 **Label**
Label the fractions and decimals as shown.

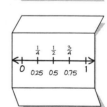

Reading and Writing As you read and study the chapter, use the foldable to help you see how fractions and decimals are related.

Study Skill

HOW TO...
Take Good Notes

VENN DIAGRAMS

The notebook you carry to class helps you stay organized. One way to organize your thoughts in math class is to use a diagram like a Venn diagram.

A *Venn diagram* uses overlapping circles to show the similarities and differences of two groups of items. Any item that is located where the circles overlap has a characteristic of both circles.

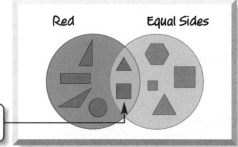

These are red shapes with equal sides.

You can also make a Venn diagram using numbers. The Venn diagram below shows the factors of 12 in one circle and the factors of 18 in the second circle.

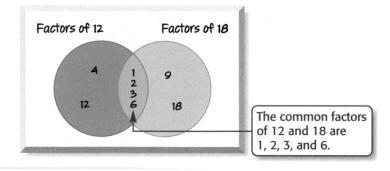

The common factors of 12 and 18 are 1, 2, 3, and 6.

SKILL PRACTICE
Make a Venn diagram that shows the factors for each pair of numbers.

1. 8, 12

2. 20, 30

3. 25, 28

4. 15, 30

5. Organize the numbers 2, 5, 9, 27, 29, 35, and 43 into a Venn diagram. Use the headings *prime numbers* and *composite numbers*. What numbers are in the overlapping circles? Explain.

Greatest Common Factor

WHEN am I ever going to use this?

What You'll LEARN

Find the greatest common factor of two or more numbers.

NEW Vocabulary

Venn diagram

greatest common factor (GCF)

REVIEW Vocabulary

factor: two or more numbers that are multiplied together to form a product **(Lesson 1-3)**

prime number: a whole number that has exactly two factors, 1 and the number itself **(Lesson 1-3)**

WATERPARKS The Venn diagram below shows which water slides Curtis and his friends rode. **Venn diagrams** use overlapping circles to show common elements.

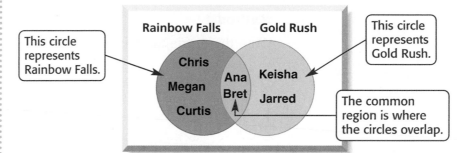

This circle represents Rainbow Falls.

This circle represents Gold Rush.

The common region is where the circles overlap.

1. Who rode Rainbow Falls?
2. Who rode Gold Rush?
3. Who rode both Rainbow Falls and Gold Rush?

Numbers often have common factors. The greatest of the common factors of two or more numbers is the **greatest common factor (GCF)** of the numbers. To find the GCF, you can make a list.

EXAMPLE   Find the GCF by Listing Factors

1 Find the GCF of 42 and 56 by making a list.

First, list the factors by pairs.

Factors of 42
1, 42
2, 21
3, 14
6, 7

Factors of 56
1, 56
2, 28
4, 14
7, 8

The common factors are 1, 2, 7, and 14. The greatest common factor or GCF of 42 and 56 is 14.

Use a Venn diagram to show the factors. Notice that the factors 1, 2, 7, and 14 are the common factors of 42 and 56 and the GCF is 14.

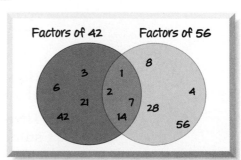

Your Turn Find the GCF of each set of numbers by making a list.

a. 35 and 60   **b.** 15 and 45   **c.** 12 and 19

You can also use prime factors to find the GCF.

EXAMPLE **Find the GCF by Using Prime Factors**

2 Find the GCF of 18 and 30 by using prime factors.

STUDY TIP

Look Back To review the divisibility rules, see Lesson 1-2.

Method 1

Write the prime factorization.

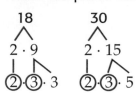

Method 2

Divide by prime numbers.

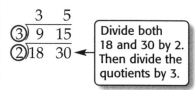

Divide both 18 and 30 by 2. Then divide the quotients by 3.

Using either method, the common prime factors are 2 and 3. So, the GCF of 18 and 30 is 2 × 3 or 6.

Your Turn Find the GCF of each set of numbers by using prime factors.

d. 12 and 66   **e.** 36 and 45   **f.** 32 and 48

Many real-life situations involve greatest common factors.

EXAMPLES **Use the GCF to Solve a Problem**

3 **MONEY** Ms. Taylor recorded the amount of money collected from sixth grade classes for a field trip. Each student paid the same amount. What is the most the field trip could cost per student?

Ms. Taylor's Class Field Trip Money	
Monday	$48
Tuesday	$40
Thursday	$24

factors of 48: 1, 2, 3, 4, 6, 8, 12, 16, 24, 48

factors of 40: 1, 2, 4, 5, 8, 10, 20, 40

factors of 24: 1, 2, 3, 4, 6, 8, 12, 24

List all the factors of each number. Then find the greatest common factor.

The GCF of 48, 40, and 24 is 8. So, the most the field trip could cost is $8 per student.

4 How many students have paid to attend the class trip if the cost of a ticket was $8?

There is a total of $48 + $40 + $24 or $112. So, the number of students that have paid for a ticket is $112 ÷ $8 or 14.

Skill and Concept Check

Writing Math
Exercises 2–4

1. **List** all of the factors of 24 and 32. Then list the common factors.

2. **Explain** three ways to find the GCF of two or more numbers.

3. **OPEN ENDED** Write two numbers whose GCF is 1. Then write a sentence explaining why their GCF is 1.

4. **Which One Doesn't Belong?** Identify the number that does not have the same greatest common factor as the other three. Explain your reasoning.

30	18	12	15

GUIDED PRACTICE

Find the GCF of each set of numbers.

5. 8 and 32   6. 12 and 21   7. 90 and 110

PLANTS For Exercises 8 and 9, use the following information.
The table lists the number of tree seedlings Emily has to sell at a school plant sale. She wants to display them in rows with the same number of each type of seedling in each row.

8. Find the greatest number of seedlings that can be placed in each row.

9. How many rows of each tree seedlings will there be?

Seedlings For Sale	
Type	**Amount**
pine	32
oak	48
maple	80

Practice and Applications

Find the GCF of each set of numbers.

10. 3 and 9   11. 4 and 16   12. 12 and 18

13. 10 and 15   14. 21 and 25   15. 37 and 81

16. 18 and 42   17. 48 and 60   18. 30 and 72

19. 45 and 75   20. 36 and 90   21. 34 and 85

22. 14, 35, 84   23. 9, 18, 42   24. 16, 52, 76

HOMEWORK HELP

For Exercises	See Examples
10–24	1, 2
30, 32–33	3, 4

Extra Practice
See pages 603, 628.

25. Write three numbers whose GCF is 15.

26. Write three numbers whose GCF is 12 with one of the numbers being greater than 100.

27. Find two even numbers between 10 and 30 whose GCF is 14.

28. Find two numbers having a GCF of 9 and a sum of 99.

29. **PATTERNS** What is the GCF of all of the numbers in the pattern 12, 24, 36, 48, . . . ?

30. **MUSIC** Elizabeth has three CD storage cases that can hold 16, 24, and 32 CDs. The cases have sections holding the same number of CDs. What is the greatest number of CDs in a section?

31. Give a counterexample for the statement written below. (A *counterexample* is an example that disproves the statement.)

    The GCF of two or more numbers cannot be equal to one of the numbers.

FOOD For Exercises 32 and 33, use the information below.
The table shows the amount of fruit Jordan picked at his family's fruit farm. The fruit must be placed in bags so that each bag contains the same number of pieces of fruit without mixing the fruit.

Fruit Picked	
Fruit	Amount
limes	45
oranges	105
tangerines	75

32. Find the greatest number of pieces of fruit that can be put in each bag.

33. How many bags of each kind of fruit will there be?

34. **CRITICAL THINKING** Is the GCF of any two even numbers always even? Is the GCF of any two odd numbers always odd? Explain your reasoning.

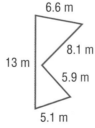

Standardized Test Practice and Mixed Review

35. **MULTIPLE CHOICE** Which number is not a factor of 48?

    Ⓐ 3   Ⓑ 6   Ⓒ 8   Ⓓ 14

36. **SHORT RESPONSE** Adam and Jocelyn are decorating the gym for the school dance. They want to cut all of the streamers the same length. One roll is 72 feet long, and the other is 64 feet long. What is the greatest length they should cut each streamer?

37. **GEOMETRY** Find the circumference of a circle with a radius of 5 meters. (Lesson 4-6)

GEOMETRY Find the perimeter of each figure. (Lesson 4-5)

38.
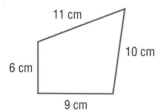
11 cm
10 cm
6 cm
9 cm

39.
6.6 m
8.1 m
13 m
5.9 m
5.1 m

GETTING READY FOR THE NEXT LESSON

PREREQUISITE SKILL Tell whether both numbers in each number pair are divisible by 2, 3, 4, 5, 6, 9, or 10. (Lesson 1-2)

40. 9 and 24   41. 15 and 25   42. 4 and 10   43. 18 and 21

Simplifying Fractions

What You'll LEARN

Use models to simplify fractions.

Materials
- ruler
- paper
- scissors
- colored pencils

To simplify a fraction using a model, you can use paper folding.

ACTIVITY *Work in groups.*

Use a model to simplify $\frac{6}{8}$.

The factors of 8 are 1, 2, 4, and 8. Cut out three rectangles of the same size to model the factors 2, 4 and 8. You do not need a rectangle for 1.

To model six-eighths, fold one rectangle into eight equal parts and then shade six parts.

Fold the remaining rectangles into sections to show the factors 2 and 4.

$\frac{1.5}{2}$

$\frac{3}{4}$

Place the model that shows six eighths under each model. Trace the shading. Choose the model that gives a whole number as the numerator.

The simplified form of $\frac{6}{8}$ is $\frac{3}{4}$.

Your Turn Use a model to simplify each fraction.

a. $\frac{2}{8}$

b. $\frac{4}{6}$

c. $\frac{3}{9}$

Writing Math

1. **Explain** how the model at the right shows the simplified form of $\frac{3}{5}$.

2. **MAKE A CONJECTURE** Write a rule that you could use to write a fraction in simplest form.

Simplifying Fractions

What You'll LEARN

Express fractions in simplest form.

NEW Vocabulary

equivalent fractions
simplest form

Link to READING

Everyday meaning of Equivalent: equal in amount or value

WHEN am I ever going to use this?

SURVEYS The results of a favorite magazine survey are shown.

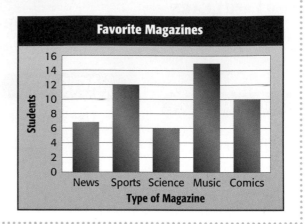

Favorite Magazines

1. How many students were surveyed?

2. How many students prefer music magazines?

In the graph above, you can compare the students who prefer music magazines to the total number of students by using a fraction.

$$\frac{15}{50} \quad \leftarrow \text{ students who prefer music magazines}$$
$$\leftarrow \text{ total number of students}$$

Equivalent fractions are fractions that name the same number. The models at the right are the same size, and the same part or fraction is shaded. So, the fractions are equivalent. That is, $\frac{15}{50} = \frac{3}{10}$.

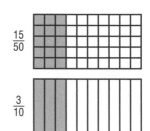

$\frac{15}{50}$

$\frac{3}{10}$

To find equivalent fractions, you can multiply or divide the numerator and denominator by the same nonzero number.

$$\frac{15 \div 5}{50 \div 5} = \frac{3}{10}$$

EXAMPLES Write Equivalent Fractions

Replace each ● with a number so the fractions are equivalent.

1 $\frac{5}{7} = \frac{●}{21}$

Since $7 \times 3 = 21$, multiply the numerator and denominator by 3.

$$\overset{\times 3}{\frac{5}{7} = \frac{●}{21}}, \text{ so } \frac{5}{7} = \frac{15}{21}.$$
$$\underset{\times 3}{}$$

2 $\frac{12}{16} = \frac{6}{●}$

Since $12 \div 2 = 6$, divide the numerator and denominator by 2.

$$\overset{\div 2}{\frac{12}{16} = \frac{6}{●}}, \text{ so } \frac{12}{16} = \frac{6}{8}.$$
$$\underset{\div 2}{}$$

A fraction is in **simplest form** when the GCF of the numerator and denominator is 1.

> **Key Concept**   **Simplest Form**
>
> To write a fraction in simplest form, you can either:
> - divide the numerator and denominator by common factors until the only common factor is 1, or
> - divide the numerator and denominator by the GCF.

EXAMPLE   Write Fractions in Simplest Form

3 Write $\frac{18}{24}$ in simplest form.

Method 1

Divide by common factors.

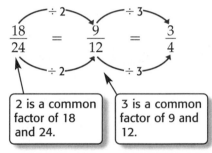

$$\frac{18}{24} = \frac{9}{12} = \frac{3}{4}$$

| 2 is a common factor of 18 and 24. | 3 is a common factor of 9 and 12. |

Since 3 and 4 have no common factor greater than 1, the fraction $\frac{3}{4}$ is in simplest form.

Method 2

Divide by the GCF.

factors of 18: 1, 2, 3, 6, 9, 18

factors of 24: 1, 2, 3, 4, 6, 8, 12, 24

The GCF of 18 and 24 is 6.

$$\frac{18}{24} = \frac{3}{4}$$ Divide the numerator and denominator by the GCF, 6.

Since the GCF of 3 and 4 is 1, the fraction $\frac{3}{4}$ is in simplest form.

Your Turn Write each fraction in simplest form. If the fraction is already in simplest form, write *simplest form*.

a. $\frac{21}{24}$ b. $\frac{9}{15}$ c. $\frac{2}{3}$

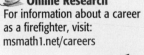

REAL-LIFE MATH

How Does a Firefighter Use Math?
Firefighters use math to calculate the amount of water they are running through their fire hoses.

Online Research
For information about a career as a firefighter, visit: msmath1.net/careers

EXAMPLE   Express Fractions in Simplest Form

4 **FIREFIGHTERS** Approximately 36 out of 40 firefighters work in city or county fire departments. Express the fraction $\frac{36}{40}$ in simplest form.

$$\frac{\overset{9}{\cancel{36}}}{\underset{10}{\cancel{40}}} = \frac{9}{10}$$ Simplify. Mentally divide both the numerator and denominator by the GCF, 4.

In simplest form, the fraction $\frac{36}{40}$ is written $\frac{9}{10}$. So $\frac{9}{10}$ of firefighters work in city or county fire departments.

Skill and Concept Check

Writing Math
Exercise 2

1. **OPEN ENDED** Write three fractions that are equivalent.

2. **Which One Doesn't Belong?** Identify the fraction that is not equivalent to the other three. Explain your reasoning.

$$\frac{8}{12} \qquad \frac{12}{18} \qquad \frac{9}{12} \qquad \frac{32}{48}$$

GUIDED PRACTICE

Replace each ● with a number so the fractions are equivalent.

3. $\frac{3}{8} = \frac{●}{24}$

4. $\frac{4}{5} = \frac{●}{50}$

5. $\frac{15}{25} = \frac{3}{●}$

Write each fraction in simplest form. If the fraction is already in simplest form, write *simplest form*.

6. $\frac{2}{10}$

7. $\frac{8}{25}$

8. $\frac{10}{38}$

Practice and Applications

Replace each ● with a number so the fractions are equivalent.

9. $\frac{1}{2} = \frac{●}{8}$

10. $\frac{1}{3} = \frac{●}{27}$

11. $\frac{●}{5} = \frac{9}{15}$

12. $\frac{●}{6} = \frac{20}{24}$

13. $\frac{7}{9} = \frac{14}{●}$

14. $\frac{12}{16} = \frac{3}{●}$

15. $\frac{30}{35} = \frac{●}{7}$

16. $\frac{36}{45} = \frac{●}{5}$

HOMEWORK HELP	
For Exercises	See Examples
9–16	1, 2
17–30, 32	3, 4

Extra Practice
See pages 603, 628.

Write each fraction in simplest form. If the fraction is already in simplest form, write *simplest form*.

17. $\frac{6}{9}$

18. $\frac{4}{10}$

19. $\frac{5}{8}$

20. $\frac{11}{12}$

21. $\frac{18}{21}$

22. $\frac{32}{40}$

23. $\frac{10}{38}$

24. $\frac{27}{54}$

25. $\frac{19}{37}$

26. $\frac{32}{85}$

27. $\frac{28}{77}$

28. $\frac{15}{100}$

29. **RECREATION** Of the 32 students in a class, 24 students own scooters. Write this fraction in simplest form.

30. **FOOD** Twenty out of two dozen cupcakes are chocolate cupcakes. Write this amount as a fraction in simplest form.

GAMES For Exercises 31 and 32, use the table. The 100 tiles in a popular word game are labeled as shown. Two of the tiles are blank tiles.

Letter Tiles			
A-9	B-2	C-2	D-4
E-12	F-2	G-3	H-2
I-9	J-1	K-1	L-4
M-2	N-6	O-8	P-2
Q-1	R-6	S-4	T-6
U-4	V-2	W-2	X-1
Y-2	Z-1		

31. Write a fraction that compares the number of tiles containing the letter E to the total number of tiles.

32. Write this fraction in simplest form.

STATISTICS For Exercises 33–35, use the following information and the table shown.

In a frequency table, the *relative frequency* of a category is the fraction of the data that falls in that class.

Favorite Pet Survey		
Pet	Tally	Frequency
Dog	⊪⊪ I	6
Cat	⊪⊪	5
Hamster	III	3
Gerbil	IIII	4
Turtle	II	2

33. What is the total number of items?

34. Write the relative frequency of each category as a fraction in simplest form.

35. Write a sentence describing what fraction prefer cats as pets.

36. **CRITICAL THINKING** A fraction is equivalent to $\frac{3}{4}$, and the sum of the numerator and denominator is 84. What is the fraction?

EXTENDING THE LESSON
Another way you can find equivalent fractions is to think of the patterns on a multiplication table.

Use the table at the right to find three equivalent fractions for each fraction listed.

37. $\frac{5}{7}$ **38.** $\frac{2}{9}$ **39.** $\frac{4}{6}$

×	1	2	3	4	5	6	7	8	9	10
1	1	2	3	4	5	6	7	8	9	10
2	2	4	6	8	10	12	14	16	18	20
3	3	6	9	12	15	18	21	24	27	30
4	4	8	12	16	20	24	28	32	36	40
5	5	10	15	20	25	30	35	40	45	50
6	6	12	18	24	30	36	42	48	54	60
7	7	14	21	28	35	42	49	56	63	70
8	8	16	24	32	40	48	56	64	72	80
9	9	18	27	36	45	54	63	72	81	90
10	10	20	30	40	50	60	70	80	90	100

You can see the equivalent fractions for $\frac{3}{10}$ as you look across the table.

Standardized Test Practice and Mixed Review

40. **MULTIPLE CHOICE** Which two models show equivalent fractions?

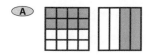

41. **SHORT RESPONSE** Write and model a fraction that is equivalent to $\frac{5}{8}$.

Find the GCF of each set of numbers. (Lesson 5-1)

42. 40 and 36   **43.** 45 and 75   **44.** 120 and 150

45. **GEOMETRY** Find the circumference of a circle with a diameter of 13.4 meters. (Lesson 4-6)

GETTING READY FOR THE NEXT LESSON

PREREQUISITE SKILL Divide. Round to the nearest tenth. (Page 591)

46. $8 \div 3$ **47.** $19 \div 6$ **48.** $52 \div 8$ **49.** $67 \div 9$

Mixed Numbers and Improper Fractions

What You'll LEARN

Write mixed numbers as improper fractions and vice versa.

NEW Vocabulary

mixed number
improper fraction

HANDS-ON Mini Lab

Work with a partner.

STEP 1 Draw and shade a rectangle to represent the whole number 1.

STEP 2 Draw another rectangle the same size as the first. Divide it into three equal parts to show thirds. Shade one part to represent $\frac{1}{3}$.

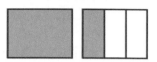

STEP 3 Divide the whole number portion into thirds.

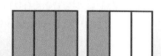

1. How many shaded $\frac{1}{3}$s are there?

2. What fraction is equivalent to $1\frac{1}{3}$?

Make a model to show each number.

3. the number of fourths in $2\frac{3}{4}$ 4. the number of halves in $4\frac{1}{2}$

A number like $1\frac{1}{3}$ is an example of a mixed number. A **mixed number** indicates the sum of a whole number and a fraction.

$$1\frac{1}{3} = 1 + \frac{1}{3}$$

The numbers below the number line show two groups of fractions. Notice how the numerators differ from the denominators.

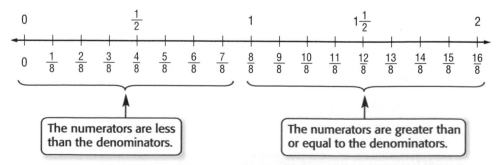

The numerators are less than the denominators.

The numerators are greater than or equal to the denominators.

The values of the fractions to the right of 1 are greater than or equal to 1. These fractions are **improper fractions**. Mixed numbers can be written as improper fractions.

READING Math

Mixed Numbers The mixed number $4\frac{1}{6}$ is read *four and one-sixth*.

1 Draw a model for $4\frac{1}{6}$. Then write $4\frac{1}{6}$ as an improper fraction.

Since the mixed number is greater than 4, draw five models that are divided into six equal sections to show sixths. Then shade four wholes and one sixth.

There are twenty-five $\frac{1}{6}$s. So, $4\frac{1}{6}$ can be written as $\frac{25}{6}$.

You can also write mixed numbers as improper fractions using mental math.

EXAMPLE **Mixed Numbers as Improper Fractions**

REAL-LIFE MATH

BIRDS Male bald eagles generally measure 3 feet from head to tail, weigh 7 to 10 pounds, and have a wingspan of about $6\frac{1}{2}$ feet.

Source: U.S. Fish and Wildlife Service

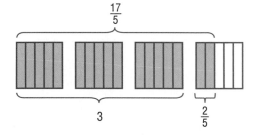

2 **BIRDS** Use the information at the left. Write the length of the male bald eagle's wingspan as an improper fraction.

$$6\frac{1}{2} \Rightarrow 6 \quad \frac{1}{2} + = \frac{(6 \times 2) + 1}{2} = \frac{13}{2}$$

Multiply the whole number and denominator. Then add the numerator.

Your Turn Write each mixed number as an improper fraction.

a. $2\frac{1}{4}$ b. $1\frac{2}{5}$ c. $4\frac{3}{4}$

Improper fractions can also be written as mixed numbers. Divide the numerator by the denominator.

EXAMPLE **Improper Fractions as Mixed Numbers**

3 Write $\frac{17}{5}$ as a mixed number.

Divide 17 by 5.

$$\begin{array}{r} 3\frac{2}{5} \\ 5\overline{)17} \\ -15 \\ \hline 2 \end{array}$$

Use the remainder as the numerator of the fraction.

$\frac{17}{5}$

3         $\frac{2}{5}$

So, $\frac{17}{5}$ can be written as $3\frac{2}{5}$.

Your Turn Write each improper fraction as a mixed number.

d. $\frac{7}{3}$ e. $\frac{18}{5}$ f. $\frac{13}{2}$

1. **Draw** a model for $\frac{8}{5}$. Write the fraction as a mixed number.

2. **OPEN ENDED** Write a fraction that is equal to a whole number.

3. **NUMBER SENSE** Explain how you know whether a fraction is less than, equal to, or greater than 1.

GUIDED PRACTICE

Draw a model for each mixed number. Then write the mixed number as an improper fraction.

4. $3\frac{2}{3}$

5. $2\frac{3}{4}$

Write each mixed number as an improper fraction.

6. $4\frac{1}{8}$

7. $1\frac{4}{5}$

8. $5\frac{2}{3}$

Write each improper fraction as a mixed number.

9. $\frac{31}{6}$

10. $\frac{15}{4}$

11. $\frac{21}{8}$

12. **BATS** Write the length of the body of the vampire bat shown as an improper fraction.

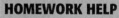
$2\frac{3}{4}$ in.

Practice and Applications

Draw a model for each mixed number. Then write the mixed number as an improper fraction.

13. $3\frac{1}{4}$

14. $1\frac{5}{6}$

15. $2\frac{7}{8}$

16. $4\frac{3}{5}$

Write each mixed number as an improper fraction.

17. $6\frac{1}{3}$

18. $8\frac{2}{3}$

19. $3\frac{2}{5}$

20. $7\frac{4}{5}$

21. $1\frac{5}{8}$

22. $1\frac{7}{8}$

23. $7\frac{1}{4}$

24. $5\frac{3}{4}$

25. $3\frac{5}{6}$

26. $4\frac{1}{6}$

27. $6\frac{6}{7}$

28. $7\frac{3}{8}$

29. Express *six and three-fifths* as an improper fraction.

30. Find the improper fraction that is equivalent to *eight and five-sixths*.

31. What mixed number is equivalent to *twenty-nine thirds*?

32. Express *thirty-three fourths* as a mixed number.

HOMEWORK HELP

For Exercises	See Examples
13–16	1
17–30, 48	2
31–47, 50	3

Extra Practice
See pages 603, 628.

Write each improper fraction as a mixed number.

33. $\dfrac{16}{5}$　　34. $\dfrac{27}{5}$　　35. $\dfrac{9}{8}$　　36. $\dfrac{19}{8}$

37. $\dfrac{16}{9}$　　38. $\dfrac{22}{9}$　　39. $\dfrac{35}{6}$　　40. $\dfrac{50}{6}$

41. $\dfrac{25}{7}$　　42. $\dfrac{68}{7}$　　43. $\dfrac{47}{10}$　　44. $\dfrac{61}{10}$

45. one hundred ninths

46. two hundred thirty-five sevenths

47. **FOOD** Sabino bought a carton of 18 eggs for his mom at the grocery store. How many dozen eggs did Sabino buy?

48. **MEASUREMENT** The width of a piece of paper is $8\frac{1}{2}$ inches. Write $8\frac{1}{2}$ as an improper fraction.

MOVIES For Exercises 49 and 50, use the table at the right that shows the running times of movies.

49. For each movie, write a fraction that compares the running time to the number of minutes in an hour.

50. Write each fraction as a mixed number in simplest form.

 Data Update What is the running time of your favorite movie? Write the mixed number that represents the running time in hours. Visit msmath1.net/data_update to learn more.

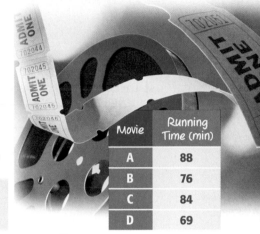

Movie	Running Time (min)
A	88
B	76
C	84
D	69

51. **CRITICAL THINKING** Jonah cannot decide whether $\dfrac{17}{17}$ is an improper fraction. How could he determine the answer?

Standardized Test Practice and Mixed Review

52. **MULTIPLE CHOICE** Which improper fraction is *not* equivalent to any of the mixed numbers listed in the table?

　Ⓐ $\dfrac{9}{5}$　　Ⓑ $\dfrac{11}{5}$　　Ⓒ $\dfrac{17}{6}$　　Ⓓ $\dfrac{7}{2}$

Ingredient	Amount
flour	$2\frac{5}{6}$ cups
butter	$1\frac{4}{5}$ cups
chocolate chips	$3\frac{1}{2}$ bags

53. **MULTIPLE CHOICE** Write $\dfrac{23}{8}$ as a mixed number.

　Ⓕ $1\frac{7}{8}$　　Ⓖ $2\frac{3}{8}$　　Ⓗ $2\frac{5}{8}$　　Ⓘ $2\frac{7}{8}$

54. Write $\dfrac{35}{42}$ in simplest form. (Lesson 5-2)

Find the GCF of each set of numbers. (Lesson 5-1)

55. 9 and 39　　56. 33 and 88　　57. 44 and 70　　58. 24, 48, and 63

GETTING READY FOR THE NEXT LESSON

PREREQUISITE SKILL Find the prime factorization of each number. (Lesson 1-3)

59. 36　　60. 76　　61. 105　　62. 145

CHAPTER 5 Mid-Chapter Practice Test

Vocabulary and Concepts

1. **Write** a fraction that is equivalent to $\frac{3}{8}$. (Lesson 5-2)

2. **OPEN ENDED** Write two numbers whose GCF is 7. (Lesson 5-1)

Skills and Applications

Find the GCF of each set of numbers. (Lesson 5-1)

3. 12 and 32

4. 27 and 45

5. 24, 40, and 72

Replace each ● with a number so that the fractions are equivalent.
(Lesson 5-2)

6. $\frac{2}{9} = \frac{●}{45}$

7. $\frac{5}{12} = \frac{●}{60}$

8. $\frac{27}{36} = \frac{●}{4}$

Write each fraction in simplest form. If the fraction is already in simplest form, write *simplest form*. (Lesson 5-2)

9. $\frac{15}{24}$

10. $\frac{12}{42}$

11. $\frac{9}{14}$

12. **FOOD** A basket contains 9 apples and 6 pears. Write a fraction in simplest form that compares the number of apples to the total number of pieces of fruit. (Lesson 5-2)

Write each mixed number as an improper fraction. (Lesson 5-3)

13. $3\frac{5}{6}$

14. $3\frac{3}{5}$

15. $8\frac{4}{9}$

Write each improper fraction as a mixed number. (Lesson 5-3)

16. $\frac{23}{4}$

17. $\frac{37}{9}$

18. $\frac{69}{8}$

Standardized Test Practice

19. **MULTIPLE CHOICE** Which fraction is *not* an improper fraction? (Lesson 5-3)

 Ⓐ $\frac{3}{8}$

 Ⓑ $\frac{6}{5}$

 Ⓒ $\frac{4}{3}$

 Ⓓ $\frac{3}{2}$

20. **SHORT RESPONSE** Write three factors of 72. (Lesson 5-1)

The Game Zone

A Place To Practice Your Math Skills

GCF Spin-Off

● **GET READY!**

Players: four
Materials: two spinners

● **GET SET!**

* Divide into teams of two players.

* One team labels the six equal sections of one spinner 6, 8, 12, 16, 18, and 24 as shown.

* The other team labels the six equal sections of another spinner 48, 54, 56, 64, 72, and 80 as shown.

● **GO!**

* One player from each team spins their spinner.
* The other player on each team tries to be the first to name the GCF of the two numbers.
* The first person to correctly name the GCF gets 5 points for their team.
* Players take turns spinning the spinners and naming the GCF.
* **Who Wins?** The first team to get 25 points wins.

5-4a Problem-Solving Strategy

A Preview of Lesson 5-4

What You'll Learn

Solve problems by making an organized list.

Make an Organized List

Malila, this year's school carnival is going to have a dart game booth, a ring toss booth, and a face-painting booth. I wonder how we could arrange the booths.

Well, Jessica, we could **make an organized list** to determine all of the different possibilities.

Explore	We know that there are three booths. We need to know the number of possible arrangements.
Plan	Make a list of all of the different possible arrangements. Use D for darts, R for ring toss, and F for face painting.
Solve	D R F R F D F R D D F R R D F F D R There are six different ways the booths can be arranged.
Examine	Check the answer by seeing if each booth is accounted for two times in the first, second, and third positions.

Analyze the Strategy

1. **Explain** how making an organized list helps you to solve a problem.

2. **Analyze** the six possible arrangements. Do you agree or disagree with the possibilities? Explain your reasoning.

3. **Think** of another approach to solving this problem. Write a short explanation of your problem-solving approach.

Solve. Use the make an organized list strategy.

4. Lourdes is having a birthday party. She wants to sit with her three best friends. How many arrangements are there for all of them to sit along one side of a rectangular table?

5. Michael needs to go to the pharmacy, the grocery store, and the video store. How many different ways can Michael make the stops?

Mixed Problem Solving

Solve. Use any strategy.

6. **MANUFACTURING** A sweater company offers 6 different styles of sweaters in 5 different colors. How many combinations of style and color are possible?

7. **NUMBER SENSE** How many different arrangements are possible for the prime factors of 30?

8. **GEOMETRY** Find the difference in the areas of the square and rectangle.

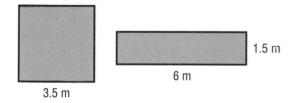

3.5 m

6 m

1.5 m

9. **SHOES** A store sells four types of tennis shoes in white, black, or blue. How many combinations of style and color are possible?

10. **FOOD** A small bag of potato chips weighs about 0.85 ounce. What is the weight of 12 bags of potato chips?

11. **MONEY** Caroline earns $258 each month delivering newspapers. How much does she earn each year?

12. **MULTI STEP** What is the total cost of a coat that is on sale for $50 if the sales tax rate is 0.065?

13. **PATTERNS** What number is missing in the pattern . . . , 234, 345, __?__, 567, . . . ?

14. **LANGUAGE ARTS** On Monday, 86 science fiction books were sold at a book sale. This is 8 more than twice the amount sold on Thursday. How many science fiction books were sold on Thursday?

15. **STANDARDIZED TEST PRACTICE**
Refer to the graph. Which is a reasonable conclusion that can be drawn from the graph?

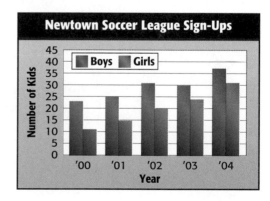

Newtown Soccer League Sign-Ups

Ⓐ More girls than boys signed up in 2003.

Ⓑ The difference between the number of boys and girls has been decreasing since 2000.

Ⓒ The number of boys who joined soccer decreased every year.

Ⓓ There were more soccer sign-ups in 2002 than any other year.

You will use the make an organized list strategy in the next lesson.

Least Common Multiple

What You'll LEARN

Find the least common multiple of two or more numbers.

NEW Vocabulary

multiple
common multiples
least common multiple (LCM)

REVIEW Vocabulary

prime factorization:
expressing a composite number as a product of prime numbers
(Lesson 1-3)

WHEN am I ever going to use this?

PATTERNS The multiplication table shown below lists the products of any two numbers from 1 to 10.

×	1	2	3	4	5	6	7	8	9	10
1	1	2	3	4	5	6	7	8	9	10
2	2	4	6	8	10	12	14	16	18	20
3	3	6	9	12	15	18	21	24	27	30
4	4	8	12	16	20	24	28	32	36	40
5	5	10	15	20	25	30	35	40	45	50
6	6	12	18	24	30	36	42	48	54	60
7	7	14	21	28	35	42	49	56	63	70
8	8	16	24	32	40	48	56	64	72	80
9	9	18	27	36	45	54	63	72	81	90
10	10	20	30	40	50	60	70	80	90	100

1. Write the products of 8 and each of the numbers 1, 2, 3, 4, 5, 6, 7, 8, 9, and 10.

2. Describe the pattern you see in the way the numbers increase.

3. Study the products of any number and the numbers from 1 to 10. Write a rule for the pattern you find. Give examples to support your rule.

A **multiple** of a number is the product of the number and any whole number. The multiples of 8 and 4 are listed below.

$$0 \times 8 = 0$$
$$1 \times 8 = 8$$
$$2 \times 8 = 16$$ multiples of 8
$$3 \times 8 = 24$$
$$4 \times 8 = 32$$
$$\vdots$$

$$0 \times 4 = 0$$
$$1 \times 4 = 4$$
$$2 \times 4 = 8$$ multiples of 4
$$3 \times 4 = 12$$
$$4 \times 4 = 16$$
$$\vdots$$

Notice that 0, 8, 16, and 24 are multiples of both 4 and 8. These are **common multiples**. The smallest number other than 0 that is a multiple of two or more whole numbers is the **least common multiple (LCM)** of the numbers.

multiples of 4: 0, 4, 8, 12, 16, 20, 24, …
multiples of 8: 0, 8, 16, 24, 32, 40, 48, …

The least of the common multiples, or LCM, of 4 and 8 is 8.

To find the LCM of two or more numbers, you can make a list.

EXAMPLE **Find the LCM by Making a List**

1 Find the LCM of 9 and 12 by making a list.

Step 1 List the nonzero multiples.

multiples of 9: 9, 18, 27, 36, 45, 54, 63, 72, …
multiples of 12: 12, 24, 36, 48, 60, 72, …

Step 2 Identify the LCM from the common multiples.
The LCM of 9 and 12 is 36.

Prime factors can also be used to find the LCM.

EXAMPLE **Find the LCM by Using Prime Factors**

2 Find the LCM of 15 and 25 by using prime factors.

Step 1 Write the prime factorization of each number.

$$15$$
$$3 \times 5$$

$$25$$
$$5 \times 5$$

Step 2 Identify all common prime factors.

$15 = 3 \times \boxed{5}$
$25 = 5 \times \boxed{5}$

Step 3 Find the product of the prime factors using each common prime factor once and any remaining factors. The LCM is $3 \times 5 \times 5$ or 75.

Your Turn Find the LCM of each set of numbers.

a. 4 and 7 **b.** 3 and 9

REAL-LIFE MATH

ANIMALS The Arctic grayling is a freshwater fish commonly found in the cold waters of the Hudson Bay and the Great Lakes. These fish can grow up to 30 inches long and weigh up to 6 pounds.

Source: eNature.com

EXAMPLE **Use the LCM to Solve a Problem**

3 **ANIMALS** The weight in pounds of each fish is shown on the scale readout. Suppose more of each type of fish weighing the same amount is added to each scale. At what weight will the scales have the same readout?

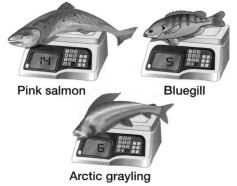

Pink salmon   Bluegill

Arctic grayling

Find the LCM using prime factors.

14   5   6
②×⑦   ⑤× 1   2 ×③

The scales will show the same weight at $2 \times 7 \times 5 \times 3$ or 210 pounds.

Skill and Concept Check

Writing Math
Exercises 2–4

1. **Tell** whether the first number is a multiple of the second number.

   a. 30; 6   b. 27; 4   c. 35; 3   d. 84; 7

2. **OPEN ENDED** Choose any two numbers less than 100. Then find the LCM of the numbers and explain your steps.

3. **FIND THE ERROR** Kurt and Ana are finding the LCM of 6 and 8. Who is correct? Explain.

Kurt
$6 = \boxed{2} \times 3$
$8 = \boxed{2} \times 2 \times 2$
The LCM of 6 and 8 is 2.

Ana
$6 = \boxed{2} \times 3$
$8 = \boxed{2} \times 2 \times 2$
The LCM of 6 and 8 is $2 \times 2 \times 2 \times 3$ or 24.

4. **NUMBER SENSE** Numbers that share no common prime factors are called *relatively prime*. Find the LCM for each pair of numbers. What do you notice about the LCMs?

   4 and 9   5 and 7   3 and 10   6 and 11

GUIDED PRACTICE

Find the LCM of each set of numbers.

5. 5 and 8   6. 3 and 14   7. 4, 6, and 9

8. **ANIMALS** Refer to Example 3 on page 195. How many of each fish would you need to reach the minimum common weight?

Practice and Applications

Find the LCM of each set of numbers.

9. 2 and 10   10. 9 and 54   11. 3 and 4

12. 7 and 9   13. 5 and 12   14. 8 and 36

15. 16 and 20   16. 15 and 12   17. 4, 8, and 10

18. 3, 9, and 18   19. 15, 25, and 75   20. 9, 12, and 15

21. Find the LCM of 3, 5, and 7.   22. What is the LCM of 2, 3, and 13?

HOMEWORK HELP

For Examples	See Exercises
9–22	1, 2
25–27	3

Extra Practice
See pages 604, 628.

23. **PATTERNS** Find the three missing common multiples from the list of common multiples for 3 and 15.

    90, 135, 180, __?__, __?__, __?__, 360, 405, . . .

24. **PATTERNS** List the next five common multiples after the LCM of 6 and 9.

25. **BICYCLES** The front bicycle gear shown has 42 teeth, and the back gear has 14 teeth. How many complete rotations must the smaller gear make for both gears to be aligned in the original starting position?

26. **SCIENCE** The approximate cycles for the appearance of cicadas and tent caterpillars are shown. Suppose the peak cycles of these two insects coincided in 1998. What will be the next year in which they will coincide?

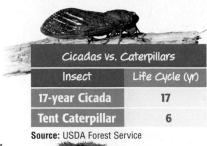

Cicadas vs. Caterpillars	
Insect	Life Cycle (yr)
17-year Cicada	17
Tent Caterpillar	6

Source: USDA Forest Service

SCHOOL For Exercises 27 and 28, use the following information.
Daniela and Tionna are on the school dance team. At practice, the dance team members always line up in even rows.

27. What is the least number of people needed to be able to line up in rows of 3, 4, or 5?

28. Describe the possible arrangements.

29. **PATTERNS** Some equivalent fractions for $\frac{5}{8}$ are $\frac{10}{16}$, $\frac{15}{24}$, $\frac{20}{32}$, and $\frac{25}{40}$. What pattern exists in the numerators and denominators?

30. **CRITICAL THINKING** Is the statement below *sometimes, always,* or *never* true? Give at least two examples to support your reasoning.
The LCM of two square numbers is another square number.

Standardized Test Practice and Mixed Review

31. **MULTIPLE CHOICE** Sahale decorated his house with two strands of holiday lights. The red lights strand blinks every 4 seconds, and the green lights strand blinks every 6 seconds. How many seconds will go by until both strands blink at the same time?

    Ⓐ 24 s   Ⓑ 20 s   Ⓒ 12 s   Ⓓ 8 s

32. **MULTIPLE CHOICE** Use the graph shown. If train A and train B both leave the station at 9:00 A.M., at what time will they next leave the station together?

    Ⓕ 9:24 A.M.   Ⓖ 9:48 A.M.

    Ⓗ 10:48 A.M.   Ⓘ 12:00 P.M.

Train Schedule	
Train	Departs
A	every 8 minutes
B	every 6 minutes

33. Write $\frac{47}{5}$ as a mixed number. (Lesson 5-3)

Replace each ● with a number so that the fractions are equivalent.
(Lesson 5-2)

34. $\frac{1}{5} = \frac{●}{25}$ 35. $\frac{3}{17} = \frac{9}{●}$ 36. $\frac{42}{48} = \frac{●}{8}$ 37. $\frac{3}{11} = \frac{●}{55}$

GETTING READY FOR THE NEXT LESSON

BASIC SKILL Choose the letter of the point that represents each fraction.

38. $\frac{1}{2}$ 39. $\frac{3}{4}$ 40. $\frac{2}{3}$

41. $\frac{1}{6}$ 42. $\frac{5}{6}$ 43. $\frac{1}{4}$

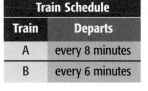

Comparing and Ordering Fractions

What You'll LEARN

Compare and order fractions.

NEW Vocabulary

least common denominator (LCD)

WHEN am I ever going to use this?

MARBLES Suppose one marble has a diameter of $\frac{5}{8}$ inch and another one has a diameter of $\frac{9}{16}$ inch.

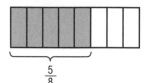 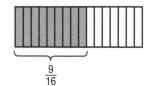

$\frac{5}{8}$ $\frac{9}{16}$

1. Use the models to determine which marble is larger.

To compare two fractions without using models, you can write them as fractions with the same denominator.

Key Concept   Compare Two Fractions

To compare two fractions,

- Find the **least common denominator (LCD)** of the fractions. That is, find the least common multiple of the denominators.
- Rewrite each fraction as an equivalent fraction using the LCD.
- Compare the numerators.

EXAMPLE   Compare Fractions

1 Replace ● with <, >, or = to make $\frac{5}{8}$ ● $\frac{9}{16}$ true.

- First, find the LCD; that is, the LCM of the denominators. The LCM of 8 and 16 is 16. So, the LCD is 16.

- Next, rewrite each fraction with a denominator of 16.

$$\frac{5}{8} = \frac{●}{16}, \text{ so } \frac{5}{8} = \frac{10}{16}. \qquad \frac{9}{16} = \frac{9}{16}$$

(×2)

- Then compare. Since $10 > 9$, $\frac{10}{16} > \frac{9}{16}$. So, $\frac{5}{8} > \frac{9}{16}$.

Your Turn Replace each ● with <, >, or = to make a true sentence.

a. $\frac{2}{3}$ ● $\frac{4}{9}$ b. $\frac{5}{12}$ ● $\frac{7}{8}$ c. $\frac{1}{2}$ ● $\frac{8}{16}$

You can use what you have learned about comparing fractions to order fractions.

EXAMPLE Order Fractions

2 Order the fractions $\frac{1}{2}$, $\frac{9}{14}$, $\frac{3}{4}$, and $\frac{5}{7}$ from least to greatest.

The LCD of the fractions is 28. So, rewrite each fraction with a denominator of 28.

$$\overset{\times 14}{\frac{1}{2} = \frac{\bullet}{28}}, \text{ so } \frac{1}{2} = \frac{14}{28}. \qquad \overset{\times 7}{\frac{3}{4} = \frac{\bullet}{28}}, \text{ so } \frac{3}{4} = \frac{21}{28}.$$

$$\overset{\times 2}{\frac{9}{14} = \frac{\bullet}{28}}, \text{ so } \frac{9}{14} = \frac{18}{28}. \qquad \overset{\times 4}{\frac{5}{7} = \frac{\bullet}{28}}, \text{ so } \frac{5}{7} = \frac{20}{28}.$$

The order of the fractions from least to greatest is $\frac{1}{2}$, $\frac{9}{14}$, $\frac{5}{7}$, $\frac{3}{4}$.

Your Turn Order the fractions from least to greatest.

d. $\frac{1}{2}$, $\frac{5}{6}$, $\frac{2}{3}$, $\frac{3}{5}$ e. $\frac{4}{5}$, $\frac{3}{4}$, $\frac{2}{5}$, $\frac{1}{4}$ f. $\frac{5}{6}$, $\frac{2}{3}$, $\frac{3}{5}$, $\frac{1}{5}$

Standardized tests often contain questions that involve comparing and ordering fractions in real-life situations.

Standardized Test Practice

EXAMPLE Compare and Order Fractions

3 MULTIPLE-CHOICE TEST ITEM
According to the survey data, what did most people say should be done with the penny?

  Ⓐ get rid of the penny

  Ⓑ undecided

  Ⓒ keep the penny

  Ⓓ cannot tell from the data

Read the Test Item You need to compare the fractions.

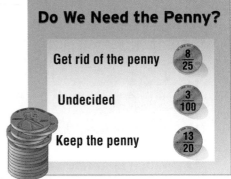

Do We Need the Penny?

Get rid of the penny $\frac{8}{25}$

Undecided $\frac{3}{100}$

Keep the penny $\frac{13}{20}$

Source: Coinstar

Solve the Test Item Rewrite the fractions with the LCD, 100.

$$\overset{\times 4}{\frac{8}{25} = \frac{32}{100}} \qquad \overset{\times 1}{\frac{3}{100} = \frac{3}{100}} \qquad \overset{\times 5}{\frac{13}{20} = \frac{65}{100}}$$

So, $\frac{13}{20}$ is the greatest fraction, and the answer is C.

Test-Taking Tip
The Princeton Review

Writing Equivalent Fractions Any common denominator can be used, but using the least common denominator usually makes the computation easier.

Skill and Concept Check

1. **Determine** the LCD of each pair of fractions.

   a. $\frac{1}{2}$ and $\frac{3}{8}$ b. $\frac{3}{5}$ and $\frac{3}{4}$ c. $\frac{5}{6}$ and $\frac{2}{9}$

2. **OPEN ENDED** Write two fractions whose LCD is 24.

3. **NUMBER SENSE** Match the fractions $\frac{3}{5}$, $\frac{9}{10}$, $\frac{3}{4}$, and $\frac{9}{20}$ to the models below. What is the order from least to greatest?

   a.    b. 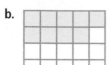   c. 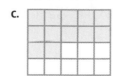   d.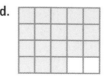

4. **NUMBER SENSE** How can you compare $\frac{1}{5}$ and $\frac{7}{8}$ without using their LCD?

GUIDED PRACTICE

Replace each ● with $<$, $>$, or $=$ to make a true sentence.

5. $\frac{3}{7}$ ● $\frac{1}{4}$ 6. $\frac{5}{7}$ ● $\frac{15}{21}$ 7. $\frac{9}{16}$ ● $\frac{5}{8}$

Order the fractions from least to greatest.

8. $\frac{4}{5}$, $\frac{1}{2}$, $\frac{9}{10}$, $\frac{3}{4}$ 9. $\frac{3}{8}$, $\frac{1}{4}$, $\frac{5}{6}$, $\frac{2}{3}$

Practice and Applications

Replace each ● with $<$, $>$, or $=$ to make a true sentence.

10. $\frac{1}{3}$ ● $\frac{3}{5}$ 11. $\frac{7}{8}$ ● $\frac{5}{6}$ 12. $\frac{6}{9}$ ● $\frac{2}{3}$ 13. $\frac{3}{4}$ ● $\frac{9}{16}$

14. $\frac{7}{12}$ ● $\frac{1}{2}$ 15. $\frac{14}{18}$ ● $\frac{7}{9}$ 16. $\frac{4}{5}$ ● $\frac{13}{15}$ 17. $\frac{5}{8}$ ● $\frac{20}{32}$

HOMEWORK HELP	
For Exercises	See Examples
10–21	1
22–27	2
28–30	3
Extra Practice See pages 604, 628.	

18. Which is less, *three-fifths* or *three-twentieths*?

19. Which is greater, *fifteen twenty-fourths* or *five-eighths*?

20. **TIME** What is shorter, $\frac{2}{3}$ of an hour or $\frac{8}{15}$ of an hour?

21. **MONEY** Which is more, $\frac{1}{2}$ of a dollar or $\frac{3}{4}$ of a dollar?

Order the fractions from least to greatest.

22. $\frac{1}{2}$, $\frac{2}{3}$, $\frac{1}{4}$, $\frac{5}{6}$ 23. $\frac{2}{3}$, $\frac{2}{9}$, $\frac{5}{6}$, $\frac{11}{18}$ 24. $\frac{1}{6}$, $\frac{2}{5}$, $\frac{3}{7}$, $\frac{3}{5}$ 25. $\frac{5}{8}$, $\frac{3}{4}$, $\frac{1}{2}$, $\frac{9}{16}$

26. Order the fractions $\frac{5}{8}$, $\frac{1}{6}$, $\frac{7}{8}$, and $\frac{3}{4}$ from greatest to least.

27. Write the fractions $\frac{1}{4}$, $\frac{3}{8}$, $\frac{5}{16}$, and $\frac{4}{5}$ in order from greatest to least.

28. **ART** Three paint brushes have widths of $\frac{3}{8}$-inch, $\frac{1}{2}$-inch, and $\frac{3}{4}$-inch. What is the measure of the brush with the greatest width?

29. **STATISTICS** The graphic shows the smelliest household odors. Do more people think pet odors or tobacco smoke is smellier?

What are the Smelliest Household Odors?

$\frac{7}{20}$ Pet Odors $\frac{1}{10}$ Cooking Odors $\frac{2}{5}$ Tobacco Smoke $\frac{9}{100}$ Other

Source: Aprilaire Fresh Air Exchangers

30. **TOOLS** Marissa is using three different-sized wood screws to build a cabinet. The sizes are $1\frac{1}{4}$ inch, $1\frac{5}{8}$ inch, and $1\frac{1}{2}$ inch. What is the size of the longest wood screws?

31. **CRITICAL THINKING** A fraction is in simplest form. It is not improper. Its numerator and denominator have a difference of two. The sum of the numerator and denominator is equal to a dozen. What is the fraction?

Standardized Test Practice and Mixed Review

32. **SHORT RESPONSE** Order the fractions $\frac{7}{9}$, $\frac{2}{3}$, $\frac{2}{6}$, $\frac{2}{5}$, and $\frac{4}{9}$ from greatest to least.

33. **MULTIPLE CHOICE** The table shows the fraction of drivers and passengers that are wearing seat belts in four states. Which state has the greatest fraction of drivers and passengers wearing seat belts?

   Ⓐ Arizona   Ⓑ Iowa

   Ⓒ Montana   Ⓓ Ohio

State	Drivers/Passengers Wearing Seat Belts
Arizona	$\frac{3}{4}$
Iowa	$\frac{39}{50}$
Montana	$\frac{19}{25}$
Ohio	$\frac{13}{20}$

Source: National Safety Council

Find the LCM of each set of numbers. (Lesson 5-4)

34. 5 and 13

35. 4 and 6

36. 6, 9, and 21

37. 8, 12, and 27

38. Express $5\frac{3}{8}$ as an improper fraction. (Lesson 5-3)

39. **ANIMALS** Sarah has four rabbits, three cats, two dogs, two parakeets, and one duck. Write a fraction in simplest form that compares the number of two-legged animals to the total number of animals. (Lesson 5-2)

GETTING READY FOR THE NEXT LESSON

PREREQUISITE SKILL Write each decimal in standard form. (Lesson 3-1)

40. seven tenths

41. four and six tenths

42. eighty-nine hundredths

43. twenty-five thousandths

Writing Decimals as Fractions

5-6

What You'll LEARN

Write decimals as fractions or mixed numbers in simplest form.

WHEN am I ever going to use this?

STATISTICS The graph shows the data results of a favorite summer treat survey.

1. Write the word form of the decimal that represents the amount of people who prefer Italian ice.

2. Write this decimal as a fraction.

3. Repeat Exercises 1 and 2 with each of the other decimals.

Favorite Summer Treats	
Treat	**People Who Prefer the Treat**
Ice cream	0.64
Italian ice	0.15
Popsicle	0.14
Other	0.07

Source: Opinion Research Corporation

Decimals like 0.64, 0.15, 0.14, and 0.07 can be written as fractions with denominators of 10, 100, 1,000, and so on.

Key Concept — Write Decimals as Fractions

To write a decimal as a fraction, you can follow these steps.
- Identify the place value of the last decimal place.
- Write the decimal as a fraction using the place value as the denominator.
- If necessary, simplify the fraction.

EXAMPLES — Write Decimals as Fractions

Write each decimal as a fraction in simplest form.

1 0.6

The place-value chart shows that the place value of the last decimal place is tenths. So, 0.6 means six tenths.

$$0.6 = \frac{6}{10} \quad \text{0.6 means six tenths.}$$

$$= \frac{\overset{3}{\cancel{6}}}{\underset{5}{\cancel{10}}} \quad \begin{array}{l}\text{Simplify. Divide the} \\ \text{numerator and denominator by the GCF, 2.}\end{array}$$

$$= \frac{3}{5}$$

Place-Value Chart

1,000	100	10	1	0.1	0.01	0.001	0.0001
thousands	hundreds	tens	ones	tenths	hundredths	thousandths	ten-thousandths
O	O	O	O .	6	O	O	O

STUDY TiP

Look Back To review **decimal place value**, see Lesson 3-1.

So, in simplest form, 0.6 is written as $\frac{3}{5}$.

Mental Math
Here are some commonly used decimal-fraction equivalencies:

$0.2 = \dfrac{1}{5}$

$0.25 = \dfrac{1}{4}$

$0.5 = \dfrac{1}{2}$

$0.75 = \dfrac{3}{4}$

2 0.45

$0.45 = \dfrac{45}{100}$ 0.45 means forty-five hundredths.

$= \dfrac{\cancel{45}\,9}{\cancel{100}\,20}$ Simplify. Divide by the GCF, 5.

$= \dfrac{9}{20}$

Place-Value Chart

1,000	100	10	1	0.1	0.01	0.001	0.0001
thousands	hundreds	tens	ones	tenths	hundredths	thousandths	ten-thousandths
O	O	O	O .	4	5	O	O

3 0.375

$0.375 = \dfrac{375}{1,000}$ 0.375 means three hundred seventy-five thousandths.

$= \dfrac{\cancel{375}\,3}{\cancel{1,000}\,8}$ Simplify. Divide by the GCF, 125.

$= \dfrac{3}{8}$

Place-Value Chart

1,000	100	10	1	0.1	0.01	0.001	0.0001
thousands	hundreds	tens	ones	tenths	hundredths	thousandths	ten-thousandths
O	O	O	O .	3	7	5	O

Your Turn Write each decimal as a fraction or mixed number in simplest form.

a. 0.8 **b.** 0.28 **c.** 0.125

d. 2.75 **e.** 5.12 **f.** 9.35

Decimals like 3.25, 26.82, and 125.54 can be written as mixed numbers in simplest form.

EXAMPLE **Write Decimals as Mixed Numbers**

4 FOOD According to the graphic, 8.26 billion pounds of cheese were produced in the United States in 2000. Write this amount as a mixed number in simplest form.

$8.26 = 8\dfrac{26}{100}$

$= 8\dfrac{\cancel{26}\,13}{\cancel{100}\,50}$ Simplify.

$= 8\dfrac{13}{50}$

So, $8.26 = 8\dfrac{13}{50}$.

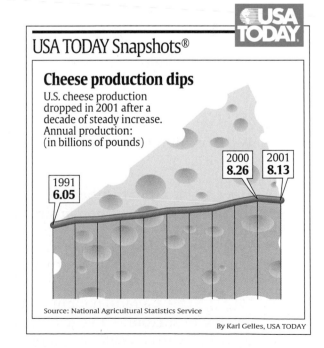

USA TODAY Snapshots®

Cheese production dips

U.S. cheese production dropped in 2001 after a decade of steady increase. Annual production: (in billions of pounds)

1991 **6.05**

2000 **8.26**

2001 **8.13**

Source: National Agricultural Statistics Service

By Karl Gelles, USA TODAY

Skill and Concept Check

Writing Math
Exercise 2

1. **OPEN ENDED** Write a decimal that, when written as a fraction in simplest form, has a denominator of 25.

2. **FIND THE ERROR** Miguel and Halley are writing 3.72 as a mixed number. Who is correct? Explain.

Miguel	Halley
$3.72 = 3\frac{72}{1,000}$ or $3\frac{9}{125}$	$3.72 = 3\frac{72}{100}$ or $3\frac{18}{25}$

GUIDED PRACTICE

Write each decimal as a fraction or mixed number in simplest form.

3. 0.4   4. 0.68   5. 0.525   6. 13.5

7. **WEATHER** The newspaper reported that it rained 1.25 inches last month. Write 1.25 as a mixed number in simplest form.

Practice and Applications

Write each decimal as a fraction or mixed number in simplest form.

8. 0.3   9. 0.7   10. 0.2   11. 0.5

12. 0.33   13. 0.21   14. 0.65   15. 0.82

16. 0.875   17. 0.425   18. 0.018   19. 0.004

20. 14.06   21. 7.08   22. 50.605   23. 65.234

24. fifty-two thousandths   25. thirteen and twenty-eight hundredths

26. **STOCK MARKET** One share of ABZ stock fell 0.08 point yesterday. Write 0.08 as a fraction in simplest form.

ROAD SIGNS For Exercises 27–29, use the road sign.

27. What fraction of a mile is each landmark from the exit?

28. How much farther is the zoo than the hotels? Write the distance as a fraction in simplest form.

29. **WRITE A PROBLEM** Write another problem that can be solved by using the road sign.

30. **MULTI STEP** Gerado bought 13.45 yards of fencing to surround a vegetable garden. He used 12.7 yards. Write the amount remaining as a fraction.

HOMEWORK HELP	
For Exercises	See Examples
8–19, 24, 26–28, 30, 32	1–3
20–23, 25, 31–32	4

Extra Practice
See pages 604, 628.

Zoo	0.8 mi ←
Camping	0.5 mi →
Hotels	0.2 mi ←

31. **MULTI STEP** Tamara ran 4.6 miles on Tuesday, 3.45 miles on Wednesday, and 3.75 miles on Friday. Write the total amount as a mixed number.

32. **BUTTERFLIES** The average wingspan of a western tailed-blue butterfly shown at the right is between 0.875 and 1.125 inches. Find two lengths that are within the given span. Write them as fractions in simplest form.

33. **RESEARCH** Use the Internet or another source to find the wingspan of one species of butterfly. How does the wingspan compare to the western tailed-blue butterfly?

34. **CRITICAL THINKING** Tell whether the following statement is *true* or *false*. Explain your reasoning.

Decimals like 0.8, 0.75 and 3.852 can be written as a fraction with a denominator that is divisible by 2 and 5.

Standardized Test Practice and Mixed Review

35. **MULTIPLE CHOICE** Write *eight hundred seventy-five thousandths* as a fraction.

    A $\frac{5}{13}$ **B** $\frac{5}{9}$ **C** $\frac{7}{8}$ **D** $\frac{11}{12}$

36. **SHORT RESPONSE** The distance between the Wisconsin cities of Sturgeon Bay and Sister Bay is about 27.125 miles. What fraction represents this distance?

Replace each ● with <, >, or = to make a true sentence. (Lesson 5-5)

37. $\frac{1}{3}$ ● $\frac{2}{7}$ 38. $\frac{5}{9}$ ● $\frac{6}{11}$ 39. $\frac{3}{5}$ ● $\frac{12}{20}$ 40. $\frac{4}{15}$ ● $\frac{8}{27}$

41. Find the LCM of 15, 20, and 25. (Lesson 5-4)

Find the mean for each set of data. (Lesson 2-6)

42. 14, 73, 25, 25, 53   43. 84, 88, 78, 79, 84, 84, 86, 83, 81

44. **STATISTICS** Choose an appropriate scale for a frequency table for the following set of data: 24, 67, 11, 9, 52, 38, 114, and 98. (Lesson 2-1)

45. **SCHOOL** The number of students at Aurora Junior High is half the number it was twenty years ago. If there were 758 students twenty years ago, how many students are there now? (Lesson 1-1)

GETTING READY FOR THE NEXT LESSON

PREREQUISITE SKILL Divide. (Lesson 4-3)

46. $2.0 \div 5$ 47. $3.0 \div 4$ 48. $1.0 \div 8$ 49. $1.0 \div 4$

Writing Fractions as Decimals

What You'll LEARN

Write fractions as terminating and repeating decimals.

NEW Vocabulary

terminating decimal
repeating decimal

Link to READING

Everyday meaning of **terminate:** end

WHEN am I ever going to use this?

GARDENING The graphic shows what fraction of home gardens in the U.S. grow each vegetable.

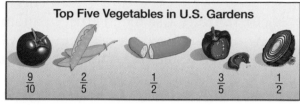

Top Five Vegetables in U.S. Gardens

$\frac{9}{10}$ $\frac{2}{5}$ $\frac{1}{2}$ $\frac{3}{5}$ $\frac{1}{2}$

Source: National Gardening Association

1. Write the decimal for $\frac{9}{10}$.

2. Write the fraction equivalent to $\frac{1}{2}$ with 10 as the denominator.

3. Write the decimal for $\frac{5}{10}$.

4. Write $\frac{3}{5}$ and $\frac{2}{5}$ as decimals.

Fractions with denominators of 10, 100, or 1,000 can be written as a decimal using place value. Any fraction can be written as a decimal using division. Decimals like 0.25 and 0.75 are **terminating decimals** because the division ends, or terminates, when the remainder is zero.

EXAMPLE Write Fractions as Terminating Decimals

1 Write $\frac{7}{8}$ as a decimal.

Method 1

Use paper and pencil.

$$\frac{7}{8} \rightarrow 8\overline{)7.000}$$

$$\begin{array}{r} 0.875 \\ 8\overline{)7.000} \\ -64 \\ \hline 60 \\ -56 \\ \hline 40 \\ -40 \\ \hline 0 \end{array}$$ Divide 7 by 8.

Therefore, $\frac{7}{8} = 0.875$.

Method 2

Use a calculator.

7 ÷ 8 ENTER 0.875

So, $\frac{7}{8} = 0.875$.

 Your Turn Write each fraction or mixed number as a decimal.

a. $\frac{1}{8}$ b. $\frac{1}{2}$ c. $1\frac{3}{5}$ d. $2\frac{3}{5}$

READING Math

Bar Notation The notation $0.\overline{8}$ indicates that the digit 8 repeats forever.

Not all decimals are terminating decimals. A decimal like 0.2222222 . . . is called a **repeating decimal** because the digits repeat. Other examples of repeating decimals are shown below.

$0.8888888 \ldots = 0.\overline{8}$	The digit 8 repeats.
$2.8787878 \ldots = 2.\overline{87}$	The digits 8 and 7 repeat.
$20.1939393 \ldots = 20.1\overline{93}$	The digits 9 and 3 repeat.

EXAMPLES **Write Fractions as Repeating Decimals**

2 Write $\frac{5}{11}$ as a decimal.

Method 1

Use paper and pencil.

$$\frac{5}{11} \rightarrow 11\overline{)5.0000} \quad \text{Divide 5 by 11.}$$

$$
\begin{array}{r}
0.4545 \\
11\overline{)5.0000} \\
-44 \\
\hline
60 \\
-55 \\
\hline
50 \\
-44 \\
\hline
60 \\
-55 \\
\hline
5 \\
\end{array}
$$
The pattern will continue.

Therefore, $\frac{5}{11} = 0.\overline{45}$.

Method 2

Use a calculator.

5 ÷ 11 ENTER = 0.45454545 . . .

So, $\frac{5}{11} = 0.\overline{45}$.

REAL-LIFE MATH

INSECTS There are about 9,000 different species of grasshoppers. The two-striped grasshopper shown has two dark brown stripes extending backward from its eyes. This grasshopper can measure up to $2\frac{1}{6}$ inches.

Source: www.enature.com

3 **INSECTS** Use the information at the left. Write the maximum length of the grasshopper as a decimal.

You can use a calculator to write $2\frac{1}{6}$ as a decimal.

2 + 1 ÷ 6 ENTER = 2.1666666 . . .

The maximum length of the grasshopper is $2.1\overline{6}$ inches.

Your Turn Write each fraction or mixed number as a decimal.

e. $\frac{4}{9}$ f. $\frac{7}{11}$ g. $3\frac{2}{3}$

Concept Summary   **Common Repeating Decimals**

The list at the right contains some commonly used fractions that are repeating decimals.

Fraction	Decimal	Fraction	Decimal
$\frac{1}{3}$	0.33333...	$\frac{1}{6}$	0.16666...
$\frac{2}{3}$	0.66666...	$\frac{5}{6}$	0.83333...

Skill and Concept Check

Writing Math
Exercises 2–4

1. **Write** each decimal using appropriate bar notation.
   a. 0.5555555 . . .
   b. 4.345345345 . . .
   c. 13.6767676 . . .
   d. 25.1545454 . . .

2. **Explain** how to write $\frac{5}{8}$ as a decimal.

3. **OPEN ENDED** Give an example of a terminating decimal and of a repeating decimal. Explain the difference between these two types of decimals.

4. **Which One Doesn't Belong?** Identify the fraction that is not equal to a repeating decimal. Explain your reasoning.

   $\frac{1}{9}$ $\frac{3}{11}$ $\frac{3}{20}$ $\frac{7}{12}$

GUIDED PRACTICE

Write each fraction or mixed number as a decimal.

5. $\frac{3}{4}$ 6. $\frac{2}{9}$ 7. $5\frac{4}{5}$ 8. $3\frac{4}{11}$

9. **MEASUREMENT** A half-ounce is $\frac{1}{16}$ of a cup. Write this measurement as a decimal.

Practice and Applications

Write each fraction or mixed number as a decimal.

10. $\frac{7}{10}$ 11. $\frac{13}{100}$ 12. $\frac{1}{4}$ 13. $\frac{9}{16}$

14. $\frac{8}{15}$ 15. $\frac{5}{12}$ 16. $\frac{11}{12}$ 17. $\frac{1}{11}$

18. $8\frac{1}{2}$ 19. $10\frac{9}{10}$ 20. $15\frac{3}{16}$ 21. $4\frac{9}{20}$

22. $8\frac{5}{6}$ 23. $6\frac{4}{15}$ 24. $9\frac{5}{11}$ 25. $12\frac{5}{9}$

26. $\frac{7}{1,000}$ 27. $\frac{25}{1,000}$ 28. $9\frac{8}{10,000}$ 29. $12\frac{32}{10,000}$

HOMEWORK HELP

For Exercises	See Examples
10–17, 26–27, 32–37	1, 2
18–25, 28–31, 38	3

Extra Practice
See pages 605, 628.

30. **SPORTS** Oleta ran $3\frac{2}{5}$ miles. What decimal does this represent?

31. **MEASUREMENT** The distance 12,540 feet is $2\frac{3}{8}$ miles. Write this mileage as a decimal.

32. Write *seven fourths* as a decimal.

33. What is the decimal equivalent of *eight thirds*?

Replace each ● with <, >, or = to make a true sentence.

34. $\frac{4}{9}$ ● $\frac{2}{5}$ **35.** $\frac{4}{5}$ ● $\frac{5}{6}$ **36.** $\frac{18}{25}$ ● $\frac{8}{11}$

37. Order $\frac{1}{6}, \frac{2}{9}, \frac{1}{5}, \frac{3}{20}$ from least to greatest.

38. Write $2\frac{2}{3}, 2\frac{6}{11}, 2\frac{5}{9}$, and $2\frac{1}{2}$ in order from greatest to least.

NATURE For Exercises 39–42, use the table that shows the sizes for different seashells found along the Atlantic coast.

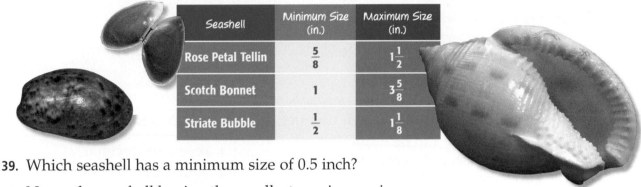

Seashell	Minimum Size (in.)	Maximum Size (in.)
Rose Petal Tellin	$\frac{5}{8}$	$1\frac{1}{2}$
Scotch Bonnet	1	$3\frac{5}{8}$
Striate Bubble	$\frac{1}{2}$	$1\frac{1}{8}$

39. Which seashell has a minimum size of 0.5 inch?

40. Name the seashell having the smallest maximum size.

41. Which seashell has the greatest minimum size?

42. **WRITE A PROBLEM** Write a problem that can be solved using the data in the table.

43. **CRITICAL THINKING** Look for a pattern in the prime factorization of the denominators of the fractions at the right. Write a rule you could use to determine whether a fraction in simplest form is a terminating decimal or a repeating decimal.

Fractions that are Terminating Decimals							
$\frac{1}{2}$	$\frac{3}{4}$	$\frac{2}{5}$	$\frac{5}{8}$	$\frac{7}{10}$	$\frac{5}{16}$	$\frac{9}{32}$	$\frac{17}{20}$

Standardized Test Practice and Mixed Review

44. **MULTIPLE CHOICE** It rained $\frac{11}{16}$ inch overnight. Which decimal represents this amount?

  Ⓐ 0.68   Ⓑ 0.6875   Ⓒ $0.\overline{6875}$ Ⓓ $0.6\overline{8}$

45. **MULTIPLE CHOICE** Which decimal represents the shaded portion of the figure shown at the right?

  Ⓕ 0.2   Ⓖ 0.202   Ⓗ 0.22   Ⓘ $0.\overline{2}$

Write each decimal as a fraction or mixed number in simplest form. (Lesson 5-6)

46. 0.25   **47.** 0.73   **48.** 8.118   **49.** 11.14

50. Which fraction is greater, $\frac{13}{40}$ or $\frac{3}{7}$? (Lesson 5-5)

Vocabulary and Concept Check

common multiples (p. 194)
equivalent fraction (p. 182)
greatest common factor (GCF) (p. 177)
improper fraction (p. 186)
least common denominator (LCD) (p. 198)
least common multiple (LCM) (p. 194)

mixed number (p. 186)
multiple (p. 194)
repeating decimal (p. 207)
simplest form (p. 183)
terminating decimal (p. 206)
Venn diagram (p. 177)

Choose the letter of the term or number that best matches each phrase.

1. a decimal that can be written as a fraction with a denominator of 10, 100, 1,000 and so on
2. the least common multiple of a set of denominators
3. fractions that name the same number
4. the least number other than 0 that is a multiple of two or more whole numbers
5. a decimal such as 0.8888888…
6. a fraction whose value is greater than 1

a. repeating decimal
b. equivalent fractions
c. terminating decimal
d. least common multiple
e. improper fraction
f. least common denominator

Lesson-by-Lesson Exercises and Examples

 Greatest Common Factor (pp. 177–180)

Find the GCF of each set of numbers.
7. 15 and 18   8. 30 and 36
9. 28 and 70   10. 26, 52, and 65

11. **CANDY** A candy shop has 45 candy canes, 72 lollipops, and 108 chocolate bars. The candy must be placed in bags so that each bag contains the same number of each type of candy. What is the greatest amount of each type of candy that can be put in each bag?

Example 1 Find the GCF of 36 and 54.

To find the GCF, you can make a list.

Factors of 36	Factors of 54
1, 36	1, 54
2, 18	2, 27
3, 12	3, 18
4, 9	6, 9
6, 6	

The common factors of 36 and 54 are 1, 2, 3, 6, 9, and 18.
So, the GCF of 36 and 54 is 18.

5-2 Simplifying Fractions (pp. 182–185)

Replace each ● with a number so that the fractions are equivalent.

12. $\dfrac{2}{3} = \dfrac{●}{24}$ 13. $\dfrac{5}{8} = \dfrac{35}{●}$

14. $\dfrac{●}{6} = \dfrac{12}{24}$ 15. $\dfrac{7}{●} = \dfrac{63}{81}$

Write each fraction in simplest form. If the fraction is already in simplest form, write *simplest form*.

16. $\dfrac{21}{24}$ 17. $\dfrac{15}{80}$

18. $\dfrac{14}{23}$ 19. $\dfrac{42}{98}$

Example 2 Replace the ● with a number so that $\dfrac{4}{9}$ and $\dfrac{●}{27}$ are equivalent.

Since $9 \times 3 = 27$, multiply the numerator and denominator by 3.

$$\overset{\times 3}{\dfrac{4}{9}} = \dfrac{●}{27} \quad \text{So, } \dfrac{4}{9} = \dfrac{12}{27}.$$

5-3 Mixed Numbers and Improper Fractions (pp. 186–189)

Write each mixed number as an improper fraction.

20. $3\dfrac{1}{4}$ 21. $5\dfrac{3}{8}$

22. $2\dfrac{5}{6}$ 23. $7\dfrac{2}{9}$

Write each improper fraction as a mixed number.

24. $\dfrac{23}{4}$ 25. $\dfrac{48}{9}$

26. $\dfrac{38}{7}$ 27. $\dfrac{49}{2}$

28. **SNACKS** Akia bought six $\dfrac{1}{4}$-cup packages of peanuts. How many cups of peanuts does she have?

Example 3 Write $4\dfrac{2}{5}$ as an improper fraction.

$$4\dfrac{2}{5} = \dfrac{(4 \times 5) + 2}{5} = \dfrac{22}{5}$$

> Multiply the whole number and denominator. Then add the numerator.

Example 4 Write $\dfrac{49}{6}$ as a mixed number.

Divide 49 by 6.

$$6\overline{)49}\,\,\,8\dfrac{1}{6}$$
$$\underline{-48}$$
$$1$$

So, $\dfrac{49}{6}$ can be written as $8\dfrac{1}{6}$.

5-4 Least Common Multiple (pp. 194–197)

Find the LCM of each set of numbers.

29. 10 and 25   30. 28 and 35

31. 12 and 18   32. 6 and 8

33. Find the LCM of 8, 12, and 16.

34. What is the LCM of 12, 15, and 20?

Example 5 Find the LCM of 8 and 18.

multiples of 8:   8, 16, 24, 32, 40, 48, 56, 64, 72, …

multiples of 18:   18, 36, 54, 72, …

So, the LCM of 8 and 18 is 72.

5-5 Comparing and Ordering Fractions (pp. 198–201)

Replace each ● with <, >, or = to make a true sentence.

35. $\frac{2}{5}$ ● $\frac{4}{9}$

36. $\frac{12}{15}$ ● $\frac{4}{5}$

37. $\frac{3}{8}$ ● $\frac{4}{10}$

38. $\frac{7}{12}$ ● $\frac{5}{9}$

Order the fractions from least to greatest.

39. $\frac{2}{3}, \frac{3}{4}, \frac{1}{2}, \frac{5}{9}$

40. $\frac{7}{12}, \frac{5}{8}, \frac{5}{6}, \frac{3}{4}$

41. **BAKING** Which is more, $\frac{2}{3}$ cup of flour or $\frac{5}{6}$ cup of flour?

Example 6 Replace ● with <, >, or = to make $\frac{2}{5}$ ● $\frac{3}{8}$ true.

First, find the LCD. The LCM of 5 and 8 is 40. So, the LCD is 40.

Next, rewrite both fractions with a denominator of 40.

$\frac{2}{5} \overset{\times 8}{=} \frac{16}{40}$ and $\frac{3}{8} \overset{\times 3}{=} \frac{15}{40}$

Since $16 > 15$, $\frac{16}{40} > \frac{15}{40}$. So, $\frac{2}{5} > \frac{3}{8}$.

5-6 Writing Decimals as Fractions (pp. 202–205)

Write each decimal as a fraction or mixed number in simplest form.

42. 0.9

43. 0.35

44. 0.72

45. 0.125

46. 3.006

47. 9.315

48. 2.64

49. 0.048

50. **PETS** Michael's dog weighs 8.75 pounds. Write 8.75 as a mixed number in simplest form.

Example 7 Write 0.85 as a fraction in simplest form.

$0.85 = \frac{85}{100}$ 0.85 means 85 hundredths.

$= \frac{\overset{17}{\cancel{85}}}{\underset{20}{\cancel{100}}}$ Simplify. Divide the numerator and denominator by the GCF, 5.

$= \frac{17}{20}$

So, 0.85 can be written as $\frac{17}{20}$.

5-7 Writing Fractions as Decimals (pp. 206–209)

Write each fraction or mixed number as a decimal.

51. $\frac{7}{8}$

52. $\frac{9}{15}$

53. $\frac{5}{12}$

54. $4\frac{2}{9}$

55. $12\frac{3}{4}$

56. $8\frac{9}{16}$

57. **INSECTS** Carpenter bees generally measure $\frac{3}{4}$ inch long. Write this length as a decimal.

Example 8 Write $\frac{5}{8}$ as a decimal.

$$
\begin{array}{r}
0.625 \\
8\overline{)5.000} \quad \text{Divide 5 by 8.} \\
-\,48 \\
\hline
20 \\
-\,16 \\
\hline
40 \\
-\,40 \\
\hline
0
\end{array}
$$

So, $\frac{5}{8} = 0.625$.

Vocabulary and Concepts

1. **OPEN ENDED** Write two fractions whose least common denominator is 20.

2. **Define** *least common multiple* of two numbers.

Skills and Applications

Find the GCF of each set of numbers.

3. 27 and 45

4. 16 and 24

5. 24, 48, and 84

6. **MARBLES** Ben has 6 red marbles, 5 yellow marbles, 4 blue marbles, and 9 green marbles. Write a fraction in simplest form that compares the number of red marbles to the total number of marbles in Ben's collection.

Write each mixed number as an improper fraction.

7. $2\frac{5}{7}$

8. $4\frac{2}{3}$

Find the LCM of each set of numbers.

9. 8 and 12

10. 6 and 15

11. 4, 9, and 18

12. **FOOD** Which is more, $\frac{3}{8}$ of a pizza or $\frac{1}{3}$ of a pizza?

Write each decimal as a fraction or mixed number in simplest form.

13. 0.7

14. 0.32

15. 4.008

Write each fraction or mixed number as a decimal.

16. $\frac{3}{5}$

17. $\frac{9}{16}$

18. $8\frac{2}{3}$

Standardized Test Practice

(A) (B) (C) (D)

19. **MULTIPLE CHOICE** Which fraction represents *two hundred seventy-five thousandths* in simplest form?

(A) $\frac{7}{9}$

(B) $\frac{29}{52}$

(C) $\frac{5}{13}$

(D) $\frac{11}{40}$

20. **MULTIPLE CHOICE** Which list of fractions is ordered from least to greatest?

(F) $\frac{5}{8}, \frac{2}{3}, \frac{3}{5}$

(G) $\frac{2}{3}, \frac{5}{8}, \frac{3}{5}$

(H) $\frac{5}{8}, \frac{3}{5}, \frac{2}{3}$

(I) $\frac{3}{5}, \frac{5}{8}, \frac{2}{3}$

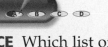

msmath1.net/chapter_test

PART 1 Multiple Choice

Record your answers on the answer sheet provided by your teacher or on a sheet of paper.

1. Refer to the table below. Which list orders the lengths of the movies from greatest to least? (Prerequisite Skill, p. 588)

Movie	Length (min)
1	68
2	72
3	74
4	66

(A) 1, 2, 3, 4
(B) 2, 3, 4, 1
(C) 3, 2, 1, 4
(D) 4, 1, 2, 3

2. Which operation should be done first in the expression $3 + 6 - 4 \div 2 \times 3$? (Lesson 1-5)

(F) $3 + 6$
(G) 2×3
(H) $6 - 4$
(I) $4 \div 2$

3. The frequency table shows which sports the students in a classroom prefer. Which statement correctly describes the data? (Lesson 2-1)

Sport	Tally	Frequency
baseball	ЖЖ III	8
basketball	ЖЖ II	7
football	IIII	4
hockey	II	2

(A) Twice as many students prefer football as hockey.
(B) More students prefer basketball than baseball.
(C) The least preferred sport is football.
(D) There were 20 students who participated in the survey.

4. Q Brand jeans come in six styles, priced at $25, $15, $50, $75, $35, and $40. What is the mean price of the six styles of jeans? (Lesson 2-6)

(F) $37.50
(G) $40
(H) $50
(I) $240

5. Which pair of numbers are equivalent? (Lesson 3-1)

(A) 0.06 and $\frac{1}{10}$
(B) 0.06 and $\frac{6}{10}$
(C) 0.6 and $\frac{1}{10}$
(D) 0.6 and $\frac{6}{10}$

6. Which number is the greatest common factor of 18 and 24? (Lesson 5-1)

(F) 4
(G) 6
(H) 18
(I) 24

7. Sheldon has a total of 200 marbles. Of these, 50 are red. What fraction of Sheldon's marbles are *not* red? (Lesson 5-2)

(A) $\frac{1}{25}$
(B) $\frac{1}{40}$
(C) $\frac{1}{4}$
(D) $\frac{3}{4}$

8. For which of the following figures is the greatest fraction of its area *not* shaded? (Lesson 5-5)

(F)
(G)
(H)
(I)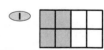

TEST-TAKING TIP

Question 6 Use the answer choices to help find a solution. To find the GCF, divide 18 and 24 by each possible answer choice. The greatest value that divides evenly into both numbers is the solution.

PART 2   Short Response/Grid In

Record your answers on the answer sheet provided by your teacher, or on a sheet of paper.

9. Terri planted a garden made up of two small squares. The area of a square can be found using the expression s^2, where s is the length of one side. What is the total area of Terri's garden, in square units? (Lesson 1-8)

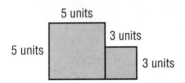

5 units

5 units

3 units

3 units

10. The restaurant manager placed an order for 15.2 pounds of ham, 6.8 pounds of turkey, and 8.4 pounds of roast beef. Find the total amount of meat ordered. (Lesson 3-5)

11. What number is the greatest common factor of 16 and 40? (Lesson 5-1)

12. Which is greater, $\frac{1}{5}$ or $\frac{1}{2}$? (Lesson 5-5)

13. Mrs. Cardona bought wood in the following lengths.

Type of Wood	Amount (ft)
Balsam	$6\frac{1}{5}$
Cedar	$6\frac{2}{3}$
Oak	$6\frac{7}{15}$
Pine	$6\frac{11}{30}$

Which type of wood did she purchase in the *greatest* amount? (Lesson 5-5)

14. The Hope Diamond is a rare, deep blue diamond that weighs 45.52 carats. Write this weight as a mixed number in simplest form. (Lesson 5-6)

15. Write the mixed number in simplest form that represents $20.25. (Lesson 5-6)

16. To write the decimal equivalent of $\frac{1}{8}$, should 8 be divided by 1 or should 1 be divided by 8? (Lesson 5-7)

17. What is the decimal equivalent of $\frac{2}{5}$? (Lesson 5-7)

18. A whippet is a type of a greyhound. It can run $35\frac{1}{2}$ miles per hour. What decimal does this represent? (Lesson 5-7)

19. The table shows the lengths of certain insects. Which insect's length is less than $\frac{1}{2}$ inch? (Lesson 5-7)

Insect	Length (in.)
Hornet	$\frac{5}{8}$
Queen Honey Bee	$\frac{3}{4}$
Lady Beetle	$\frac{1}{4}$

PART 3   Extended Response

Record your answers on a sheet of paper. Show your work.

20. Copy the hexagon and the rectangle. Both shapes have the same area. (Lessons 5-2 and 5-6)

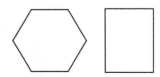

a. Shade $\frac{1}{3}$ of the hexagon.

b. Shade 0.25 of the rectangle.

c. Which figure has the greater shaded area? Explain your answer.

6 Adding and Subtracting Fractions

"How do you use math when you cook?"

The recipe for your favorite chocolate chip cookies is written using fractions of cups and teaspoons. If you were making two batches of cookies, you might estimate with fractions to see whether you have enough of each ingredient.

You will solve problems about fractions and cooking in Lessons 6-1 and 6-2.

GETTING STARTED

Take this quiz to see whether you are ready to begin Chapter 6. Refer to the lesson number in parentheses if you need more review.

▶ **Vocabulary Review**

State whether each sentence is *true* or *false*. If *false*, replace the underlined word to make a true sentence.

1. The fraction $\frac{13}{4}$ is an example of an <u>improper</u> fraction. (Lesson 5-3)

2. LCD stands for least common <u>divisor</u>. (Lesson 5-5)

▶ **Prerequisite Skills**

Estimate using rounding. (Lesson 3-4)

3. $1.2 + 6.6$

4. $9.6 - 2.3$

5. $8.25 - 4.8$

6. $5.85 + 7.1$

7. $10.3 + 9.9$

8. $8.6 - 2.9$

Write each fraction in simplest form.
(Lesson 5-2)

9. $\frac{3}{18}$

10. $\frac{21}{28}$

11. $\frac{16}{40}$

12. $\frac{6}{38}$

Write each improper fraction as a mixed number. (Lesson 5-3)

13. $\frac{11}{10}$

14. $\frac{14}{5}$

15. $\frac{7}{5}$

16. $\frac{15}{9}$

17. $\frac{25}{20}$

18. $\frac{26}{12}$

Find the least common denominator of each pair of fractions. (Lesson 5-5)

19. $\frac{1}{3}$ and $\frac{4}{9}$

20. $\frac{3}{8}$ and $\frac{2}{9}$

Fractions Make this Foldable to help you organize your notes. Begin with two sheets of $8\frac{1}{2}"\times 11"$ paper, four index cards, and glue.

STEP 1 **Fold**
Fold one sheet in half widthwise.

STEP 2 **Open and Fold Again**
Fold the bottom to form a pocket. Glue edges.

STEP 3 **Repeat Steps 1 and 2**
Glue the back of one piece to the front of the other to form a booklet.

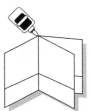

STEP 4 **Label**
Label each left-hand pocket *What I Know* and each right-hand pocket *What I Need to Know*. Place an index card in each pocket.

Reading and Writing Write examples of what you know and what you need to know about adding fractions on the first two index cards. On the remaining index cards, repeat for subtracting fractions.

Rounding Fractions

What You'll LEARN

Round fractions to 0, $\frac{1}{2}$, and 1.

Materials

• grid paper
• colored paper

In Lesson 3-3, you learned to round decimals. You can use a similar method to round fractions.

ACTIVITY *Work with a partner.*

Round each fraction to the nearest half.

① $\frac{4}{20}$ **②** $\frac{4}{10}$ **③** $\frac{4}{5}$

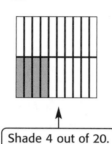

Shade 4 out of 20.

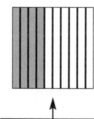

Shade 4 out of 10.

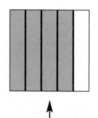

Shade 4 out of 5.

Very few sections are shaded. So, $\frac{4}{20}$ rounds to 0.

About one half of the sections are shaded. So, $\frac{4}{10}$ rounds to $\frac{1}{2}$.

Almost all of the sections are shaded. So, $\frac{4}{5}$ rounds to 1.

Your Turn **Round each fraction to the nearest half.**

a. $\frac{13}{20}$ b. $\frac{98}{100}$ c. $\frac{9}{10}$ d. $\frac{1}{5}$ e. $\frac{37}{50}$

f. $\frac{2}{25}$ g. $\frac{6}{10}$ h. $\frac{17}{20}$ i. $\frac{1}{8}$ j. $\frac{28}{50}$

Writing Math

1. Sort the fractions in Exercises a–j into three groups: those that round to 0, those that round to $\frac{1}{2}$, and those that round to 1.

2. Compare the numerators and denominators of the fractions in each group. **Make a conjecture** about how to round any fraction to the nearest half.

3. Test your conjecture by repeating Exercise 1 using the fractions $\frac{3}{5}$, $\frac{3}{17}$, $\frac{16}{20}$, $\frac{2}{13}$, $\frac{6}{95}$, $\frac{7}{15}$, $\frac{7}{9}$, and $\frac{9}{11}$.

6-1 Rounding Fractions and Mixed Numbers

What You'll LEARN

Round fractions and mixed numbers.

REVIEW Vocabulary

mixed number: the sum of a whole number and a fraction (Lesson 5-3)

HANDS-ON Mini Lab

Work with a partner.

The line segment is about $1\frac{7}{8}$ inches long. The segment is between $1\frac{1}{2}$ and 2 inches long. It is closer to 2 inches. So, to the nearest half inch, the length of the segment is 2 inches.

Materials

- inch ruler
- pencil
- paper

STEP 1 Draw five short line segments. Exchange segments with your partner.

STEP 2 Measure each segment in eighths and record its length. Then measure to the nearest half inch and record.

STEP 3 Sort the measures into three columns, those that round up to the next greater whole number, those that round to a half inch, and those that round down to the smaller whole number.

1. Compare the numerators and denominators of the fractions in each list. How do they compare?

2. Write a rule about how to round to the nearest half inch.

To round to the nearest half, you can use these guidelines.

Concept Summary — Rounding to the Nearest Half

Round Up	**Round to $\frac{1}{2}$**	**Round Down**
If the numerator is almost as large as the denominator, round the number up to the next whole number.	If the numerator is about half of the denominator, round the fraction to $\frac{1}{2}$.	If the numerator is much smaller than the denominator, round the number down to the whole number.
Example	**Example**	**Example**
$\frac{7}{8}$ rounds to 1.	$2\frac{3}{8}$ rounds to $2\frac{1}{2}$.	$\frac{1}{8}$ rounds to 0.
7 is almost as large as 8.	3 is about half of 8.	1 is much smaller than 8.

EXAMPLES **Round to the Nearest Half**

1 **Round $3\frac{1}{6}$ to the nearest half.**

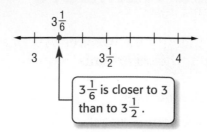

$3\frac{1}{6}$ is closer to 3 than to $3\frac{1}{2}$.

The numerator of $\frac{1}{6}$ is much smaller than the denominator. So, $3\frac{1}{6}$ rounds to 3.

2 **Find the length of the segment to the nearest half inch.**

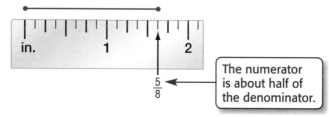

The numerator is about half of the denominator.

To the nearest half inch, $1\frac{5}{8}$ rounds to $1\frac{1}{2}$.

To the nearest half inch, the segment length is $1\frac{1}{2}$ inches.

Your Turn **Round each number to the nearest half.**

a. $8\frac{1}{12}$　　　　b. $2\frac{9}{10}$　　　　c. $\frac{2}{9}$

d. $\frac{5}{12}$　　　　e. $1\frac{2}{5}$　　　　f. $4\frac{3}{7}$

Sometimes you should round a number down when it is better for a measure to be too small than too large. Other times you should round up despite what the rule says.

EXAMPLE **Use Rounding to Solve a Problem**

3 **COOKING** A recipe for tacos calls for $1\frac{1}{4}$ pounds of ground beef. Should you buy a $1\frac{1}{2}$-pound package or a 1-pound package?

One pound is less than $1\frac{1}{4}$ pounds. So, in order to have enough ground beef, you should round up and buy the $1\frac{1}{2}$-pound package.

msmath1.net/extra_examples

Skill and Concept Check

Writing
Math

Exercise 2

1. **OPEN ENDED** Describe a situation where it would make sense to round a fraction up to the nearest unit.

2. **Which One Doesn't Belong?** Identify the number that does not round to the same number as the other three. Explain.

$$5\frac{1}{5} \qquad 4\frac{5}{6} \qquad 5\frac{6}{7} \qquad 4\frac{11}{12}$$

GUIDED PRACTICE

Round each number to the nearest half.

3. $\frac{7}{8}$ 4. $3\frac{1}{10}$ 5. $\frac{3}{8}$ 6. $6\frac{2}{3}$ 7. $\frac{1}{5}$

NUMBER SENSE Tell whether each number should be rounded up or down. Explain your reasoning.

8. the time needed to get to school

9. the width of a piece of paper that will fit in an $8\frac{1}{2}$-inch wide binder pocket

10. **MEASUREMENT** Find the length of the rubber band to the nearest half inch.

Practice and Applications

Round each number to the nearest half.

11. $\frac{5}{6}$ 12. $2\frac{4}{5}$ 13. $4\frac{2}{9}$ 14. $9\frac{1}{6}$ 15. $3\frac{2}{9}$

16. $3\frac{1}{12}$ 17. $6\frac{11}{12}$ 18. $8\frac{6}{7}$ 19. $2\frac{2}{11}$ 20. $6\frac{1}{7}$

21. $\frac{1}{3}$ 22. $5\frac{3}{10}$ 23. $\frac{7}{12}$ 24. $3\frac{2}{3}$ 25. $4\frac{4}{9}$

26. $6\frac{2}{5}$ 27. $\frac{13}{16}$ 28. $6\frac{5}{16}$ 29. $9\frac{7}{24}$ 30. $4\frac{19}{32}$

HOMEWORK HELP

For Exercises	See Examples
11–30	1
31–35	3
36–38	2

Extra Practice
See pages 605, 629.

NUMBER SENSE Tell whether each number should be rounded up or down. Explain your reasoning.

31. the amount of air to put into a balloon

32. the number of rolls of wallpaper to buy for your room

33. the size of box you need for a gift that is $14\frac{3}{8}$ inches tall

34. the width of blinds to fit inside a window opening that is $24\frac{3}{4}$ inches wide

35. the height of shelves that will fit in a room with an $8\frac{1}{4}$-foot ceiling

Find the length of each item to the nearest half inch.

36.

37.

38. **CRAFTS** Marina is making birthday cards. She is using envelopes that are $6\frac{3}{4}$ inches by $4\frac{5}{8}$ inches. To the nearest half inch, how large can she make her cards?

SCHOOL For Exercises 39 and 40, use the histogram. It shows the results of a survey of 49 students about the number of board games they own.

A *histogram* uses bars to represent the frequency of numerical data organized in intervals.

39. Write a fraction to indicate the part of students in each interval.

40. Are more than half of the students represented by any one interval?

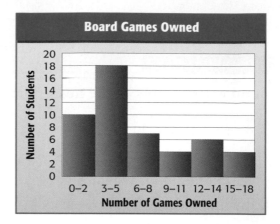

41. **CRITICAL THINKING** Name three mixed numbers that round to $4\frac{1}{2}$.

Standardized Test Practice and Mixed Review

42. **MULTIPLE CHOICE** What is $6\frac{4}{7}$ rounded to the nearest half?

Ⓐ 6   Ⓑ $6\frac{1}{2}$ Ⓒ 7   Ⓓ $7\frac{1}{2}$

43. **MULTIPLE CHOICE** A punch bowl holds $4\frac{13}{16}$ gallons. Round the mixed number to the nearest half.

Ⓕ 4   Ⓖ $4\frac{1}{2}$ Ⓗ 5   Ⓘ $5\frac{1}{2}$

Write each fraction or mixed number as a decimal. (Lesson 5-7)

44. $\frac{2}{3}$ 45. $\frac{1}{6}$ 46. $2\frac{5}{11}$ 47. $6\frac{1}{12}$

48. **EARTH SCIENCE** Mercury moves in its orbit at a speed of 29.75 miles per second. Write this speed as a mixed number in simplest form. (Lesson 5-6)

GETTING READY FOR THE NEXT LESSON

PREREQUISITE SKILL Estimate using rounding. (Lesson 3-4)

49. $0.8 + 0.9$ 50. $5.75 + 8.2$ 51. $2.2 + 6.8 + 3.1$

52. $1.8 - 0.9$ 53. $10.02 - 4.25$ 54. $6.2 - 3.852$

Estimating Sums and Differences

What You'll LEARN

Estimate sums and differences of fractions and mixed numbers.

WHEN am I ever going to use this?

WORLD RECORDS The table lists a few world records written as fractions.

1. To the nearest whole number, how tall is the largest cowboy boot?

2. To the nearest whole number, how tall is the largest tricycle wheel?

3. About how much taller is the tricycle wheel than the cowboy boots?

World Records	
tallest snowman	$113\frac{2}{3}$ ft
largest cowboy boots	$4\frac{7}{12}$ ft
largest tricycle wheel	$15\frac{1}{4}$ ft

Source: guinnessworldrecords.com

You can estimate sums and differences of fractions by rounding each fraction to the nearest half. Then add or subtract.

EXAMPLES Fraction Sums and Differences

1 Estimate $\frac{5}{8} + \frac{11}{12}$.

$\boxed{\frac{5}{8} \text{ rounds to } \frac{1}{2}.}$ $\boxed{\frac{11}{12} \text{ rounds to } 1.}$

$\frac{5}{8} + \frac{11}{12}$ is about $\frac{1}{2} + 1$ or $1\frac{1}{2}$.

2 Estimate $\frac{7}{12} - \frac{1}{6}$.

$\boxed{\frac{7}{12} \text{ rounds to } \frac{1}{2}.}$ $\boxed{\frac{1}{6} \text{ rounds to } 0.}$

$\frac{7}{12} - \frac{1}{6}$ is about $\frac{1}{2} - 0$ or $\frac{1}{2}$.

Your Turn Estimate.

a. $\frac{2}{5} + \frac{7}{8}$ b. $\frac{5}{12} + \frac{4}{9}$ c. $\frac{9}{10} - \frac{4}{9}$ d. $\frac{5}{8} - \frac{3}{5}$

You can estimate sums and differences of mixed numbers by rounding each number to the nearest whole number.

EXAMPLE Mixed Number Sum

3 Estimate $4\frac{5}{6} + 2\frac{1}{5}$.

$4\frac{5}{6}$ rounds to 5. $2\frac{1}{5}$ rounds to 2.

Estimate $5 + 2 = 7$

So, $4\frac{5}{6} + 2\frac{1}{5}$ is about 7.

EXAMPLE **Mixed Number Difference**

4 Estimate $2\frac{5}{8} - \frac{2}{5}$.

Round $2\frac{5}{8}$ to $2\frac{1}{2}$. Round $\frac{2}{5}$ to $\frac{1}{2}$.

Estimate $2\frac{1}{2} - \frac{1}{2} = 2$

So, $2\frac{5}{8} - \frac{2}{5}$ is about 2.

Your Turn Estimate.

e. $3\frac{1}{8} + 1\frac{5}{6}$
f. $2\frac{5}{6} - 1\frac{7}{8}$
g. $4\frac{5}{9} + \frac{1}{3}$

Sometimes you need to round all fractions up.

STUDY TIP

Look Back You can review **perimeter** in Lesson 4-5.

EXAMPLE **Round Up to Solve a Problem**

5 **DECORATING** Carmen plans to put a wallpaper border around the top of her room. The room is $9\frac{1}{4}$ feet wide and $12\frac{1}{6}$ feet long. About how much border does she need?

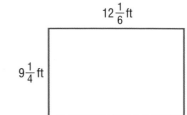

Carmen wants to make sure she buys enough border. So, she rounds up.
Round $9\frac{1}{4}$ to 10 and $12\frac{1}{6}$ to 13.

Estimate $10 + 13 + 10 + 13 = 46$ Add the lengths of all four sides.

So, Carmen needs about 46 feet of border.

Skill and Concept Check

Writing Math

Exercises 1 & 2

1. **OPEN ENDED** Write a problem in which you would need to estimate the difference between $4\frac{3}{4}$ and $2\frac{1}{6}$.

2. **NUMBER SENSE** Without calculating, replace ● with < or > to make a true sentence. Explain your reasoning.

   a. $\frac{3}{8} + \frac{9}{10}$ ● 1
   b. $2\frac{1}{6} + 4\frac{7}{8}$ ● 6
   c. 2 ● $3\frac{9}{10} - 1\frac{1}{5}$

GUIDED PRACTICE

Estimate.

3. $\frac{7}{8} - \frac{2}{5}$
4. $\frac{9}{10} + \frac{3}{5}$
5. $4\frac{1}{8} + 5\frac{11}{12}$
6. $6\frac{5}{8} - 2\frac{1}{3}$

7. **BAKING** A cookie recipe uses $3\frac{1}{4}$ cups of flour. A bread recipe uses $2\frac{2}{3}$ cups of flour. About how much flour is needed for both?

Practice and Applications

Estimate.

8. $\dfrac{5}{8} + \dfrac{11}{12}$

9. $\dfrac{9}{10} + \dfrac{1}{2}$

10. $\dfrac{3}{8} - \dfrac{1}{9}$

11. $\dfrac{5}{6} - \dfrac{3}{8}$

12. $1\dfrac{5}{9} - \dfrac{3}{7}$

13. $1\dfrac{9}{10} - 1\dfrac{1}{8}$

14. $8\dfrac{7}{8} + 2\dfrac{1}{3}$

15. $4\dfrac{1}{3} + 1\dfrac{5}{6}$

16. $7\dfrac{1}{6} - \dfrac{7}{9}$

17. $3\dfrac{2}{3} - 1\dfrac{7}{8}$

18. $4\dfrac{1}{10} + 3\dfrac{4}{5} + 7\dfrac{2}{3}$

19. $6\dfrac{1}{4} + 2\dfrac{5}{6} + 3\dfrac{3}{8}$

HOMEWORK HELP

For Exercises	See Examples
8–11	1, 2
12–20	3, 4
21–22	5

Extra Practice
See pages 605, 629.

20. **GEOMETRY** Estimate the perimeter of the rectangle.

$12\dfrac{7}{8}$ in.

$2\dfrac{1}{4}$ in.

21. **FRAMING** Luis wants to make a square frame that is $5\dfrac{3}{4}$ inches on each side. About how much framing material should he buy?

22. **MULTI STEP** Nina needs $1\dfrac{1}{4}$ yards of ribbon to trim a pillow and $4\dfrac{5}{8}$ yards of the same ribbon for curtain trim. She has 5 yards of ribbon. Does she have enough ribbon? Explain.

23. **CRITICAL THINKING** The estimate for the sum of two fractions is 1. If 2 is added to each fraction, the estimate for the sum of the two mixed numbers is 4. What are examples of the fractions?

Standardized Test Practice and Mixed Review

24. **MULTIPLE CHOICE** About how much taller than Rose 3 did Rose 1 grow?

    Ⓐ $1\dfrac{1}{2}$ in.

    Ⓑ $2\dfrac{1}{2}$ in.

    Ⓒ 3 in.

    Ⓓ 23 in.

Height of Nicole's Roses		
Rose 1	Rose 2	Rose 3
$12\dfrac{5}{6}$ in.	$10\dfrac{1}{3}$ in.	$9\dfrac{5}{8}$ in.

25. **MULTIPLE CHOICE** Which is the best estimate for $6\dfrac{5}{6} + 2\dfrac{1}{12}$?

    Ⓕ 7   Ⓖ 8   Ⓗ 9   Ⓘ 10

26. Tell whether the amount of money you need for fast food should be rounded up or down. Explain your reasoning. (Lesson 6-1)

Write each fraction or mixed number as a decimal. (Lesson 5-7)

27. $\dfrac{5}{6}$

28. $\dfrac{7}{9}$

29. $4\dfrac{1}{3}$

30. $1\dfrac{5}{18}$

GETTING READY FOR THE NEXT LESSON

PREREQUISITE SKILL Write each improper fraction as a mixed number. (Lesson 5-3)

31. $\dfrac{6}{5}$

32. $\dfrac{7}{4}$

33. $\dfrac{12}{9}$

34. $\dfrac{10}{6}$

35. $\dfrac{18}{12}$

What You'll Learn
Solve problems by acting them out.

Act It Out

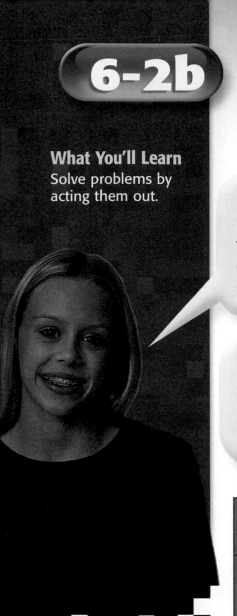

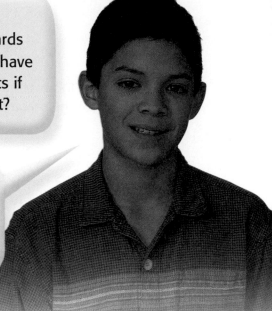

Ricardo, this roll of craft paper had $11\frac{1}{4}$ yards on it. But we used $2\frac{5}{8}$ yards for the art project. Are we going to have enough paper for four more projects if each project uses the same amount?

Let's **act it out**, Kelsey. We can measure $11\frac{1}{4}$ yards in the hall and use the completed project as a pattern to determine if there is enough paper for four more.

Explore	We know the roll of paper had $11\frac{1}{4}$ yards on it and $2\frac{5}{8}$ yards were used. We need to see whether there is enough for four more projects.
Plan	Let's start by marking the floor to show a length of $11\frac{1}{4}$ yards. Then mark off the amount used in the first project and continue until there are a total of five projects marked.
Solve	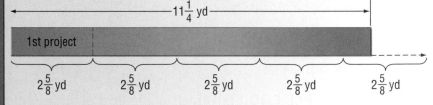 There is not enough for 4 additional projects.
Examine	We can estimate. Round $2\frac{5}{8}$ to $2\frac{1}{2}$. Then $2\frac{1}{2} + 2\frac{1}{2} + 2\frac{1}{2} + 2\frac{1}{2} + 2\frac{1}{2} = 12\frac{1}{2}$. So, $11\frac{1}{4}$ yards will not be enough for five projects.

Analyze the Strategy

1. **Explain** how this strategy could help determine if your answer after completing the calculations was reasonable.

2. **Write** a problem that could be solved by using the act it out strategy. Then explain how you would act it out to find the solution.

Solve. Use the act it out strategy.

3. **FITNESS** Dante runs 10 yards forward and then 5 yards backward. How many sets will he run to reach the end of the 100-yard field?

4. **BANNERS** The Spirit Club is making a banner using three sheets of paper. How many different banners can they make using their school colors of green, gold, and white one time each? Show the possible arrangements.

Mixed Problem Solving

Solve. Use any strategy.

5. **FLOORING** A kitchen floor measuring 12 feet by 10 feet needs to be tiled. There are 4 boxes of tiles with 24 in each box measuring 12 inches by 12 inches. Will this be enough tile to cover the floor?

6. **FOOD** Carlota bought three packages of ground turkey that weighed $2\frac{3}{4}$ pounds, $1\frac{7}{8}$ pounds, and $2\frac{1}{3}$ pounds. About how much ground turkey did she buy?

7. **BIRTHDAYS** Oscar took a survey of the dates of birth in his classroom. He listed them in a stem-and-leaf plot. Which is greater for this set of data, the mode or the median?

Stem	Leaf
0	1 1 2 3 5 5 5 8 9
1	1 2 3 3 7 8 8
2	0 3 5 5 6 7 7 7
3	0 0 1 $1|4 = 14^{th}$ day of the month

8. **CLOTHES** You can buy school uniforms through an online catalog. Boys can order either navy blue or khaki pants with a red, white, or blue shirt. How many uniform combinations are there online for boys?

9. **MONEY** The table gives admission costs for a health fair. Twelve people paid a total of $50 for admission. If 8 children attended the health fair, how many adults and senior citizens attended?

Health Fair Admission Costs	
Adults	$6
Children	$4
Senior Citizens	$3

10. **SHOPPING** Jhan has $95 to spend on athletic shoes. The shoes he wants cost $59.99. If you buy one pair, you get a second pair for half price. About how much money will he have left if he purchases two pairs of the shoes?

11. **TIME** School is out at 3:45 P.M., band practice is $2\frac{1}{2}$ hours, dinner takes 45 minutes, and you go to bed at 10:00 P.M. How much free time will you have if you study for 2 hours for a math exam?

12. **FOOD** About how much more money is spent on strawberry and grape jelly than the other types of jelly?

Yearly Jelly Sales (thousands)	
strawberry and grape	$366.2
all others	$291.5

Source: Nielsen Marketing Research

13. **STANDARDIZED TEST PRACTICE** Mrs. Samuelson had $350 to spend on a field trip for herself and 18 students. Admission was $12.50 per person and lunch cost about $5.00 per person. Which sentence best describes the amount of money left after the trip?

Ⓐ $c = 350 + 19(12.50 + 5.00)$

Ⓑ $c = 350 + 19(12.50 - 5.00)$

Ⓒ $c = 350 - 19(12.50) - 19(5.00)$

Ⓓ $c = 350 - 19(12.50 - 5.00)$

6-3 Adding and Subtracting Fractions with Like Denominators

What You'll LEARN

Add and subtract fractions with like denominators.

NEW Vocabulary

like fractions

REVIEW Vocabulary

simplest form: the form of a fraction when the GCF of the numerator and denominator is 1 **(Lesson 5-2)**

HANDS-ON Mini Lab

Materials
• grid paper
• markers

Work with a partner.

You can use grid paper to model adding fractions such as $\frac{3}{12}$ and $\frac{2}{12}$.

STEP 1 On grid paper, draw a rectangle like the one shown. Since the grid has 12 squares, each square represents $\frac{1}{12}$.

STEP 2 With a marker, color three squares to represent $\frac{3}{12}$. With a different marker, color two more squares to represent $\frac{2}{12}$.

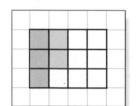

STEP 3 Five of the 12 squares are colored. So, the sum of $\frac{3}{12}$ and $\frac{2}{12}$ is $\frac{5}{12}$.

Find each sum using grid paper.

1. $\frac{4}{12} + \frac{3}{12}$ 2. $\frac{1}{6} + \frac{1}{6}$ 3. $\frac{3}{10} + \frac{5}{10}$

4. What patterns do you notice with the numerators?

5. What patterns do you notice with the denominator?

6. Explain how you could find the sum $\frac{3}{8} + \frac{1}{8}$ without using grid paper.

Fractions with the same denominator are called **like fractions**. You add and subtract the numerators of like fractions the same way you add and subtract whole numbers. The denominator names the units being added or subtracted.

Key Concept

Adding Like Fractions

Words	To add fractions with the same denominators, add the numerators. Use the same denominator in the sum.
Symbols	Arithmetic — $\frac{2}{5} + \frac{1}{5} = \frac{2+1}{5}$ or $\frac{3}{5}$ Algebra — $\frac{a}{c} + \frac{b}{c} = \frac{a+b}{c}$

EXAMPLE **Add Like Fractions**

1 Find the sum of $\frac{4}{5}$ and $\frac{3}{5}$.

$\frac{4}{5}$ + $\frac{3}{5}$

Estimate $1 + \frac{1}{2} = 1\frac{1}{2}$

$1\frac{2}{5}$

$\frac{4}{5} + \frac{3}{5} = \frac{4 + 3}{5}$ Add the numerators.

$= \frac{7}{5}$ Simplify.

$= 1\frac{2}{5}$ Write the improper fraction as a mixed number.

Compared to the estimate, the answer is reasonable.

The rule for subtracting fractions is similar to the rule for adding fractions.

Key Concept   **Subtracting Like Fractions**

Words   To subtract fractions with the same denominators, subtract the numerators. Use the same denominator in the difference.

Symbols   Arithmetic   Algebra

$\frac{3}{5} - \frac{1}{5} = \frac{3 - 1}{5}$ or $\frac{2}{5}$ $\frac{a}{c} - \frac{b}{c} = \frac{a - b}{c}$

EXAMPLE **Subtract Like Fractions**

2 Find $\frac{7}{8} - \frac{5}{8}$. Write in simplest form.

$\frac{7}{8} - \frac{5}{8} = \frac{7 - 5}{8}$ Subtract the numerators.

$= \frac{2}{8}$ or $\frac{1}{4}$ Simplify.

Your Turn Add or subtract. Write in simplest form.

a. $\frac{4}{7} + \frac{6}{7}$ **b.** $\frac{1}{9} + \frac{5}{9}$ **c.** $\frac{5}{9} - \frac{2}{9}$ **d.** $\frac{11}{12} - \frac{5}{12}$

EXAMPLE **Use Fractions to Solve a Problem**

3 **GEOGRAPHY** About $\frac{12}{100}$ of the population of the United States lives in California. Another $\frac{7}{100}$ lives in Texas. How much more of the population lives in California than in Texas?

$\frac{12}{100} - \frac{7}{100} = \frac{12 - 7}{100}$ Subtract the numerators.

$= \frac{5}{100}$ or $\frac{1}{20}$ Simplify.

About $\frac{1}{20}$ more of the population of the United States lives in California than in Texas.

1. **State** a simple rule for adding and subtracting like fractions.

2. **OPEN ENDED** Write two fractions whose sum is $\frac{7}{8}$.

3. **FIND THE ERROR** Three-eighths quart of pineapple juice was added to a bowl containing some orange juice to make seven-eighths quart of mixed juice. Della and Nikki determined how much orange juice was in the bowl. Who is correct? Explain.

Della	Nikki
$\frac{7}{8} - \frac{3}{8} = \frac{4}{8}$ or $\frac{1}{2}$ quart	$\frac{3}{8} + \frac{7}{8} = \frac{10}{8}$ or $1\frac{1}{4}$ quart

GUIDED PRACTICE

Add or subtract. Write in simplest form.

4. $\frac{3}{5} + \frac{1}{5}$ 5. $\frac{3}{8} - \frac{1}{8}$ 6. $\frac{7}{8} - \frac{3}{8}$ 7. $\frac{3}{4} + \frac{3}{4}$

8. **MEASUREMENT** How much more is $\frac{3}{4}$ gallon than $\frac{1}{4}$ gallon?

Practice and Applications

Add or subtract. Write in simplest form.

9. $\frac{4}{5} + \frac{2}{5}$ 10. $\frac{5}{7} + \frac{6}{7}$ 11. $\frac{3}{8} + \frac{7}{8}$ 12. $\frac{1}{9} + \frac{5}{9}$

13. $\frac{6}{6} - \frac{1}{6}$ 14. $\frac{5}{9} - \frac{2}{9}$ 15. $\frac{9}{10} - \frac{3}{10}$ 16. $\frac{5}{8} - \frac{3}{8}$

17. $\frac{5}{6} + \frac{5}{6}$ 18. $\frac{4}{5} + \frac{1}{5}$ 19. $\frac{7}{12} - \frac{2}{12}$ 20. $\frac{5}{8} - \frac{1}{8}$

21. $\frac{5}{9} - \frac{5}{9}$ 22. $\frac{9}{10} - \frac{9}{10}$ 23. $\frac{13}{14} + \frac{5}{14}$ 24. $\frac{15}{16} + \frac{7}{16}$

HOMEWORK HELP

For Exercises	See Examples
9–24	1, 2
25–32	3

Extra Practice
See pages 606, 629.

25. **MEASUREMENT** How much longer than $\frac{5}{16}$ inch is $\frac{13}{16}$ inch?

26. **MEASUREMENT** How much is $\frac{2}{3}$ cup plus $\frac{2}{3}$ cup?

27. **INSECTS** A mosquito's proboscis, the part that sucks blood, is the first $\frac{1}{3}$ of its body's length. The rest of the mosquito is made up of the head, thorax, and abdomen. How much of a mosquito is the head, thorax, and abdomen?

28. **WEATHER** One week, the rainfall for Monday through Saturday was $\frac{3}{8}$ inch. By Sunday evening, the total rainfall for the week was $\frac{7}{8}$ inch. How much rain fell on Sunday?

GEOGRAPHY For Exercises 29 and 30, use the following information.

About $\frac{3}{10}$ of the Earth's surface is land. The rest is covered by water.

29. How much of the Earth is covered by water?

30. How much more of the Earth is covered by water than by land?

FOOD For Exercises 31 and 32, use the circle graph.

31. What part of the school population likes their donuts filled, glazed, or frosted?

32. How much larger is the part that prefers glazed donuts than the part that prefers plain donuts?

33. **CRITICAL THINKING** Find the sum $\frac{1}{20} + \frac{19}{20} + \frac{2}{20} + \frac{18}{20} + \frac{3}{20} + \frac{17}{20} + \dots + \frac{10}{20} + \frac{10}{20}$. Look for a pattern to help you.

Genoa Middle School's Favorite Donuts

Plain $\frac{1}{10}$
Frosted $\frac{2}{10}$
Glazed $\frac{4}{10}$
Filled $\frac{3}{10}$

Standardized Test Practice and Mixed Review

34. **MULTIPLE CHOICE** A recipe calls for $\frac{2}{3}$ cup water, $\frac{1}{3}$ cup oil, and $\frac{1}{3}$ cup milk. How much liquid is used in the recipe?

    (A) $1\frac{1}{3}$ c   (B) $2\frac{1}{3}$ c   (C) 3 c   (D) $3\frac{1}{3}$ c

35. **MULTIPLE CHOICE** Kelli spent $\frac{3}{8}$ hour studying for her math test and $\frac{6}{8}$ hour studying for her Spanish test. How much time did Kelli spend studying for both tests?

    (F) 1 h   (G) $1\frac{1}{8}$ h   (H) $1\frac{2}{8}$ h   (I) $1\frac{3}{8}$ h

36. Estimate $18\frac{3}{11} + 4\frac{1}{9}$. (Lesson 6-2)

Round each number to the nearest half. (Lesson 6-1)

37. $\frac{9}{11}$ 38. $3\frac{3}{8}$ 39. $2\frac{1}{8}$ 40. $\frac{11}{12}$ 41. $\frac{5}{9}$

Add or subtract. (Lesson 3-5)

42. $14 + 23.5$ 43. $83.4 - 29.7$ 44. $17.3 + 33.5$ 45. $105.6 - 39.8$

46. **WEATHER** The average annual precipitation in Georgia is 48.61 inches. About how many inches of precipitation does Georgia average each month? (Lesson 2-6)

GETTING READY FOR THE NEXT LESSON

PREREQUISITE SKILL Find the least common denominator for each pair of fractions. (Lesson 5-5)

47. $\frac{3}{4}$ and $\frac{5}{8}$ 48. $\frac{2}{3}$ and $\frac{1}{2}$ 49. $\frac{3}{10}$ and $\frac{3}{4}$ 50. $\frac{4}{5}$ and $\frac{2}{9}$ 51. $\frac{4}{25}$ and $\frac{49}{20}$

Vocabulary and Concepts

1. **Define** *like fractions*. (Lesson 6-3)
2. **Describe** the numerator and denominator of a fraction that would be rounded up to a whole number. (Lesson 6-1)

Skills and Applications

Round each number to the nearest half. (Lesson 6-1)

3. $\frac{7}{8}$

4. $3\frac{2}{7}$

5. $6\frac{3}{4}$

Estimate. (Lesson 6-2)

6. $\frac{5}{9} + \frac{6}{7}$

7. $4\frac{1}{8} - 1\frac{11}{12}$

8. $6\frac{3}{4} + 2\frac{8}{10}$

9. **CARPENTRY** A board that is $63\frac{5}{8}$ inches long is about how much longer than a board that is $62\frac{1}{4}$ inches long? (Lesson 6-2)

10. **GEOMETRY** Estimate the perimeter of the rectangle. (Lesson 6-2)

$4\frac{1}{8}$ in.

$2\frac{3}{4}$ in.

11. **FOOD** Malinda is making a punch that calls for $\frac{3}{4}$ quart of grapefruit juice, $\frac{3}{4}$ quart of orange juice, and $\frac{3}{4}$ quart of pineapple juice. How much punch does the recipe make? (Lesson 6-3)

Add or subtract. Write in simplest form. (Lesson 6-3)

12. $\frac{5}{9} + \frac{7}{9}$

13. $\frac{9}{11} - \frac{5}{11}$

14. $\frac{1}{6} + \frac{5}{6}$

Standardized Test Practice

15. **MULTIPLE CHOICE** Which of the following has a sum that is less than 1? (Lesson 6-2)

   Ⓐ $1 + \frac{3}{4}$

   Ⓑ $\frac{2}{5} + \frac{4}{5}$

   Ⓒ $\frac{7}{11} + \frac{3}{11}$

   Ⓓ $\frac{1}{2} + \frac{1}{2}$

16. **SHORT RESPONSE** Sean lives $\frac{9}{10}$ mile from school. He has jogged $\frac{4}{10}$ mile from home toward the school. How much farther does he have to jog? (Lesson 6-3)

The GameZone

A Place To Practice Your Math Skills

Math Skill
Adding Fractions

Fraction Rummy

● **GET READY!**

Players: two
Materials: 27 index cards

● **GET SET!**

Cut the index cards in half. Then label the cards.

- 4 cards each: $\frac{1}{2}, \frac{1}{3}, \frac{1}{4}$

- 3 cards each: $\frac{1}{5}, \frac{2}{5}, \frac{3}{5}, \frac{4}{5}, \frac{1}{6}, \frac{5}{6}, \frac{1}{10}, \frac{3}{10}, \frac{7}{10}, \frac{9}{10}, \frac{1}{12}, \frac{5}{12}, \frac{7}{12}, \frac{11}{12}$

● **GO!**

- Choose one person to be the dealer.
- The dealer shuffles the cards and deals seven cards to each person, one card at a time, facedown.
- The dealer places the remaining cards facedown, turns the top card faceup, and places it next to the deck to start a discard pile.
- The players look at their cards and try to form sets of two or more cards whose sum is one. For example, $\frac{3}{10} + \frac{7}{10} = \frac{10}{10}$ or 1. Cards forming sums of one are placed in front of the player.
- Players take turns drawing the top card from the facedown deck, or the discard pile, and trying to find sums of one, which they place in front of them. To finish a turn, the player discards one card face up on the discard pile.
- **Who Wins?** The first person to discard his or her last card wins.

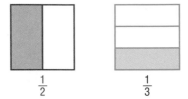

6-4a HANDS-ON LAB

A Preview of Lesson 6-4

Common Denominators

In this Activity, you will add or subtract fractions with *unlike* denominators.

What You'll LEARN

Add and subtract fractions with unlike denominators.

Materials

- paper squares
- ruler
- markers

ACTIVITY *Work with a partner.*

Add $\frac{1}{2} + \frac{1}{3}$.

STEP 1 First, model each fraction.

$\frac{1}{2}$ $\quad$ $\frac{1}{3}$

STEP 2 Next, find a common denominator.

The LCD of $\frac{1}{2}$ and $\frac{1}{3}$ is 6. Divide each square into sixths as shown.

$\frac{1}{2}$ $\quad$ $\frac{1}{3}$

STEP 3 Then, combine the fractional parts on one model.

Shade 3 of the sixths blue to represent $\frac{3}{6}$ or $\frac{1}{2}$.

Shade 2 of the sixths green to represent $\frac{2}{6}$ or $\frac{1}{3}$.

The combined shading is $\frac{5}{6}$ of the square. So, $\frac{1}{2} + \frac{1}{3} = \frac{5}{6}$.

Your Turn Use fraction models to add or subtract.

a. $\frac{1}{4} + \frac{2}{3}$ b. $\frac{4}{5} + \frac{1}{2}$ c. $\frac{5}{6} - \frac{1}{2}$ d. $\frac{4}{5} - \frac{1}{3}$

Writing Math

1. **Explain** why you need a common denominator to add or subtract fractions with unlike denominators.

2. **MAKE A CONJECTURE** What is the relationship between common multiples and adding and subtracting unlike fractions?

Adding and Subtracting Fractions with Unlike Denominators

What You'll LEARN

Add and subtract fractions with unlike denominators.

REVIEW Vocabulary

least common denominator (LCD): the LCM of the denominators of two or more fractions **(Lesson 5-5)**

HANDS-ON Mini Lab

Materials
• pennies and nickels

Work with a partner.

To name a group of different items, you need to find a common name for them.

1. How can you describe the sum of 3 pennies and 2 nickels using a common name?

2. How can you describe the sum of 2 pens and 2 pencils using a common name?

3. Explain why you need a common unit name to find the sum.

4. What do you think you need to do to find the sum of $\frac{1}{2}$ and $\frac{3}{4}$?

In the Mini Lab, you found common unit names for a group of unlike objects. When you work with fractions with different, or unlike, denominators, you do the same thing.

To find the sum or difference of two fractions with unlike denominators, rename the fractions using the least common denominator (LCD). Then add or subtract and simplify.

EXAMPLE   Add Unlike Fractions

1 Find $\frac{1}{2} + \frac{1}{4}$.

The LCD of $\frac{1}{2}$ and $\frac{1}{4}$ is 4.

Write the problem.

$$\begin{array}{r} \frac{1}{2} \\ + \frac{1}{4} \\ \hline \end{array}$$

→

Rename $\frac{1}{2}$ as $\frac{2}{4}$.

$$\frac{1}{2} \times \frac{2}{2} = \frac{2}{4}$$

$$\frac{1}{4}$$

→

Add the fractions.

$$\begin{array}{r} \frac{2}{4} \\ + \frac{1}{4} \\ \hline \frac{3}{4} \end{array}$$

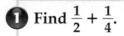 **Your Turn** Add. Write in simplest form.

a. $\frac{1}{6} + \frac{2}{3}$　　　　b. $\frac{9}{10} + \frac{1}{2}$　　　　c. $\frac{1}{4} + \frac{3}{8}$

Subtract Unlike Fractions

2 Find $\frac{2}{3} - \frac{1}{2}$.

The LCD of $\frac{2}{3}$ and $\frac{1}{2}$ is 6.

Write the problem.	Rename $\frac{2}{3}$ as $\frac{4}{6}$ and $\frac{1}{2}$ as $\frac{3}{6}$.	Subtract the fractions.
$\begin{array}{r} \frac{2}{3} \\ -\frac{1}{2} \\ \hline \end{array}$ →	$\frac{2}{3} \times \frac{2}{2} = \frac{4}{6}$ $\frac{1}{2} \times \frac{3}{3} = \frac{3}{6}$ →	$\begin{array}{r} \frac{4}{6} \\ -\frac{3}{6} \\ \hline \frac{1}{6} \end{array}$

Your Turn Subtract. Write in simplest form.

d. $\frac{5}{8} - \frac{1}{4}$ **e.** $\frac{3}{4} - \frac{1}{3}$ **f.** $\frac{1}{2} - \frac{2}{5}$

EXAMPLE **Use Fractions to Solve a Problem**

3 **BLOOD TYPES** Use the table to find the fraction of the population that have type O or type A blood.

Find $\frac{11}{25} + \frac{21}{50}$.   The LCD of $\frac{11}{25}$ and $\frac{22}{50}$ is 50.

	Blood Type Frequencies			
ABO Type	O	A	B	AB
Fraction	$\frac{11}{25}$	$\frac{21}{50}$	$\frac{1}{10}$	$\frac{1}{25}$

Source: anthro.palomar.edu

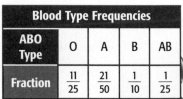

Write the problem.	Rename $\frac{11}{25}$ as $\frac{22}{50}$.	Add the fractions.
$\begin{array}{r} \frac{11}{25} \\ +\frac{21}{50} \\ \hline \end{array}$ →	$\begin{array}{r} \frac{22}{50} \\ \frac{21}{50} \\ \end{array}$ →	$\begin{array}{r} \frac{22}{50} \\ +\frac{21}{50} \\ \hline \frac{43}{50} \end{array}$

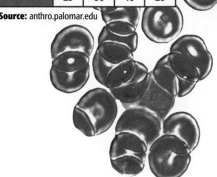

So, $\frac{43}{50}$ of the population has type O or type A blood.

You can evaluate algebraic expressions with fractions.

EXAMPLE **Evaluate an Expression with Fractions**

4 **ALGEBRA** Evaluate $a - b$ if $a = \frac{3}{4}$ and $b = \frac{1}{6}$.

$$a - b = \frac{3}{4} - \frac{1}{6} \qquad \text{Replace } a \text{ with } \frac{3}{4} \text{ and } b \text{ with } \frac{1}{6}.$$

$$= \frac{3}{4} \times \frac{3}{3} - \frac{1}{6} \times \frac{2}{2} \qquad \text{Rename } \frac{3}{4} \text{ and } \frac{1}{6} \text{ using the LCD, 12.}$$

$$= \frac{9}{12} - \frac{2}{12} \qquad \text{Simplify.}$$

$$= \frac{7}{12} \qquad \text{Subtract the numerators.}$$

1. **OPEN ENDED** Write two fractions with unlike denominators. Write them as equivalent fractions using the least common denominator.

2. **FIND THE ERROR** Victor and Seki are finding $\frac{3}{4} + \frac{1}{2}$. Who is correct? Explain.

Victor
$$\frac{3}{4} + \frac{1}{2} = \frac{3+1}{4+2}$$
$$= \frac{4}{6} \text{ or } \frac{2}{3}$$

Seki
$$\frac{3}{4} + \frac{1}{2} = \frac{3}{4} + \frac{2}{4}$$
$$= \frac{3+2}{4}$$
$$= \frac{5}{4} \text{ or } 1\frac{1}{4}$$

GUIDED PRACTICE

Add or subtract. Write in simplest form.

3. $\frac{2}{3} + \frac{2}{9}$ 4. $\frac{1}{4} + \frac{5}{8}$ 5. $\frac{2}{3} - \frac{1}{2}$ 6. $\frac{3}{5} - \frac{1}{2}$

7. What is $\frac{3}{4}$ minus $\frac{1}{8}$?   8. Find the sum of $\frac{3}{10}$ and $\frac{2}{5}$.

9. **ALGEBRA** Evaluate $c + d$ if $c = \frac{5}{6}$ and $d = \frac{7}{12}$.

Practice and Applications

Add or subtract. Write in simplest form.

10. $\frac{3}{8}$ 11. $\frac{2}{5}$ 12. $\frac{9}{10}$ 13. $\frac{5}{8}$
   $+\frac{1}{4}$ $+\frac{1}{2}$ $-\frac{1}{2}$ $-\frac{1}{4}$

14. $\frac{1}{6}$ 15. $\frac{1}{4}$ 16. $\frac{5}{6}$ 17. $\frac{3}{4}$
   $+\frac{3}{4}$ $+\frac{2}{3}$ $-\frac{7}{10}$ $-\frac{2}{5}$

HOMEWORK HELP

For Exercises	See Examples
10–25	1, 2
26–29, 32–36	3
30–31	4

Extra Practice
See pages 606, 629.

18. $\frac{8}{9} + \frac{1}{2}$ 19. $\frac{7}{8} - \frac{3}{4}$ 20. $\frac{7}{8} + \frac{3}{4}$ 21. $\frac{7}{12} + \frac{2}{3}$

22. $\frac{9}{10} + \frac{2}{3}$ 23. $\frac{7}{12} + \frac{5}{8}$ 24. $\frac{9}{10} - \frac{2}{3}$ 25. $\frac{15}{16} - \frac{1}{3}$

26. How much more is $\frac{3}{4}$ cup than $\frac{2}{3}$ cup?   27. How much longer than $\frac{9}{16}$ inch is $\frac{7}{8}$ inch?

28. What is the sum of $\frac{3}{4}$, $\frac{5}{12}$, and $\frac{1}{6}$?   29. What is the sum of $\frac{5}{8}$, $\frac{3}{4}$, and $\frac{2}{5}$?

30. **ALGEBRA** Evaluate $a + b$ if $a = \frac{7}{10}$ and $b = \frac{5}{6}$.

31. **ALGEBRA** Evaluate $c - d$ if $c = \frac{7}{8}$ and $d = \frac{3}{5}$.

GEOGRAPHY For Exercises 32–34, use the circle graph.

Portion of Earth's Landmass

Antarctica, Europe, Australia, Oceania

?

Asia $\frac{3}{10}$

South America — $\frac{1}{8}$

North America — $\frac{1}{6}$

Africa $\frac{1}{5}$

Source: *Oxford Atlas of the World*

32. What portion of the Earth's landmass is Asia and Africa?

33. How much more is the landmass of North America than South America?

34. **MULTI STEP** What portion of Earth's landmass is Antarctica, Europe, Australia, and Oceania?

STATISTICS For Exercises 35 and 36, use the table shown and refer to Exercises 33–35 on page 185.

35. Copy and complete the table by finding the relative frequency of each category.

36. Find the sum of the relative frequencies.

Favorite Type of Book			
Type	Tally	Frequency	Relative Frequency
Fiction	IIII III	8	?
Mystery	IIII I	6	?
Romance	II	2	?
Nonfiction	IIII	4	?

CRITICAL THINKING Tell whether each sentence is *sometimes*, *always*, or *never* true. Explain.

37. The sum of two fractions that are less than 1 is less than 1.

38. The difference of two fractions is less than both fractions.

Standardized Test Practice and Mixed Review

Ⓐ Ⓑ Ⓒ Ⓓ

39. **MULTIPLE CHOICE** What is the perimeter of the figure?

ⓐ $\frac{16}{24}$ in. ⓑ $\frac{16}{8}$ in. ⓒ 2 in. ⓓ $2\frac{1}{2}$ in.

$\frac{1}{4}$ in. $\frac{3}{4}$ in. $\frac{7}{8}$ in. $\frac{5}{8}$ in.

40. **SHORT RESPONSE** Hannah has 1 quarter, 3 dimes, and 5 nickels. What part of a dollar does she have?

41. **GARDENING** John finished filling a $\frac{7}{8}$-gallon watering can by pouring $\frac{5}{8}$ of a gallon of water into the can. How much water was already in the can? (Lesson 6-3)

Estimate. (Lesson 6-2)

42. $4\frac{3}{7} + \frac{1}{8}$ 43. $3\frac{4}{5} + 1\frac{1}{7}$ 44. $8\frac{7}{8} - 3\frac{5}{6}$ 45. $13\frac{2}{5} - 9\frac{6}{7}$

GETTING READY FOR THE NEXT LESSON

PREREQUISITE SKILL Replace each ● with a number so that the fractions are equivalent. (Lesson 5-2)

46. $\frac{3}{4} = \frac{●}{12}$ 47. $\frac{1}{8} = \frac{●}{24}$ 48. $\frac{1}{3} = \frac{●}{12}$ 49. $\frac{5}{6} = \frac{●}{18}$

msmath1.net/self_check_quiz

Study Skill

HOW TO...

Read Math Problems

SUBTRACTION

You know that one meaning of subtraction is *to take away*. But there are other meanings too. Look for these meanings when you're solving a word problem.

Subtraction has three meanings.

- **To take away**

  Chad found $\frac{5}{8}$ of a pizza in the refrigerator. He ate $\frac{1}{8}$ of the original pizza. How much of the original pizza is left?

- **To find a missing addend**

  Heather made a desktop by gluing a sheet of oak veneer to a sheet of $\frac{3}{4}$-inch plywood. The total thickness of the desktop is $\frac{13}{16}$ inch. What was the thickness of the oak veneer?

- **To compare the size of two sets**

  Yesterday, it rained $\frac{7}{8}$ inch. Today, it rained $\frac{1}{4}$ inch. How much more did it rain yesterday than today?

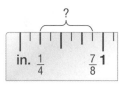

SKILL PRACTICE

1. Solve each problem above.

Identify the meaning of subtraction shown in each problem. Then solve the problem.

2. Marcus opened a carton of milk and drank $\frac{1}{4}$ of it. How much of the carton of milk is left?

3. How much bigger is a $\frac{15}{16}$-inch wrench than a $\frac{3}{8}$-inch wrench?

4. Part of a hiking trail is $\frac{3}{4}$ mile long. When you pass the $\frac{1}{8}$-mile marker, how much farther is it until the end of the trail?

Adding and Subtracting Mixed Numbers

What You'll LEARN

Add and subtract mixed numbers.

REVIEW Vocabulary

mixed number: the sum of a whole number and a fraction (Lesson 5-3)

HANDS-ON **Mini Lab**

Work with a partner.

You can use paper plates to add and subtract mixed numbers.

Materials

• paper plates
• scissors

STEP 1 Cut a paper plate into fourths and another plate into halves.

STEP 2 Use one whole plate and three fourths of a plate to show the mixed number $1\frac{3}{4}$.

STEP 3 Use two whole plates and one half of a plate to show $2\frac{1}{2}$.

STEP 4 Make as many whole paper plates as you can.

1. How many whole paper plates can you make?

2. What fraction is represented by the leftover pieces?

3. What is the sum $1\frac{3}{4} + 2\frac{1}{2}$?

The Mini Lab suggests the following rule.

Key Concept   Adding and Subtracting Mixed Numbers

To add or subtract mixed numbers, first add or subtract the fractions. Then add or subtract the whole numbers. Rename and simplify if necessary.

EXAMPLE   Subtract Mixed Numbers

1 Find $4\frac{5}{6} - 2\frac{1}{6}$.   **Estimate** $5 - 2 = 3$

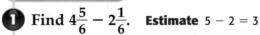

Subtract the fractions.

$$
\begin{array}{r}
4\frac{5}{6} \\
- 2\frac{1}{6} \\
\hline
\frac{4}{6}
\end{array}
$$

→

Subtract the whole numbers.

$$
\begin{array}{r}
4\frac{5}{6} \\
- 2\frac{1}{6} \\
\hline
2\frac{4}{6} \text{ or } 2\frac{2}{3}
\end{array}
$$

Compare to the estimate.

EXAMPLE **Add Mixed Numbers**

2 Find $5\frac{1}{4} + 10\frac{2}{3}$. **Estimate** $5 + 11 = 16$

The LCM of 4 and 3 is 12.	Rename the fractions.	Add the fractions.	Add the whole numbers.

$$5\frac{1}{4} \times \frac{3}{3}$$
$$+ 10\frac{2}{3} \times \frac{4}{4}$$
$\rightarrow$
$$5\frac{3}{12}$$
$$+ 10\frac{8}{12}$$
$\rightarrow$
$$5\frac{3}{12}$$
$$+ 10\frac{8}{12}$$
$$\frac{11}{12}$$
$\rightarrow$
$$5\frac{3}{12}$$
$$+ 10\frac{8}{12}$$
$$15\frac{11}{12}$$

Your Turn Add or subtract. Write in simplest form.

a. $5\frac{2}{8} + 3\frac{1}{8}$ b. $4\frac{3}{8} + 7\frac{1}{8}$ c. $5\frac{1}{2} - 2\frac{1}{3}$

Standardized Test Practice

EXAMPLE **Use Mixed Numbers to Solve a Problem**

3 **MULTIPLE-CHOICE TEST ITEM**
Refer to the diagram. How far will Mieko travel if she walks from her home to the library and then to the bagel shop?

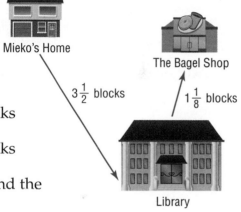

Mieko's Home

The Bagel Shop

$3\frac{1}{2}$ blocks $1\frac{1}{8}$ blocks

Library

   Ⓐ $2\frac{3}{8}$ blocks   Ⓑ $4\frac{1}{2}$ blocks

   Ⓒ $4\frac{5}{8}$ blocks   Ⓓ $5\frac{1}{8}$ blocks

Read the Test Item You need to find the distance Mieko will walk.

Solve the Test Item
First use the LCD to rename the fractions. Then add.

$$3\frac{1}{2}$$
$$+ 1\frac{1}{8}$$
$\rightarrow$
$$3\frac{4}{8}$$
$$+ 1\frac{1}{8}$$
$$4\frac{5}{8}$$

Mieko will walk $4\frac{5}{8}$ blocks.

The answer is C.

Test-Taking Tip
The Princeton Review

Eliminating Choices
You can estimate to eliminate some choices. By estimating $3\frac{1}{2} + 1\frac{1}{8}$, you know the distance must be greater than 4 blocks and less than 5 blocks.

EXAMPLE **Evaluate an Expression**

4 **ALGEBRA** Evaluate $x - y$ if $x = 4\frac{2}{3}$ and $y = 2\frac{1}{6}$.

$x - y = 4\frac{2}{3} - 2\frac{1}{6}$ Replace x with $4\frac{2}{3}$ and y with $2\frac{1}{6}$.

$= 4\frac{4}{6} - 2\frac{1}{6}$ Rename $4\frac{2}{3}$ as $4\frac{4}{6}$.

$= 2\frac{3}{6}$ or $2\frac{1}{2}$ Simplify.

Skill and Concept Check

Writing Math
Exercises 1 & 2

1. **OPEN ENDED** Write a problem where you need to subtract $1\frac{1}{4}$ from $3\frac{3}{8}$.

2. **NUMBER SENSE** Is the sum of two mixed numbers always a mixed number? Explain. If not, give a counterexample.

GUIDED PRACTICE

Add or subtract. Write in simplest form.

3. $5\frac{3}{4}$
    $-1\frac{1}{4}$

4. $2\frac{3}{8}$
    $+4\frac{1}{8}$

5. $14\frac{3}{5}$
    $-6\frac{3}{10}$

6. $4\frac{3}{5}$
    $+4\frac{1}{2}$

7. $6\frac{9}{10} + 8\frac{1}{4}$

8. $3\frac{2}{3} - \frac{3}{5}$

9. $4\frac{5}{6} + 3\frac{3}{4}$

10. **ALGEBRA** Evaluate $m + n$ if $m = 2\frac{3}{4}$ and $n = 3\frac{3}{5}$.

Practice and Applications

Add or subtract. Write in simplest form.

11. $3\frac{5}{6}$
    $+4\frac{1}{6}$

12. $4\frac{5}{8}$
    $-2\frac{3}{8}$

13. $9\frac{4}{5}$
    $-4\frac{2}{5}$

14. $4\frac{5}{12}$
    $+6\frac{7}{12}$

15. $10\frac{3}{4}$
    $-3\frac{1}{2}$

16. $8\frac{1}{3}$
    $-3\frac{1}{6}$

17. $3\frac{1}{2}$
    $+2\frac{2}{3}$

18. $11\frac{3}{10}$
    $+9\frac{1}{4}$

19. $6\frac{3}{5} + \frac{4}{5}$

20. $3\frac{3}{8} + 6\frac{5}{8}$

21. $7\frac{7}{9} - 4\frac{1}{3}$

22. $6\frac{6}{7} - 4\frac{5}{14}$

23. $6\frac{5}{8} + 7\frac{1}{4}$

24. $4\frac{5}{6} + 15\frac{3}{8}$

25. $5\frac{11}{12} - 1\frac{7}{10}$

26. $13\frac{7}{10} + 4\frac{1}{6}$

HOMEWORK HELP	
For Exercises	See Examples
11–26, 29	1, 2
30–33	4
27–28, 34–37	3
Extra Practice See pages 606, 629.	

27. **MEASUREMENT** How much longer is $35\frac{1}{2}$ seconds than $30\frac{3}{10}$ seconds?

28. **BASEBALL** The table shows the standings in the American League East. How many more games behind 1st place is Baltimore than Tampa Bay?

 Data Update Find the baseball standings for the end of the last baseball season. Visit msmath1.net/data_update to learn more.

—Daily Sports Review—

Team	W	L	GB*
Boston	8	4	–
New York	8	7	$1^{1}/_{2}$
Toronto	6	7	$2^{1}/_{2}$
Tampa Bay	5	7	3
Baltimore	4	9	$4^{1}/_{2}$

*games behind 1st place

29. Find the sum of $3\frac{1}{5}$, $1\frac{7}{8}$, and $6\frac{7}{10}$.

ALGEBRA Evaluate each expression if $a = 2\frac{1}{6}$, $b = 4\frac{3}{4}$, and $c = 5\frac{2}{3}$.

30. $a + b$ **31.** $c - a$ **32.** $b + c$ **33.** $a + b + c$

GEOMETRY Find the perimeter of each figure.

34.
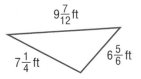
$9\frac{7}{12}$ ft $7\frac{1}{4}$ ft $6\frac{5}{6}$ ft

35.
$4\frac{1}{6}$ ft $8\frac{3}{4}$ ft

36. SPACE Mercury's average distance from the Sun is 36 million miles. Venus averages $67\frac{1}{2}$ million miles from the Sun. How much greater is Venus' average distance from the Sun than Mercury's?

37. MULTI STEP To win horse racing's Triple Crown, a 3-year-old horse must win the Kentucky Derby, Preakness Stakes, and Belmont Stakes. The lengths of the tracks are shown. How much longer is the longest race than the shortest?

Triple Crown	
Race	Length (mi)
Kentucky Derby	$1\frac{1}{4}$
Preakness Stakes	$1\frac{3}{16}$
Belmont Stakes	$1\frac{1}{2}$

38. CRITICAL THINKING Use the digits 1, 1, 2, 2, 3, and 4 to create two mixed numbers whose sum is $4\frac{1}{4}$.

Standardized Test Practice and Mixed Review

39. MULTIPLE CHOICE Mrs. Matthews bought $3\frac{1}{4}$ pounds of caramels and $2\frac{1}{2}$ pounds of chocolate. How many pounds of candy did she buy altogether?

   Ⓐ $5\frac{3}{4}$ lb   Ⓑ $5\frac{2}{6}$ lb   Ⓒ $1\frac{1}{2}$ lb   Ⓓ none of these

40. MULTIPLE CHOICE In the school-wide recycling program, Robert Frost Middle School recycled $89\frac{3}{8}$ pounds of paper this year. They recycled $77\frac{1}{3}$ pounds last year. How many more pounds did the school recycle this year than last year?

   Ⓕ $11\frac{1}{8}$ lb   Ⓖ $11\frac{11}{24}$ lb   Ⓗ $12\frac{1}{24}$ lb   Ⓘ $166\frac{17}{24}$ lb

Add or subtract. Write in simplest form. (Lessons 6-3 and 6-4)

41. $\frac{1}{3} + \frac{1}{3}$ **42.** $\frac{9}{10} - \frac{3}{10}$ **43.** $\frac{5}{6} + \frac{4}{6}$ **44.** $\frac{7}{9} - \frac{5}{9}$

45. $\frac{4}{5} - \frac{3}{4}$ **46.** $\frac{7}{9} + \frac{5}{12}$ **47.** $\frac{9}{10} + \frac{3}{4}$ **48.** $\frac{5}{6} - \frac{1}{8}$

GETTING READY FOR THE NEXT LESSON

PREREQUISITE SKILL Write each mixed number as an improper fraction. (Lesson 5-3)

49. $1\frac{2}{5}$ **50.** $1\frac{4}{9}$ **51.** $1\frac{3}{8}$ **52.** $2\frac{5}{6}$ **53.** $2\frac{1}{12}$

Subtracting Mixed Numbers with Renaming

What You'll LEARN

Subtract mixed numbers involving renaming.

REVIEW Vocabulary

circumference: the distance around a circle (Lesson 4-6)

WHEN am I ever going to use this?

SPORTS The table shows some differences between the softballs and baseballs used in the Olympics.

Sport	Circumference (inches)	Weight (ounces)
Softball	$11\frac{7}{8}$ to $12\frac{1}{8}$	$6\frac{1}{4}$ to 7
Baseball	9 to $9\frac{1}{4}$	5 to $5\frac{1}{4}$

1. Which sport's ball has the greater weight?

2. Explain how you could find the difference between the greatest weights allowed for a softball and a baseball.

To find the difference between the greatest weight allowed for a softball and a baseball, subtract $5\frac{1}{4}$ from 7. Sometimes it is necessary to rename the fraction part of a mixed number as an improper fraction in order to subtract.

EXAMPLES Rename to Subtract

1 Find $7 - 5\frac{1}{4}$.

$$7 \rightarrow 6\frac{4}{4}$$
$$-5\frac{1}{4} \rightarrow -5\frac{1}{4}$$
$$\overline{\phantom{-5\frac{1}{4}}1\frac{3}{4}}$$

So, $7 - 5\frac{1}{4} = 1\frac{3}{4}$.

Rename 7 as $6\frac{4}{4}$. Then cross out $5\frac{1}{4}$. $\rightarrow$

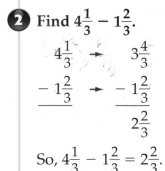

$1\frac{3}{4}$ remains.

2 Find $4\frac{1}{3} - 1\frac{2}{3}$.

$$4\frac{1}{3} \rightarrow 3\frac{4}{3}$$
$$-1\frac{2}{3} \rightarrow -1\frac{2}{3}$$
$$\overline{\phantom{-1\frac{2}{3}}2\frac{2}{3}}$$

So, $4\frac{1}{3} - 1\frac{2}{3} = 2\frac{2}{3}$.

Rename $4\frac{1}{3}$ as $3\frac{4}{3}$. Then cross out $1\frac{2}{3}$. $\rightarrow$

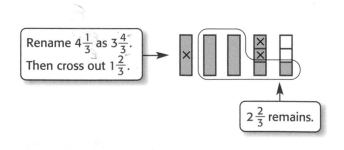

$2\frac{2}{3}$ remains.

Your Turn Subtract. Write in simplest form.

a. $5 - 3\frac{1}{2}$ b. $3\frac{1}{4} - 2\frac{3}{4}$ c. $6\frac{2}{5} - 3\frac{3}{5}$

3 Find $12\frac{1}{8} - 9\frac{1}{4}$.

Step 1 $12\frac{1}{8}$ → $12\frac{1}{8}$

        $-9\frac{1}{4}$ → $-9\frac{2}{8}$

> The LCM of 8 and 4 is 8.

Step 2 $12\frac{1}{8}$ → $11\frac{9}{8}$

> Rename $12\frac{1}{8}$ as $11\frac{9}{8}$.

        $-9\frac{2}{8}$ → $-9\frac{2}{8}$

                   $2\frac{7}{8}$

So, $12\frac{1}{8} - 9\frac{1}{4} = 2\frac{7}{8}$.

Your Turn Subtract. Write in simplest form.

d. $6\frac{1}{2} - 2\frac{3}{4}$ **e.** $10\frac{1}{6} - 7\frac{1}{3}$ **f.** $8\frac{7}{10} - 6\frac{3}{4}$

EXAMPLE **Use Renaming to Solve a Problem**

4 GEOGRAPHY
Kayla can see Mt. Rainier from her home near Seattle. Mt. Rainier is about $2\frac{3}{4}$ miles high. Use the graph to compare Mt. Rainier to the highest mountain in the world.

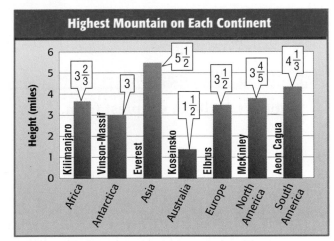

Highest Mountain on Each Continent

Height (miles)

Kilimanjaro — Africa: $3\frac{2}{3}$
Vinson-Massif — Antarctica: 3
Everest — Asia: $5\frac{1}{2}$
Kossinsko — Australia: $1\frac{1}{2}$
Elbrus — Europe: $3\frac{1}{2}$
McKinley — North America: $3\frac{4}{5}$
Aeon Cagua — South America: $4\frac{1}{3}$

Source: *Oxford Atlas of the World*

Mt. Rainier: $2\frac{3}{4}$ miles high

Mt. Everest: $5\frac{1}{2}$ miles high

Find $5\frac{1}{2} - 2\frac{3}{4}$.   **Estimate** $6 - 3 = 3$

    $5\frac{1}{2}$ → $5\frac{2}{4}$ → $4\frac{6}{4}$

> Rename $5\frac{2}{4}$ as $4\frac{6}{4}$.

    $-2\frac{3}{4}$ → $-2\frac{3}{4}$ → $-2\frac{3}{4}$

                          $2\frac{3}{4}$

So, Mt. Everest is $2\frac{3}{4}$ miles higher than Mt. Rainier.

1. **Draw** a model or use paper plates to show how to find $3\frac{1}{2} - 1\frac{5}{8}$.

2. **OPEN ENDED** Write a problem that can be solved by finding $1\frac{1}{4} - \frac{1}{2}$. Solve the problem.

3. **Complete.**
   a. $10 = 9\frac{\bullet}{8}$ b. $9 = 8\frac{\bullet}{6}$ c. $5\frac{3}{4} = 4\frac{\bullet}{4}$ d. $8\frac{2}{3} = 7\frac{\bullet}{3}$

4. **FIND THE ERROR** Jeremy and Tom are finding $9\frac{1}{2} - 3$. Who is correct? Explain.

   Jeremy
   $9\frac{1}{2} - 3 = 9\frac{1}{2} - 2\frac{2}{2}$
   $= 7\frac{1}{2}$

   Tom
   $9\frac{1}{2} - 3 = 6\frac{1}{2}$

5. **Which One Doesn't Belong?** Identify the subtraction problem that does not need to be renamed before subtracting. Explain your reasoning.

   $7 - 2\frac{1}{2}$ $9\frac{1}{3} - 4\frac{2}{3}$ $6\frac{3}{10} - 2\frac{7}{10}$ $3\frac{5}{8} - 2\frac{3}{8}$

GUIDED PRACTICE

Subtract. Write in simplest form.

6. $\quad 4$
   $\underline{-\ 3\frac{1}{6}}$

7. $\quad 3\frac{1}{8}$
   $\underline{-\ 1\frac{3}{8}}$

8. $6\frac{1}{2} - 2\frac{3}{4}$

9. $5\frac{1}{4} - 3\frac{3}{5}$

10. **TRANSPORTATION** The U.S. Department of Transportation prohibits a truck driver from driving more than 70 hours in any 8-day period. Mr. Galvez has driven $53\frac{3}{4}$ hours in the last 6 days. How many more hours is he allowed to drive during the next 2 days?

Practice and Applications

Subtract. Write in simplest form.

11. $\quad 7$
    $\underline{-\ 5\frac{1}{2}}$

12. $\quad 9$
    $\underline{-\ 3\frac{3}{5}}$

13. $\quad 4\frac{1}{4}$
    $\underline{-\ 2\frac{3}{4}}$

14. $\quad 9\frac{3}{8}$
    $\underline{-\ 6\frac{5}{8}}$

HOMEWORK HELP	
For Exercises	See Examples
11–22	1–3
23–28	4
Extra Practice See pages 607, 629.	

15. $12\frac{1}{5} - 5\frac{3}{10}$

16. $8\frac{1}{3} - 1\frac{5}{6}$

17. $14\frac{3}{8} - 5\frac{3}{4}$

18. $10\frac{5}{9} - 3\frac{2}{3}$

19. $7\frac{1}{3} - 3\frac{1}{2}$

20. $5\frac{1}{6} - 1\frac{3}{8}$

21. $12\frac{3}{4} - \frac{4}{5}$

22. $9\frac{5}{9} - \frac{5}{6}$

23. **MEASUREMENT** How much longer is $1\frac{1}{2}$ inches than $\frac{3}{4}$ inch?

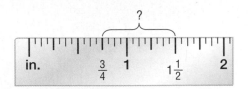

24. **MEASUREMENT** How much more is 8 quarts than $5\frac{1}{2}$ quarts?

25. **SPORTS** Use the table on the top of page 244 to find the difference between the circumference of the biggest baseball and the biggest softball.

26. **TRAVEL** The pilot of your flight said it will take $2\frac{1}{2}$ hours to reach your destination. You have been in flight $1\frac{3}{4}$ hours. How much longer is your flight?

GEOGRAPHY **For Exercises 27 and 28, use the graph on page 245.**

27. How much higher is Mt. Everest than Mt. Kilimanjaro?

28. **RESEARCH** Use the Internet or another source to find the height of Mt. Olympus in Greece. How much higher is Mt. Elbrus?

29. **CRITICAL THINKING** Write a problem where you must subtract by renaming. The difference should be between $\frac{1}{3}$ and $\frac{1}{2}$.

Standardized Test Practice and Mixed Review

30. **MULTIPLE CHOICE** The table lists differences between the court size for basketball in the Olympics and in the National Basketball Association (NBA). How much longer is the court in the NBA than in the Olympics?

Sport	Length of Court (feet)	Width of Court (feet)
Olympic Basketball	$91\frac{5}{6}$	$49\frac{1}{6}$
NBA Basketball	94	50

Ⓐ $3\frac{2}{3}$ ft

Ⓑ $2\frac{5}{6}$ ft

Ⓒ $2\frac{1}{6}$ ft

Ⓓ $\frac{5}{6}$ ft

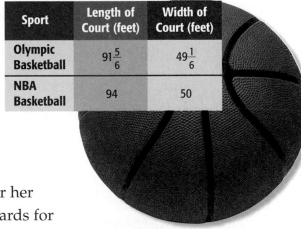

31. **SHORT RESPONSE** Melika was making a quilt for her room. She had $10\frac{1}{4}$ yards of material. It took $7\frac{1}{2}$ yards for the quilt. How much material was *not* used for the quilt?

ALGEBRA **Evaluate each expression if** $a = 1\frac{1}{6}$, $b = 4\frac{3}{4}$, **and** $c = 6\frac{5}{8}$. (Lesson 6-5)

32. $b + c$

33. $b - a$

34. $c - a$

35. $a + b + c$

36. Find the sum of $\frac{7}{8}$ and $\frac{5}{6}$. Write in simplest form. (Lesson 6-4)

Find the prime factorization of each number. (Lesson 1-3)

37. 45

38. 73

39. 57

40. 72

Vocabulary and Concept Check

like fractions (p. 228)

Choose the correct term or number to complete each sentence.

1. To the nearest half, $5\frac{1}{5}$ rounds to $\left(5, 5\frac{1}{2}\right)$.

2. When you subtract fractions with like denominators, you subtract the (denominators, numerators).

3. The (LCD, GCF) is the least common multiple of the denominators of unlike fractions.

4. The LCD of $\frac{1}{8}$ and $\frac{3}{10}$ is (80, 40).

5. The mixed number $9\frac{1}{4}$ can be renamed as $\left(8\frac{3}{4}, 8\frac{5}{4}\right)$.

6. When finding the sum of fractions with unlike denominators, rename the fractions using the (LCD, GCF).

Lesson-by-Lesson Exercises and Examples

6-1 Rounding Fractions and Mixed Numbers (pp. 219–222)

Round each number to the nearest half.

7. $\frac{4}{5}$

8. $4\frac{1}{3}$

9. $6\frac{6}{14}$

10. $\frac{11}{20}$

11. $2\frac{2}{11}$

12. $9\frac{4}{9}$

13. **MEASUREMENT** Find the length of the key to the nearest half inch.

Example 1 Round $\frac{5}{8}$ to the nearest half.

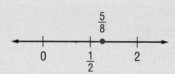

$\frac{5}{8}$ rounds to $\frac{1}{2}$.

Example 2 Round $2\frac{4}{5}$ to the nearest half.

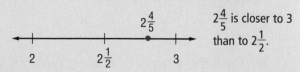

$2\frac{4}{5}$ is closer to 3 than to $2\frac{1}{2}$.

$2\frac{4}{5}$ rounds to 3.

6-2 Estimating Sums and Differences (pp. 223–225)

Estimate.

14. $\frac{3}{4} + \frac{1}{3}$

15. $\frac{7}{8} - \frac{1}{5}$

16. $3\frac{1}{9} - 1\frac{6}{7}$

17. $9\frac{3}{8} - 2\frac{4}{5}$

18. $6\frac{7}{12} - 2\frac{1}{4}$

19. $\frac{1}{9} + 2\frac{9}{10}$

20. $4\frac{9}{10} + 1\frac{2}{9}$

21. $3\frac{7}{8} - \frac{1}{12}$

Example 3

Estimate $\frac{5}{6} - \frac{4}{9}$.

$\frac{5}{6} - \frac{4}{9}$ is about $1 - \frac{1}{2}$ or $\frac{1}{2}$.

Example 4

Estimate $2\frac{4}{5} + 6\frac{1}{8}$.

$2\frac{4}{5} + 6\frac{1}{8}$ is about $3 + 6$ or 9.

6-3 Adding and Subtracting Fractions with Like Denominators (pp. 228–231)

Add or subtract. Write in simplest form.

22. $\frac{3}{8} + \frac{1}{8}$

23. $\frac{7}{12} + \frac{1}{12}$

24. $\frac{7}{10} + \frac{3}{10}$

25. $\frac{6}{7} - \frac{2}{7}$

26. $\frac{11}{12} - \frac{7}{12}$

27. $\frac{7}{9} + \frac{4}{9}$

28. **PIZZA** Tori ate $\frac{4}{12}$ of a large pizza, and Ben ate $\frac{5}{12}$. How much of the pizza did they eat in all? Write in simplest form.

Example 5

Find $\frac{3}{8} + \frac{1}{8}$. **Estimate** $\frac{1}{2} + 0 = \frac{1}{2}$

$\frac{3}{8} + \frac{1}{8} = \frac{3+1}{8}$ Add the numerators.

$= \frac{4}{8}$ or $\frac{1}{2}$ Simplify.

Example 6

Find $\frac{7}{12} - \frac{5}{12}$. **Estimate** $\frac{1}{2} - \frac{1}{2} = 0$

$\frac{7}{12} - \frac{5}{12} = \frac{7-5}{12}$ Subtract the numerators.

$= \frac{2}{12}$ or $\frac{1}{6}$ Simplify.

6-4 Adding and Subtracting Fractions with Unlike Denominators (pp. 235–238)

Add or subtract. Write in simplest form.

29. $\frac{1}{2} + \frac{2}{3}$

30. $\frac{5}{8} + \frac{1}{4}$

31. $\frac{7}{9} - \frac{1}{12}$

32. $\frac{9}{10} - \frac{1}{4}$

33. $\frac{3}{4} + \frac{5}{6}$

34. $\frac{7}{9} - \frac{1}{6}$

35. $\frac{7}{8} - \frac{1}{3}$

36. $\frac{4}{5} + \frac{2}{10}$

Example 7

Find $\frac{3}{8} + \frac{2}{3}$. **Estimate** $\frac{1}{2} + \frac{1}{2} = 1$

Rename $\frac{3}{8}$ as $\frac{9}{24}$ and $\frac{2}{3}$ as $\frac{16}{24}$.

$$\begin{array}{ccc} \frac{3}{8} & \frac{3}{3} \times \frac{3}{8} = \frac{9}{24} & \frac{9}{24} \\ +\frac{2}{3} & \frac{8}{8} \times \frac{2}{3} = \frac{16}{24} & +\frac{16}{24} \\ & & \overline{\frac{25}{24} \text{ or } 1\frac{1}{24}} \end{array}$$

6-5 Adding and Subtracting Mixed Numbers (pp. 240–243)

Add or subtract. Write in simplest form.

37. $3\frac{2}{5} + 1\frac{3}{5}$ 38. $9\frac{7}{8} - 5\frac{3}{8}$

39. $5\frac{7}{10} - 3\frac{2}{5}$ 40. $2\frac{3}{4} + 1\frac{1}{6}$

41. $1\frac{8}{9} - 1\frac{1}{3}$ 42. $6\frac{7}{12} + 4\frac{3}{8}$

43. $7\frac{5}{6} + 9\frac{3}{4}$ 44. $4\frac{3}{7} - 2\frac{5}{14}$

45. **HOMEWORK** Over the weekend, Brad spent $\frac{3}{4}$ hour on his math homework and $2\frac{1}{6}$ hours on his science paper. How much time did he spend on these two subjects?

Example 8

Find $6\frac{5}{8} - 2\frac{2}{5}$.   **Estimate** $7 - 2 = 5$

The LCM of 8 and 5 is 40.	Rename the fractions.	Subtract the fractions. Then subtract the whole numbers.
$6\frac{5}{8}$	$6\frac{25}{40}$	$6\frac{25}{40}$
$-\ 2\frac{2}{5}$	$-\ 2\frac{16}{40}$	$-\ 2\frac{16}{40}$
		$4\frac{9}{40}$

$6\frac{5}{8} \rightarrow 6\frac{25}{40} \rightarrow 6\frac{25}{40}$

6-6 Subtracting Mixed Numbers with Renaming (pp. 244–247)

Subtract. Write in simplest form.

46. $5 - 3\frac{2}{3}$ 47. $6\frac{3}{8} - 3\frac{5}{6}$

48. $9\frac{1}{6} - 4\frac{5}{6}$ 49. $7\frac{1}{2} - 6\frac{2}{3}$

50. $12\frac{2}{5} - 9\frac{2}{3}$ 51. $8\frac{5}{8} - 1\frac{3}{4}$

52. $4\frac{7}{12} - 2\frac{5}{6}$ 53. $2\frac{3}{7} - 1\frac{2}{3}$

54. **COOKING** A recipe for spaghetti sauce calls for $1\frac{2}{3}$ cups of tomato sauce. A can of tomato sauce holds $2\frac{1}{2}$ cups. How much tomato sauce will be left?

55. **CONSTRUCTION** A board measures $9\frac{2}{3}$ feet. A piece measuring $5\frac{7}{8}$ feet is cut off. Find the length of the remaining board.

Example 9

Find $3\frac{1}{5} - 1\frac{4}{5}$.   **Estimate** $3 - 2 = 1$

$3\frac{1}{5}$	Rename $3\frac{1}{5}$ as $2\frac{6}{5}$.	$2\frac{6}{5}$
$-\ 1\frac{4}{5}$		$-\ 1\frac{4}{5}$
		$1\frac{2}{5}$

Example 10

Find $5\frac{1}{4} - 3\frac{2}{3}$.   **Estimate** $5 - 4 = 1$

The LCM of 4 and 3 is 12.

$5\frac{1}{4}$	$\rightarrow$	$5\frac{3}{12}$	$\rightarrow$	$4\frac{15}{12}$	Rename $5\frac{3}{12}$ as $4\frac{15}{12}$.
$-\ 3\frac{2}{3}$	$\rightarrow$	$-\ 3\frac{8}{12}$		$-\ 3\frac{8}{12}$	
				$1\frac{7}{12}$	

Vocabulary and Concepts

1. **Explain** how to find the sum of two fractions with unlike denominators.

2. **OPEN ENDED** Write a subtraction problem with mixed numbers where you need to rename.

3. **State** the process used to subtract mixed numbers.

Skills and Applications

Round each number to the nearest half.

4. $4\frac{7}{8}$

5. $1\frac{10}{18}$

6. $8\frac{1}{5}$

7. $11\frac{1}{17}$

Estimate.

8. $\frac{8}{10} + \frac{3}{5}$

9. $5\frac{1}{7} - 2\frac{9}{11}$

10. $3\frac{6}{14} + 2\frac{2}{3}$

11. $5\frac{4}{9} - 4\frac{13}{23}$

Add or subtract. Write in simplest form.

12. $\frac{2}{9} + \frac{5}{9}$

13. $\frac{9}{10} - \frac{4}{10}$

14. $\frac{5}{6} - \frac{2}{6}$

15. $\frac{2}{9} + \frac{5}{6}$

16. $\frac{11}{12} - \frac{3}{8}$

17. $\frac{2}{5} + \frac{2}{4}$

18. $2\frac{1}{5} + 4\frac{2}{5}$

19. $6\frac{5}{8} - 4\frac{1}{2}$

20. $5\frac{7}{9} + 1\frac{3}{4}$

21. **CARPENTRY** In industrial technology class, Aiko made a plaque by gluing a piece of $\frac{3}{8}$-inch oak to a piece of $\frac{5}{8}$-inch poplar. What was the total thickness of the plaque?

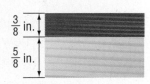

Subtract. Write in simplest form.

22. $4\frac{1}{4} - 2\frac{5}{8}$

23. $7\frac{2}{3} - 3\frac{3}{4}$

24. $11\frac{1}{2} - 7\frac{3}{5}$

Standardized Test Practice

Ⓐ Ⓑ Ⓒ Ⓓ

25. **MULTIPLE CHOICE** Sarah has $4\frac{1}{4}$ cups of flour. She is making cookies using a recipe that calls for $2\frac{5}{8}$ cups of flour. How much flour will she have left?

Ⓐ $6\frac{7}{8}$ c

Ⓑ $2\frac{1}{2}$ c

Ⓒ $2\frac{3}{8}$ c

Ⓓ $1\frac{5}{8}$ c

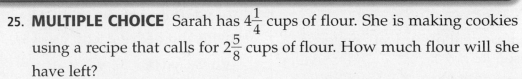

msmath1.net/chapter_test

PART 1 Multiple Choice

Record your answers on the answer sheet provided by your teacher or on a sheet of paper.

1. At the end of practice, Marcus places the tennis balls from three baskets holding 24, 19, and 31 balls into a large storage box. About how many tennis balls will be in the box? (Prerequisite Skill, p. 589)

    Ⓐ fewer than 60

    Ⓑ between 60 and 70

    Ⓒ between 70 and 80

    Ⓓ more than 90

2. In which order should the operations be performed in the expression $7 + 6 \div 2 \times 4 - 3$? (Lesson 1-5)

    Ⓕ multiply, subtract, divide, add

    Ⓖ add, divide, multiply, subtract

    Ⓗ multiply, divide, add, subtract

    Ⓘ divide, multiply, add, subtract

3. Wendy and Martin interviewed students for the school newspaper and made a graph showing the results.

    Favorite Sandwich Topping

    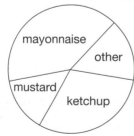

    About what percent of students prefer mayonnaise? (Lesson 2-3)

    Ⓐ 10%   Ⓑ 25%

    Ⓒ 40%   Ⓓ 50%

> **TEST-TAKING TIP**
>
> **Question 4** To round fractions quickly, divide the denominator by 2. If the numerator is the same as or greater than that number, round up. If the numerator is smaller than that number, round down.

4. Carey recorded the time she spent doing homework one week in a table. Which is the *best* estimate for the total time she spent on homework? (Lesson 6-2)

Day	Time (hours)
Monday	$2\frac{1}{5}$
Tuesday	$1\frac{2}{9}$
Wednesday	$\frac{2}{3}$
Thursday	$1\frac{3}{4}$

    Ⓕ 2 h   Ⓖ 4 h   Ⓗ 5 h   Ⓘ 6 h

5. A recipe calls for $\frac{1}{4}$ cup of brown sugar and $\frac{2}{3}$ cup of granulated sugar. What is the total amount of sugar needed? (Lesson 6-4)

    Ⓐ $\frac{1}{6}$ c   Ⓑ $\frac{3}{7}$ c

    Ⓒ $\frac{2}{3}$ c   Ⓓ $\frac{11}{12}$ c

6. What is the difference between $9\frac{5}{7}$ and $4\frac{3}{7}$? (Lesson 6-5)

    Ⓕ $5\frac{2}{7}$ Ⓖ $5\frac{2}{14}$

    Ⓗ $5\frac{2}{49}$ Ⓘ $4\frac{2}{49}$

7. David had $6\frac{3}{8}$ yards of twine. He used $2\frac{3}{4}$ yards on a craft project. How much twine did he have left? (Lesson 6-6)

    Ⓐ $3\frac{5}{8}$ yd   Ⓑ $3\frac{3}{4}$ yd

    Ⓒ 4 yd   Ⓓ $4\frac{3}{8}$ yd

PART 2   Short Response/Grid In

Record your answers on the answer sheet provided by your teacher, or on a sheet of paper.

8. Scott bought some items for his dog at Pet Mart, as shown. To the nearest dollar, how much did Scott spend?
(Lesson 3-4)

> **Pet Mart**
>
> Bone................$2.67
> Squeak Toy......$5.03
> Leash..............$7.18
> Treat................$1.01
>
> Total..............$15.89

9. If one British pound is worth 1.526 U.S. dollars, how much are 5 British pounds worth in U.S. dollars? (Lesson 4-1)

10. Ali made a drawing of the route she walks every day. The curved part of the route is in the shape of a semicircle with a radius of 10.5 meters.

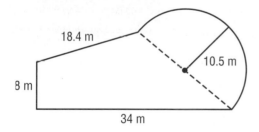

To the nearest tenth, what is the distance Ali walks every day? (Lesson 4-6)

11. Emma's teacher gave her a summer reading list. After four weeks of summer break, Emma had read 0.45 of the books on the list. What is 0.45 expressed as a fraction? (Lesson 5-6)

12. Martina measures the height of a drawing that she wants to frame. It is $16\frac{5}{8}$ inches tall. To round the height to the nearest inch, should she round up or down? Why? (Lesson 6-1)

13. What is the *best* whole number estimate for $3\frac{3}{5} + 1\frac{4}{5} + \frac{1}{3}$? (Lesson 6-2)

14. What is the value of $\frac{4}{7} - \frac{1}{2}$? (Lesson 6-4)

15. Find the difference between $6\frac{5}{7}$ and $4\frac{2}{7}$. (Lesson 6-5)

16. So far this week, Julio swam $\frac{1}{2}$ hour, $1\frac{1}{4}$ hours, and $\frac{5}{6}$ hour. If his goal is to swim for 3 hours each week, how many more hours does he need to swim to reach his goal? (Lesson 6-6)

PART 3   Extended Response

Record your answers on a sheet of paper. Show your work.

17. Keisha plans to put a wallpaper border around her room. The room dimensions are shown below. (Lessons 6-5 and 6-6)

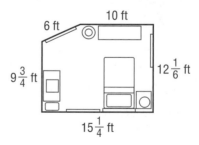

a. Explain how you can find the length of border needed to go around the entire room.

b. How many feet of border are needed?

c. A standard roll of wallpaper border measures 15 feet. How many rolls of border are needed to go around the room?

d. How many feet of border will be left over?

Multiplying and Dividing Fractions

"How is math useful on road trips?"

When traveling by car, you can calculate the gas mileage, or miles per gallon, by using division. For example, the gas mileage of a car that travels **407 miles** on $18\frac{1}{2}$ **gallons** of gasoline is **407 ÷ $18\frac{1}{2}$.** In mathematics, you will divide fractions and mixed numbers to solve many real-life problems.

You will solve problems about gas mileage in Lesson 7-5.

Take this quiz to see whether you are ready to begin Chapter 7. Refer to the lesson number in parentheses if you need more review.

▶ Vocabulary Review

Complete each sentence.

1. The GCF represents the __?__ of a set of numbers. (Lesson 5-1)

2. $\frac{7}{2}$ is a(n) __?__ because the numerator is greater than the denominator. (Lesson 5-1)

3. 6, 9, and 12 are all __?__ of 3. (Lesson 5-1)

▶ Prerequisite Skills

Use a calculator to find each product. Round to the nearest tenth. (Lesson 3-3)

4. $\pi \times 20$

5. $8 \times \pi$

6. $2 \times \pi \times 5$

7. $4 \times \pi \times 9$

Find the GCF of each set of numbers.
(Lesson 5-1)

8. 6, 24

9. 18, 12

10. 14, 8

11. 10, 20

12. 21, 9

13. 28, 32

Write each mixed number as an improper fraction. (Lesson 5-3)

14. $2\frac{3}{4}$

15. $1\frac{6}{7}$

16. $5\frac{7}{9}$

17. $3\frac{1}{8}$

Round each fraction to 0, $\frac{1}{2}$, or 1. (Lesson 6-1)

18. $\frac{1}{5}$

19. $\frac{4}{7}$

20. $\frac{11}{12}$

21. $\frac{2}{15}$

Fractions Make this Foldable to help you organize information about fractions. Begin with a sheet of $8\frac{1}{2}$" by 11" paper.

STEP 1 Fold
Fold the paper along the width, leaving a 1-inch margin at the top.

STEP 2 Fold Again
Fold in half widthwise.

STEP 3 Unfold and Cut
Unfold. Cut along the vertical fold from the bottom to the first fold.

STEP 4 Label
Label each tab as shown. In the top margin write *Fractions*, and draw arrows to the tabs.

Reading and Writing As you read and study the chapter, under each tab write notes, examples, and models that show multiplication and division of fractions.

Estimating Products

What You'll LEARN

Estimate products using compatible numbers and rounding.

NEW Vocabulary

compatible numbers

REVIEW Vocabulary

multiple of a number: the product of the number and any whole number (Lesson 5-4)

WHEN am I ever going to use this?

SPORTS Kayla made about $\frac{1}{3}$ of the 14 shots she attempted in a basketball game.

1. For the shots attempted, what is the nearest multiple of 3?

2. How many basketballs should be added to reflect the nearest multiple of 3?

3. Divide the basketballs into three groups each having the same number. How many basketballs are in each group?

4. About how many shots did Kayla make?

One way to estimate products is to use **compatible numbers**, which are numbers that are easy to divide mentally.

EXAMPLES Estimate Using Compatible Numbers

1 Estimate $\frac{1}{4} \times 13$. $\frac{1}{4} \times 13$ means $\frac{1}{4}$ of 13.

Find a number close to 13 that is a multiple of 4.

$\frac{1}{4} \times 13 \rightarrow \frac{1}{4} \times 12$ 12 and 4 are compatible numbers since 12 ÷ 4 = 3.

$\frac{1}{4} \times 12 = 3$ 12 ÷ 4 = 3.

So, $\frac{1}{4} \times 13$ is *about* 3.

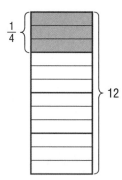

2 Estimate $\frac{2}{5} \times 11$.

Estimate $\frac{1}{5} \times 11$ first.

$\frac{1}{5} \times 11 \rightarrow \frac{1}{5} \times 10$ Use 10 since 10 and 5 are compatible numbers.

$\frac{1}{5} \times 10 = 2$ 10 ÷ 5 = 2

If $\frac{1}{5}$ of 10 is 2, then $\frac{2}{5}$ of 10 is 2 × 2 or 4.

So, $\frac{2}{5} \times 11$ is *about* 4.

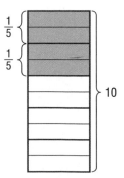

Your Turn Estimate each product.

a. $\frac{1}{5} \times 16$

b. $\frac{5}{6} \times 13$

c. $\frac{3}{4} \times 23$

EXAMPLE **Estimate by Rounding to 0, $\frac{1}{2}$, or 1**

3 Estimate $\frac{1}{3} \times \frac{7}{8}$.

$$\frac{1}{3} \times \frac{7}{8} \rightarrow \frac{1}{2} \times 1$$

$$\frac{1}{2} \times 1 = \frac{1}{2}$$

So, $\frac{1}{3} \times \frac{7}{8}$ is *about* $\frac{1}{2}$.

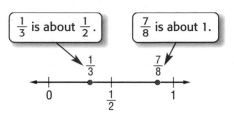

$\frac{1}{3}$ is about $\frac{1}{2}$. $\frac{7}{8}$ is about 1.

Your Turn Estimate each product.

d. $\frac{5}{8} \times \frac{9}{10}$ e. $\frac{5}{6} \times \frac{9}{10}$ f. $\frac{5}{6} \times \frac{1}{9}$

EXAMPLE **Estimate With Mixed Numbers**

STUDY TIP

Look Back You can review **area of rectangles** in Lesson 1-8.

4 **GEOMETRY** Estimate the area of the rectangle.

Round each mixed number to the nearest whole number.

$$11\frac{3}{4} \times 8\frac{1}{6} \rightarrow 12 \times 8 = 96$$

Round $11\frac{3}{4}$ to 12.   Round $8\frac{1}{6}$ to 8.

So, the area is *about* 96 square feet.

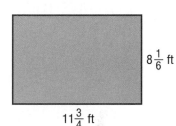

$8\frac{1}{6}$ ft

$11\frac{3}{4}$ ft

Skill and Concept Check

Writing Math

Exercises 1 & 3

1. **Explain** how to use compatible numbers to estimate $\frac{2}{3} \times 8$.

2. **OPEN ENDED** Write an example of two mixed numbers whose product is about 6.

3. **NUMBER SENSE** Is $\frac{2}{3} \times 20$ greater than 14 or less than 14? Explain.

GUIDED PRACTICE

Estimate each product.

4. $\frac{1}{8} \times 15$ 5. $\frac{3}{4} \times 21$ 6. $\frac{1}{4} \times \frac{8}{9}$

7. $\frac{5}{8} \times \frac{1}{9}$ 8. $6\frac{2}{3} \times 4\frac{1}{5}$ 9. $2\frac{9}{10} \times 10\frac{3}{4}$

10. **PAINTING** A wall measures $8\frac{1}{2}$ feet by $12\frac{3}{4}$ feet. If a gallon of paint covers about 150 square feet, will one gallon of paint be enough to cover the wall? Explain.

Estimate each product.

11. $\frac{1}{4} \times 21$ **12.** $\frac{2}{3} \times 10$ **13.** $\frac{5}{7} \times \frac{3}{4}$ **14.** $\frac{5}{6} \times \frac{8}{9}$

15. $\frac{5}{7} \times \frac{1}{9}$ **16.** $\frac{1}{10} \times \frac{7}{8}$ **17.** $\frac{11}{12} \times \frac{3}{8}$ **18.** $\frac{2}{5} \times \frac{9}{10}$

19. $4\frac{1}{3} \times 2\frac{3}{4}$ **20.** $6\frac{4}{5} \times 4\frac{1}{9}$ **21.** $5\frac{1}{8} \times 9\frac{1}{12}$ **22.** $2\frac{9}{10} \times 8\frac{5}{6}$

23. Estimate $\frac{3}{8} \times \frac{1}{11}$.   **24.** Estimate $\frac{5}{9}$ of $7\frac{7}{8}$.

HOMEWORK HELP	
For Exercises	See Examples
11–12, 25–26	1, 2
13–18, 23	3
19–22, 24	4

Extra Practice
See pages 607, 630.

25. VOLUNTEERING The circle graph shows the fraction of teens who volunteer. Suppose 100 teens were surveyed. About how many teens do *not* volunteer?

Teens Volunteering

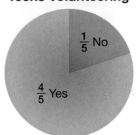

Source: 200M and Research & Consulting

26. SPORTS Barry Zito of the Oakland Athletics won about $\frac{4}{5}$ of the games for which he was the pitcher of record in 2002. If he was the pitcher of record for 28 games, about how many games did he win? Explain.

27. CRITICAL THINKING Which point on the number line could be the graph of the product of the numbers graphed at C and D?

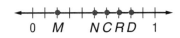

28. MULTIPLE CHOICE Which is the best estimate of the area of the rectangle?

  Ⓐ 15 in²   Ⓑ 20 in²   Ⓒ 24 in²   Ⓓ 16 in²

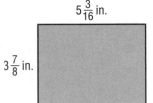
$5\frac{3}{16}$ in. $3\frac{7}{8}$ in.

29. SHORT RESPONSE Ruby has budgeted $\frac{1}{4}$ of her allowance for savings. If she receives \$25 a month, about how much will she put in savings?

30. BAKING Viho needs $2\frac{1}{4}$ cups of flour for making cookies, $1\frac{2}{3}$ cups for almond bars, and $3\frac{1}{2}$ cups for cinnamon rolls. How much flour does he need in all? (Lesson 6-6)

Subtract. Write in simplest form. (Lesson 6-5)

31. $10\frac{3}{8} - 7\frac{1}{8}$ **32.** $5\frac{1}{6} - 3\frac{8}{9}$ **33.** $8\frac{2}{3} - 3\frac{6}{7}$ **34.** $6\frac{3}{5} - 4\frac{2}{3}$

GETTING READY FOR THE NEXT LESSON

PREREQUISITE SKILL Find the GCF of each set of numbers. (Lesson 5-1)

35. 6, 9   **36.** 8, 6   **37.** 10, 4   **38.** 15, 9   **39.** 24, 16

What You'll LEARN

Multiply fractions using models.

Materials

- paper
- markers

Multiplying Fractions

In Chapter 4, you used decimal models to multiply decimals. You can use a similar model to multiply fractions.

ACTIVITY *Work with a partner.*

1 Find $\frac{1}{3} \times \frac{1}{2}$ using a model.

To find $\frac{1}{3} \times \frac{1}{2}$, find $\frac{1}{3}$ of $\frac{1}{2}$.

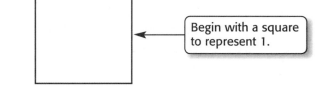

Begin with a square to represent 1.

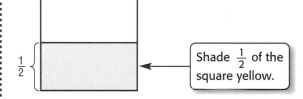

Shade $\frac{1}{2}$ of the square yellow.

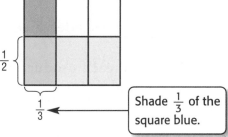

Shade $\frac{1}{3}$ of the square blue.

One sixth of the square is shaded green. So, $\frac{1}{3} \times \frac{1}{2} = \frac{1}{6}$.

Your Turn Find each product using a model.

a. $\frac{1}{4} \times \frac{1}{2}$　　　b. $\frac{1}{3} \times \frac{1}{4}$　　　c. $\frac{1}{2} \times \frac{1}{5}$

Writing Math

1. Describe how you would change the model to find $\frac{1}{2} \times \frac{1}{3}$. Is the product the same as $\frac{1}{3} \times \frac{1}{2}$? Explain.

ACTIVITY *Work with a partner.*

2 Find $\frac{3}{5} \times \frac{2}{3}$ using a model. Write in simplest form.

To find $\frac{3}{5} \times \frac{2}{3}$, find $\frac{3}{5}$ of $\frac{2}{3}$.

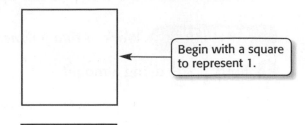

Begin with a square to represent 1.

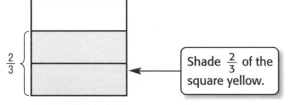

$\frac{2}{3}$

Shade $\frac{2}{3}$ of the square yellow.

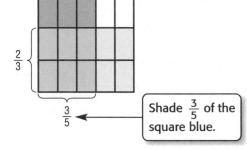

$\frac{2}{3}$

$\frac{3}{5}$

Shade $\frac{3}{5}$ of the square blue.

Six out of 15 parts are shaded green. So, $\frac{3}{5} \times \frac{2}{3} = \frac{6}{15}$ or $\frac{2}{5}$.

Your Turn Find each product using a model. Then write in simplest form.

d. $\frac{3}{4} \times \frac{2}{3}$ e. $\frac{2}{5} \times \frac{5}{6}$ f. $\frac{4}{5} \times \frac{3}{8}$

Writing Math

2. Draw a model to show that $\frac{2}{3} \times \frac{5}{6} = \frac{10}{18}$. Then explain how the model shows that $\frac{10}{18}$ simplifies to $\frac{5}{9}$.

3. Explain the relationship between the numerators of the problem and the numerator of the product. What do you notice about the denominators of the problem and the denominator of the product?

4. **MAKE A CONJECTURE** Write a rule you can use to multiply fractions.

Multiplying Fractions

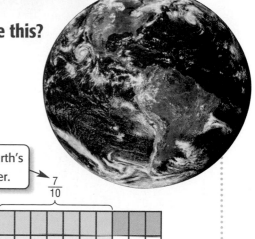

What You'll LEARN

Multiply fractions.

REVIEW Vocabulary

greatest common factor (GCF): the greatest of the common factors of two or more numbers **(Lesson 5-1)**

WHEN am I ever going to use this?

EARTH The model represents the part of Earth that is covered by water and the part that is covered by the Pacific Ocean. The overlapping area represents $\frac{1}{2}$ of $\frac{7}{10}$ or $\frac{1}{2} \times \frac{7}{10}$.

About $\frac{7}{10}$ of Earth's surface is water.

$\frac{7}{10}$

$\frac{1}{2}$

About $\frac{1}{2}$ of Earth's water surface is the Pacific Ocean.

1. What part of Earth's surface is covered by the Pacific Ocean?

2. What is the relationship between the numerators and denominators of the factors and the numerator and denominator of the product?

Key Concept   **Multiplying Fractions**

Words   To multiply fractions, multiply the numerators and multiply the denominators.

Symbols   **Arithmetic**   **Algebra**

$$\frac{2}{5} \times \frac{1}{2} = \frac{2 \times 1}{5 \times 2} \qquad \frac{a}{b} \times \frac{c}{d} = \frac{a \times c}{b \times d}, \text{ where } b \text{ and } d \text{ are not } 0.$$

EXAMPLE   **Multiply Fractions**

1 Find $\frac{1}{3} \times \frac{1}{4}$.

$\frac{1}{3} \times \frac{1}{4} = \frac{1 \times 1}{3 \times 4}$ Multiply the numerators.
Multiply the denominators.

$= \frac{1}{12}$ Simplify.

$\frac{1}{4}$

$\frac{1}{3}$

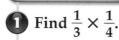 **Multiply. Write in simplest form.**

a. $\frac{1}{2} \times \frac{3}{5}$ **b.** $\frac{1}{3} \times \frac{3}{4}$ **c.** $\frac{2}{3} \times \frac{5}{6}$

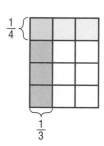
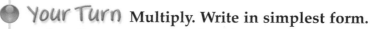

To multiply a fraction and a whole number, first write the whole number as a fraction.

EXAMPLE Multiply Fractions and Whole Numbers

2 Find $\frac{3}{5} \times 4$. **Estimate** $\frac{1}{2} \times 4 = 2$

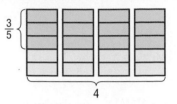

$$\frac{3}{5} \times 4 = \frac{3}{5} \times \frac{4}{1} \qquad \text{Write 4 as } \frac{4}{1}.$$

$$= \frac{3 \times 4}{5 \times 1} \qquad \text{Multiply.}$$

$$= \frac{12}{5} \text{ or } 2\frac{2}{5} \qquad \text{Simplify. Compare to the estimate.}$$

Your Turn Multiply. Write in simplest form.

d. $\frac{2}{3} \times 6$ e. $\frac{3}{4} \times 5$ f. $3 \times \frac{1}{2}$

If the numerators and the denominators have a common factor, you can simplify *before* you multiply.

EXAMPLE Simplify Before Multiplying

3 Find $\frac{3}{4} \times \frac{5}{6}$. **Estimate** $\frac{1}{2} \times 1 = \frac{1}{2}$

The numerator 3 and the denominator 6 have a common factor, 3.

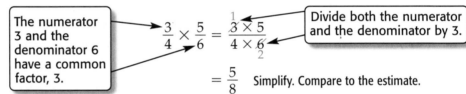

$$\frac{3}{4} \times \frac{5}{6} = \frac{\overset{1}{\cancel{3}} \times 5}{4 \times \cancel{6}_{2}}$$

Divide both the numerator and the denominator by 3.

$$= \frac{5}{8} \qquad \text{Simplify. Compare to the estimate.}$$

Your Turn Multiply. Write in simplest form.

g. $\frac{3}{4} \times \frac{4}{9}$ h. $\frac{5}{6} \times \frac{9}{10}$ i. $\frac{3}{5} \times 10$

EXAMPLE Evaluate Expressions

4 **ALGEBRA** Evaluate ab if $a = \frac{2}{3}$ and $b = \frac{3}{8}$.

$$ab = \frac{2}{3} \times \frac{3}{8} \qquad \text{Replace } a \text{ with } \frac{2}{3} \text{ and } b \text{ with } \frac{3}{8}.$$

$$= \frac{\overset{1}{\cancel{2}} \times \overset{1}{\cancel{3}}}{\underset{1}{\cancel{3}} \times \underset{4}{\cancel{8}}} \qquad \begin{array}{l}\text{The GCF of 2 and 8 is 2. The GCF of 3 and} \\ \text{3 is 3. Divide both the numerator and the} \\ \text{denominator by 2 and then by 3.}\end{array}$$

$$= \frac{1}{4} \qquad \text{Simplify.}$$

Your Turn

j. Evaluate $\frac{3}{4}c$ if $c = \frac{2}{5}$. k. Evaluate $5a$ if $a = \frac{3}{10}$.

STUDY TIP

Mental Math You can multiply some fractions mentally. For example, $\frac{1}{3}$ of $\frac{3}{8} = \frac{1}{8}$. So, $\frac{2}{3}$ of $\frac{3}{8} = \frac{2}{8}$ or $\frac{1}{4}$.

Writing Math

Exercises 3 & 4

1. **Draw** a model to show why $\frac{2}{3} \times \frac{1}{2} = \frac{1}{3}$.

2. **OPEN ENDED** Write an example of multiplying two fractions where you can simplify before you multiply.

3. **NUMBER SENSE** Natalie multiplied $\frac{2}{3}$ and 22 and got 33. Is this answer reasonable? Why or why not?

4. **NUMBER SENSE** Without multiplying, tell whether the product of $\frac{5}{9}$ and $\frac{4}{7}$ is a fraction or a mixed number. Explain.

GUIDED PRACTICE

Multiply. Write in simplest form.

5. $\frac{1}{8} \times \frac{1}{2}$

6. $\frac{4}{5} \times 10$

7. $\frac{1}{3} \times \frac{3}{4}$

8. $\frac{3}{10} \times \frac{5}{6}$

9. $\frac{3}{4} \times 12$

10. $\frac{3}{5} \times \frac{5}{6}$

11. **ALGEBRA** Evaluate xy if $x = \frac{1}{4}$ and $y = \frac{5}{6}$.

12. **HOBBIES** Suppose you are building a model car that is $\frac{1}{12}$ the size of the actual car. How long is the model if the actual car is shown at the right?

16 ft

Practice and Applications

Multiply. Write in simplest form.

13. $\frac{1}{3} \times \frac{2}{5}$

14. $\frac{1}{8} \times \frac{3}{4}$

15. $\frac{3}{4} \times 2$

16. $\frac{2}{3} \times 4$

17. $\frac{2}{3} \times \frac{1}{4}$

18. $\frac{3}{5} \times \frac{5}{7}$

19. $\frac{4}{9} \times \frac{3}{8}$

20. $\frac{2}{5} \times \frac{5}{6}$

21. $\frac{3}{4} \times \frac{5}{8}$

22. $\frac{5}{6} \times 15$

23. $\frac{1}{2} \times \frac{4}{9}$

24. $\frac{7}{8} \times \frac{2}{3}$

25. $\frac{1}{2} \times \frac{1}{3} \times \frac{1}{4}$

26. $\frac{2}{3} \times \frac{3}{4} \times \frac{2}{3}$

27. $\frac{1}{2} \times \frac{2}{5} \times \frac{15}{16}$

28. $\frac{2}{3} \times \frac{9}{10} \times \frac{5}{9}$

HOMEWORK HELP

For Exercises	See Examples
13–14	1
15–16, 33–35	2
17–24, 36	3
29–32	4

Extra Practice
See pages 607, 630.

ALGEBRA Evaluate each expression if $a = \frac{3}{5}$, $b = \frac{1}{2}$, and $c = \frac{1}{3}$.

29. ab

30. bc

31. $\frac{1}{3}a$

32. ac

33. **LIFE SCIENCE** About $\frac{7}{10}$ of the human body is water. How many pounds of water are in a person weighing 120 pounds?

34. MALLS The area of a shopping mall is 700,000 square feet. About $\frac{1}{9}$ of the area is for stores that are food related. About how many square feet in the mall are for food-related stores?

35. GEOGRAPHY Rhode Island's area is 1,545 square miles. Water makes up about $\frac{1}{3}$ of the area of the state. About how many square miles of water does Rhode Island have?

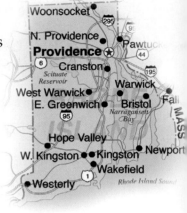

 Data Update What part of the area of your state is water? Visit msmath1.net/data_update to learn more.

36. FLAGS In a recent survey, $\frac{4}{5}$ of Americans said they were displaying the American flag. Five-eighths of these displayed the flag on their homes. What fraction of Americans displayed a flag on their home?

37. CRITICAL THINKING Find $\frac{1}{2} \times \frac{2}{3} \times \frac{3}{4} \times \frac{4}{5} \times \ldots \times \frac{99}{100}$.

Standardized Test Practice and Mixed Review

38. MULTIPLE CHOICE Evaluate ab if $a = \frac{7}{8}$ and $b = \frac{2}{3}$.

   Ⓐ $\frac{21}{16}$ Ⓑ $\frac{9}{11}$ Ⓒ $\frac{72}{83}$ Ⓓ $\frac{7}{12}$

39. SHORT RESPONSE On a warm May day, $\frac{6}{7}$ of the students at West Middle School wore short-sleeved T-shirts, and $\frac{2}{3}$ of those students wore shorts. What fraction of the students wore T-shirts and shorts to school?

Estimate each product. (Lesson 7-1)

40. $\frac{1}{6} \times 29$ **41.** $1\frac{8}{9} \times 5\frac{1}{6}$ **42.** $\frac{1}{7} \times 3\frac{5}{6}$ **43.** $\frac{4}{9} \times \frac{8}{9}$

44. HEALTH Joaquin is $65\frac{1}{2}$ inches tall. Juan is $61\frac{3}{4}$ inches tall. How much taller is Joaquin than Juan? (Lesson 6-6)

45. SPORTS The table shows the finishing times for four runners in a 100-meter race. In what order did the runners cross the finish line? (Lesson 3-2)

Runner	Time
Sarah	14.31 s
Camellia	13.84 s
Fala	13.97 s
Debbie	13.79 s

GETTING READY FOR THE NEXT LESSON

PREREQUISITE SKILL Write each mixed number as an improper fraction. (Lesson 5-3)

46. $4\frac{1}{2}$ **47.** $3\frac{1}{4}$ **48.** $5\frac{2}{3}$ **49.** $2\frac{5}{7}$ **50.** $9\frac{3}{4}$ **51.** $6\frac{5}{8}$

Multiplying Mixed Numbers

What You'll LEARN

Multiply mixed numbers.

REVIEW Vocabulary

improper fraction: a fraction with a numerator that is greater than or equal to the denominator (Lesson 5-3)

WHEN am I ever going to use this?

EXERCISE Jasmine walks 3 days a week, $2\frac{1}{2}$ miles each day. The number line shows the miles she walks in a week.

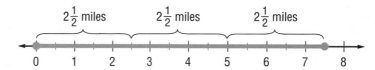

1. How many miles does Jasmine walk in a week?

2. Write a multiplication sentence that shows the total miles walked in a week.

3. Write the multiplication sentence using improper fractions.

Use a number line and improper fractions to find each product.

4. $2 \times 1\frac{1}{3}$

5. $2 \times 2\frac{1}{4}$

6. $3 \times 1\frac{3}{4}$

7. Describe how multiplying mixed numbers is similar to multiplying fractions.

Multiplying mixed numbers is similar to multiplying fractions.

Key Concept — Multiplying Mixed Numbers

To multiply mixed numbers, write the mixed numbers as improper fractions and then multiply as with fractions.

EXAMPLE — Multiply a Fraction and a Mixed Number

1 Find $\frac{1}{4} \times 4\frac{4}{5}$. **Estimate** Use compatible numbers → $\frac{1}{4} \times 4 = 1$

$$\frac{1}{4} \times 4\frac{4}{5} = \frac{1}{4} \times \frac{24}{5} \qquad \text{Write } 4\frac{4}{5} \text{ as } \frac{24}{5}.$$

$$= \frac{1 \times \overset{6}{24}}{\underset{1}{4} \times 5} \qquad \text{Divide 24 and 4 by their GCF, 4.}$$

$$= \frac{6}{5} \text{ or } 1\frac{1}{5} \qquad \text{Simplify. Compare to the estimate.}$$

Your Turn Multiply. Write in simplest form.

a. $\frac{2}{3} \times 2\frac{1}{2}$

b. $\frac{3}{8} \times 3\frac{1}{3}$

c. $3\frac{1}{2} \times \frac{1}{3}$

EXAMPLE   Multiply Mixed Numbers

2 BAKING Jessica is making $2\frac{1}{2}$ batches of chocolate chip cookies for a bake sale. If one batch uses $2\frac{1}{4}$ cups of flour, how much flour will she need?

Estimate $3 \times 2 = 6$

Each batch uses $2\frac{1}{4}$ cups of flour. So, multiply $2\frac{1}{2}$ by $2\frac{1}{4}$.

$$2\frac{1}{2} \times 2\frac{1}{4} = \frac{5}{2} \times \frac{9}{4}$$ First, write mixed numbers as improper fractions.

$$= \frac{5}{2} \times \frac{9}{4}$$ Then, multiply the numerators and multiply the denominators.

$$= \frac{45}{8} \text{ or } 5\frac{5}{8}$$ Simplify.

Jessica will need $5\frac{5}{8}$ cups of flour.   Compare this to the estimate.

EXAMPLE   Evaluate Expressions

3 ALGEBRA If $c = 1\frac{7}{8}$ and $d = 3\frac{1}{3}$, what is the value of cd?

$$cd = 1\frac{7}{8} \times 3\frac{1}{3}$$ Replace c with $1\frac{7}{8}$ and d with $3\frac{1}{3}$.

$$= \frac{\overset{5}{\cancel{15}}}{\underset{4}{\cancel{8}}} \times \frac{\overset{5}{\cancel{10}}}{\underset{1}{\cancel{3}}}$$ Divide the numerator and denominator by 3 and by 2.

$$= \frac{25}{4} \text{ or } 6\frac{1}{4}$$ Simplify.

Skill and Concept Check

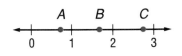

Writing Math

Exercises 1–3

1. **Describe** how to multiply mixed numbers.

2. **OPEN ENDED** Write a problem that can be solved by multiplying mixed numbers. Explain how to find the product.

3. **NUMBER SENSE** Without multiplying, tell whether the product $2\frac{1}{2} \times \frac{2}{3}$ is located on the number line at point A, B, or C. Explain your reasoning.

GUIDED PRACTICE

Multiply. Write in simplest form.

4. $\frac{1}{2} \times 2\frac{3}{8}$

5. $1\frac{1}{2} \times \frac{2}{3}$

6. $2 \times 6\frac{1}{4}$

7. $1\frac{3}{4} \times 2\frac{4}{5}$

8. $3\frac{1}{3} \times 1\frac{1}{5}$

9. $\frac{3}{8} \times 1\frac{1}{4}$

10. **ALGEBRA** If $x = \frac{9}{10}$ and $y = 1\frac{1}{3}$, find xy.

Practice and Applications

HOMEWORK HELP

For Exercises	See Examples
11–14, 27–30	1
15–20	2
23–26	3

Extra Practice
See pages 608, 630.

Multiply. Write in simplest form.

11. $\frac{1}{2} \times 2\frac{1}{3}$ **12.** $\frac{3}{4} \times 2\frac{5}{6}$ **13.** $1\frac{7}{8} \times \frac{4}{5}$ **14.** $1\frac{4}{5} \times \frac{5}{6}$

15. $1\frac{1}{3} \times 1\frac{1}{4}$ **16.** $3\frac{1}{5} \times 3\frac{1}{6}$ **17.** $3\frac{3}{4} \times 2\frac{2}{5}$ **18.** $4\frac{1}{2} \times 2\frac{5}{6}$

19. $6\frac{2}{3} \times 3\frac{3}{10}$ **20.** $3\frac{3}{5} \times 5\frac{5}{12}$ **21.** $\frac{3}{4} \times 2\frac{1}{2} \times \frac{4}{5}$ **22.** $1\frac{1}{2} \times \frac{2}{3} \times \frac{3}{5}$

ALGEBRA Evaluate each expression if $a = \frac{2}{3}$, $b = 3\frac{1}{2}$, and $c = 1\frac{3}{4}$.

23. ab **24.** $\frac{1}{2}c$ **25.** bc **26.** $\frac{1}{8}a$

MUSIC For Exercises 27–30, use the following information.
A dot following a music note (o•) means that the note gets $1\frac{1}{2}$ times as many beats as the same note without a dot (o). How many beats does each note get?

27. dotted whole note
28. dotted quarter note
29. dotted eighth note
30. dotted half note

Name	Note	Number of Beats
Whole Note	o	4
Half Note	♩	2
Quarter Note	♩	1
Eighth Note	♪	$\frac{1}{2}$

31. CRITICAL THINKING Is the product of two mixed numbers *always*, *sometimes*, or *never* less than 1? Explain.

Standardized Test Practice and Mixed Review

32. MULTIPLE CHOICE To find the area of a parallelogram, multiply the length of the base by the height. What is the area of this parallelogram?

Ⓐ $5\frac{3}{4}$ ft²   Ⓑ $6\frac{1}{4}$ ft²   Ⓒ $6\frac{3}{4}$ ft²   Ⓓ $8\frac{1}{4}$ ft²

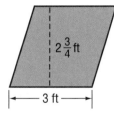

$2\frac{3}{4}$ ft

3 ft

33. SHORT RESPONSE A bag of apples weighs $3\frac{1}{2}$ pounds. How much do $1\frac{1}{2}$ bags weigh?

Multiply. Write in simplest form. (Lesson 7-2)

34. $\frac{5}{7} \times \frac{3}{4}$ **35.** $\frac{2}{3} \times \frac{1}{6}$ **36.** $\frac{3}{8} \times \frac{2}{5}$ **37.** $\frac{1}{2} \times \frac{4}{7}$

38. RECREATION There are about 7 million pleasure boats in the United States. About $\frac{2}{3}$ of these boats are motorboats. About how many motorboats are in the United States? (Lesson 7-1)

GETTING READY FOR THE NEXT LESSON

PREREQUISITE SKILL Multiply. Write in simplest form. (Lesson 7-2)

39. $\frac{1}{4} \times \frac{3}{8}$ **40.** $\frac{2}{7} \times \frac{3}{4}$ **41.** $\frac{1}{2} \times \frac{1}{6}$ **42.** $\frac{2}{5} \times \frac{5}{6}$

Vocabulary and Concepts

1. **Explain** how to multiply fractions. (Lesson 7-2)

2. **State** the first step you should do when multiplying mixed numbers. (Lesson 7-3)

Skills and Applications

Estimate each product. (Lesson 7-1)

3. $\frac{1}{3} \times 22$

4. $\frac{8}{9} \times \frac{2}{15}$

5. $3\frac{2}{3} \times 5\frac{1}{9}$

6. $\frac{1}{9} \times 44$

7. $7\frac{3}{4} \times 3\frac{1}{8}$

8. $\frac{11}{12} \times \frac{3}{5}$

Multiply. Write in simplest form. (Lesson 7-2)

9. $\frac{1}{4} \times \frac{4}{9}$

10. $\frac{3}{5} \times \frac{2}{9}$

11. $\frac{5}{8} \times \frac{4}{7}$

12. $\frac{6}{7} \times \frac{14}{15}$

13. **ALGEBRA** Evaluate ab if $a = \frac{5}{6}$ and $b = \frac{1}{2}$. (Lesson 7-2)

Multiply. Write in simplest form. (Lesson 7-3)

14. $\frac{3}{8} \times 2\frac{2}{3}$

15. $1\frac{4}{5} \times 3$

16. $12\frac{1}{2} \times \frac{1}{5}$

17. $3\frac{3}{8} \times 1\frac{4}{9}$

18. **GEOMETRY** To find the area of a parallelogram, use the formula $A = bh$, where b is the length of the base and h is the height. Find the area of the parallelogram. (Lesson 7-3)

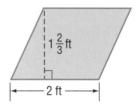

$1\frac{2}{3}$ ft

2 ft

Standardized Test Practice

Ⓐ Ⓑ Ⓒ Ⓓ

19. **MULTIPLE CHOICE** What is the product of $4\frac{1}{2}$ and $2\frac{2}{3}$? (Lesson 7-3)

Ⓐ $1\frac{11}{16}$

Ⓑ $7\frac{1}{6}$

Ⓒ $8\frac{1}{3}$

Ⓓ 12

20. **MULTIPLE CHOICE** Which expression is equal to $\frac{21}{48}$? (Lesson 7-2)

Ⓕ $\frac{1}{2} \times \frac{11}{48}$

Ⓖ $\frac{3}{48} \times \frac{7}{48}$

Ⓗ $\frac{7}{12} \times \frac{3}{4}$

Ⓘ $\frac{10}{6} \times \frac{11}{8}$

The Game Zone

A Place To Practice Your Math Skills

Multiplication Chaos

- ● **GET READY!**

    Players: two, three, or four
    Materials: poster board, straightedge, 2 number cubes

- ● **GET SET!**

    • Draw a large game board on your poster board like the one shown.

- ● **GO!**

    • Place the game board on the floor.

    • Each player rolls the number cubes. The person with the highest total starts.

    • The first player rolls the number cubes onto the game board. If a number cube rolls off the board or lands on a line, roll it again.

    • The player then multiplies the two numbers on which the number cubes land and simplifies the product. Each correct answer is worth 1 point.

    • Then the next player rolls the number cubes and finds the product.

    • **Who Wins?** The first player to score 10 points wins.

The game board shows fractions and whole numbers arranged in concentric rings:

Outer ring: $\frac{3}{8}$, 3, $\frac{5}{7}$, $\frac{6}{7}$, $\frac{7}{8}$, 2, $\frac{5}{6}$, $\frac{5}{8}$

Middle ring: $\frac{2}{3}$, $\frac{5}{8}$, $\frac{2}{5}$, $\frac{3}{4}$, $\frac{4}{5}$, $\frac{3}{5}$, $\frac{3}{7}$, $\frac{3}{8}$

Inner ring: $\frac{1}{4}$, $\frac{1}{3}$, $\frac{1}{8}$, $\frac{1}{5}$

Center: $\frac{1}{2}$

Dividing Fractions

What You'll LEARN

Divide fractions using models.

Materials

• paper
• colored pencils
• scissors

There are 8 pieces of candy that are given away 2 at a time. How many people will get candy?

1. How many 2s are in 8? Write as a division expression.

Suppose there are two granola bars divided equally among 8 people. What part of a granola bar will each person get?

2. What part of 8 is in 2? Write as a division expression.

ACTIVITY *Work with a partner.*

1 Find $1 \div \frac{1}{5}$ using a model.

STEP 1 Make a model of the dividend, 1.

THINK How many $\frac{1}{5}$s are in 1?

STEP 2 Rename 1 as $\frac{5}{5}$ so the numbers have common denominators. So, the problem is $\frac{5}{5} \div \frac{1}{5}$. Redraw the model to show $\frac{5}{5}$.

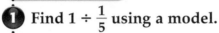

How many $\frac{1}{5}$s are in $\frac{5}{5}$?

STEP 3 Circle groups that are the size of the divisor $\frac{1}{5}$.

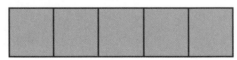

There are five $\frac{1}{5}$s in $\frac{5}{5}$.

So, $1 \div \frac{1}{5} = 5$.

Your Turn Find each quotient using a model.

a. $2 \div \frac{1}{5}$ b. $3 \div \frac{1}{3}$ c. $3 \div \frac{2}{3}$ d. $2 \div \frac{3}{4}$

A model can also be used to find the quotient of two fractions.

ACTIVITY *Work with a partner.*

2 Find $\frac{3}{4} \div \frac{3}{8}$ using a model.

STEP 1 Rename $\frac{3}{4}$ as $\frac{6}{8}$ so the fractions have common denominators. So, the problem is $\frac{6}{8} \div \frac{3}{8}$. Draw a model of the dividend, $\frac{6}{8}$.

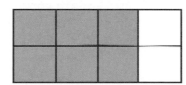

THINK How many $\frac{3}{8}$s are in $\frac{6}{8}$?

STEP 2 Circle groups that are the size of the divisor, $\frac{3}{8}$.

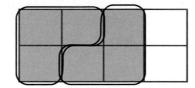

There are two $\frac{3}{8}$s in $\frac{6}{8}$.

So, $\frac{3}{4} \div \frac{3}{8} = 2$.

Your Turn Find each quotient using a model.

e. $\frac{4}{10} \div \frac{1}{5}$ f. $\frac{3}{4} \div \frac{1}{2}$ g. $\frac{4}{5} \div \frac{1}{5}$ h. $\frac{1}{6} \div \frac{1}{3}$

Writing Math

Use *greater than*, *less than*, or *equal to* to complete each sentence. Then give an example to support your answer.

1. When the dividend is equal to the divisor, the quotient is __?__ 1.

2. When the dividend is greater than the divisor, the quotient is __?__ 1.

3. When the dividend is less than the divisor, the quotient is __?__ 1.

4. You know that multiplication is commutative because the product of 3×4 is the same as 4×3. Is division commutative? Give examples to explain your answer.

Dividing Fractions

What You'll LEARN

Divide fractions.

NEW Vocabulary

reciprocal

HANDS-ON Mini Lab

Materials
- paper
- pencil

Work with a partner.

Kenji and his friend Malik made 4 pizzas. They estimate that a $\frac{1}{2}$-pizza will serve one person.

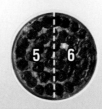

1. How many $\frac{1}{2}$-pizza servings are there?

2. The model shows $4 \div \frac{1}{2}$. What is $4 \div \frac{1}{2}$?

Draw a model to find each quotient.

3. $3 \div \frac{1}{4}$ 4. $2 \div \frac{1}{6}$ 5. $4 \div \frac{1}{2}$

The Mini Lab shows that $4 \div \frac{1}{2} = 8$. Notice that dividing by $\frac{1}{2}$ gives the same result as multiplying by 2.

$$4 \div \frac{1}{2} = 8 \qquad 4 \times 2 = 8$$

Notice that $\frac{1}{2} \times 2 = 1$.

The numbers $\frac{1}{2}$ and 2 have a special relationship. Their product is 1. Any two numbers whose product is 1 are called **reciprocals**.

EXAMPLES Find Reciprocals

1 **Find the reciprocal of 5.**

Since $5 \times \frac{1}{5} = 1$, the reciprocal of 5 is $\frac{1}{5}$.

2 **Find the reciprocal of $\frac{2}{3}$.**

Since $\frac{2}{3} \times \frac{3}{2} = 1$, the reciprocal of $\frac{2}{3}$ is $\frac{3}{2}$.

STUDY TIP

Mental Math To find the reciprocal of a fraction, *invert* the fraction. That is, switch the numerator and denominator.

You can use reciprocals to divide fractions.

Key Concept
Dividing Fractions

Words   To divide by a fraction, multiply by its reciprocal.

Symbols   Arithmetic   Algebra

$$\frac{1}{2} \div \frac{2}{3} = \frac{1}{2} \times \frac{3}{2} \qquad \frac{a}{b} \div \frac{c}{d} = \frac{a}{b} \times \frac{d}{c}, \text{ where } b, c, \text{ and } d \neq 0$$

EXAMPLE Divide by a Fraction

③ Find $\frac{1}{8} \div \frac{3}{4}$.

$$\frac{1}{8} \div \frac{3}{4} = \frac{1}{8} \times \frac{4}{3} \qquad \text{Multiply by the reciprocal, } \frac{4}{3}.$$

$$= \frac{1 \times \overset{1}{\cancel{4}}}{\underset{2}{\cancel{8}} \times 3} \qquad \text{Divide 8 and 4 by the GCF, 4.}$$

$$= \frac{1}{6} \qquad \begin{array}{l}\text{Multiply numerators.}\\ \text{Multiply denominators.}\end{array}$$

Your Turn Divide. Write in simplest form.

a. $\frac{1}{4} \div \frac{3}{8}$ b. $\frac{2}{3} \div \frac{3}{8}$ c. $4 \div \frac{3}{4}$

EXAMPLE Divide Fractions to Solve a Problem

④ **PAINTBALL** It costs $5 to play paintball for one-half hour. How many five-dollar bills do you need to play paintball for 3 hours?

Divide 3 by $\frac{1}{2}$ to find the number of half hours in 3 hours.

$$3 \div \frac{1}{2} = \frac{3}{1} \times \frac{2}{1} \qquad \text{Multiply by the reciprocal of } \frac{1}{2}.$$

$$= \frac{6}{1} \text{ or } 6 \qquad \text{Simplify.}$$

So, you need 6 five-dollar bills or $30 to play for 3 hours.

Standardized Test Practice

EXAMPLE Divide by a Whole Number

⑤ **GRID-IN TEST ITEM** A neighborhood garden that is $\frac{2}{3}$ of an acre is to be divided into 4 equal-size areas. What is the size of each area?

Read the Test Item
You need to find the size of each area. To do so, divide $\frac{2}{3}$ into 4 equal parts.

Solve the Test Item

$$\frac{2}{3} \div 4 = \frac{2}{3} \times \frac{1}{4} \qquad \text{Multiply by the reciprocal.}$$

$$= \frac{\overset{1}{\cancel{2}}}{3} \times \frac{1}{\underset{2}{\cancel{4}}} \qquad \begin{array}{l}\text{Divide 2 and 4}\\ \text{by the GCF, 2.}\end{array}$$

$$= \frac{1}{6} \qquad \text{Simplify.}$$

Each area is $\frac{1}{6}$ acre.

Fill in the Grid

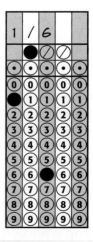

Writing Math
Exercises 2 & 4

1. **Draw** a model that shows $2 \div \frac{1}{3} = 6$.

2. **Explain** why $\frac{1}{2} \div \frac{2}{3} = \frac{1}{2} \times \frac{3}{2} = \frac{3}{4}$. Use a model in your explanation.

3. **OPEN ENDED** Write two fractions that are reciprocals of each other.

4. **FIND THE ERROR** Ryan and Joshua are solving $\frac{2}{3} \div 4$. Who is correct? Explain.

Ryan
$$\frac{2}{3} \div 4 = \frac{2}{3} \times \frac{4}{1}$$
$$= \frac{8}{3} \text{ or } 2\frac{2}{3}$$

Joshua
$$\frac{2}{3} \div 4 = \frac{2}{3} \times \frac{1}{4}$$
$$= \frac{2}{12} \text{ or } \frac{1}{6}$$

GUIDED PRACTICE

Find the reciprocal of each number.

5. $\frac{2}{3}$
6. $\frac{1}{7}$
7. $\frac{2}{5}$
8. 4

Divide. Write in simplest form.

9. $\frac{1}{4} \div \frac{1}{2}$
10. $\frac{5}{6} \div \frac{1}{3}$
11. $\frac{4}{5} \div 2$

12. $2 \div \frac{1}{3}$
13. $\frac{5}{8} \div \frac{3}{4}$
14. $\frac{3}{4} \div \frac{2}{5}$

15. **FOOD** Mrs. Cardona has $\frac{2}{3}$ of a pan of lasagna left for dinner. She wants to divide the lasagna into 6 equal pieces for her family. What part of the original pan of lasagna will each person get?

Practice and Applications

Find the reciprocal of each number.

16. $\frac{1}{4}$
17. $\frac{1}{10}$
18. $\frac{5}{6}$
19. $\frac{2}{5}$

20. $\frac{7}{9}$
21. 8
22. 1
23. $\frac{3}{8}$

HOMEWORK HELP	
For Exercises	See Examples
16–23	1, 2
24–33, 36	3
34–35, 37	5
42–44	4

Extra Practice
See pages 608, 630.

Divide. Write in simplest form.

24. $\frac{1}{8} \div \frac{1}{2}$
25. $\frac{1}{2} \div \frac{2}{3}$
26. $\frac{1}{3} \div \frac{1}{9}$
27. $\frac{1}{4} \div \frac{1}{8}$

28. $\frac{5}{8} \div \frac{1}{4}$
29. $\frac{3}{4} \div \frac{2}{3}$
30. $\frac{3}{4} \div \frac{9}{10}$
31. $3 \div \frac{3}{4}$

32. $2 \div \frac{3}{5}$
33. $\frac{2}{3} \div \frac{2}{5}$
34. $\frac{5}{6} \div 5$
35. $\frac{5}{8} \div 2$

36. If you divide $\frac{1}{2}$ by $\frac{1}{8}$, what is the quotient?

37. If you divide $\frac{6}{7}$ by 3, what is the quotient?

ALGEBRA Find the value of each expression if $a = \frac{2}{3}$, $b = \frac{3}{4}$, and $c = \frac{1}{2}$.

38. $a \div b$ **39.** $b \div c$ **40.** $a \div c$ **41.** $c \div b$

42. DOGS Maria works at a kennel and uses 30-pound bags of dog food to feed the dogs. If each dog gets $\frac{2}{5}$ pound of food, how many dogs can she feed with one bag?

43. WRITE A PROBLEM Write two real-life problems that involve the fraction $\frac{1}{2}$ and the whole number 3. One problem should involve multiplication, and the other should involve division.

44. MULTI STEP Lena has painted $\frac{3}{4}$ of a room. She has used $1\frac{1}{2}$ gallons of paint. How much paint will she need to finish the job?

45. CRITICAL THINKING Solve mentally.

a. $\frac{2{,}345}{1{,}015} \times \frac{12}{11} \div \frac{2{,}345}{1{,}015}$ b. $\frac{2{,}345}{11} \times \frac{12}{1{,}015} \div \frac{2{,}345}{1{,}015}$

Standardized Test Practice and Mixed Review

46. MULTIPLE CHOICE The table shows the weight factors of other planets relative to Earth. For example, an object on Jupiter is 3 times heavier than on Earth. About how many times heavier is an object on Venus than on Mercury?

   Ⓐ $1\frac{3}{5}$ Ⓑ $2\frac{1}{16}$ Ⓒ $2\frac{7}{10}$ Ⓓ $3\frac{1}{8}$

Planets' Weight Factors	
Planet	**Weight Factor**
Mercury	$\frac{1}{3}$
Venus	$\frac{9}{10}$
Jupiter	3

Source: www.factmonster.com

47. MULTIPLE CHOICE Which of the following numbers, when divided by $\frac{1}{2}$, gives a result *less than* $\frac{1}{2}$?

   Ⓕ $\frac{2}{8}$ Ⓖ $\frac{7}{12}$ Ⓗ $\frac{2}{3}$ Ⓘ $\frac{5}{24}$

Multiply. Write in simplest form. (Lesson 7-3)

48. $2\frac{2}{5} \times 3\frac{1}{3}$ **49.** $1\frac{5}{6} \times 2\frac{3}{4}$ **50.** $3\frac{3}{7} \times 2\frac{3}{8}$ **51.** $4\frac{4}{9} \times 5\frac{1}{4}$

52. VOLUNTEERING According to a survey, nine in 10 teens volunteer at least once a year. Of these, about $\frac{1}{3}$ help clean up their communities. What fraction of teens volunteer by helping clean up their communities? (Lesson 7-2)

GETTING READY FOR THE NEXT LESSON

PREREQUISITE SKILL Write each mixed number as an improper fraction. Then find the reciprocal of each. (Lesson 5-3)

53. $1\frac{2}{3}$ **54.** $1\frac{5}{9}$ **55.** $4\frac{1}{2}$ **56.** $3\frac{3}{4}$ **57.** $6\frac{4}{5}$

Dividing Mixed Numbers

REVIEW Vocabulary

mixed number: the sum of a whole number and a fraction (**Lesson 5-3**)

> **WHEN** **am I ever going to use this?**
>
> **DESIGNER** Suppose you are going to cut pieces of fabric $1\frac{3}{4}$ yards long from a bolt containing $5\frac{1}{2}$ yards of fabric.
>
> 1. To the nearest yard, how long is each piece?
>
> 2. To the nearest yard, how long is the fabric on the bolt?
>
> 3. About how many pieces can you cut?

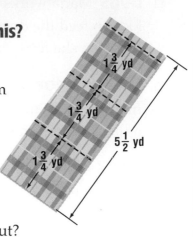

When you multiply mixed numbers, you write each mixed number as an improper fraction. The same is true with division.

EXAMPLE **Divide by a Mixed Number**

1 Find $5\frac{1}{2} \div 1\frac{3}{4}$.   **Estimate** $6 \div 2 = 3$

$$5\frac{1}{2} \div 1\frac{3}{4} = \frac{11}{2} \div \frac{7}{4}$$ Write mixed numbers as improper fractions.

$$= \frac{11}{2} \times \frac{4}{7}$$ Multiply by the reciprocal.

$$= \frac{11}{\underset{1}{2}} \times \frac{\overset{2}{4}}{7}$$ Divide 2 and 4 by the GCF, 2.

$$= \frac{22}{7} \text{ or } 3\frac{1}{7}$$ Compare to the estimate.

Your Turn Divide. Write in simplest form.

a. $4\frac{1}{5} \div 2\frac{1}{3}$ b. $8 \div 2\frac{1}{2}$ c. $1\frac{5}{9} \div 2\frac{1}{3}$

EXAMPLE **Evaluate Expressions**

2 **ALGEBRA** Find $m \div n$ if $m = 1\frac{3}{4}$ and $n = \frac{2}{5}$.

$$m \div n = 1\frac{3}{4} \div \frac{2}{5}$$ Replace m with $1\frac{3}{4}$ and n with $\frac{2}{5}$.

$$= \frac{7}{4} \div \frac{2}{5}$$ Write the mixed number as an improper fraction.

$$= \frac{7}{4} \times \frac{5}{2}$$ Multiply by the reciprocal.

$$= \frac{35}{8} \text{ or } 4\frac{3}{8}$$ Simplify.

STUDY TIP

Estimation
$$1\frac{3}{4} \div \frac{2}{5} \approx 2 \div \frac{1}{2}$$
$$\approx 4$$
Compare the actual quotient to the estimate.

EXAMPLES Solve Problems with Mixed Numbers

3 **WEATHER** A tornado traveled 100 miles in $1\frac{1}{2}$ hours. How many miles per hour did it travel? **Estimate** $100 \div 2 = 50$

$$100 \div 1\frac{1}{2} = \frac{100}{1} \div \frac{3}{2}$$ Write the mixed number as an improper fraction.

$$= \frac{100}{1} \times \frac{2}{3}$$ Multiply by the reciprocal.

$$= \frac{200}{3}$$ Simplify.

$$= 66\frac{2}{3}$$ Compare to the estimate.

So, the tornado traveled $66\frac{2}{3}$ miles per hour.

4 How far would the tornado travel in $\frac{1}{2}$ hour at the same speed?

Estimate $\frac{1}{2}$ of $70 = 35$

$$\frac{1}{2} \times 66\frac{2}{3} = \frac{1}{2} \times \frac{200}{3}$$ Write the mixed number as an improper fraction.

$$= \frac{1}{\cancel{2}} \times \frac{\cancel{200}^{100}}{3}$$ Divide 2 and 100 by their GCF, 2.

$$= \frac{100}{3} \text{ or } 33\frac{1}{3}$$ Simplify.

So, the tornado would travel $33\frac{1}{3}$ miles in $\frac{1}{2}$ hour.

Skill and Concept Check

Writing Math

1. **OPEN ENDED** Write about a real-life situation that is represented by $12\frac{3}{4} \div 2\frac{1}{2}$.   *Exercises 1 & 2*

2. **Which One Doesn't Belong?** Identify the expression whose quotient is less than 1. Explain your reasoning.

$2\frac{1}{2} \div 1\frac{1}{3}$	$4\frac{1}{3} \div 2\frac{2}{5}$	$2\frac{1}{8} \div 3\frac{1}{3}$	$3\frac{1}{2} \div 1\frac{3}{5}$

GUIDED PRACTICE

Divide. Write in simplest form.

3. $3\frac{1}{2} \div 2$ 4. $8 \div 1\frac{1}{3}$ 5. $3\frac{1}{5} \div \frac{2}{7}$

6. **ALGEBRA** What is the value of $c \div d$ if $c = \frac{3}{8}$ and $d = 1\frac{1}{2}$?

7. **BAKING** Jay is cutting a roll of cookie dough into slices that are $\frac{3}{8}$ inch thick. If the roll is $10\frac{1}{2}$ inches long, how many slices can he cut?

Practice and Applications

Divide. Write in simplest form.

8. $5\frac{1}{2} \div 2$

9. $4\frac{1}{6} \div 10$

10. $3 \div 4\frac{1}{2}$

11. $6 \div 2\frac{1}{4}$

12. $15 \div 3\frac{1}{8}$

13. $18 \div 2\frac{2}{5}$

14. $6\frac{1}{2} \div \frac{3}{4}$

15. $7\frac{4}{5} \div \frac{1}{5}$

16. $\frac{11}{12} \div 3\frac{1}{2}$

17. $1\frac{1}{4} \div \frac{5}{6}$

18. $6\frac{1}{2} \div 3\frac{1}{4}$

19. $8\frac{3}{4} \div 2\frac{1}{6}$

20. $3\frac{3}{5} \div 1\frac{4}{5}$

21. $3\frac{3}{4} \div 5\frac{5}{8}$

22. $4\frac{2}{3} \div 2\frac{2}{9}$

23. $6\frac{3}{5} \div 2\frac{3}{4}$

24. $4\frac{3}{8} \div 1\frac{2}{3}$

25. $5\frac{1}{3} \div 2\frac{2}{5}$

HOMEWORK HELP

For Exercises	See Examples
8–25	1
26–27, 34–38	3, 4
28–33	2

Extra Practice
See pages 608, 630.

26. **FOOD** How many $\frac{1}{4}$-pound hamburgers can be made from $2\frac{1}{2}$ pounds of ground beef?

27. **MEASUREMENT** Suppose you are designing the layout for your school yearbook. If a student photograph is $1\frac{3}{8}$ inches wide, how many photographs will fit across a page that is $6\frac{7}{8}$ inches wide?

ALGEBRA Evaluate each expression if $a = 4\frac{4}{5}$, $b = \frac{2}{3}$, $c = 6$, and $d = 1\frac{1}{2}$.

28. $12 \div a$

29. $b \div 1\frac{2}{9}$

30. $a \div b$

31. $a \div c$

32. $c \div d$

33. $c \div (ab)$

34. **SLED DOG RACING** In 2001, Doug Swingley won the Iditarod Trail Sled Dog Race for the fourth time. He completed the 1,100-mile course in $9\frac{5}{6}$ days. How many miles did he average each day?

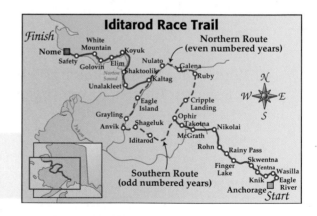

Iditarod Race Trail

 Data Update Find the winning time of the Iditarod for the current year. What was the average number of miles per day? Visit **msmath1.net/data_update** to learn more.

OCEANS **For Exercises 35 and 36, use the following information.**
A tsunami is a tidal wave in the Pacific Ocean. Suppose a tsunami traveled 1,400 miles from a point in the Pacific Ocean to the Alaskan coastline in $2\frac{1}{2}$ hours.

35. How many miles per hour did the tsunami travel?

36. How far would the tsunami travel in $1\frac{1}{2}$ hours at the same speed?

TRAVEL For Exercises 37 and 38, use the following information.
The Days drove their car from Nashville, Tennessee, to Orlando, Florida. They filled the gas tank before leaving home. They drove 407 miles before filling the gas tank with $18\frac{1}{2}$ gallons of gasoline.

37. How many miles per gallon did they get on that portion of their trip?

38. How much did they pay for the gasoline if it cost $1.12 per gallon?

39. **CRITICAL THINKING** Tell whether $\frac{8}{10} \div 1\frac{2}{3}$ is greater than or less than $\frac{8}{10} \div 1\frac{3}{4}$. Explain your reasoning.

Standardized Test Practice and Mixed Review

40. **SHORT RESPONSE** The width of 10 blooms in a test of a new marigold variety are shown. What is the average (mean) bloom width?

Marigold Bloom Width (in.)				
$3\frac{1}{4}$	$2\frac{3}{4}$	3	$2\frac{3}{4}$	$2\frac{1}{2}$
$3\frac{1}{4}$	$3\frac{1}{2}$	$3\frac{1}{4}$	3	$3\frac{1}{4}$

41. **MULTIPLE CHOICE** There are $18\frac{2}{3}$ cups of juice to be divided among a group of children. If each child gets $\frac{2}{3}$ cup of juice, how many children are there?

   Ⓐ 25   Ⓑ 26   Ⓒ 27   Ⓓ 28

MEASUREMENT For Exercises 42 and 43, use the graphic at the right and the information below. (Lesson 7-4)
One U.S. ton equals $\frac{9}{10}$ metric ton. So, you can use $t \div \frac{9}{10}$ to convert t metric tons to U.S. tons.

42. Write a division expression to represent the U.S. tons of gold that were produced in South Africa. Then simplify.

43. How many U.S. tons of gold were produced in Europe?

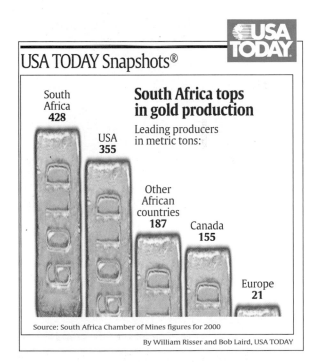

USA TODAY Snapshots®

South Africa tops in gold production
Leading producers in metric tons:
South Africa 428
USA 355
Other African countries 187
Canada 155
Europe 21

Source: South Africa Chamber of Mines figures for 2000
By William Risser and Bob Laird, USA TODAY

Multiply. Write in simplest form. (Lesson 7-3)

44. $\frac{4}{5} \times 1\frac{3}{4}$

45. $2\frac{5}{8} \times \frac{2}{7}$

46. $1\frac{1}{8} \times 5\frac{1}{3}$

47. $3\frac{1}{3} \times 2\frac{1}{2}$

GETTING READY FOR THE NEXT LESSON

PREREQUISITE SKILL What number should be added to the first number to get the second number? (Lesson 6-6)

48. $8\frac{1}{2}$, 10

49. 9, $12\frac{1}{2}$

50. $1\frac{2}{3}$, $2\frac{1}{3}$

51. $7\frac{3}{4}$, $9\frac{1}{4}$

Problem-Solving Strategy
A Preview of Lesson 7-6

What You'll Learn
Solve problems by looking for a pattern.

Look for a Pattern

> Emelia, do you know what time your brother's bus will get here? He said he would be on the first bus after 8:00 P.M.

> Buses arrive at the terminal every 50 minutes. The first bus arrives at 3:45 P.M. We can figure out when his bus will get here by **looking for a pattern**.

Explore	We know that the first bus arrives at 3:45 P.M. and they arrive every 50 minutes. We need to find the first bus after 8:00 P.M.
Plan	Let's start with the time of the first bus and look for a pattern.
Solve	3:45 P.M. + 50 minutes → 4:35 P.M. 4:35 P.M. + 50 minutes → 5:25 P.M. 5:25 P.M. + 50 minutes → 6:15 P.M. 6:15 P.M. + 50 minutes → 7:05 P.M. 7:05 P.M. + 50 minutes → 7:55 P.M. 7:55 P.M. + 50 minutes → 8:45 P.M. So, the first bus to arrive after 8:00 P.M. will be the 8:45 P.M. bus.
Examine	Write the times using fractions. 50 minutes = $\frac{50}{60}$ or $\frac{5}{6}$ of an hour The first bus would arrive at $3\frac{3}{4}$. Add 6 groups of $\frac{5}{6}$. $3\frac{3}{4} + 6\left(\frac{5}{6}\right) = 3\frac{3}{4} + 5$ or $8\frac{3}{4}$, which is 8:45 P.M.

Analyze the Strategy

1. **Describe** another pattern that you could use to find the time the bus arrives.

2. **Explain** when you would use the look for a pattern strategy to solve a problem.

3. **Write** a problem that can be solved by looking for a pattern. Then write the steps you would take to find the solution to your problem.

Solve. Use the look for a pattern strategy.

4. **NUMBER SENSE** Describe the pattern below. Then find the missing number.

    30, 300, __?__, 30,000

5. **GEOMETRY** Draw the next two figures in the sequence.

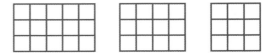

Solve. Use any strategy.

6. **GEOMETRY** Use the pattern below to find the perimeter of the eighth figure.

    Figure 1   Figure 2   Figure 3

7. **MONEY** In 1997, Celina earned $18,000 per year, and Roger earned $14,500. Each year Roger received a $1,000 raise, and Celina received a $500 raise. In what year will they earn the same amount of money? How much will it be?

8. **HEIGHT** Fernando is 2 inches taller than Jason. Jason is 1.5 inches shorter than Kendra and 1 inch taller than Nicole. Hao, who is 5 feet 10 inches tall, is 2.5 inches taller than Fernando. How tall is each student?

9. **PHYSICAL SCIENCE**
A cup of marbles hangs from a rubber band. The length of the rubber band is measured as shown in the graph. Predict the approximate length of the rubber band if 5 marbles are in the cup.

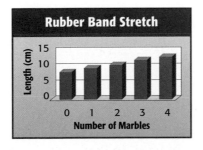

10. **MONEY** What was the price of the sweatshirt before taxes?

Sweatshirt Price	Tax	Total Cost
?	$2.50	$42.49

11. **NUMBER THEORY** The numbers below are called *triangular numbers*. Find the next three triangular numbers.

    1   3   6

12. **STANDARDIZED TEST PRACTICE**
Jody and Lazaro are cycling in a 24-mile race. Jody is cycling at an average speed of 8 miles per hour. Lazaro is cycling at an average speed of 6 miles per hour. Which of the following statements is *not* true?

   Ⓐ If Lazaro has a 6-mile head start, they will finish at the same time.

   Ⓑ Lazaro will finish the race one hour after Jody.

   Ⓒ Jody is 4 miles ahead of Lazaro after two hours.

   Ⓓ Jody will finish the race one hour after Lazaro.

> You will use the look for a pattern strategy in the next lesson.

Sequences

What You'll LEARN

Recognize and extend sequences.

NEW Vocabulary

sequence

WHEN am I ever going to use this?

MUSIC The diagram shows the most common notes used in music. The names of the first four notes are whole note, half note, quarter note, and eighth note.

$1 \quad \frac{1}{2} \quad \frac{1}{4} \quad \frac{1}{8}$

1. What are the names of the next three notes?
2. Write the fraction that represents each of the next three notes.
3. Identify the pattern in the numbers.

A **sequence** is a list of numbers in a specific order. By determining the pattern, you can find additional numbers in the sequence. The numbers $1, \frac{1}{2}, \frac{1}{4}$, and $\frac{1}{8}$ are an example of a sequence.

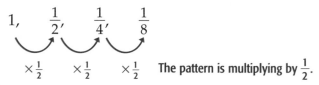

$1, \quad \frac{1}{2}, \quad \frac{1}{4}, \quad \frac{1}{8}$

$\times \frac{1}{2} \quad \times \frac{1}{2} \quad \times \frac{1}{2}$ The pattern is multiplying by $\frac{1}{2}$.

The next number in this sequence is $\frac{1}{8} \times \frac{1}{2}$ or $\frac{1}{16}$.

EXAMPLE **Extend a Sequence by Adding**

1 Describe the pattern in the sequence 16, 24, 32, 40, … .
Then find the next two numbers in the sequence.

$16, \quad 24, \quad 32, \quad 40, \dots$ Each number is 8 more than the number before it.
$+8 \quad +8 \quad +8$

In this sequence, 8 is added to each number. The next two numbers are 40 + 8, or 48, and 48 + 8, or 56.

Your Turn Describe each pattern. Then find the next two numbers in each sequence.

a. $1\frac{1}{2}, 3, 4\frac{1}{2}, 6, \dots$ b. $20, 16, 12, 8, \dots$ c. $27\frac{1}{2}, 25, 22\frac{1}{2}, 20, \dots$

In some sequences, the numbers are found by multiplying by the same number.

EXAMPLE Extend a Sequence by Multiplying

2 Describe the pattern in the sequence 5, 15, 45, 135, ... Then find the next two numbers in the sequence.

5,   15,   45,   135, ...
  $\times 3$ $\times 3$ $\times 3$

Each number is multiplied by 3.

The next two numbers in the sequence are 405 and 1,215.

EXAMPLE Use Sequences to Solve a Problem

3 **SPORTS** The NCAA basketball tournament starts with 64 teams. The second round consists of 32 teams, and the third round consists of 16 teams. How many teams are in the fifth round?

Write the sequence. Find the fifth number.

64,   32,   16,   8,   4
  $\times \frac{1}{2}$ $\times \frac{1}{2}$ $\times \frac{1}{2}$ $\times \frac{1}{2}$

There are 4 teams in the fifth round.

Your Turn Describe each pattern. Then find the next two numbers in each sequence.

d. 3, 12, 48, 192, ...   e. 125, 25, 5, 1, ...

Skill and Concept Check

Writing Math
Exercises 1 & 3

1. Tell how the numbers are related in the sequence 9, 3, 1, $\frac{1}{3}$.

2. **OPEN ENDED** Write a sequence in which $1\frac{1}{4}$ is added to each number.

3. **FIND THE ERROR** Meghan and Drake are finding the missing number in the sequence 3, $4\frac{1}{2}$, ___?___, $7\frac{1}{2}$, ... Who is correct? Explain.

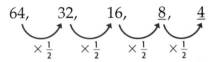

Meghan
3, $4\frac{1}{2}$, $5\frac{1}{2}$, $7\frac{1}{2}$

Drake
3, $4\frac{1}{2}$, 6, $7\frac{1}{2}$, ...

GUIDED PRACTICE

Describe each pattern. Then find the next two numbers in the sequence.

4. $7\frac{1}{2}$, 6, $4\frac{1}{2}$, 3, ...   5. 3, 6, 12, 24, ...   6. 32, 8, 2, $\frac{1}{2}$, ...

7. Find the missing number in the sequence 13, 21, ___?___, 37.

Practice and Applications

HOMEWORK HELP

For Exercises	See Examples
8–10	1
11–13	2
14–18	3

Extra Practice
See pages 609, 630.

Describe each pattern. Then find the next two numbers in the sequence.

8. $2, 3\frac{1}{2}, 5, 6\frac{1}{2}, \ldots$

9. $20, 16, 12, 8, \ldots$

10. $90, 75, 60, 45, \ldots$

11. $8, 16, 32, 64, \ldots$

12. $12, 6, 3, 1\frac{1}{2}, \ldots$

13. $162, 54, 18, 6, \ldots$

Find the missing number in each sequence.

14. $7, \underline{\ ?\ }, 16, 20\frac{1}{2}, \ldots$

15. $30, \underline{\ ?\ }, 19, 13\frac{1}{2}, \ldots$

16. $\underline{\ ?\ }, 16, 4, 1, \ldots$

17. $\underline{\ ?\ }, 1, 3, 9, \ldots$

18. **TOOLS** Mr. Black's drill bit set includes the following sizes (in inches).

$$\ldots, \frac{13}{64}, \frac{7}{32}, \frac{15}{64}, \frac{1}{4}, \ldots$$

What are the next two smaller bits?

19. **CRITICAL THINKING** The largest square at the right represents 1.

a. Find the first ten numbers of the sequence represented by the model. The first number is $\frac{1}{2}$.

b. Estimate the sum of the first ten numbers without actually adding. Explain.

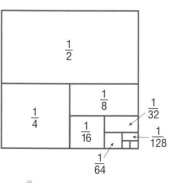

Standardized Test Practice and Mixed Review

20. **MULTIPLE CHOICE** What number is missing from the sequence 14, 56, $\underline{\ ?\ }$, 896, 3,584?

Ⓐ 284   Ⓑ 194   Ⓒ 334   Ⓓ 224

21. **SHORT RESPONSE** What is the next term in the sequence $x, x^2, x^3, x^4, \ldots$?

22. Find $2\frac{4}{5} \div \frac{7}{10}$. (Lesson 7-5)

23. **FOOD** Each serving of an apple pie is $\frac{1}{10}$ of the pie. If $\frac{1}{2}$ of the pie is left, how many servings are left? (Lesson 7-4)

WebQuest **Interdisciplinary Project**

Cooking Up a Mystery!

It's time to complete your project. Use the volcano you've created and the data you have gathered about volcanoes to prepare a class demonstration. Be sure to include a graph of real volcanic eruptions with your project.

msmath1.net/webquest

Study Guide and Review

Vocabulary and Concept Check

compatible numbers (p. 256)	reciprocal (p. 272)	sequence (p. 282)

Determine whether each sentence is *true* or *false*. If *false*, replace the underlined word or number to make a true sentence.

1. Any two numbers whose product is 1 are called <u>opposites</u>.
2. When dividing by a fraction, <u>multiply</u> by its reciprocal.
3. To multiply fractions, multiply the numerators and <u>add</u> the denominators.
4. A list of numbers in a specific order is called a <u>sequence</u>.
5. Any whole number can be written as a fraction with a denominator of <u>1</u>.
6. The reciprocal of $\frac{8}{3}$ is $\frac{3}{8}$.
7. To divide mixed numbers, first write each mixed number as a <u>decimal</u>.
8. The missing number in the sequence 45, 41, 37, __?__, 29, 25 is <u>32</u>.

Lesson-by-Lesson Exercises and Examples

7-1   Estimating Products (pp. 256–258)

Estimate each product.

9. $\frac{1}{5} \times 21$ 10. $10 \times 2\frac{3}{4}$

11. $\frac{5}{6} \times 13$ 12. $7\frac{3}{4} \times \frac{1}{4}$

13. $4\frac{5}{6} \times 8\frac{3}{10}$ 14. $\frac{3}{7} \times \frac{11}{12}$

15. Estimate $\frac{5}{6}$ of 35.

Example 1   Estimate $\frac{1}{7} \times 41$.

$\frac{1}{7} \times 41 \rightarrow \frac{1}{7} \times 42$ 42 and 7 are compatible numbers since $42 \div 7 = 6$.

$\frac{1}{7} \times 42 = 6$ $\frac{1}{7}$ of 42 is 6.

So, $\frac{1}{7} \times 41$ is *about* 6.

7-2   Multiplying Fractions (pp. 261–264)

Multiply. Write in simplest form.

16. $\frac{1}{3} \times \frac{1}{4}$ 17. $\frac{3}{5} \times \frac{2}{9}$

18. $\frac{7}{8} \times \frac{4}{21}$ 19. $\frac{5}{6} \times 9$

Example 2   Find $\frac{3}{10} \times \frac{4}{9}$.

$\frac{3}{10} \times \frac{4}{9} = \frac{\overset{1}{3} \cdot \overset{2}{4}}{\underset{5}{10} \cdot \underset{3}{9}}$ Divide the numerator and denominator by the GCF.

$= \frac{2}{15}$ Simplify.

Multiplying Mixed Numbers (pp. 265–267)

Multiply. Write in simplest form.

20. $\frac{2}{3} \times 4\frac{1}{2}$

21. $6\frac{5}{8} \times 4$

22. $1\frac{1}{5} \times 1\frac{2}{3}$

23. $3\frac{3}{4} \times 1\frac{1}{5}$

24. $3\frac{1}{8} \times 2\frac{2}{5}$

25. $2\frac{1}{4} \times 6\frac{2}{3}$

Example 3 Find $3\frac{1}{2} \times 4\frac{2}{3}$.

$3\frac{1}{2} \times 4\frac{2}{3} = \frac{7}{2} \times \frac{14}{3}$ — Write the numbers as improper fractions.

$= \frac{7}{2} \times \frac{\overset{7}{14}}{3}$ — Divide 2 and 14 by their GFC, 2.

$= \frac{49}{3}$ or $16\frac{1}{3}$ — Simplify.

Dividing Fractions (pp. 272–275)

Divide. Write in simplest form.

26. $\frac{2}{3} \div \frac{4}{5}$

27. $\frac{1}{8} \div \frac{3}{4}$

28. $5 \div \frac{4}{9}$

29. $\frac{3}{8} \div 6$

Example 4 Find $\frac{3}{8} \div \frac{2}{3}$.

$\frac{3}{8} \div \frac{2}{3} = \frac{3}{8} \times \frac{3}{2}$ — Multiply by the reciprocal of $\frac{2}{3}$.

$= \frac{9}{16}$ — Multiply the numerators and multiply the denominators.

Dividing Mixed Numbers (pp. 276–279)

Divide. Write in simplest form.

30. $2\frac{4}{5} \div 5\frac{3}{5}$

31. $8 \div 2\frac{1}{2}$

32. **PIZZA** Bret has $1\frac{1}{2}$ pizzas. The pizzas are to be divided evenly among 6 friends. How much of a pizza will each friend get?

Example 5 Find $5\frac{1}{2} \div 1\frac{5}{6}$.

$5\frac{1}{2} \div 1\frac{5}{6} = \frac{11}{2} \div \frac{11}{6}$ — Rewrite as improper fractions.

$= \frac{11}{2} \times \frac{6}{11}$ — Multiply by the reciprocal.

$= \frac{\overset{1}{11}}{\underset{1}{2}} \times \frac{\overset{3}{6}}{\underset{1}{11}}$ — Divide by the GCF.

$= \frac{3}{1}$ or 3 — Simplify.

Sequences (pp. 282–284)

Describe each pattern. Then find the next two numbers in the sequence.

33. 6, 12, 24, 48, …

34. 20, $17\frac{1}{2}$, 15, $12\frac{1}{2}$, …

35. 5000, 1000, 200, 40, …

36. 11, 21, 31, 41, …

Example 6 Describe the pattern. Then find the next two numbers in the sequence 625, 125, 25, 5, … .

Each number is multiplied by $\frac{1}{5}$.

$5 \times \frac{1}{5} = 1 \qquad 1 \times \frac{1}{5} = \frac{1}{5}$

The next two numbers are 1 and $\frac{1}{5}$.

Practice Test

Vocabulary and Concepts

1. **Explain** how to multiply a fraction and a whole number.
2. **Define** *sequence*.
3. **Compare and contrast** dividing two fractions and multiplying two fractions.

Skills and Applications

Estimate each product.

4. $38 \times \frac{1}{4}$

5. $6\frac{7}{8} \times 8\frac{1}{6}$

6. $\frac{1}{6} \times 22$

Multiply. Write in simplest form.

7. $\frac{9}{10} \times \frac{5}{8}$

8. $6 \times \frac{5}{24}$

9. $\frac{7}{12} \times \frac{3}{28}$

10. $1\frac{4}{5} \times 2\frac{2}{3}$

11. $\frac{1}{6} \times 3\frac{3}{8}$

12. $3\frac{1}{5} \times 1\frac{1}{4}$

GEOMETRY Find the area of each rectangle.

13. 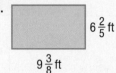 $6\frac{2}{5}$ ft
$9\frac{3}{8}$ ft

14. 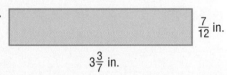 $\frac{7}{12}$ in.
$3\frac{3}{7}$ in.

Divide. Write in simplest form.

15. $\frac{1}{8} \div \frac{3}{4}$

16. $\frac{2}{5} \div 4$

17. $6 \div 1\frac{4}{5}$

18. $5\frac{3}{4} \div 1\frac{1}{2}$

19. $8\frac{1}{3} \div 2\frac{1}{2}$

20. $3\frac{5}{8} \div 4$

21. **KITES** Latanya works at a kite store. To make a kite tail, she needs $2\frac{1}{4}$ feet of fabric. If Latanya has $29\frac{1}{4}$ feet of fabric, how many kite tails can she make?

Describe each pattern. Then find the next two numbers in the sequence.

22. 14, 19, 24, 29, ...

23. 243, 81, 27, ...

24. 71, 60, 49, 38, ...

Standardized Test Practice

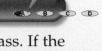

25. **SHORT RESPONSE** There are 24 students in Annie's math class. If the total number of students at her school is $21\frac{3}{8}$ times the number of students in her math class, how many students attend Annie's school?

PART 1 Multiple Choice

Record your answers on the answer sheet provided by your teacher or on a sheet of paper.

1. Marta recorded the number of seeds she planted in flowerpots and the plants that grew.

Number of Seeds (s)	Number of Plants (p)
2	1
4	3
5	4
6	5

Which expression describes the relationship between the number of seeds s and the number of plants p? (Lesson 1-5)

Ⓐ $p = 3 + s$ Ⓑ $p = s + 1$

Ⓒ $p = s - 1$ Ⓓ $p = 2s$

2. A city has radio stations with the frequencies 100.8, 101.7, 101.3, and 100.1. Which shows the frequencies ordered from least to greatest? (Lesson 3-2)

Ⓕ 100.8, 101.7, 101.3, 100.1

Ⓖ 100.1, 100.8, 101.3, 101.7

Ⓗ 100.1, 101.3, 101.7, 100.8

Ⓘ 101.7, 101.3, 100.8, 100.1

3. What is the circumference of the coin below? Use 3.14 for π and round to the nearest tenth. (Lesson 4-6)

Ⓐ 462.0 mm

Ⓑ 152.4 mm

Ⓒ 76.2 mm

Ⓓ 38.1 mm

←— 24.26 mm —→

4. Blaine finished 17 out of 30 questions. Which is the *best* estimate of the fraction of questions he finished? (Lesson 6-1)

Ⓕ $\frac{1}{4}$ Ⓖ $\frac{1}{3}$ Ⓗ $\frac{1}{2}$ Ⓘ $\frac{7}{8}$

5. Kelly uses $9\frac{1}{8}$ inches of yarn to make a tassle. Which is the *best* estimate for the amount of yarn that she will need for 16 tassles? (Lesson 7-1)

Ⓐ 10 in. Ⓑ 80 in.

Ⓒ 150 in. Ⓓ 180 in.

6. Which model shows $\frac{1}{2}$ of $\frac{1}{2}$? (Lesson 7-2)

Ⓕ 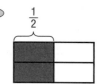 Ⓖ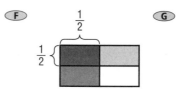

Ⓗ 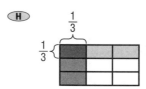 Ⓘ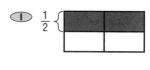

7. What is the value of $2\frac{1}{4} \times 3\frac{1}{3}$? (Lesson 7-3)

Ⓐ $3\frac{1}{3}$ Ⓑ $6\frac{1}{12}$ Ⓒ $7\frac{1}{2}$ Ⓓ $9\frac{1}{3}$

8. Which rule can be used to find the next number in the sequence below? (Lesson 7-6)

12, 21, 30, 39, ___?___

Ⓕ Add 9. Ⓖ Divide by 3.

Ⓗ Multiply by 6. Ⓘ Subtract 9.

TEST-TAKING TIP

Question 7 Round each mixed number down to a whole number and then multiply. Then round each up to a whole number and then multiply. The answer is between the two products.

PART 2 Short Response/Grid In

Record your answers on the answer sheet provided by your teacher or on a sheet of paper.

9. A roller coaster car holds 4 people. How many people can ride at the same time if there are 50 cars? (Prerequisite Skill, p. 590)

10. The stem-and-leaf plot shows the average lengths of fifteen species of poisonous snakes in Texas.

Stem	Leaf
2	0 0 4 4 6 6
3	6 6 6 6 6
4	2 2 8
5	
6	0 3|6 = 36 inches

What is the length of the longest poisonous snake? (Lesson 2-5)

11. Melanie rented a mountain bike for 2 days from Speedy's Bike Rentals. What did it cost to rent the mountain bike each day? (Lesson 4-3)

Speedy's Bike Rentals

Type of Bike	Cost for 2 Days
mountain bike	$42.32
racing bike	$48.86

12. Write $1\frac{5}{6}$ as an improper fraction. (Lesson 5-3)

13. A recipe for a two-layer, 8-inch cake calls for a box of cake mix, 2 eggs, and $1\frac{1}{3}$ cups of water. How much of each ingredient is needed to make a three-layer, 8-inch cake? (Lesson 7-3)

14. Colin is walking on a track that is $\frac{1}{10}$ mile long. How many laps should he walk if he wants to walk a total distance of 2 miles? (Lesson 7-4)

15. The pizzas below are to be divided equally among 3 people.

What fraction of a whole pizza will each person get? (Lesson 7-5)

16. If the pattern below continues, what is the perimeter of the sixth square? (Lesson 7-6)

Square	Side Length (cm)
1	3
2	6
3	12
4	24
5	48

PART 3 Extended Response

Record your answers on a sheet of paper. Show your work.

17. Mr. Williams and Ms. Ling each teach science to 50 students. Two-fifths of Mr. Williams' students signed up for a field trip to the museum. About two-thirds of Ms. Ling's students signed up for the same trip. (Lessons 7-1 and 7-2)

a. About how many of Ms. Ling's students signed up to go to the museum?

b. How many of Mr. Williams' students signed up to go to the museum?

c. One fourth of the students who attended the trip from Mr. Williams' class bought souvenirs. What fraction of his total students attended the trip and bought souvenirs?

 msmath1.net/standardized_test

UNIT 4
Algebra

To accurately describe our world, you will need more than just whole numbers, fractions, and decimals. In this unit, you will use negative numbers to describe many real-life situations and solve equations using them.

WebQuest INTERDISCIPLINARY PROJECT — MATH and SCIENCE

WEATHER WATCHERS

Some nasty weather is brewing! You've been selected to join an elite group of weather watchers on a whirlwind adventure. You'll be gathering and charting data about weather patterns in your own state and doing other meteorological research. You'll also be tracking the path of a tornado that's just touched down. Hurry! Pack your algebra tool kit and your raincoat. There's no time to waste! There are more storms on the horizon!

Log on to msmath1.net/webquest to begin your WebQuest.

Algebra: Integers

"What does football have to do with math?"

Have you ever heard a sports announcer use the phrase "yards gained" or "yards lost" when referring to football? In a football game, one team tries to move the football forward to score a touchdown. On each play, yards are either gained or lost. **In mathematics, phrases such as "a gain of three yards" and "a loss of 2 yards" can be represented using positive and negative integers.**

You will solve problems about football in Lesson 8-2.

GETTING STARTED

Take this quiz to see whether you are ready to begin Chapter 8. Refer to the page number in parentheses if you need more review.

▶ Vocabulary Review

Choose the correct term to complete each sentence.

1. The (sum, product) of 3 and 4 is 12.
   (Page 590)

2. The result of dividing two numbers is called the (difference, quotient).
   (Page 591)

▶ Prerequisite Skills

Add. (Page 589)

3. $12 + 15$ 4. $3 + 4$

5. $5 + 7$ 6. $16 + 9$

7. $8 + 13$ 8. $5 + 17$

Subtract. (Page 589)

9. $14 - 6$ 10. $9 - 4$

11. $11 - 5$ 12. $8 - 3$

13. $7 - 5$ 14. $10 - 6$

Multiply. (Page 590)

15. 7×6 16. 10×2

17. 5×9 18. 8×3

19. 4×4 20. 6×8

Divide. (Page 591)

21. $32 \div 4$ 22. $63 \div 7$

23. $21 \div 3$ 24. $18 \div 9$

25. $72 \div 9$ 26. $45 \div 3$

FOLDABLES™ Study Organizer

Integers Make this Foldable to help you organize information about integers. Begin with a sheet of 11" × 17" unlined paper.

STEP 1 **Fold**
Fold the short sides so they meet in the middle.

STEP 2 **Fold Again**
Fold the top to the bottom.

STEP 3 **Cut**
Unfold and cut along the second fold to make four tabs.

STEP 4 **Label**
Label each tab as shown.

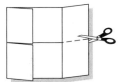

Reading and Writing As you read and study the chapter, write examples of addition, subtraction, multiplication, and division problems under each tab.

Integers

What You'll LEARN

Identify, graph, compare, and order integers.

NEW Vocabulary

integer
negative integer
positive integer
graph
opposites

MATH Symbols

$<$ is less than
$>$ is greater than

WHEN am I ever going to use this?

MONEY The number line shows the amounts of money Molly, Kevin, Blake, and Jenna either have in their wallets or owe one of their parents. A value of -4 represents a debt of 4 dollars.

1. What does a value of -6 represent?
2. Who has the most money?
3. Who owes the most money?
4. What number represents having 8 dollars in a wallet?

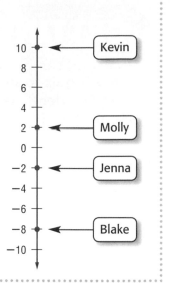

The numbers 8 and -6 are integers. An **integer** is any number from the set $\{..., -4, -3, -2, -1, 0, 1, 2, 3, 4, ...\}$ where ... means *continues without end*.

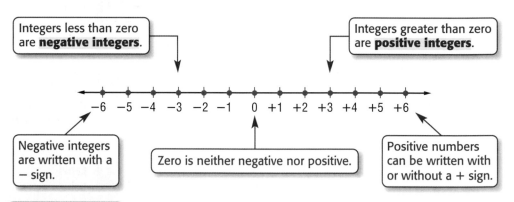

Integers less than zero are **negative integers**.

Integers greater than zero are **positive integers**.

Negative integers are written with a $-$ sign.

Zero is neither negative nor positive.

Positive numbers can be written with or without a $+$ sign.

EXAMPLES Write Integers for Real-Life Situations

Write an integer to describe each situation.

1 **FOOTBALL** a gain of 5 yards on the first down

The word *gain* represents an increase.

The integer is $+5$ or 5.

2 **WEATHER** a temperature of 10 degrees below zero

Any number that is *below zero* is a negative number.

The integer is -10.

READING Math

Positive and Negative Signs The integer $+5$ is read *positive 5* or *5*. The integer -2 is read *negative two*.

Your Turn Write an integer to describe each situation.

a. lost 6 points

b. 12 feet above sea level

To **graph** an integer on a number line, draw a dot at the location on the number line that corresponds to the integer.

EXAMPLE Graph an Integer on a Number Line

3 Graph −3 on a number line.

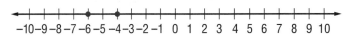

> Draw a number line. Then draw a dot at the location that represents −3.

Your Turn Graph each integer on a number line.

c. −1   d. 4   e. 0

A number line can also be used to compare and order integers. On a number line, the number to the left is always less than the number to the right.

READING Math

Inequality Symbols
The sentence
−6 < −4 is read
negative 6 is less than negative 4.
The sentence
−4 > −6 is read
negative 4 is greater than negative 6.

EXAMPLE Compare Integers

4 Replace the ● in −6 ● −4 with <, >, or = to make a true sentence.

Graph −6 and −4 on a number line. Then compare.

Since −6 is to the left of −4, −6 < −4.

Your Turn Replace each ● with <, >, or = to make a true sentence.

f. −3 ● −5   g. −5 ● 0   h. 6 ● −1

EXAMPLE Order Integers

5 **SCIENCE** The average surface temperatures of Jupiter, Mars, Earth, and the Moon are shown in the table. Order the temperatures from least to greatest.

Name	Average Surface Temperature (°F)
Jupiter	−162
Moon	−10
Mars	−81
Earth	59

Source: *The World Almanac*

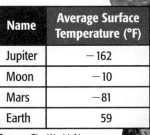

First, graph each integer. Then, write the integers as they appear on the number line from left to right.

STUDY TIP

Ordering Integers
When ordering integers from greatest to least, write the integers as they appear from right to left.

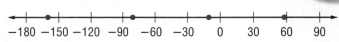

The order from least to greatest is −162°, −81°, −10°, and 59°.

Opposites are numbers that are the same distance from zero in opposite directions on the number line.

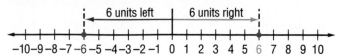 **EXAMPLE** Find the Opposite of an Integer

6 Write the opposite of +6.

```
        6 units left    6 units right
◄─┼──┼──┼──┼──┼──┼──┼──┼──┼──┼──┼──┼──┼──┼──┼──┼──┼──┼──┼──┼──┼──►
   -10-9-8-7-6-5-4-3-2-1  0  1  2  3  4  5  6  7  8  9  10
```

The opposite of +6 is −6.

Your Turn Write the opposite of each integer.

  i. −4   **j.** +8   **k.** −9

Skill and Concept Check

Writing Math

Exercises 1 & 2

1. **Explain** how to list integers from greatest to least.

2. **OPEN ENDED** Write about a situation in which integers would be compared or ordered.

3. **NUMBER SENSE** Describe three characteristics of the set of integers.

GUIDED PRACTICE

Write an integer to describe each situation.

  4. gain 3 pounds   **5.** withdraw $15

Draw a number line from −10 to 10. Then graph each integer on the number line.

  6. −7   **7.** 10   **8.** −4   **9.** 3

Replace each ● with <, >, or = to make a true sentence.

  10. −4 ● −8   **11.** −1 ● 3   **12.** −5 ● 0

13. Order −4, 3, 0, and −5 from least to greatest.

Write the opposite of each integer.

  14. +7   **15.** +5   **16.** −1   **17.** −5

18. HISTORY The timeline shows the years various cities were established. Which city was established after Rome but before London?

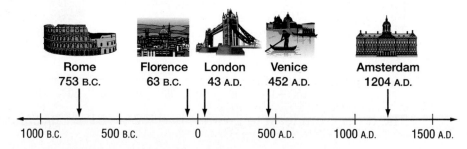

```
    Rome        Florence  London   Venice        Amsterdam
   753 B.C.      63 B.C.  43 A.D.  452 A.D.       1204 A.D.
      │             │  │       │              │
      ▼             ▼  ▼       ▼              ▼
◄──┼──────────┼───────────┼───────────┼──────────┼──────────┼──►
  1000 B.C.  500 B.C.     0      500 A.D.   1000 A.D.   1500 A.D.
```

Practice and Applications

HOMEWORK HELP

For Exercises	See Examples
19–22, 48, 49	1, 2
23–30	3
31–39, 50	4, 5
40–47	6

Extra Practice
See pages 609, 631.

Write an integer to describe each situation.

19. 5 feet below sea level   **20.** deposit $30 into an account

21. move ahead 4 spaces   **22.** the stock market lost 6 points

Draw a number line from −10 to 10. Then graph each integer on the number line.

23. 1   **24.** 5   **25.** 3   **26.** 7

27. −5   **28.** −10   **29.** −6   **30.** −8

Replace each ● with <, >, or = to make a true sentence.

31. −2 ● −4   **32.** 2 ● 4   **33.** 1 ● −3

34. −6 ● 3   **35.** 5 ● 0   **36.** −3 ● 2

37. Order −7, 4, −5, and 6 from least to greatest.

38. Write −2, 5, 0, and −3 from greatest to least.

39. Graph 8, $\frac{1}{2}$, 0.6, $−5$, and $6\frac{3}{4}$ on a number line. Then order the numbers from least to greatest.

Write the opposite of each integer.

40. +1   **41.** +4   **42.** −6   **43.** −10

44. −8   **45.** −3   **46.** +9   **47.** +2

48. GEOGRAPHY New Orleans, Louisiana, is 8 feet below sea level. Write this number as an integer.

49. GEOGRAPHY Jacksonville, Florida, is at sea level. Write this elevation as an integer.

STATISTICS For Exercises 50 and 51, refer to the table.

50. Which state had the lowest temperature?

51. What is the median coldest recorded temperature for the states listed in the table?

Coldest Temperature on Record (°F)

Alaska	−80
California	−45
Florida	−2
Ohio	−39
Montana	−70

Source: *The World Almanac*

52. LIFE SCIENCE Some sea creatures live near the surface while others live in the depths of the ocean. Make a drawing showing the relative habitats of the following creatures.

- blue marlin: 0 to 600 feet below the surface

- lantern fish: 3,300 to 13,200 feet below the surface

- ribbon fish: 600 to 3,300 feet below the surface

FINANCE For Exercises 53–55, use the bar graph.

53. Order the percents on a number line.

54. Which type of bankruptcy showed the greater decline, Chapter 12 or Chapter 7? Explain your answer.

55. Which type of bankruptcy showed the greatest change? Explain your answer.

EXTENDING THE LESSON *Absolute value* is the distance a number is from zero on the number line. The number line below shows that −3 and 3 have the same absolute value, 3. The symbol for absolute value is $|n|$.

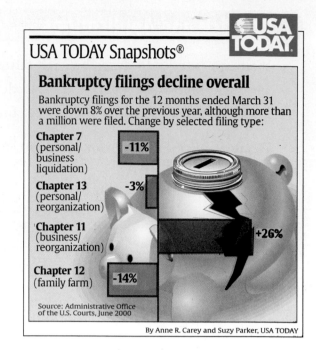

USA TODAY Snapshots®

Bankruptcy filings decline overall

Bankruptcy filings for the 12 months ended March 31 were down 8% over the previous year, although more than a million were filed. Change by selected filing type:

Chapter 7 (personal/business liquidation) **-11%**

Chapter 13 (personal/reorganization) **-3%**

Chapter 11 (business/reorganization) **+26%**

Chapter 12 (family farm) **-14%**

Source: Administrative Office of the U.S. Courts, June 2000

By Anne R. Carey and Suzy Parker, USA TODAY

$|3| = 3$ The absolute value of 3 is 3.

$|-3| = 3$ The absolute value of −3 is 3.

Evaluate each expression.

56. $|5|$ 57. $|-2|$ 58. $|7|$ 59. $|-4|$

60. **CRITICAL THINKING** Explain why any negative integer is less than any positive integer.

Standardized Test Practice and Mixed Review

61. **MULTIPLE CHOICE** What integer is 3 units less than 1?

Ⓐ −2   Ⓑ −1   Ⓒ 0   Ⓓ 2

62. **MULTIPLE CHOICE** Which point on the number line represents −3?

Ⓕ point A Ⓖ point B Ⓗ point C Ⓘ point D

Find the missing number in each sequence. (Lesson 7-6)

63. 2, 4, ___?___, 16, …   64. 5, ___?___, $1\frac{4}{5}$, $1\frac{2}{25}$, …   65. 8, $6\frac{2}{3}$, ___?___, 4, …

66. Find $2\frac{3}{5}$ divided by $1\frac{2}{3}$. Write the quotient in simplest form. (Lesson 7-5)

GETTING READY FOR THE NEXT LESSON

PREREQUISITE SKILL Add or subtract. (Page 589)

67. $6 + 4$ 68. $6 - 4$ 69. $10 + 3$ 70. $10 - 3$

Zero Pairs

What You'll LEARN

Use models to understand zero pairs.

Materials
- counters
- integer mat

Counters can be used to help you understand integers. A yellow counter ⊕ represents the integer $+1$. A red counter ⊖ represents the integer -1. When one yellow counter is paired with one red counter, the result is zero. This pair of counters is called a **zero pair**.

ACTIVITY *Work with a partner.*

1 Use counters to model $+4$ and -4. Then form as many zero pairs as possible to find the sum $+4 + (-4)$.

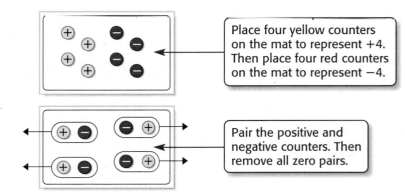

Place four yellow counters on the mat to represent $+4$. Then place four red counters on the mat to represent -4.

Pair the positive and negative counters. Then remove all zero pairs.

There are no counters on the mat. So, $+4 + (-4) = 0$.

Your Turn Use counters to model each pair of integers. Then form zero pairs to find the sum of the integers.

a. $+3, -3$ b. $+5, -5$ c. $-7, +7$

Writing Math

1. What is the value of a zero pair? Explain your reasoning.

2. Suppose there are 5 zero pairs on an integer mat. What is the value of these zero pairs? Explain.

3. **Explain** the effect of removing a zero pair from the mat. What effect does this have on the remaining counters?

4. Integers like $+4$ and -4 are called *opposites*. What is the sum of any pair of opposites?

5. **Write** a sentence describing how zero pairs are used to find the sum of any pair of opposites.

6. **MAKE A CONJECTURE** How do you think you could find $+5 + (-2)$ using counters?

Adding Integers

What You'll LEARN

Add integers.

WHEN am I ever going to use this?

GAMES Gracia and Conner are playing a board game. The cards Gracia selected on each of her first three turns are shown in order from left to right.

Move Ahead 5 Spaces
Move Ahead 6 Spaces
Move Back 3 Spaces

1. After Gracia's third turn, how many spaces from the start is her game piece?

To add integers, you can use counters or a number line.

EXAMPLES Add Integers with the Same Sign

1 Find +3 + (+2).

Method 1 Use counters.

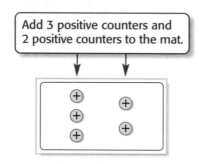

Add 3 positive counters and 2 positive counters to the mat.

So, +3 + (+2) = +5 or 5.

Method 2 Use a number line.

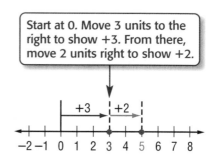

Start at 0. Move 3 units to the right to show +3. From there, move 2 units right to show +2.

2 Find −2 + (−4).

Method 1 Use counters.

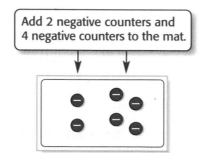

Add 2 negative counters and 4 negative counters to the mat.

So, −2 + (−4) = −6.

Method 2 Use a number line.

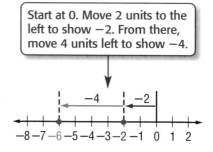

Start at 0. Move 2 units to the left to show −2. From there, move 4 units left to show −4.

Your Turn Add. Use counters or a number line if necessary.

a. +1 + (+4) b. −3 + (−4) c. −7 + (−4)

To add two integers with different signs, it is necessary to remove any zero pairs. A *zero pair* is a pair of counters that includes one positive counter and one negative counter.

EXAMPLES **Add Integers with Different Signs**

READING Math

Positive Integers
A number without a sign is assumed to be positive.

3 Find $1 + (-5)$.

Method 1 Use counters.

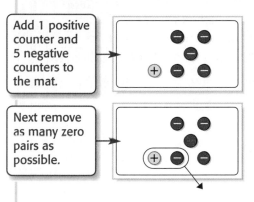

Add 1 positive counter and 5 negative counters to the mat.

Next remove as many zero pairs as possible.

So, $1 + (-5) = -4$.

Method 2 Use a number line.

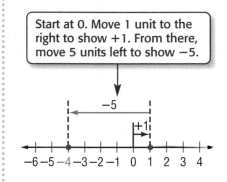

Start at 0. Move 1 unit to the right to show $+1$. From there, move 5 units left to show -5.

4 Find $-8 + 6$.

Method 1 Use counters.

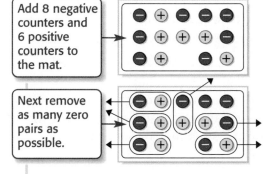

Add 8 negative counters and 6 positive counters to the mat.

Next remove as many zero pairs as possible.

So, $-8 + 6 = -2$.

Method 2 Use a number line.

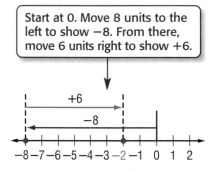

Start at 0. Move 8 units to the left to show -8. From there, move 6 units right to show $+6$.

Your Turn Add. Use counters or a number line if necessary.

d. $+7 + (-3)$ e. $4 + (-4)$ f. $-6 + 3$

The following rules are often helpful when adding integers.

Key Concept	**Adding Integers**
Words	The sum of two positive integers is always positive. The sum of two negative integers is always negative.
Examples	$5 + 1 = 6$ $-5 + (-1) = -6$
Words	The sum of a positive integer and a negative integer is sometimes positive, sometimes negative, and sometimes zero.
Examples	$5 + (-1) = 4$ $-5 + 1 = -4$ $-5 + 5 = 0$

1. **OPEN ENDED** Write an addition problem that involves one positive integer and one negative integer whose sum is −2.

2. **FIND THE ERROR** Savannah and Cesar are finding $4 + (−6)$. Who is correct? Explain.

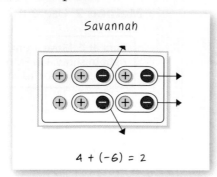

Savannah

$4 + (−6) = 2$

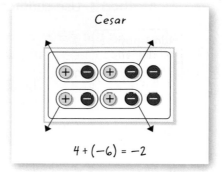

Cesar

$4 + (−6) = −2$

3. **NUMBER SENSE** Tell whether each sum is *positive*, *negative*, or *zero* without adding.

   a. $−8 + (−8)$ b. $−2 + 2$ c. $6 + (+6)$

   d. $−5 + 2$ e. $−3 + 8$ f. $−2 + (−6)$

GUIDED PRACTICE

Add. Use counters or a number line if necessary.

4. $+3 + (+1)$ 5. $4 + (+2)$ 6. $−3 + (−5)$

7. $−6 + (−4)$ 8. $+2 + (−5)$ 9. $−4 + 9$

10. **GAMES** You and a friend are playing a card game. The scores for the games played so far are shown at the right. If the player with the most points wins, who is winning?

Game	You	Friend
1	+10	+8
2	−5	−10
3	+15	+13

Practice and Applications

Add. Use counters or a number line if necessary.

11. $+2 + (+1)$ 12. $+5 + (+1)$ 13. $6 + (+2)$

14. $3 + (+4)$ 15. $+8 + 3$ 16. $+9 + 4$

17. $−4 + (−1)$ 18. $−5 + (−4)$ 19. $−2 + (−4)$

20. $−3 + (−3)$ 21. $+2 + (−3)$ 22. $+6 + (−10)$

23. $−7 + (+5)$ 24. $−3 + (+3)$ 25. $−2 + 6$

26. $−9 + 6$ 27. $4 + (−3)$ 28. $1 + (−4)$

29. $−12 + 7$ 30. $15 + (−6)$ 31. $8 + (−18)$

32. $5 + 2 + (−8)$ 33. $3 + (−5) + 7$ 34. $(−8) + 1 + 5$

HOMEWORK HELP

For Exercises	See Examples
11–20	1, 2
21–31	3, 4

Extra Practice
See pages 609, 631.

35. What is the sum of positive seven and negative three?

36. Find the result when negative nine is added to positive four.

For Exercises 37 and 38, write an addition problem that represents the situation. Then solve.

37. WEATHER The temperature outside is −5°F. If the temperature drops another 4 degrees, what temperature will it be?

38. DIVING A scuba diver descends 25 feet below the surface of the water and then swims 12 feet up towards the surface. What is the location of the diver?

39. FOOTBALL The table shows the yards gained or yards lost on each of the first three plays of a football game. What is the total number of yards gained or lost on the three plays?

Play	Yards Lost or Gained
1	+8
2	−3
3	+2

40. WRITE A PROBLEM Write a problem that can be solved by adding two integers.

41. SCIENCE At night, the surface temperature on Mars can reach a low of −120°F. During the day, the temperature rises at most 100°F. What is the maximum surface temperature on Mars during the day?

 Data Update What are the nighttime and daytime surface temperatures on Mercury? Visit **msmath1.net/data_update** to learn more.

CRITICAL THINKING For Exercises 42 and 43, write an addition sentence that satisfies each statement.

42. All addends are negative integers, and the sum is −7.

43. At least one addend is a positive integer, and the sum is −7.

Standardized Test Practice and Mixed Review

44. MULTIPLE CHOICE Paco has $150 in the bank and withdraws $25. Which addition sentence represents this situation?

    Ⓐ −150 + (−25)   Ⓑ 150 + 25

    Ⓒ −150 + (−25)   Ⓓ 150 + (−25)

45. SHORT RESPONSE What number added to −8 equals −13?

Draw a number line from −10 to 10. Then graph each integer on the number line. (Lesson 8-1)

46. −8   **47.** 0   **48.** +7

49. PATTERNS Find the next two numbers in the sequence 160, 80, 40, 20, ... (Lesson 7-6)

GETTING READY FOR THE NEXT LESSON

PREREQUISITE SKILL Subtract. (Page 589)

50. 5 − 3   **51.** 6 − 4   **52.** 9 − 5   **53.** 10 − 3

Subtracting Integers

What You'll LEARN

Subtract integers.

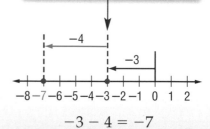

HANDS-ON Mini Lab

Materials
• number lines

Work with a partner.

The number lines below model the subtraction problems 8 − 2 and −3 − 4.

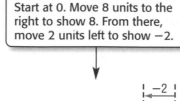

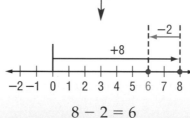

Start at 0. Move 8 units to the right to show 8. From there, move 2 units left to show −2.

Start at 0. Move 3 units to the left to show −3. From there, move 4 units left to show −4.

$$8 - 2 = 6$$

$$-3 - 4 = -7$$

1. Model $8 + (-2)$ using a number line.

2. Compare this model to the model for $8 - 2$. How is $8 - 2$ related to $8 + (-2)$?

3. Use a number line to model $-3 + (-4)$.

4. Compare this model to the model for $-3 - 4$. How is $-3 - 4$ related to $-3 + (-4)$?

The Mini Lab shows that when you subtract a number, the result is the same as adding the opposite of the number.

The numbers 2 and −2 are opposites.

The numbers 4 and −4 are opposites.

$$8 - 2 = 6 \qquad 8 + (-2) = 6 \qquad\qquad -3 - 4 = -7 \qquad -3 + (-4) = -7$$

Notice that the result is the same.

Notice that the result is the same.

To subtract integers, you can use counters or the following rule.

Key Concept   **Subtracting Integers**

Words   To subtract an integer, add its opposite.

Examples $5 - 2 = 5 + (-2)$

$-3 - 4 = -3 + (-4)$

$-1 - (-2) = -1 + 2$

EXAMPLES   Subtract Positive Integers

1 Find $3 - 1$.

Method 1 Use counters.

Place 3 positive counters on the mat to show +3. Remove 1 positive counter.

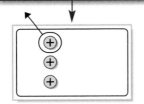

So, $3 - 1 = 2$.

Check   Use a number line to find $3 + (-1)$.

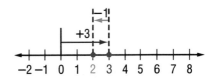

Method 2 Use the rule.

$3 - 1 = 3 + (-1)$ To subtract 1, add −1.

$= 2$ Simplify.

The difference of 3 and 1 is 2.

STUDY TIP

Check by Adding
In Example 1, you can check $3 - 1 = 2$ by adding.
$2 + 1 = 3$
In Example 2, you can check $-5 - (-3) = -2$ by adding.
$-2 + (-3) = -5$

Your Turn   Subtract. Use counters if necessary.

a. $6 - 4$ b. $+5 - 2$ c. $9 - 6$

EXAMPLES   Subtract Negative Integers

2 Find $-5 - (-3)$.

Method 1 Use counters.

Place 5 negative counters on the mat to show −5. Remove 3 negative counters.

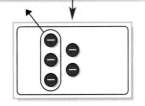

So, $-5 - (-3) = -2$.

Check   Use a number line to find $-5 + 3$.

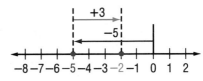

Method 2 Use the rule.

$-5 - (-3) = -5 + 3$ To subtract −3, add 3.

$= -2$ Simplify.

The difference of −5 and −3 is −2.

Your Turn   Subtract. Use counters if necessary.

d. $-8 - (-2)$ e. $-6 - (-1)$ f. $-5 - (-4)$

Sometimes you need to add zero pairs before you can subtract. Recall that when you add zero pairs, the values of the integers on the mat do not change.

EXAMPLES **Subtract Integers Using Zero Pairs**

3 Find $-2 - 3$.

Method 1 Use counters.

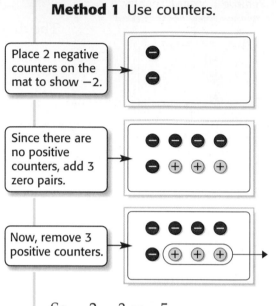

Place 2 negative counters on the mat to show -2.

Since there are no positive counters, add 3 zero pairs.

Now, remove 3 positive counters.

Method 2 Use the rule.

$-2 - 3 = -2 + (-3)$ To subtract 3, add -3.

$\qquad = -5$ Simplify.

The difference of -2 and 3 is -5.

Check Use a number line to find $-2 + (-3)$.

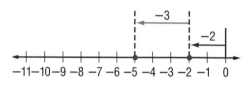

So, $-2 - 3 = -5$.

Your Turn Subtract. Use counters if necessary.

g. $-8 - 3$ h. $-2 - 5$ i. $-9 - 5$

Skill and Concept Check

Writing Math
Exercises 2 & 3

1. **OPEN ENDED** Write a subtraction sentence that contains two integers in which the result is –3.

2. **Which One Doesn't Belong?** Identify the sentence that is not equal to -2. Explain.

| $-6 - (-4)$ | $-6 + 4$ | $4 - 6$ | $-4 + 6$ |

3. **NUMBER SENSE** Explain how $6 - 3$ is related to $6 + (-3)$.

GUIDED PRACTICE

Subtract. Use counters if necessary.

4. $7 - 5$ 5. $+4 - 1$ 6. $-9 - (-4)$

7. $-6 - (-6)$ 8. $-1 - 5$ 9. $-2 - (-3)$

10. **ALGEBRA** Evaluate $a - b$ if $a = 5$ and $b = 7$.

11. **STATISTICS** Find the range of the temperatures $-5°F$, $12°F$, $-21°F$, $4°F$, and $0°F$.

Practice and Applications

Subtract. Use counters if necessary.

12. $8 - 3$ **13.** $6 - 5$ **14.** $+6 - 2$

15. $+8 - 1$ **16.** $-7 - (-5)$ **17.** $-8 - (-4)$

18. $-10 - (-5)$ **19.** $-12 - (-7)$ **20.** $-9 - (-9)$

21. $-2 - (-2)$ **22.** $-4 - 8$ **23.** $-5 - 4$

24. $5 - (-4)$ **25.** $6 - (-2)$ **26.** $-4 - (-2)$

27. $-9 - (-6)$ **28.** $-12 - (3)$ **29.** $-15 - (5)$

HOMEWORK HELP

For Exercises	See Examples
12–15, 28–29	1
16–21, 24–27	2
22–23	3

Extra Practice
See pages 610, 631.

30. ALGEBRA Find the value of $m - n$ if $m = -3$ and $n = 4$.

31. WEATHER One morning, the temperature was $-5°$F. By noon, it was $10°$F. Find the change in temperature.

32. SOCCER The difference between the number of goals scored by a team (goals for), and the number of goals scored by the opponent (goals against) is called the *goal differential*. Use the table to find each team's goal differential.

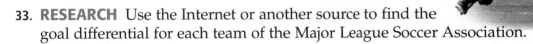

Women's United Soccer Association, 2002		
Team	Goals For	Goals Against
Carolina	40	30
San Jose	34	30
San Diego	28	42
New York	31	62

33. RESEARCH Use the Internet or another source to find the goal differential for each team of the Major League Soccer Association.

CRITICAL THINKING Tell whether each statement is *sometimes, always,* or *never* true. Give an example or a counterexample for each.

34. positive $-$ positive $=$ positive   **35.** negative $-$ negative $=$ negative

36. negative $-$ positive $=$ negative   **37.** positive $-$ negative $=$ negative

Standardized Test Practice and Mixed Review

38. MULTIPLE CHOICE At noon, the temperature was $-3°$F. If the temperature decreases $9°$F, what will the new temperature be?

   A $12°$F   **B** $6°$F   **C** $-6°$F   **D** $-12°$F

39. SHORT RESPONSE Find the value of $7 - (-6)$.

Add. Use counters or a number line if necessary. (Lesson 8-2)

40. $3 + (-4)$ **41.** $4 + (-4)$ **42.** $-2 + (-3)$

43. Graph $-6, 2, 0, -2, 4$ on the same number line. (Lesson 8-1)

GETTING READY FOR THE NEXT LESSON

PREREQUISITE SKILL Multiply. (Page 590)

44. 5×6 **45.** 8×7 **46.** 9×4 **47.** 8×9

Vocabulary and Concepts

1. **Write** the integer that describes 4 miles below sea level. (Lesson 8-1)
2. **Draw** a model to show $4 + (-3)$. (Lesson 8-2)
3. **Explain** how to subtract integers. (Lesson 8-3)

Skills and Applications

Draw a number line from −10 to 10. Then graph each integer on the number line. (Lesson 8-1)

4. -8 5. 6 6. -9

Replace each ● with $<$, $>$, or $=$ to make a true sentence. (Lesson 8-1)

7. $-7 ● -3$ 8. $4 ● -2$ 9. $-8 ● 6$

10. Order –5, –7, 4, –3, and –2 from greatest to least. (Lesson 8-1)

11. **GOLF** In golf, the lowest score wins. The table shows the scores of the winners of the U.S. Open from 2000 to 2002. What was the lowest winning score? (Lesson 8-1)

Year	Winner	Score
2002	Tiger Woods	−3
2001	Retief Goosen	−4
2000	Tiger Woods	−12

Source: www.golfonline.com

Add. Use counters or a number line if necessary.
(Lesson 8-2)

12. $+8 + (-3)$ 13. $-6 + 2$ 14. $-4 + (-7)$

Subtract. Use counters if necessary. (Lesson 8-3)

15. $9 - (+3)$ 16. $-3 - 5$ 17. $8 - (-2)$ 18. $-4 - (-7)$

Standardized Test Practice

19. **MULTIPLE CHOICE** Find the value of d that makes $d - (-5) = 17$ a true sentence. (Lesson 8-3)

    (A) 22
    (B) −12
    (C) 12
    (D) −22

20. **SHORT RESPONSE** A mole is trying to crawl out of its burrow that is 12 inches below ground. The mole crawls up 6 inches and then slides down 2 inches. What integer represents the mole's location in relation to the surface? (Lesson 8-2)

The GameZone

A Place To Practice Your Math Skills

Falling Off the Ends

● **GET READY!**

Players: two or four
Materials: 2 different colored number cubes,
2 different colored counters

● **GET SET!**

• Draw a game board as shown below.

| -6 | -5 | -4 | -3 | -2 | -1 | 0 | 1 | 2 | 3 | 4 | 5 | 6 |

● **GO!**

• Let one number cube represent positive integers and the other represent negative integers.

• Each player places a counter on zero.

• One player rolls both number cubes and adds the numbers shown. If the sum is positive, the player moves his or her counter right the number of spaces indicated by the sum. If the sum is negative, the player moves the counter left the number of spaces indicated by the sum.

• Players take turns rolling the number cubes and moving the counters.

• **Who Wins?** The first player to go off the board in either direction wins the game.

Multiplying Integers

What You'll LEARN

Multiply integers.

HANDS-ON Mini Lab

Materials
- counters
- integer mat

Work with a partner.

The models show 3×2 and $3 \times (-2)$.

For 3×2, place 3 sets of 2 positive counters on the mat.

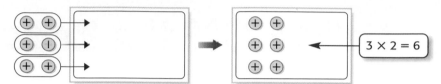

$3 \times 2 = 6$

For $3 \times (-2)$, place 3 sets of 2 negative counters on the mat.

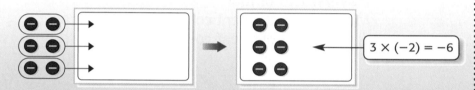

$3 \times (-2) = -6$

1. Use counters to find $4 \times (-3)$ and $5 \times (-2)$.

2. **Make a conjecture** as to the sign of the product of a positive and negative integer.

To find the sign of products like -3×2 and $-3 \times (-2)$, you can use patterns.

$3 \times 2 = 6$	$3 \times (-2) = -6$
$2 \times 2 = 4$	$2 \times (-2) = -4$
$1 \times 2 = 2$	$1 \times (-2) = -2$
$0 \times 2 = 0$	$0 \times (-2) = 0$
$-1 \times 2 = -2$	$-1 \times (-2) = 2$
$-2 \times 2 = -4$	$-2 \times (-2) = 4$
$-3 \times 2 = ?$	$-3 \times (-2) = ?$

By extending the number pattern, you find that $-3 \times 2 = -6$.

By extending the number pattern, you find that $-3 \times (-2) = 6$.

To multiply integers, the following rules apply.

Key Concept   **Multiplying Integers**

Words	The product of two integers with different signs is negative.
Examples	$3 \times (-2) = -6$ $-3 \times 2 = -6$
Words	The product of two integers with the same sign is positive.
Examples	$3 \times 2 = 6$ $-3 \times (-2) = 6$

Multiply Integers with Different Signs

Multiply.

1 $4 \times (-2)$

$4 \times (-2) = -8$ The integers have different signs. The product is negative.

2 -8×3

$-8 \times 3 = -24$ The integers have different signs. The product is negative.

Your Turn Multiply.

a. $6 \times (-3)$ b. $3(-1)$ c. -4×5 d. $-2(7)$

EXAMPLES **Multiply Integers with Same Signs**

Multiply.

3 4×8

$4 \times 8 = 32$ The integers have the same sign. The product is positive.

4 $-5 \times (-6)$

$-5 \times (-6) = 30$ The integers have the same sign. The product is positive.

Your Turn Multiply.

e. 3×3 f. $6(8)$ g. $-5 \times (-3)$ h. $-4(-3)$

REAL-LIFE MATH

How Does a Marine Biologist Use Math?
A marine biologist uses math to calculate the rate of ascent during a dive.

🖱 **Online Research**
For information about a career as a marine biologist, visit:
msmath1.net/careers

Many real-life situations can be solved by multiplying integers.

EXAMPLES **Use Integers to Solve a Problem**

5 **DOLPHINS** A dolphin dives from the surface of the water at a rate of 3 feet per second. Where will the dolphin be in relation to the surface after 5 seconds?

To find the location of the dolphin after 5 seconds, you can multiply 5 by the amount of change per second, -3 feet.

> Start at 0. Move 3 feet below the water's surface for every second.

$5 \times (-3) = -15$

So, after 5 seconds, the dolphin will be -15 feet from the surface or 15 feet *below* the surface.

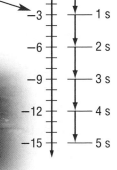

Skill and Concept Check

1. **Model** the product of 4 and −6.

2. **OPEN ENDED** Write two integers whose product is negative.

3. **Which One Doesn't Belong?** Identify the product that is not positive. Explain your reasoning.

3 x 4	−5 x (−2)	−8(6)	−3(−7)

GUIDED PRACTICE

Multiply.

4. $4 \times (-7)$ 5. $8(-7)$ 6. 4×4

7. $9(3)$ 8. $-1 \times (-7)$ 9. $-7(-6)$

10. **SCIENCE** For each kilometer above Earth's surface, the temperature decreases 7°C. If the temperature at Earth's surface is 0°C, what will the temperature be 2 kilometers above the surface?

Practice and Applications

Multiply.

11. $6 \times (-6)$ 12. $9 \times (-1)$ 13. $7(-3)$

14. $2(-10)$ 15. -7×5 16. -2×9

17. $-5(6)$ 18. $-6(9)$ 19. 8×7

20. $9(4)$ 21. $-5 \times (-8)$ 22. $-9 \times (-7)$

23. $-6(-10)$ 24. $-1(-9)$ 25. -7×7

26. Find the product of −10 and 10.

27. **ALGEBRA** Evaluate st if $s = -4$ and $t = 9$.

28. **ALGEBRA** Find the value of ab if $a = -12$ and $b = -5$.

29. **SCHOOL** A school district loses 30 students per year due to student transfers. If this pattern continues for the next 4 years, what will be the loss in relation to the original enrollment?

PATTERNS Find the next two numbers in the pattern. Then describe the pattern.

30. $2, -4, 8, -16, \ldots$ 31. $-2, -6, -18, -54, \ldots$

32. **WRITE A PROBLEM** Write a problem that can be solved by multiplying a positive integer and a negative integer.

HOMEWORK HELP	
For Exercises	See Examples
11–18, 26–27	1, 2
19–25, 28	3, 4
29, 33	5
Extra Practice See pages 610, 631.	

33. **SCIENCE** In 1990, Sue Hendrickson found the T-Rex fossil that now bears her name. To excavate the fossil, her team had to remove about 3 cubic meters of dirt each day from the site. Write an integer to represent the change in the amount of soil at the site at the end of 5 days.

MULTI STEP Find the value of each expression.

34. $3(-4 - 7)$ 35. $-2(3)(-4)$ 36. $-4(5 + (-9))$

Tell whether each sentence is *sometimes*, *always*, or *never* true. Give an example or counterexample for each.

37. The product of two positive integers is negative.

38. A negative integer multiplied by a negative integer is positive.

39. The product of a positive integer and a negative integer is negative.

40. The product of any three negative integers is negative.

41. **CRITICAL THINKING** The product of 1 times any number is the number itself. What is the product of -1 times any number?

Standardized Test Practice and Mixed Review

42. **MULTIPLE CHOICE** Which value of m makes $-4m = 24$ a true statement?

   Ⓐ 6   Ⓑ 4   Ⓒ -4 Ⓓ -6

43. **MULTIPLE CHOICE** A scuba diver descends from the ocean's surface at the rate of 4 meters per minute. Where will the diver be in relation to the surface after 8 minutes?

   Ⓕ 32 m   Ⓖ -12 m   Ⓗ -32 m   Ⓘ -48 m

Subtract. Use counters if necessary. (Lesson 8-3)

44. $9 - 2$ 45. $+3 - 1$ 46. $-5 - (-8)$ 47. $-7 - 6$

Add. Use counters or a number line if necessary. (Lesson 8-2)

48. $-7 + 2$ 49. $+4 + (-3)$ 50. $-2 + (-2)$ 51. $7 + (-8)$

52. **GOLF** The table shows the scores for the top five players of the 2002 LPGA Championship. Find the median of the data. (Lessons 8-1 and 2-7)

53. Estimate the product of $\frac{3}{4}$ and 37. (Lesson 7-1)

Players	Score
Se Ri Pak	-5
Juli Inkster	$+1$
Beth Daniel	-2
Karrie Webb	$+1$
Annika Sorenstam	0

Source: www.lpga.com

GETTING READY FOR THE NEXT LESSON

PREREQUISITE SKILL Divide. (Page 591)

54. $9 \div 3$ 55. $12 \div 2$ 56. $63 \div 7$ 57. $81 \div 9$

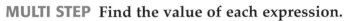

What You'll Learn
Solve problems by working backward.

Work Backward

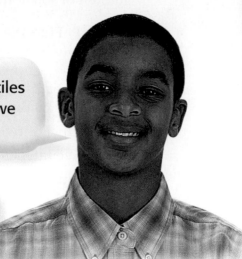

Hey Emilia, here's a puzzle about the tiles in our favorite board game. How can we find the number of pink squares?

Well Wesley, we could start from the last clue and **work backward.**

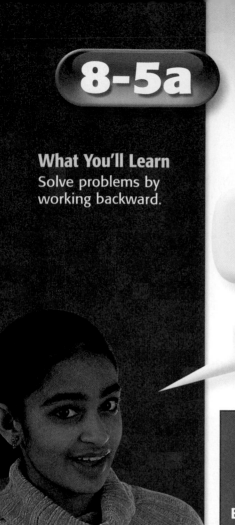

	We know the number of gray squares. We need to find the number of pink squares.
Explore	• The squares on the game board are pink, dark blue, light blue, red or gray. • There are 8 more light blue than pink. • There are twice as many light blue as dark blue. • There are 6 times as many red as dark blue. • There are 28 fewer red than gray. • There are 100 gray squares on the board.
Plan	To find the number of pink squares, let's start at the last clue and reverse the clues.
Solve	$100 - 28 = 72 \rightarrow$ number of red squares $72 \div 6 = 12 \rightarrow$ number of dark blue squares $12 \times 2 = 24 \rightarrow$ number of light blue squares $24 - 8 = 16 \rightarrow$ number of pink squares
Examine	Look at the clues. Start with 16 pink squares and follow the clues to be sure you end with 100 gray squares.

Analyze the Strategy

1. **Explain** how using the work backward strategy helped the students find the number of pink squares.

2. **Compare and contrast** the words from the game clues with the operations the students used to find the number of pink squares.

3. **Write** a problem that can be solved using the work backward strategy. Then tell the steps you would take to find the solution of the problem.

314 **Chapter 8** Algebra: Integers

Solve. Use the work backward strategy.

4. **NUMBER SENSE** A number is divided by 5. Next, 4 is subtracted from the quotient. Then, 6 is added to the difference. If the result is 10, what is the number?

5. **NUMBER SENSE** A number is multiplied by −2, and then 6 is added to the product. The result is 12. What is the number?

6. **TIME** Marta and Scott volunteer at the food bank at 9:00 A.M. on Saturdays. It takes 30 minutes to get from Scott's house to the food bank. It takes Marta 15 minutes to get to Scott's house. If it takes Marta 45 minutes to get ready in the morning, what is the latest time she should get up?

Mixed Problem Solving

Solve. Use any strategy.

7. **MONEY** Chet has $4.50 in change after purchasing a skateboard for $62.50 and a helmet for $32. How much money did Chet have originally?

8. **MONEY** Mrs. Perez wants to save an average of $150 per week over six weeks. Find how much she must save during the sixth week to meet her goal.

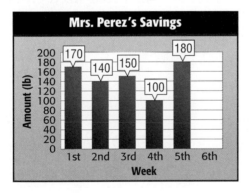

9. **NUMBER SENSE** What is the least positive number that you can divide by 7 and get a remainder of 4, divide by 8 and get a remainder of 5, and divide by 9 and get a remainder of 6?

10. **SCIENCE** A certain bacteria doubles its population every 12 hours. After 3 days, there are 1,600 bacteria. How many bacteria were there at the beginning of the first day?

11. **PUZZLES** In a magic square, each row, column, and diagonal have the same sum. Copy and complete the magic square.

−2	?	?
−3	−1	1
?	?	?

12. **SCHOOL** A multiple choice test has 10 questions. The scoring is +3 for correct answers, −1 for incorrect answers, and 0 for questions not answered. Meg scored 23 points on the test. She answered all but one of the questions. How many did she answer correctly? How many did she answer incorrectly?

13. **DESIGN** A designer wants to arrange 12 glass bricks into a rectangular shape with the least perimeter possible. How many blocks will be in each row?

14. **STANDARDIZED TEST PRACTICE**
Refer to the table. If the pattern continues, what will be the output when the input is 3?

Input	Output
0	5
1	8
2	11
3	?

Ⓐ 12   Ⓑ 13
Ⓒ 14   Ⓓ 15

You will use the work backward strategy in the next lesson.

Dividing Integers

HANDS-ON Mini Lab

Materials
• counters

Work with a partner.

You can use counters to model $-12 \div 3$.

STEP 1 Place 12 negative counters on the mat to represent -12.

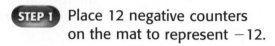

STEP 2 Separate the 12 negative counters into three equal-sized groups. There are 3 groups of 4 negative counters.

So, $-12 \div 3 = -4$.

1. Explain how you would model $-9 \div 3$.
2. What would you do differently to model $8 \div 2$?

Division means to separate the number into equal-sized groups.

EXAMPLES Divide Integers

Divide.

1 $-8 \div 4$

Separate 8 negative counters into 4 equal-sized groups.

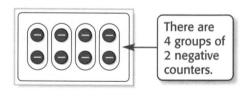

There are 4 groups of 2 negative counters.

So, $-8 \div 4 = -2$.

2 $15 \div 3$

Separate 15 positive counters into 3 equal-sized groups.

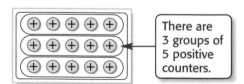

There are 3 groups of 5 positive counters.

So, $15 \div 3 = 5$.

Your Turn Divide. Use counters if necessary.

a. $-4 \div 2$ b. $-20 \div 4$ c. $18 \div 3$

You can also divide integers by working backward. For example, to find $48 \div 6$, ask yourself "what number times 6 equals 48?"

EXAMPLES Divide Integers

Divide.

3 $-10 \div 2$

Since $-5 \times 2 = -10$, it follows that $-10 \div 2 = -5$. ← The quotient is negative.

4 $14 \div (-7)$

Since $-2 \times (-7) = 14$, it follows that $14 \div (-7) = -2$. ← The quotient is negative.

5 $-24 \div (-6)$

Since $4 \times (-6) = -24$, it follows that $-24 \div (-6) = 4$. ← The quotient is positive.

Your Turn Divide. Work backward if necessary.

d. $-16 \div 4$ e. $32 \div (-8)$ f. $-21 \div (-3)$

Study the signs of the quotients in the previous examples. The following rules apply.

Key Concept   **Dividing Integers**

Words	The quotient of two integers with different signs is negative.
Examples	$8 \div (-2) = -4$ $-8 \div 2 = -4$
Words	The quotient of two integers with the same sign is positive.
Examples	$8 \div 2 = 4$ $-8 \div (-2) = 4$

EXAMPLES Divide Integers

Standardized Test Practice

6 **MULTIPLE-CHOICE TEST ITEM** A football team was penalized a total of 15 yards in 3 plays. If the team was penalized an equal number of yards on each play, which integer gives the yards for each penalty?

   A 5   **B** -3 **C** -5 **D** -6

Read the Test Item
You need to find the number of yards for each penalty.

Solve the Test Item
$-15 \div 3 = -5$

The answer is C.

Test-Taking Tip

The Princeton Review

Look for words that indicate negative numbers. A penalty is a loss, so it would be represented by a negative integer.

 msmath1.net/extra_examples

Skill and Concept Check

Writing Math
Exercise 3

1. **Write** a division problem represented by the model at the right.

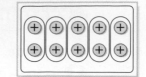

2. **OPEN ENDED** Write a division sentence whose quotient is -4.

3. **FIND THE ERROR** Emily and Mateo are finding $-42 \div 6$. Who is correct? Explain.

   | Emily |
   | $-42 \div 6 = -7$ |

   | Mateo |
   | $-42 \div 6 = 7$ |

4. **NUMBER SENSE** Is $-24 \div 6$ greater than, less than, or equal to $24 \div (-4)$?

GUIDED PRACTICE

Divide.

5. $-6 \div 2$ 6. $-12 \div 3$ 7. $15 \div 3$

8. $-25 \div 5$ 9. $72 \div (-9)$ 10. $-36 \div (-4)$

11. **GARDENING** Jacob is planting vegetable seeds in holes that are each 2 inches deep. If Jacob has dug a total of 28 inches, how many holes has he dug?

Practice and Applications

Divide.

12. $-8 \div 2$ 13. $-12 \div 4$ 14. $-18 \div 6$

15. $-32 \div 4$ 16. $21 \div 7$ 17. $35 \div 7$

18. $-40 \div 8$ 19. $-45 \div 5$ 20. $63 \div (-9)$

21. $81 \div (-9)$ 22. $-48 \div (-6)$ 23. $-54 \div (-6)$

HOMEWORK HELP	
For Exercises	See Examples
12–25	1–5
26, 29–31	6

Extra Practice
See pages 610, 631.

24. **ALGEBRA** Find the value of $c \div d$ if $c = -22$ and $d = 11$.

25. **ALGEBRA** For what value of m makes $48 \div m = -16$ true?

26. **OCEANOGRAPHY** A submarine starts at the surface of the water and then travels 95 meters below sea level in 19 seconds. If the submarine traveled an equal distance each second, what integer gives the distance traveled each second?

Find the value of each expression.

27. $\dfrac{-3 + (-7)}{2}$

28. $\dfrac{(4 + (-6)) \times (-1 + 7)}{-3}$

SCHOOL For Exercises 29 and 30, use the table shown.

A teacher posted the sign shown below so that students would know the process of earning points to have a year-end class pizza party.

Positive Behavior	Points	Negative Behavior	Points
Complete Homework	+5	Incomplete Homework	−5
Having School Supplies	+3	Not Having School Supplies	−3
Being Quiet	+2	Talking	−2
Staying in Seat	+1	Out of Seat	−1
Paying Attention	+1	Not Paying Attention	−1
Being Cooperative	+4	Not Following Directions	−4

29. After 5 days, a student has −15 points. On average, how many points is the student losing each day?

30. **MULTI STEP** Suppose a student averages −2 points every day for 4 days. How many positive points would the student need to earn on the fifth day to have a +5?

31. **MULTI STEP** A study revealed that 6,540,000 acres of coastal wetlands have been lost over the past 50 years due to draining, dredging, land fills, and spoil disposal. If the loss continues at the same rate, how many acres will be lost in the next 10 years?

32. **CRITICAL THINKING** List all the numbers by which −24 is divisible.

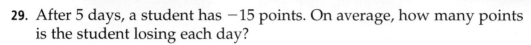

Standardized Test Practice and Mixed Review

33. **MULTIPLE CHOICE** Julia needs to plant 232 seedlings in flowerpots with 8 plants in each pot. How many flowerpots will be needed to plant the seedlings?

    Ⓐ 32   Ⓑ 29   Ⓒ 27   Ⓓ 19

34. **MULTIPLE CHOICE** Find the value of $-145 \div (-29)$.

    Ⓕ −4   Ⓖ −5   Ⓗ 5   Ⓘ 6

35. Find the product of −3 and −7. (Lesson 8-4)

36. **ALGEBRA** Find the value of $s - t$ if $s = 6$ and $t = -5$. (Lesson 8-3)

Find the missing number in each sequence. (Lesson 7-6)

37. $3\frac{7}{8}, 4\frac{1}{4}, \underline{\ ?\ }, 5, 5\frac{3}{8}$ 38. 34, 36.5, 39, 41.5, $\underline{\ ?\ }$

GETTING READY FOR THE NEXT LESSON

PREREQUISITE SKILL Draw a number line from −10 to 10. Then graph each point on the number line. (Lesson 8-1)

39. 3   40. 0   41. −2   42. −6

The Coordinate Plane

What You'll LEARN

Graph ordered pairs of numbers on a coordinate plane.

NEW Vocabulary

coordinate system
coordinate plane
x-axis
y-axis
origin
quadrants
ordered pair
x-coordinate
y-coordinate

Link to READING

Prefix quadr-: four, as in quadruplets

WHEN am I ever going to use this?

READING MAPS A street map is shown.

1. Use a letter and a number to identify where on the map Autumn Court meets Timberview Drive.

2. Location G2 is closest to the end of which street?

3. Write a sentence that explains how to locate a specific place on the map shown.

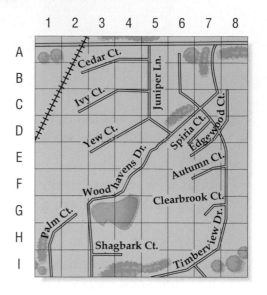

On a map, towns, streets, and points of interest are often located on a grid. In mathematics, we use a grid called a **coordinate system**, or **coordinate plane**, to locate points.

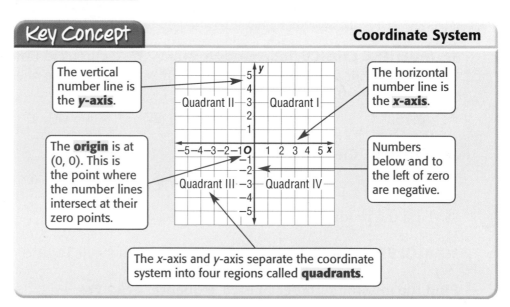

Key Concept   Coordinate System

The vertical number line is the **y-axis**.

The horizontal number line is the **x-axis**.

The **origin** is at (0, 0). This is the point where the number lines intersect at their zero points.

Numbers below and to the left of zero are negative.

The *x*-axis and *y*-axis separate the coordinate system into four regions called **quadrants**.

You can use an **ordered pair** to locate any point on the coordinate system. The first number is the **x-coordinate**. The second number is the **y-coordinate**.

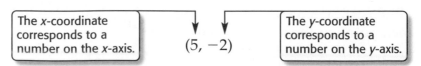

The *x*-coordinate corresponds to a number on the *x*-axis.

$(5, -2)$

The *y*-coordinate corresponds to a number on the *y*-axis.

EXAMPLES Identify Ordered Pairs

Identify the ordered pair that names each point. Then identify its quadrant.

1 point *B*

Step 1 Start at the origin. Move right on the *x*-axis to find the *x*-coordinate of point *B*, which is 4.

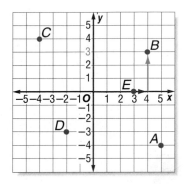

Step 2 Move up the *y*-axis to find the *y*-coordinate, which is 3.

Point *B* is named by (4, 3). Point *B* is in the first quadrant.

2 point *D*

Step 1 Start at the origin. Move left on the *x*-axis to find the *x*-coordinate of point *D*, which is −2.

Step 2 Move down the *y*-axis to find the *y*-coordinate, which is −3.

Point *D* is named by (−2, −3). Point *D* is in the third quadrant.

Your Turn Write the ordered pair that names each point. Then identify the quadrant in which each point is located.

a. *A*　　　　　　**b.** *C*　　　　　　**c.** *E*

STUDY TIP

Ordered Pairs A point located on the *x*-axis will have a *y*-coordinate of 0. A point located on the *y*-axis will have an *x*-coordinate of 0. Points located on an axis are not in any quadrant.

You can graph an ordered pair. To do this, draw a dot at the point that corresponds to the ordered pair. The coordinates are the directions to locate the point.

EXAMPLES Graph Ordered Pairs

3 Graph point *M* at (−3, 5).

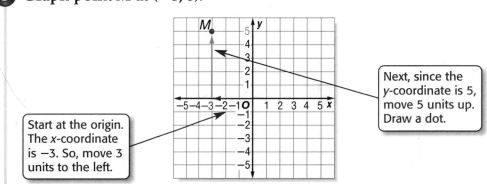

Start at the origin. The *x*-coordinate is −3. So, move 3 units to the left.

Next, since the *y*-coordinate is 5, move 5 units up. Draw a dot.

Your Turn Graph and label each point on a coordinate plane.

d. *N*(2, 4)　　　　**e.** *P*(0, 4)　　　　**f.** *Q*(5, 1)

REAL-LIFE MATH

BASKETBALL In basketball, each shot made from outside the 3-point line scores 3 points. The expression $3x$ represents the total number of points scored where x is the number of 3-point shots made.

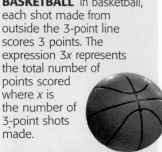

EXAMPLES Graph Real-Life Data

4 BASKETBALL Use the information at the left. List the ordered pairs (3-point shots made, total number of points) for 0, 1, 2, and 3 shots made.

Evaluate the expression $3x$ for 0, 1, 2, and 3. Make a table.

x (shots)	3x	y (points)	(x, y)
0	3(0)	0	(0, 0)
1	3(1)	3	(1, 3)
2	3(2)	6	(2, 6)
3	3(3)	9	(3, 9)

5 Graph the ordered pairs from Example 4. Then describe the graph.

The points appear to fall in a line.

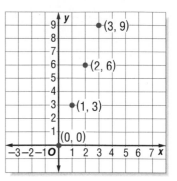

Skill and Concept Check

Writing Math

Exercise 2

1. **Draw** a coordinate plane. Then label the origin, x-axis, and y-axis.

2. **Explain** how to graph a point using an ordered pair.

3. **OPEN ENDED** Give the coordinates of a point located on the y-axis.

GUIDED PRACTICE

For Exercises 4–9, use the coordinate plane at the right. Identify the point for each ordered pair.

4. $(3, 1)$ 5. $(-1, 0)$ 6. $(3, -4)$

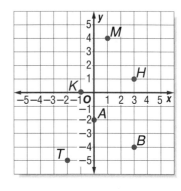

Write the ordered pair that names each point. Then identify the quadrant where each point is located.

7. M 8. A 9. T

Graph and label each point on a coordinate plane.

10. $D(2, 1)$ 11. $K(-3, 3)$ 12. $N(0, -1)$

MONEY For Exercises 13 and 14, use the following information.
Lindsey earns $4 each week in allowance. The expression $4w$ represents the total amount earned where w is the number of weeks.

13. List the ordered pairs (weeks, total amount earned) for 0, 1, 2, and 3 weeks.

14. Graph the ordered pairs. Then describe the graph.

For Exercises 15–26, use the coordinate plane at the right. Identify the point for each ordered pair.

15. (2, 2)
16. (1, 3)
17. (−4, 2)
18. (−2, 1)
19. (−3, −2)
20. (−4, −5)

HOMEWORK HELP

For Exercises	See Examples
15–26	1, 2
27–35	3
36, 37	4, 5

Extra Practice
See pages 611, 631.

Write the ordered pair that names each point. Then identify the quadrant where each point is located.

21. *G*
22. *C*
23. *A*
24. *D*
25. *P*
26. *M*

Graph and label each point on a coordinate plane.

27. $N(1, 2)$
28. $T(0, 0)$
29. $B(−3, 4)$
30. $F(5, −2)$
31. $H(−4, −1)$
32. $K(−2, −5)$
33. $J\left(2\frac{1}{2}, −2\frac{1}{2}\right)$
34. $A\left(4\frac{3}{4}, −1\frac{1}{4}\right)$
35. $D(−1.5, 2.5)$

SCOOTERS For Exercises 36 and 37, use the following information.
If it takes Ben 5 minutes to ride his scooter once around a bike path, then $5t$ represents the total time where t is the number of times around the path.

36. List the ordered pairs (number of times around the path, total time) for 0, 1, 2, and 3 times around the path.

37. Graph the ordered pairs. Then describe the graph.

CRITICAL THINKING If the graph of a set of ordered pairs forms a line, the graph is *linear*. Otherwise, the graph is *nonlinear*.

Graph each set of points on a coordinate plane. Then tell whether the graph is linear or nonlinear.

38. (−1, −1), (−2, −2), (−3, −3), (−4, −4)
39. (−1, 1), (2, −2), (−3, 3), (−4, −4)

40. **MULTIPLE CHOICE** Which ordered pair represents point *H* on the coordinate grid?

   Ⓐ (4, 1)   Ⓑ (1, 3)   Ⓒ (1, 2)   Ⓓ (2, 4)

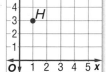

41. **SHORT RESPONSE** Graph point *D* at (4, 3) on a coordinate plane.

42. **ALGEBRA** Find the value of $x \div y$ if $x = −15$ and $y = −3$. (Lesson 8-5)

Multiply. (Lesson 8-4)

43. $−5 \times 4$
44. $−9 \times (−6)$
45. $3(−8)$
46. $−7(−7)$

Vocabulary and Concept Check

coordinate plane (p. 320)	opposites (p. 296)	x-axis (p. 320)
coordinate system (p. 320)	ordered pair (p. 320)	x-coordinate (p. 320)
graph (p. 295)	origin (p. 320)	y-axis (p. 320)
integer (p. 294)	positive integer (p. 294)	y-coordinate (p. 320)
negative integer (p. 294)	quadrant (p. 320)	

Choose the letter of the term that best matches each phrase.

1. the sign of the sum of two positive integers
2. the horizontal number line of a coordinate system
3. the sign of the quotient of a positive integer and a negative integer
4. the second number in an ordered pair
5. the first number in an ordered pair
6. numbers that are the same distance from 0 on a number line
7. a pair of counters that includes one positive counter and one negative counter

a. *y*-coordinate
b. positive
c. negative
d. *x*-axis
e. *x*-coordinate
f. opposites
g. zero pair

Lesson-by-Lesson Exercises and Examples

 Integers (pp. 294–298)

Graph each integer on a number line.

8. -2 9. 5 10. -6

Replace each ● with <, >, or = to make a true sentence.

11. $-4 ● 0$ 12. $6 ● 2$
13. $-1 ● 1$ 14. $7 ● -6$

15. Order $-2, 4, 0, -1,$ and 3 from least to greatest.

16. Order $8, -7, -5,$ and -12 from greatest to least.

17. **MONEY** Write an integer to describe owing your friend $5.

Example 1   Graph -4 on a number line.

Draw a number line. Then draw a dot at the location that represents -4.

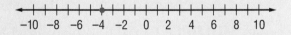

Example 2   Replace the ● in $-5 ● -2$ with <, >, or = to make a true sentence.

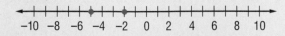

Since -5 is to the left of -2, $-5 < -2$.

msmath1.net/vocabulary_review

8-2 Adding Integers (pp. 300–303)

Add. Use counters or a number line if necessary.

18. $+4 + (+3)$
19. $7 + (+2)$
20. $-8 + (-5)$
21. $-4 + (-2)$
22. $+8 + (-3)$
23. $6 + (-9)$
24. $-10 + 4$
25. $-9 + 5$
26. $-6 + 6$
27. $7 - 7$
28. $9 + (-5)$
29. $15 + (-3)$

30. What is the sum of negative 13 and positive 4?

31. Find the sum of positive 18 and negative 3.

32. **ALGEBRA** Find the value of $m + n$ if $m = 6$ and $n = -9$.

33. **WEATHER** Suppose the temperature outside is 4°F. If the temperature drops 7 degrees, what temperature will it be?

Example 3 Find $-3 + 2$.

Method 1 Use counters.

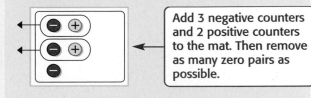

Add 3 negative counters and 2 positive counters to the mat. Then remove as many zero pairs as possible.

So, $-3 + 2 = -1$.

Method 2 Use a number line.

Start at 0. Move 3 units to the left to show -3. From there, move 2 units right to show $+2$.

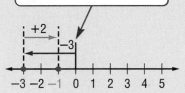

8-3 Subtracting Integers (pp. 304–307)

Subtract. Use counters if necessary.

34. $6 - 4$
35. $8 - 5$
36. $+9 - 3$
37. $+7 - 1$
38. $-4 - (-9)$
39. $-12 - (-8)$
40. $-8 - (-3)$
41. $-6 - (-4)$
42. $-2 - 5$
43. $-5 - 8$
44. $-3 - (-8)$
45. $-1 - (-2)$

46. **ALGEBRA** Evaluate $m - n$ if $m = 6$ and $n = 9$.

47. **WEATHER** In Antarctica, the temperature can be −4°F. With the windchill, this feels like −19°F. Find the difference between the actual temperature and the windchill temperature.

Example 4 Find $-2 - 3$.

Method 1 Use counters.

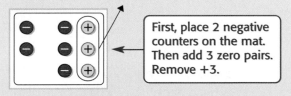

First, place 2 negative counters on the mat. Then add 3 zero pairs. Remove +3.

So, $-2 - 3 = -5$.

Method 2 Use the rule.

$-2 - 3 = -2 + (-3)$ To subtract 3, add −3.

$\quad\quad\quad = -5$ Simplify.

So, $-2 - 3 = -5$.

8-4 Multiplying Integers (pp. 310–313)

Multiply.

48. -3×5 49. $6(-4)$

50. $-2 \times (-8)$ 51. $-7(-5)$

52. $9 \times (-1)$ 53. $-8(4)$

54. $6(5)$ 55. $8(4)$

56. **MONEY** Write the integer that represents the amount lost if you lost $2 each month for 3 months.

Example 5 Find -4×3.

$-4 \times 3 = -12$ The integers have different signs. The product is negative.

Example 6 Find $-8 \times (-3)$.

$-8 \times (-3) = 24$ The integers have the same sign. The product is positive.

8-5 Dividing Integers (pp. 316–319)

Divide.

57. $8 \div (-2)$ 58. $-56 \div (-8)$

59. $-81 \div -9$ 60. $-36 \div (-3)$

61. $24 \div (-8)$ 62. $-72 \div 6$

63. $-21 \div (-3)$ 64. $42 \div (-7)$

65. **ALGEBRA** What is the value of $k \div j$ if $k = -28$ and $j = 7$?

Example 7 Find $-6 \div 3$.

$-2 \times 3 = -6$. So, $-6 \div 3 = -2$.

Example 8 Find $-20 \div (-5)$.

$4 \times -5 = -20$. So, $-20 \div (-5) = 4$.

8-6 The Coordinate Plane (pp. 320–323)

Write the ordered pair that names each point. Then identify the quadrant where each point is located.

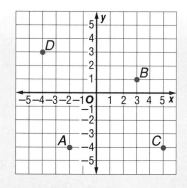

66. A 67. B

68. C 69. D

Graph and label each point on a coordinate plane.

70. $E(2, -1)$ 71. $F(-3, -2)$

72. $G(4, 1)$ 73. $H(-2, 4)$

Example 9 Graph $M(-3, -4)$.

Start at 0. Since the x-coordinate is -3, move 3 units left. Then since the y-coordinate is -4, move 4 units down.

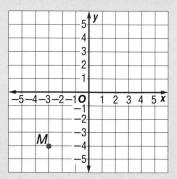

Vocabulary and Concepts

1. **Write** the rule for adding two negative integers.

2. **Explain** how to determine the sign of the quotient of two integers.

Skills and Applications

Write the opposite of each number.

3. 12   4. -3 5. 6

6. Write the integer that represents withdrawing $75.

Draw a number line from -10 to 10. Then graph each integer on the number line.

7. -9 8. 1   9. -4

Add or subtract. Use counters or a number line if needed.

10. $-5 + (-7)$ 11. $6 - (-19)$ 12. $-13 + 10$

13. $-4 - (-9)$ 14. $-7 - 3$ 15. $6 - (-5)$

16. **WEATHER** The temperature at 6:00 A.M. was $-5°F$. Find the temperature at 8:00 A.M. if it was 7 degrees warmer.

Multiply or divide.

17. $4(-9)$ 18. $24 ÷ (-4)$ 19. $-2(-7)$

20. **HEALTH** Avery's temperature dropped $1°F$ each hour for 3 hours. What was the change from his original temperature?

Name the ordered pair for each point and identify its quadrant.

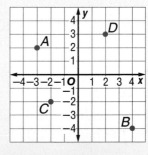

21. *A*   22. *B*

Graph and label each point on a coordinate plane.

23. $C(-4, -2)$ 24. $D(5, 3)$

Standardized Test Practice

25. **SHORT RESPONSE** If it takes Karen 4 minutes to walk around her neighborhood block, then $4t$ gives the total time where t is the number of times around the block. Write the ordered pairs (the number of times around the block, time) for 1, 3, and 5 times.

Standardized Test Practice

PART 1   Multiple Choice

Record your answers on the answer sheet provided by your teacher or on a sheet of paper.

1. Which list correctly orders the numbers from least to greatest? (Prerequisite Skill, p. 588)

   Ⓐ 124, 223, 238, 276

   Ⓑ 223, 124, 276, 238

   Ⓒ 238, 276, 124, 223

   Ⓓ 276, 238, 223, 124

2. Which of the following numbers is evenly divisible by 2 and 3? (Lesson 1-2)

   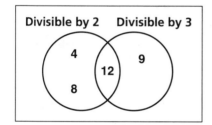

   Ⓕ 4   Ⓖ 8   Ⓗ 9   Ⓘ 12

3. What is the product 0.9×7? (Lesson 4-1)

   Ⓐ 0.063   Ⓑ 0.63

   Ⓒ 6.3   Ⓓ 63.0

4. The following fractions are all equivalent *except* which one? (Lesson 5-2)

   Ⓕ $\frac{1}{2}$ Ⓖ $\frac{3}{4}$ Ⓗ $\frac{3}{6}$ Ⓘ $\frac{10}{20}$

5. Round the length of the ribbon shown below to the nearest half inch. (Lesson 6-1)

   Ⓐ 1 in.   Ⓑ $1\frac{1}{2}$ in.

   Ⓒ 2 in.   Ⓓ $2\frac{1}{2}$ in.

6. What is the value of $\frac{4}{8} - \frac{1}{3}$? (Lesson 6-4)

   Ⓕ $\frac{1}{6}$ Ⓖ $\frac{1}{3}$ Ⓗ $\frac{1}{2}$ Ⓘ 1

7. Which list orders the integers from greatest to least? (Lesson 8-1)

   Ⓐ $-12, -15, -29, -41$

   Ⓑ $-12, -29, -15, -41$

   Ⓒ $-41, -12, -15, -29$

   Ⓓ $-41, -29, -15, -12$

8. The temperature outside was 4°C one morning. That evening the temperature fell 15°C. What was the evening temperature? (Lesson 8-3)

   Ⓕ $-19°C$ Ⓖ $-15°C$

   Ⓗ $-11°C$ Ⓘ $11°C$

9. Find -6×8. (Lesson 8-4)

   Ⓐ -48 Ⓑ -14 Ⓒ 2   Ⓓ 48

10. Which ordered pair names point X? (Lesson 8-6)

    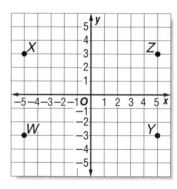

    Ⓕ $(-5, -3)$

    Ⓖ $(-5, 3)$

    Ⓗ $(5, -3)$

    Ⓘ $(5, 3)$

TEST-TAKING TIP

Question 9 When multiplying or dividing two integers, you can eliminate the choices with the wrong signs. If the signs are the same, then the correct answer choice must be positive. If the signs are different, the correct answer choice must be negative.

PART 2 Short Response/Grid In

Record your answers on the answer sheet provided by your teacher or on a sheet of paper.

11. The table shows the average height achieved by different brands of kites. Which type of graph, a bar graph or a line graph, would best display the data? Explain. (Lesson 2-2)

Brand	Average Height Achieved (ft)
Best Kites	25
Carson Kites	45
Flies Right	35

12. On his last seven quizzes, Garrett scored 10, 6, 7, 10, 9, 9, and 8 points. What is the range of the scores? (Lesson 2-7)

13. Alisha caught a fish at the shore. If her fishing basket is 36 centimeters long, how much room will be left in the basket after she packs the fish? (Lesson 3-5)

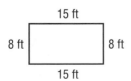

← 28.4 cm →

14. What is the perimeter of the figure shown? (Lesson 4-6)

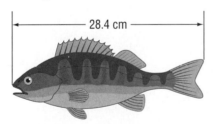

15 ft

8 ft   8 ft

15 ft

15. To write 0.27 as a fraction, which number is the numerator and which number is the denominator? (Lesson 5-6)

16. Macario works $5\frac{1}{4}$ hours each day. Estimate the number of hours he works over a 5-day period. (Lesson 7-1)

For Exercises 17–19, use the figure shown. (Lesson 8-6)

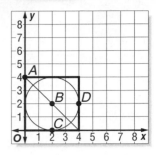

17. What are the coordinates of the intersection of the square and the diagonal of the square?

18. Write the coordinate of the point located on the circle, on the square, and on the x-axis.

19. What is the length of the radius of the circle?

PART 3 Extended Response

Record your answers on a sheet of paper. Show your work.

20. The locations of the pool, horse barn, cabin, and cafeteria at a summer camp are shown on the coordinate plane. (Lesson 8-6)

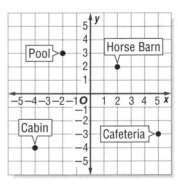

a. Which ordered pair names the location of the pool?

b. An exercise facility is to be built at $(-4, -2)$. What existing facility would be closest to the new exercise facility?

c. Describe three possible locations for a new pavilion if it is to be built at a location given by an ordered pair in which $y = 3$.

Algebra: Solving Equations

"What do field trips have to do with math?"

Millions of students take class field trips every year. Teachers and students need to plan how much it will cost the class so they can plan fund-raisers and help families prepare for the trip. They are also responsible for staying within a budget.

You will use the Distributive Property to determine what will be spent on a trip in Lesson 9-1.

GETTING STARTED

Take this quiz to see whether you are ready to begin Chapter 9. Refer to the lesson number in parentheses if you need more review.

▶ ## Vocabulary Review

State whether each sentence is *true* or *false*. If *false*, replace the underlined word to make a true sentence.

1. A letter used to represent a number is called a <u>variable</u>. (Lesson 1-6)

2. An <u>expression</u> is a mathematical sentence that contains an equals sign. (Lesson 1-7)

▶ ## Prerequisite Skills

Find the value of each expression.
(Lesson 1-4)

3. $3(5) - 9$
4. $8(2) + 4$
5. $1 + 6(4)$
6. $17 - 2(3)$

Add or subtract. (Lessons 8-2 and 8-3)

7. $2 - 4$
8. $-4 + 8$
9. $-3 + 9$
10. $6 - 9$
11. $7 - 8$
12. $-1 + 5$

Divide. (Lesson 8-5)

13. $-32 \div 4$
14. $56 \div 2$
15. $72 \div 8$
16. $-18 \div 3$
17. $36 \div (-9)$
18. $-24 \div (-6)$

Evaluate each expression for $a = -2$, $a = 1$, and $a = 2$. (Lessons 8-2 through 8-5)

19. $3a$
20. $4 + a$
21. $a - 6$
22. $8 \div a$

Solving Equations Make this Foldable to help you organize your strategies for solving problems. Begin with a piece of 11" × 17" paper.

 Fold
Fold lengthwise so the sides meet in the middle.

 Fold Again
Fold in 3 equal sections widthwise.

 Cut
Unfold. Cut to make three tabs on each side. Cut the center flaps as shown to leave a rectangle showing in the center.

 Rotate and Label
Label the center "Solving Equations." Label each tab as shown.

Reading and Writing As you read and study the chapter, write notes and examples for each lesson under the appropriate tab.

The Distributive Property

What You'll LEARN

Illustrate the Distributive Property using models.

To find the area of a rectangle, multiply the length and width. To find the area of a rectangle formed by two smaller rectangles, you can use either one of two methods.

ACTIVITY *Work with a partner.*

Find the area of the blue and yellow rectangles.

Method 1 Add the lengths. Then multiply.

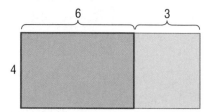

$4(6 + 3) = 4(9)$ Add.

$\quad\quad\quad\quad = 36$ Simplify.

Method 2 Find each area. Then add.

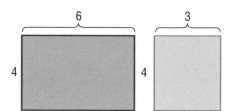

$4 \cdot 6 + 4 \cdot 3 = 24 + 12$ Multiply.

$\quad\quad\quad\quad\quad = 36$ Simplify.

In Method 1, you found that $4(6 + 3) = 36$. In Method 2, you found that $4 \cdot 6 + 4 \cdot 3 = 36$. So, $4(6 + 3) = 4 \cdot 6 + 4 \cdot 3$.

Your Turn Draw a model showing that each equation is true.

a. $2(4 + 6) = (2 \cdot 4) + (2 \cdot 6)$ b. $4(3 + 2) = (4 \cdot 3) + (4 \cdot 2)$

Writing Math

1. **Write** two expressions for the total area of the rectangle at the right.

2. **OPEN ENDED** Draw any two rectangles that have the same width. Find the total area in two ways.

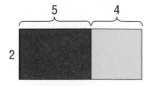

3. **MAKE A CONJECTURE** Write an expression that has the same value as $2(4 + 3)$. Explain your reasoning.

Properties

What You'll LEARN

Use the Commutative, Associative, Identity, and Distributive Properties.

Link to READING

Everyday Meaning of Distribute: to divide among several people or things

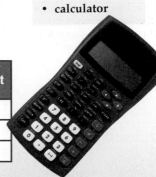

HANDS-ON Mini Lab

Work with a partner.

1. Copy and complete the table below.

	multiplication expression	product	multiplication expression	product
a.	5 × 8		6 × 12	
b.	5 × 40		6 × 200	
c.	5 × 48		6 × 212	

2. What do you notice about each set of expressions?

3. How does each product in row **c** compare to the sum of the products in rows **a** and **b**?

The expressions 5(40 + 8) and 5(40) + 5(8) illustrate how the **Distributive Property** combines addition and multiplication.

Key Concept   Distributive Property

Words	To multiply a sum by a number, multiply each addend of the sum by the number outside the parentheses.

Symbols

Arithmetic	Algebra
$2(7 + 4) = 2 \times 7 + 2 \times 4$	$a(b + c) = ab + ac$
$(5 + 6)3 = 5 \times 3 + 6 \times 3$	$(b + c)a = ba + ca$

You can use the Distributive Property to solve some multiplication problems mentally.

EXAMPLE Use the Distributive Property

1 Find 4 × 58 mentally using the Distributive Property.

$$4 \times 58 = 4(50 + 8) \qquad \text{Write 58 as 50 + 8.}$$
$$= 4(50) + 4(8) \qquad \text{Distributive Property}$$
$$= 200 + 32 \qquad \text{Multiply 4 and 50 mentally.}$$
$$= 232 \qquad \text{Add 200 and 32 mentally.}$$

Your Turn Rewrite each expression using the Distributive Property. Then find each product mentally.

a. 5 × 84   b. 12 × 32   c. 2 × 3.6

EXAMPLE **Apply the Distributive Property**

2 **FIELD TRIPS** Suppose admission to a museum costs $5 and bus tickets are $2.50 per student. What is the cost for 30 students?

Method 1 Find the cost of 30 admissions and 30 bus tickets. Then add.

$$30(\$5) + 30(\$2.50)$$

cost of 30 admissions ↑   ↑ cost of 30 bus tickets

Method 2 Find the cost for 1 person. Then multiply by 30.

$$30(\$5 + \$2.50)$$

↑ cost for 1 person

Evaluate either expression.

$$30(5 + 2.50) = 30(5) + 30(2.50) \quad \text{Distributive Property}$$
$$= 150 + 75 \quad \text{Multiply.}$$
$$= 225 \quad \text{Add.}$$

The total cost is $225.

REAL-LIFE MATH

FIELD TRIPS Nearly 400,000 children in school groups and youth organizations visited the Chicago Museum of Science and Industry in one year.

Source: *MSI, Chicago*

Other properties of addition and multiplication are given below.

Key Concept	**Properties of Addition and Multiplication**
Commutative Property The order in which numbers are added or multiplied does not change the sum or product.	
Examples $4 + 3 = 3 + 4 \quad 5 \times 4 = 4 \times 5$	
Associative Property The way in which numbers are grouped when added or multiplied does not change the sum or product.	
Examples $(3 + 4) + 5 = 3 + (4 + 5) \quad (2 \times 3) \times 4 = 2 \times (3 \times 4)$	
Additive Identity The sum of any number and 0 is the number.	
Examples $5 + 0 = 5 \quad a + 0 = a$	
Multiplicative Identity The product of any number and 1 is the number.	
Examples $7 \times 1 = 7 \quad 1 \times n = n$	

EXAMPLES **Identify Properties**

Identify the property shown by each equation.

3 $25 \times 15 = 15 \times 25$

The order in which the numbers are multiplied changes. This is the Commutative Property of Multiplication.

4 $55 + (5 + 12) = (55 + 5) + 12$

The grouping of the numbers to be added changes. This is the Associative Property of Addition.

You can use properties to find sums and products mentally.

EXAMPLE Apply Properties

5 Find 15 + 28 + 25 mentally.

Since you can easily add 25 and 15, change the order.

15 + 28 + 25 = 15 + 25 + 28   Commutative Property

Now group the numbers. The parentheses tell you which to perform first.

15 + 25 + 28 = (15 + 25) + 28   Associative Property

= 40 + 28   Add 15 and 25 mentally.

= 68   Add 40 and 28 mentally.

Your Turn Find each sum or product mentally.

d. $5 \times 26 \times 2$ e. $37 + 98 + 63$

Skill and Concept Check

Writing Math
Exercises 1–4

1. **Explain** how to use the Distributive Property to find a product mentally.

2. **OPEN ENDED** Write four equations that show each of the Commutative and Associative Properties of Addition and Multiplication.

3. **Determine** if the Commutative and Associate Properties of Addition are true for fractions. Explain using examples or counterexamples.

4. **FIND THE ERROR** Brian and Courtney are using the Distributive Property to simplify 5(4 + 2). Who is correct? Explain.

Brian
$(5 \times 4) + (5 \times 2)$

Courtney
$(5 + 4) \times (5 + 2)$

GUIDED PRACTICE

Find each product mentally. Use the Distributive Property.

5. 5×84 6. 10×2.3 7. 4.2×4

Rewrite each expression using the Distributive Property. Then evaluate.

8. $3(20 + 4)$ 9. $(60 + 5)5$ 10. $(12.5 \times 10) + (12.5 \times 8)$

Identify the property shown by each equation.

11. $17 \times 2 = 2 \times 17$ 12. $(3 + 6) + 10 = 3 + (6 + 10)$

13. $24 \times 1 = 24$ 14. $(6 + 16) + 0 = (6 + 16)$

Find each sum or product mentally.

15. $35 + 8 + 5$ 16. $86 + 28 + 14$ 17. $6 \times 8 \times 5$ 18. $5 \times 30 \times 4$

Practice and Applications

HOMEWORK HELP

For Exercises	See Examples
19–28	1
29–36	3, 4
37–42	5
43	2

Extra Practice
See pages 611, 632.

Find each product mentally. Use the Distributive Property.

19. 7×15

20. 3×72

21. 25×12

22. 15×11

23. 30×7.2

24. 60×2.5

Rewrite each expression using the Distributive Property. Then evaluate.

25. $7(30 + 6)$

26. $12(40 + 7)$

27. $(50 + 4)2$

28. $(30 + 8)13$

Identify the property shown by each equation.

29. $90 + 2 = 2 + 90$

30. $8 \times 4 = 4 \times 8$

31. $(19 + 76) + 24 = 19 + (76 + 24)$

32. $9 \times (10 \times 6) = (9 \times 10) \times 6$

33. $55 + 0 = 55$

34. $40 \times 1 = 40$

35. $5 + (85 + 16) = (85 + 16) + 5$

36. $(3 \times 15)4 = 3(15 \times 4)$

Find each sum or product mentally.

37. $15 + 9 + 35$

38. $12 + 45 + 18$

39. $4 \times 7 \times 25$

40. $2 \times 34 \times 5$

41. $115 + 20 + 15$

42. $5 \times 87 \times 20$

FOOD For Exercises 43 and 44, use the table.

43. What is the total price of 25 burgers at each fast food restaurant?

44. **MULTI STEP** Which fast food restaurant would be a better deal for 25 students if everyone ordered a burger, fries, and a soda?

	Fast Food A	Fast Food B
Burger	$2.00	$2.25
Fries	$1.50	$1.25
Soda	$1.19	$0.99

45. **CRITICAL THINKING** Evaluate each expression.

a. $0.2(2 - 0.4)$

b. $0.1(1 - 0.5)$

c. $0.8(10 - 0.25)$

Standardized Test Practice and Mixed Review

46. **MULTIPLE CHOICE** Which expression is equivalent to $(2 \times 4) + (6 \times 4)$?

Ⓐ $4(2 \times 6)$ Ⓑ $6(2 + 4)$ Ⓒ $4(2 + 6)$ Ⓓ $2(4 + 6)$

47. **SHORT RESPONSE** What property can be used to find the missing number in $10 \times (5 \times 25) = (\blacksquare \times 5) \times 25$?

48. **GEOMETRY** Graph $X(2, -3)$ and $Y(-3, 2)$. (Lesson 8-6)

Divide. (Lesson 8-5)

49. $10 \div (-2)$

50. $-24 \div 6$

51. $-36 \div (-6)$

52. $-81 \div (-9)$

GETTING READY FOR THE NEXT LESSON

PREREQUISITE SKILL Subtract. (Lesson 8-3)

53. $2 - 3$

54. $4 - 7$

55. $6 - 8$

56. $2 - 9$

Solving Addition Equations Using Models

What You'll LEARN

Solve addition equations using models.

Materials

- cups
- counters
- equation mat

REVIEW Vocabulary

equation: a sentence that contains an equals sign, = **(Lesson 1-7)**

An equation is like a balance scale. The quantity on the left side of the equals sign is *balanced* with the quantity on the right. When you solve an equation, you need to keep the equation *balanced*.

To solve an equation using cups and counters, remember to add or subtract the same number of counters from each side of the mat, so that it remains *balanced*.

ACTIVITY *Work with a partner.*

1 Solve $x + 5 = 9$ using models.

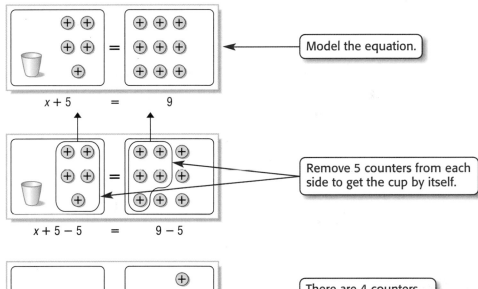

Model the equation.

Remove 5 counters from each side to get the cup by itself.

There are 4 counters remaining on the right side, so $x = 4$.

The solution is 4.

Check Replace x with 4 in the original equation.

$$x + 5 = 9$$
$$4 + 5 \stackrel{?}{=} 9$$
$$9 = 9 \checkmark \qquad \text{So, the solution is correct.}$$

Your Turn Solve each equation using models.

a. $1 + x = 8$ b. $x + 2 = 7$ c. $9 = x + 3$

Sometimes you will use zero pairs to solve equations. You can add or subtract a zero pair from each side of the mat without changing its value because the value of a zero pair is zero.

STUDY TIP

Look Back To review **zero pairs**, see Lesson 8-2.

ACTIVITY *Work with a partner.*

2 Solve $x + 2 = -5$ using models.

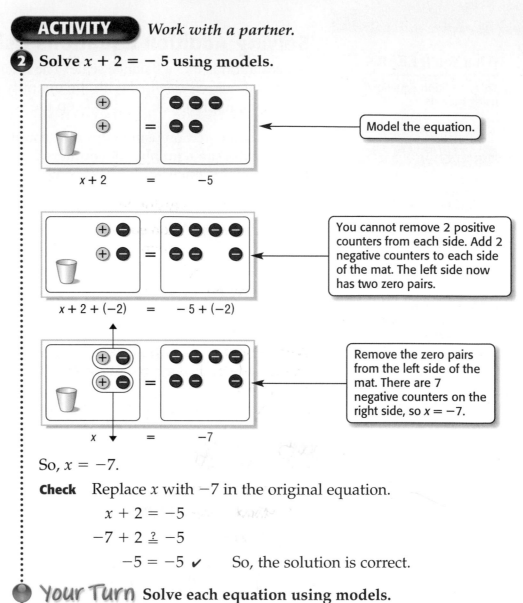

Model the equation.

$x + 2 \quad = \quad -5$

You cannot remove 2 positive counters from each side. Add 2 negative counters to each side of the mat. The left side now has two zero pairs.

$x + 2 + (-2) \quad = \quad -5 + (-2)$

Remove the zero pairs from the left side of the mat. There are 7 negative counters on the right side, so $x = -7$.

$x \quad = \quad -7$

So, $x = -7$.

Check Replace x with -7 in the original equation.

$$x + 2 = -5$$
$$-7 + 2 \stackrel{?}{=} -5$$
$$-5 = -5 \ \checkmark \quad \text{So, the solution is correct.}$$

Your Turn Solve each equation using models.

d. $x + 3 = -7$ **e.** $2 + x = -5$ **f.** $-3 = x + 3$

Writing Math

1. **Explain** how you decide how many counters to add or subtract from each side.

2. **Write** an equation in which you need to remove zero pairs in order to solve it.

3. **Model** the equation *some number plus 5 is equal to −2*. Then solve the equation.

4. **MAKE A CONJECTURE** Write a rule that you can use to solve an equation like $x + 3 = 6$ without using models.

Solving Addition Equations

What You'll LEARN

Solve addition equations.

NEW Vocabulary

inverse operations

REVIEW Vocabulary

solve: find the value of the variable that results in a true sentence (Lesson 1-7)

WHEN am I ever going to use this?

WEATHER A forecaster reported that although an additional 3 inches of rain had fallen, the total rainfall was still 9 inches below normal for the year. This is shown on the number line.

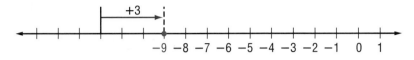

1. Write an expression to represent 3 more inches of rain.

2. Write an addition equation you could use to find the rainfall before the additional 3 inches.

3. You could solve the addition equation by counting back on the number line. What operation does counting back suggest?

When you solve an equation, the goal is to get the variable by itself on one side of the equation. One way to do this is to use inverse operations. **Inverse operations** *undo* each other. For example, to solve an addition equation, use subtraction.

EXAMPLE Solve an Equation By Subtracting

1 Solve $x + 3 = 8$.

Method 1 Use models.

Model the equation.

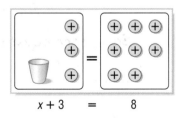

$$x + 3 = 8$$

Remove 3 counters from each side of the mat.

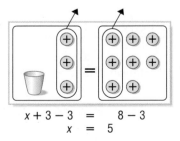

$$x + 3 - 3 = 8 - 3$$
$$x = 5$$

The solution is 5.

Method 2 Use symbols.

$x + 3 = 8$ Write the equation.

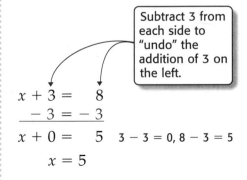

Subtract 3 from each side to "undo" the addition of 3 on the left.

$$
\begin{aligned}
x + 3 &= 8 \\
- 3 &= -3 \\
\hline
x + 0 &= 5 \quad 3 - 3 = 0, 8 - 3 = 5 \\
x &= 5
\end{aligned}
$$

EXAMPLE **Solve an Equation Using Zero Pairs**

2 Solve $b + 5 = 2$. Check your solution.

Method 1 Use models.

Model the equation.

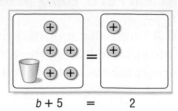

$$b + 5 \quad = \quad 2$$

Add 3 zero pairs to the right side of the mat so there are 5 positive counters on the right.

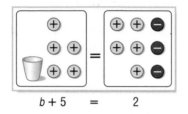

$$b + 5 \quad = \quad 2$$

Remove 5 positive counters from each side.

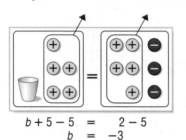

$$b + 5 - 5 \quad = \quad 2 - 5$$
$$b \quad = \quad -3$$

Method 2 Use symbols.

$b + 5 = 2$ Write the equation.

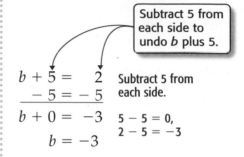

Subtract 5 from each side to undo b plus 5.

$\begin{array}{rl} b + 5 = & 2 \\ -5 = & -5 \end{array}$ Subtract 5 from each side.

$b + 0 = -3$ $5 - 5 = 0,$
$b = -3$ $2 - 5 = -3$

Check

$b + 5 = 2$ Write the equation.

$-3 + 5 \overset{?}{=} 2$ Replace b with -3.

$2 = 2$ ✔   This sentence is true.

The solution is -3.

Your Turn Solve each equation. Use models if necessary.

a. $c + 2 = 5$ b. $3 + y = 12$ c. $2 = x + 6$

d. $c + 4 = 3$ e. $x + 3 = -2$ f. $2 + g = -4$

When you solve an equation by subtracting the same number from each side of the equation, you are using the **Subtraction Property of Equality**.

Key Concept   **Subtraction Property of Equality**

Words	If you subtract the same number from each side of an equation, the two sides remain equal.	
Symbols	**Arithmetic**	**Algebra**
	$\begin{array}{rl} 5 = & 5 \\ -3 = & -3 \\ \hline 2 = & 2 \end{array}$	$\begin{array}{rl} x + 2 = & 3 \\ -2 = & -2 \\ \hline x \quad = & 1 \end{array}$

Skill and Concept Check

Writing Math
Exercises 1–3

1. **Show** how to model the equation $x + 4 = 2$. Then explain how to solve the equation using models.

2. **OPEN ENDED** Write a problem that can be represented by the equation $x + 2 = 7$. Explain the meaning of the equation.

3. **NUMBER SENSE** Without solving, tell whether the solution to $a + 14 = -2$ will be positive or negative. Explain your answer.

GUIDED PRACTICE

Solve each equation. Use models if necessary. Check your solution.

4. $x + 3 = 5$ 5. $2 + m = 7$ 6. $c + 6 = -3$ 7. $-4 = 6 + e$

8. Find the value of n if $n + 12 = 6$.

9. **HOT AIR BALLOONING** A hot air balloon is 200 feet in the air. A few minutes later it ascends to 450 feet. Write and solve an addition equation to find the change of altitude of the hot air balloon.

Practice and Applications

Solve each equation. Use models if necessary. Check your solution.

HOMEWORK HELP

For Exercises	See Examples
10–23	1, 2

Extra Practice
See pages 611, 632.

10. $y + 7 = 10$ 11. $x + 5 = 11$ 12. $9 = 2 + x$
13. $7 = 4 + y$ 14. $9 + a = 7$ 15. $6 + g = 5$
16. $d + 3 = -5$ 17. $x + 4 = -2$ 18. $-5 = 3 + f$
19. $-1 = g + 7$ 20. $b + 4 = -3$ 21. $h + (-4) = 2$

22. Find the value of x if $x + 3 = 7$.   23. If $c + 6 = 2$, what is the value of c?

Solve each equation. Check your solution.

24. $t + 1.9 = 3.8$ 25. $1.8 + n = -0.3$ 26. $a + 6.1 = -2.3$ 27. $c + 2.5 = -4.2$
28. $7.8 = x + 1.5$ 29. $5.6 = y + 2.7$ 30. $m + \dfrac{1}{3} = \dfrac{2}{3}$ 31. $t + \dfrac{1}{4} = -\dfrac{1}{2}$

32. **PETS** Zane and her dog weigh 108 pounds. Zane weighs 89 pounds. Write and solve an addition equation to find the dog's weight.

33. **EXERCISE** On average, men burn 180 more Calories per hour running than women do. If a man burns 600 Calories per hour running, write and solve an addition equation to find how many Calories a woman burns running one hour.

34. **GAMES** In the card game Clubs, it is possible to have a negative score. Suppose your friend had a score of -5 in the second hand. This made her total score after two hands equal to -2. What was her score in the first hand?

msmath1.net/self_check_quiz

35. **ROADS** A typical log truck weighs 30,000 pounds empty. What is the maximum weight of lumber that the truck can carry and not exceed the weight limit?

80,000 lb
Weight Limit

36. **PROPERTIES** How does the Subtraction Property of Equality help you solve the equation $x + 8 = 13$?

CRITICAL THINKING The solution of the equation $x + 7 = -3$ is shown. Match each step with the property used.

$$x + 7 = -3$$

37. $x + 7 - 7 = -3 - 7$ **a.** Associative Property of Addition

38. $x + 0 = -10$ **b.** Additive Identity

39. $x = -10$ **c.** Subtraction Property of Equality

Standardized Test Practice and Mixed Review

40. **MULTIPLE CHOICE** It was 3°F before an Arctic cold front came through and dropped the temperature to -9°F on New Year's Eve. The equation $3 + d = -9$ is used to find how many degrees the temperature dropped. What is the value of d?

   A $-12°$ **B** $-6°$ **C** $6°$ **D** $12°$

41. **SHORT RESPONSE** Sabrina collected 6 silver dollars. Her friend Logan gave her some more, and then she had 15. To find out how many silver dollars she was given, Sabrina wrote $s + 6 = 15$. What is the value of s?

Rewrite each expression using the Distributive Property. Then evaluate. (Lesson 9-1)

42. $6(20 + 4)$

43. $(30 \times 4) + (30 \times 0.5)$

Refer to the coordinate plane to identify the point for each ordered pair. (Lesson 8-6)

44. $(4, 2)$ 45. $(-3, 0)$ 46. $(-1, -4)$

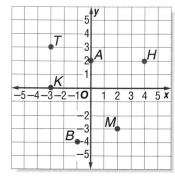

Refer to the coordinate plane to write the ordered pair that names each point. Then identify the quadrant where each point is located. (Lesson 8-6)

47. T 48. M 49. A

Multiply or divide. (Lessons 8-4 and 8-5)

50. -4×9 51. $-8(-3)$ 52. $-18 \div (-6)$ 53. $25 \div (-5)$

GETTING READY FOR THE NEXT LESSON

PREREQUISITE SKILL Add. (Lesson 8-2)

54. $-2 + 6$ 55. $-9 + 3$ 56. $-8 + 5$ 57. $-7 + 9$

Solving Subtraction Equations Using Models

What You'll LEARN

Solve subtraction equations using models.

Materials

- cups
- counters
- equation mat

Recall that subtracting an integer is the same as adding its opposite. For example, $4 - 7 = 4 + (-7)$ or $x - 3 = x + (-3)$.

ACTIVITY *Work with a partner.*

Solve $x - 3 = -2$ using models.

$x - 3 = -2 \rightarrow x + (-3) = -2$ ◄ Rewrite the equation.

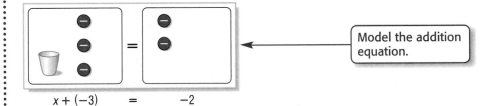

$x + (-3) \quad = \quad -2$

◄ Model the addition equation.

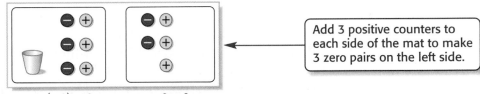

$x + (-3) + 3 \quad = \quad -2 + 3$

◄ Add 3 positive counters to each side of the mat to make 3 zero pairs on the left side.

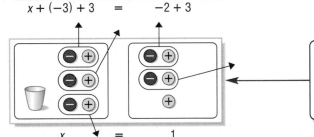

$x \quad = \quad 1$

◄ Remove 3 zero pairs from the left side and 2 zero pairs from the right side. There is one positive counter on the right side. So, $x = 1$.

The solution is 1.   **Check** $1 - 3 = 1 + (-3)$ or -2 ✔

Your Turn Solve each equation using models.

a. $x - 4 = 2$ b. $-3 = x - 1$ c. $x - 5 = -1$

Writing Math

1. **Explain** why it is helpful to rewrite a subtraction problem as an addition problem when solving equations using models.

2. **MAKE A CONJECTURE** Write a rule for solving equations like $x - 7 = -5$ without using models.

Solving Subtraction Equations

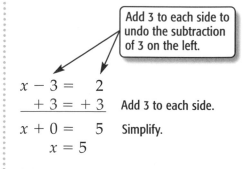

What You'll LEARN

Solve subtraction equations.

WHEN am I ever going to use this?

GROWTH Luis is 5 inches shorter than his brother Lucas. Luis is 59 inches tall.

1. Let h represent Lucas' height. Write an expression for *5 inches shorter than Lucas.*

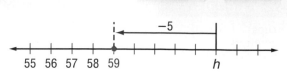

2. Write an equation for *5 inches shorter than Lucas is equal to 59 inches.*

3. You could find Lucas' height by counting forward. What operation does counting forward suggest?

4. How tall is Lucas?

Addition and subtraction are inverse operations. So, you can solve a subtraction equation by adding.

EXAMPLE Solve an Equation by Adding

1 Solve $x - 3 = 2$.

Method 1 Use models.

Model the equation.

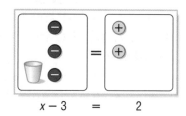

$$x - 3 = 2$$

Add 3 positive counters to each side of the mat. Remove the zero pairs.

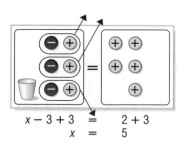

$$x - 3 + 3 = 2 + 3$$
$$x = 5$$

The solution is 5.

Method 2 Use symbols.

$x - 3 = 2$ Write the equation.

> Add 3 to each side to undo the subtraction of 3 on the left.

$$x - 3 = 2$$
$$\underline{+\ 3 = +\ 3}$$ Add 3 to each side.
$$x + 0 = 5$$ Simplify.
$$x = 5$$

STUDY TIP

Using Counters
The expression $x - 3$ is the same as $x + (-3)$. To model the expression, use one cup and three negative counters.

When you solve an equation by adding the same number to each side of the equation, you are using the **Addition Property of Equality**.

Key Concept   **Addition Property of Equality**

Words	If you add the same number to each side of an equation, the two sides remain equal.

Symbols

Arithmetic	Algebra
$5 = 5$	$x - 2 = 3$
$+ 3 = + 3$	$+ 2 = + 2$
$8 = 8$	$x = 5$

EXAMPLE   Solve a Subtraction Equation

2 Solve $y - 4 = -10$. Check your solution.

$y - 4 = -10$ Write the equation.
$\underline{+ 4 = + \ 4}$ Add 4 to each side.
$y \quad\quad = -6$ Simplify.

Check $y - 4 = -10$ Write the original equation.

$\quad\quad\quad -6 - 4 \overset{?}{=} -10$ Replace y with -6.

$\quad\quad\quad\quad\quad\quad -10 = -10$ ✔   This sentence is true.

The solution is -6.

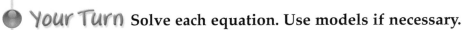 **Your Turn** Solve each equation. Use models if necessary.

   a. $b - 4 = -2$ **b.** $-5 = t - 5$ **c.** $c - 2 = -6$

Ⓐ Ⓑ Ⓒ Ⓓ
Standardized·········
Test Practice

EXAMPLE   Use an Equation to Solve a Problem

3 **GRID-IN TEST ITEM** The difference between the record high and low temperatures in Indiana is 152°F. The record low temperature is −36°F. What is the record high temperature in degrees Fahrenheit?

Read the Test Item

You need to find the record high temperature. Write and solve an equation. Let x represent the high temperature.

Test-Taking Tip
The Princeton Review

Filling in the Grid
You may start filling in the grid in the first column as shown or align the last digit on the right.

Solve the Test Item

$x - (-36) = 152$ Write the equation.

$x \quad + 36 = 152$ Definition of subtraction

$\underline{\quad\quad - 36 = - 36}$ Subtract 36 from each side.

$x \quad\quad\quad = 116$ Simplify.

The record high temperature is 116°.

Fill in the Grid

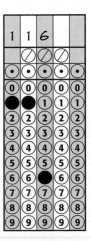

Skill and Concept Check

1. **Tell** how to check your solution to an equation.

2. **OPEN ENDED** Write two different subtraction equations that have 5 as the solution.

3. **FIND THE ERROR** Diego and Marcus are explaining how to solve the equation $d - 6 = 4$. Who is correct? Explain.

> Diego
> Subtract 6 from each side.

> Marcus
> Add 6 to each side.

4. **NUMBER SENSE** Without solving the equation, what do you know about the value of x in $x - 5 = -3$? Is x greater than 5 or less than 5? Explain your reasoning.

GUIDED PRACTICE

Solve each equation. Use models if necessary. Check your solution.

5. $a - 5 = 9$
6. $b - 3 = 7$
7. $x - 4 = -1$
8. $4 = y - 8$
9. $x - 2 = -7$
10. $-3 = n - 2$

11. **DIVING** A diver is swimming below sea level. A few minutes later the diver descends 35 feet until she reaches a depth of 75 feet below sea level. Write and solve a subtraction equation to find the diver's original position.

Practice and Applications

Solve each equation. Use models if necessary. Check your solution.

12. $c - 1 = 8$
13. $f - 1 = 5$
14. $2 = e - 1$
15. $1 = g - 3$
16. $r - 3 = -1$
17. $t - 2 = -2$
18. $t - 4 = -1$
19. $h - 2 = -9$
20. $-3 = u - 8$
21. $-5 = v - 6$
22. $x - 3 = -5$
23. $y - 4 = -7$

24. Find the value of t if $t - 7 = -12$.

25. If $b - 10 = 5$, what is the value of b?

Solve each equation. Check your solution.

26. $-6 + a = -8$
27. $-1 + c = -8$
28. $a - 1.1 = 2.3$
29. $b - 2.7 = 1.6$
30. $-4.6 = e - 3.2$
31. $-4.3 = f - 7.8$
32. $m - \frac{1}{3} = \frac{2}{3}$
33. $n - \frac{1}{4} = -\frac{1}{2}$

34. **PETS** Mika's cat lost 3 pounds. It now weighs 12 pounds. Write and solve an equation to find its original weight.

HOMEWORK HELP

For Exercises	See Examples
12–15, 25	1
16–24	2
34, 35	3

Extra Practice
See pages 612, 632.

35. **FOOTBALL** After a play resulting in the loss of 8 yards, the Liberty Middle School team's ball was 15 yards away from the goal line. Write and solve a subtraction equation to find the position of the ball at the start of the play.

36. **WEATHER** The difference between the record high and record low temperatures for September in Bryce Canyon, Utah, is 29°F. Use the information at the right to find the record high temperature.

Bryce Canyon, Utah	
Record Temperatures in September	
High	?
Low	60°F

37. **WRITE A PROBLEM** Write a real-life problem that can be solved by using a subtraction equation.

38. **CRITICAL THINKING** Describe how you would solve $6 - x = -3$.

Standardized Test Practice and Mixed Review   (A) (B) (C) (D)

39. **MULTIPLE CHOICE** A scuba diver is 84 feet below the surface of the water when she begins to swim back up. She then stops to observe a school of fish at 39 feet below the surface of the water. How many feet did she rise before stopping?

   (A) −123 ft   (B) −45 ft   (C) 45 ft   (D) 123 ft

40. **GRID IN** Find the value of b if $b - 8 = -5$.

41. **BASEBALL** Refer to the graphic. Write and solve an addition equation to find how many more people can be seated at Dodger Stadium than at Yankee Stadium. (Lesson 9-2)

USA TODAY Snapshots®

The Vet is baseball's biggest stadium
Philadelphia's Veterans Stadium, built in 1971, is the majors' largest ballpark in terms of capacity. The "Vet" is also one of only a few stadiums used for both baseball and football. Major League Baseball stadiums with the largest capacity:

Veterans Stadium (Philadelphia)
62,409

Qualcomm Stadium (San Diego)
56,133

Dodger Stadium (Los Angeles)
56,000

Shea Stadium (New York)
55,775

Yankee Stadium (New York)
55,070

Source: Major League Baseball

By Ellen J. Horrow and Bob Laird, USA TODAY

 Data Update Find the capacity of a baseball stadium not listed in the graphic. Visit **msmath1.net/data_update** to learn more.

42. Find 8×15 mentally. Use the Distributive Property. (Lesson 9-1)

Multiply. Write in simplest form. (Lesson 7-3)

43. $2\frac{3}{4} \times 3\frac{1}{3}$ 44. $3\frac{2}{3} \times 1\frac{2}{7}$

45. $7\frac{3}{5} \times \frac{8}{19}$ 46. $1\frac{3}{7} \times 2\frac{11}{12}$

47. **LIFE SCIENCE** Jamil's leaf collection includes 15 birch, 8 willow, 5 oak, 10 maple, and 8 miscellaneous leaves. Make a bar graph of this data. (Lesson 2-2)

GETTING READY FOR THE NEXT LESSON

PREREQUISITE SKILL Divide. (Lesson 8-5)

48. $-8 \div 2$ 49. $42 \div 3$ 50. $36 \div 3$ 51. $-24 \div 8$

Vocabulary and Concepts

1. **State** the Distributive Property. (Lesson 9-1)

2. **Explain** how you can use the Associative Properties of Addition and Multiplication to find sums and products mentally. (Lesson 9-1)

3. **Explain** how to solve an addition equation. (Lesson 9-2)

Skills and Applications

Rewrite each expression using the Distributive Property. Then evaluate. (Lesson 9-1)

4. $4(12 + 9)$

5. $(7 + 30)5$

6. 7×23

Identify the property shown by each equation. (Lesson 9-1)

7. $14 \times 6 = 6 \times 14$

8. $(4 + 9) + 11 = 4 + (9 + 11)$

9. $1 \times a = a$

10. $2 \times (5 \times 9) = (2 \times 5) \times 9$

Find each sum or product mentally. (Lesson 9-1)

11. $23 + 9 + 7$

12. $20 \times 38 \times 5$

13. $34 + 76 + 19$

Solve each equation. Use models if necessary. Check your solution. (Lesson 9-2 and 9-3)

14. $w + 8 = 5$

15. $7 = 2 + p$

16. $-3 + x = 11$

17. $3.4 = y + 2.1$

18. $m - 6 = 5$

19. $-8 = d - 9$

20. $k - 4 = -2$

21. $5 = a - 10$

22. $z - 3 = -6$

23. **MONEY MATTERS** Liliana bought a backpack for $28. This was $8 less than the regular price. Write and solve a subtraction equation to find the regular price. (Lesson 9-3)

Standardized Test Practice

24. **SHORT RESPONSE** The Huskies football team gained 15 yards after a loss of 11 yards on the previous play. Write a subtraction equation to find how many yards they gained before the 11-yard loss. (Lesson 9-3)

25. **MULTIPLE CHOICE** Jason and Tia have a total of 20 tadpoles and frogs. Of these, 12 are tadpoles, and the rest are frogs. Use the equation $f + 12 = 20$ to find how many frogs they have. (Lesson 9-2)

Ⓐ 8 Ⓑ 12 Ⓒ 20 Ⓓ 32

The GameZone

A Place To Practice Your Math Skills

Math Skill
Solving Equations

Four in a Line

● **GET READY!**

Players: two to ten
Materials: 12 index cards, scissors, poster board, beans

● **GET SET!**

• Cut all 12 index cards in half. Your teacher will give you a list of 24 equations. Label each card with a different equation.

• Cut one 6-inch by 5-inch playing board for each player from the poster board.

• For each playing board, copy the grid shown. Complete each column by choosing from the solutions below so that no two cards are identical.

a	b	c	d
			Free
	Free		

Solutions

a: −9, −3, 4, 6, 11, 16
b: −3, −1, 1, 5, 10, 12
c: −6, 2, 3, 7, 8, 12
d: −3, −2, 0, 1, 3, 8

● **GO!**

• Mix the equation cards and place the deck facedown.

• After an equation card is turned up, all players solve the equation.

• If a player finds a solution on the board, he or she covers it with a bean.

• **Who Wins?** The first player to cover four spaces in a row either vertically, horizontally, or diagonally is the winner.

Solving Multiplication Equations

What You'll LEARN

Solve multiplication equations.

NEW Vocabulary

coefficient

WHEN am I ever going to use this?

BABY-SITTING Kara baby-sat for 3 hours and earned $12. How much did she make each hour?

1. Let x = the amount Kara earns each hour. Write an expression for the amount Kara earns after 3 hours.

2. Explain how the equation $3x = 12$ represents the situation.

The equation $3x = 12$ is a multiplication equation. In $3x$, 3 is the **coefficient** of x because it is the number by which x is multiplied. To solve a multiplication equation, use division.

EXAMPLE Solve a Multiplication Equation

1 Solve $3x = 12$. Check your solution.

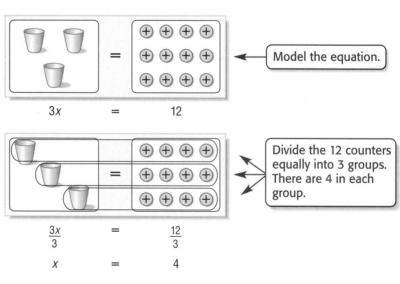

Model the equation.

$$3x = 12$$

Divide the 12 counters equally into 3 groups. There are 4 in each group.

$$\frac{3x}{3} = \frac{12}{3}$$

$$x = 4$$

Check $3x = 12$ Write the original equation.

$3(4) \stackrel{?}{=} 12$ Replace x with 4.

$12 = 12$ This sentence is true. ✔

The solution is 4.

Your Turn Solve each equation. Use models if necessary.

a. $3x = 15$ b. $8 = 4x$ c. $2x = -10$

2 Solve $-2x = 10$.

$-2x = 10$ Write the equation.

$\dfrac{-2x}{-2} = \dfrac{10}{-2}$ Divide each side by -2.

$1x = -5$ $-2 \div (-2) = 1, \; 10 \div (-2) = -5$

$x = -5$ $1x = x$

The solution is -5.   Check this solution.

STUDY TIP

Dividing Integers
Recall that the quotient of a negative integer and a negative integer is positive.

Your Turn Solve each equation.

d. $-2x = 12$ **e.** $-4t = -16$ **f.** $24 = -3c$

EXAMPLE Use an Equation to Solve a Problem

3 **GEOMETRY** The area of a rectangle is 3.6 square feet, and its width is 0.4 foot. Write an equation to find the length of the rectangle.

0.4 ft   3.6 ft² ℓ

Let ℓ = the length of the rectangle.

The area of a rectangle	is equal to	its length times its width.
3.6	=	0.4ℓ

$3.6 = 0.4\ell$ Write the equation.

$\dfrac{3.6}{0.4} = \dfrac{0.4\ell}{0.4}$ Divide each side by 0.4.

$9 = \ell$ Simplify.

The length of the rectangle is 9 feet.   **Check** $3.6 = 0.4(9)$ is true. ✔

Skill and Concept Check

Writing Math

Exercise 3

1. **Make** a model to represent the equation $2x = -12$. Then solve the equation.

2. **OPEN ENDED** Write two different multiplication equations that have 5 as the solution.

3. **Which One Doesn't Belong?** Identify the equation that does not have the same solution as the other three. Explain your reasoning.

$2x = 24$	$6a = 72$	$3c = 4$	$5y = 60$

GUIDED PRACTICE

Solve each equation. Use models if necessary.

4. $2a = 6$ **5.** $3b = 9$ **6.** $-20 = 4c$ **7.** $-16 = 8b$

8. $-4d = 12$ **9.** $-6c = 24$ **10.** $-5f = -20$ **11.** $-3g = -21$

Practice and Applications

HOMEWORK HELP

For Exercises	See Examples
12–19, 28–31	1
20–27	2
32–33, 40–42	3

Extra Practice
See pages 612, 632.

Solve each equation. Use models if necessary.

12. $5d = 30$

13. $4c = 16$

14. $36 = 6e$

15. $21 = 3g$

16. $3f = -12$

17. $4g = -24$

18. $7h = -35$

19. $9m = -72$

20. $-5a = 15$

21. $-6x = 12$

22. $-2g = 22$

23. $-3h = 12$

24. $-5t = -25$

25. $-32 = -4s$

26. $-6n = -36$

27. $-7 = -14x$

28. $2c = -7$

29. $4m = -10$

30. Solve the equation $4t = 64$.

31. What is the solution of the equation $6x = 90$?

32. Ciro's father is 3 times as old as Ciro. If Ciro's father is 39, how old is Ciro?

33. Maggie is 4 times as old as her brother. If Maggie is 12, how old is her brother?

Solve each equation. Check your solution.

34. $1.5x = 3$

35. $2.5y = 5$

36. $8.1 = 0.9a$

37. $39 = 1.3b$

38. $0.5e = 0.25$

39. $0.4g = -0.6$

40. **SCIENCE** An object on Earth weighs six times what it would weigh on the moon. If an object weighs 72 pounds on Earth, what is its weight on the moon?

 Data Update Find how an object's weight on Mars compares to its weight on Earth. Visit **msmath1.net/data_update** to learn more.

41. **BILINGUAL AMERICANS** Refer to the graphic. Write and solve an equation to find how many times more Americans speak Spanish than German.

42. **GEOMETRY** The area of a rectangle is 120 square inches, and the width is 5 inches. Write a multiplication equation to find the length of the rectangle and use it to solve the problem. Describe how you can check to be sure that your answer is correct.

43. **CRITICAL THINKING** Without solving, tell which equation below has the greater solution. Explain.

$$4x = 1{,}000 \qquad 8x = 1{,}000$$

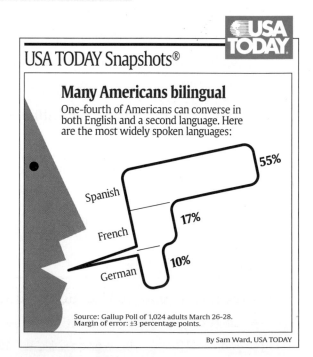

USA TODAY

USA TODAY Snapshots®

Many Americans bilingual
One-fourth of Americans can converse in both English and a second language. Here are the most widely spoken languages:

Spanish 55%

French 17%

German 10%

Source: Gallup Poll of 1,024 adults March 26-28.
Margin of error: ±3 percentage points.

By Sam Ward, USA TODAY

EXTENDING THE LESSON In the equation $\frac{a}{2} = 8$, the expression $\frac{a}{2}$ means *a divided by 2*. To solve an equation that contains division, use multiplication, which is the inverse of division.

Example Solve $\frac{a}{2} = 8$.

$$\frac{a}{2} = 8 \qquad \text{Write the equation.}$$

$$\frac{a}{2} \cdot 2 = 8 \cdot 2 \qquad \text{Multiply each side by 2.}$$

$$a = 16 \qquad \text{Simplify.}$$

Solve each equation.

44. $\frac{x}{3} = 6$ **45.** $3 = \frac{y}{4}$ **46.** $\frac{b}{2} = -3$ **47.** $\frac{c}{4} = -5$

48. $\frac{x}{-3} = 5$ **49.** $\frac{w}{-8} = 2$ **50.** $-6 = \frac{a}{-6}$ **51.** $-10 = \frac{x}{-9}$

Standardized Test Practice and Mixed Review

52. **MULTIPLE CHOICE** The Romeros are driving from New York City to Miami in three days, driving an average of 365 miles each day. What is the total distance they drive?

    Ⓐ 1,288 mi   Ⓑ 1,192 mi   Ⓒ 1,095 mi   Ⓓ 822 mi

53. **MULTIPLE CHOICE** Use the formula $A = \ell w$ to find the length of the rectangle shown.

    Ⓕ 17 ft   Ⓖ 144 ft

    Ⓗ 162 ft   Ⓘ 1,377 ft

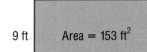

9 ft   Area = 153 ft²

Solve each equation. (Lessons 9-2 and 9-3)

54. $b - 5 = -2$ **55.** $t - 6 = 5$ **56.** $g - 6 = -7$ **57.** $a - 2 = -2$

58. $x + 4 = 9$ **59.** $p + 3 = -2$ **60.** $6 + r = 2$ **61.** $7 + q = -1$

62. Eight people borrowed a total of $56. If each borrowed the same amount, how much did each person borrow? (Lesson 8-5)

Find the circumference of each circle shown or described. Round to the nearest tenth. (Lesson 4-6)

63. 15 ft   **64.** 8 in.   **65.** $d = 0.75$ m

GETTING READY FOR THE NEXT LESSON

PREREQUISITE SKILL Find the value of each expression. (Lesson 1-5)

66. $2(4) + 6$ **67.** $4 + 3(2)$ **68.** $15 - 2(6)$ **69.** $5(4) - 6$

Materials
- cup
- counters
- equation mat

Solve Inequalities Using Models

An inequality is a sentence in which the quantity on the left side may be greater than or less than the quantity on the right side.

To solve an inequality using models, you can use these steps.

- Model the inequality on the mat.
- Follow the steps for solving equations using models.

ACTIVITY   *Work with a partner.*

Solve $x + 3 > 5$ using models.

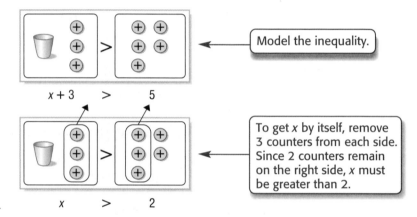

Model the inequality.

$x + 3 \quad > \quad 5$

To get x by itself, remove 3 counters from each side. Since 2 counters remain on the right side, x must be greater than 2.

$x \quad > \quad 2$

Any number greater than 2 will make the inequality $x + 3 > 5$ true.

Look at this solution on a number line.

The open dot at 2 means that 2 is not included in the solution.

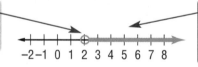

The shading tells you that all numbers greater than 2 are solutions.

Your Turn Solve each inequality using models.

a. $x + 5 > 9$ b. $x - 7 < 3$ c. $6 < x + 1$

Writing Math

1. **Compare and contrast** solving addition and subtraction inequalities with solving addition and subtraction equations.

2. **Examine** the inequality $x + 7 > 12$. Can the solution be $x = 8.5$? Explain your reasoning.

3. **MAKE A CONJECTURE** Write a rule for solving inequalities like $x - 3 > 8$ without using models.

Solving Two-Step Equations

What You'll LEARN

Solve two-step equations.

NEW Vocabulary

two-step equation

WHEN am I ever going to use this?

MONEY MATTERS Suppose you order two paperback books for a total price of $11 including shipping charges of $3. The books are the same price.

1. Let x = the cost of one book. How does the equation $2x + 3 = 11$ represent the situation?

2. Subtract 3 from each side of the equation. Write the equation that results.

3. Divide each side of the equation you wrote by 2. Write the result. What is the cost of each book?

Equations like $2x + 3 = 11$ that have two different operations are called **two-step equations**. To solve a two-step equation you need to work backward using the reverse of the order of operations.

EXAMPLE Solve a Two-Step Equation

1 Solve $2x + 3 = 11$.

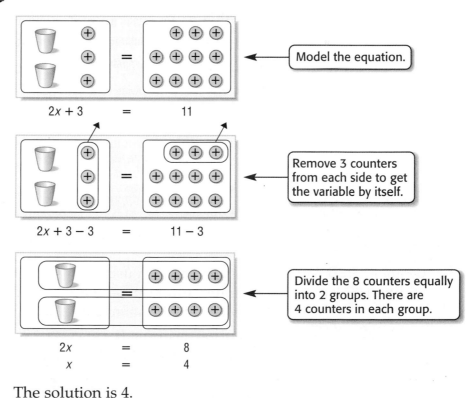

Model the equation.

$$2x + 3 = 11$$

Remove 3 counters from each side to get the variable by itself.

$$2x + 3 - 3 = 11 - 3$$

Divide the 8 counters equally into 2 groups. There are 4 counters in each group.

$$2x = 8$$
$$x = 4$$

The solution is 4.

EXAMPLE Solve a Two-Step Equation

2 Solve $3x - 2 = 7$. Check your solution.

$$3x - 2 = 7 \quad \text{Write the equation.}$$
$$\underline{+\,2 = +\,2} \quad \text{Add 2 to each side.}$$
$$3x \quad = \quad 9 \quad \text{Simplify.}$$
$$\frac{3x}{3} = \frac{9}{3} \quad \text{Divide each side by 3.}$$
$$x = 3 \quad \text{Simplify.}$$

The solution is 3.   Check this solution.

Your Turn Solve each equation. Check your solution.

a. $3a + 2 = 14$ b. $4c - 3 = 5$ c. $1 = 3a + 4$

EXAMPLE Use an Equation to Solve a Problem

3 **ICE SKATING** John and two friends went ice skating. The admission was $5 each. John brought his own skates, but his two friends had to rent skates. If they spent a total of $19 to skate, how much did each friend pay for skate rental?

Words	The cost of two skate rentals plus two admissions is $19.
Variable	Let s = cost for skate rental.

Equation	Two rentals at $$s$ each	plus admission	equals $19.
	$2s$	$+ 3(5)$	$= 19$

$$2s + 15 = 19 \quad \text{Write the equation.}$$
$$\underline{-\,15 = -\,15} \quad \text{Subtract 15 from each side.}$$
$$2s \quad = \quad 4 \quad \text{Simplify.}$$
$$\frac{2s}{2} = \frac{4}{2} \quad \text{Divide each side by 2.}$$
$$s = 2 \quad \text{Simplify.}$$

Skate rental is $2.   Is this answer reasonable?

Skill and Concept Check

1. **Tell** which operation to undo first in the equation $19 = 4 + 5x$.

   Writing Math
   Exercise 1

2. **OPEN ENDED** Write a two-step equation using multiplication and addition. Solve your equation.

GUIDED PRACTICE

Solve each equation. Use models if necessary.

3. $2a + 5 = 13$ 4. $3y + 1 = -2$ 5. $10 = 4d - 2$ 6. $-4 = 5y + 6$

7. Three times a number n plus 8 is 44. What is the value of n?

Practice and Applications

HOMEWORK HELP

For Exercises	See Examples
8–19	1, 2
20–24	3

Extra Practice
See pages 612, 632.

Solve each equation. Use models if necessary.

8. $3a + 4 = 7$

9. $2b + 6 = 12$

10. $3g + 4 = -5$

11. $-1 = 3f + 2$

12. $-8 = 6y - 2$

13. $3 = 4h - 5$

14. $4d - 1 = 11$

15. $5k - 3 = -13$

16. $2x + 3 = 9$

17. $4t + 4 = 8$

18. $10 = 2r - 8$

19. $-7 = 4s + 1$

20. Six less than twice a number is fourteen. What is the number?

21. Ten is four more than three times a number. What is the number?

22. **GEOMETRY** The perimeter of a rectangle is 48 inches. Find its length if its width is 5 inches.

23. **BOWLING** Lavone and two friends went bowling. The cost to bowl one game was $3 each. Lavone brought his own bowling shoes, but his two friends had to rent bowling shoes. If they spent a total of $15 to bowl one game, how much did each friend pay for shoe rental?

24. **TENNIS** While on vacation, Daniella played tennis. Racket rental was $7, and court time cost $27 per hour. If the total cost was $88, how many hours did Daniella play?

25. **CRITICAL THINKING** Use what you know about solving two-step equations to solve the equation $2(n - 9) = -4$.

Standardized Test Practice and Mixed Review

26. **MULTIPLE CHOICE** Seven less than four times a number is negative nineteen. What is the number?

　Ⓐ -3　　　Ⓑ -1　　　Ⓒ 1　　　Ⓓ 3

27. **MULTIPLE CHOICE** Carter bought 3 pounds of peppers, 2 pounds of onions, 1 pound of lettuce, and 4 potatoes. If he had a total of 8 pounds of vegetables, how much did the 4 potatoes weigh?

　Ⓕ $\frac{1}{2}$ lb　　　Ⓖ 2 lb　　　Ⓗ 1 lb　　　Ⓘ 4 lb

28. **MONEY MATTERS** Last week, Emilio spent 3 times as much on lunch as he spent on snacks. If he spent $12 on lunch, how much did he spend on snacks? (Lesson 9-4)

29. **ALGEBRA** Solve $y - 11 = -8$. (Lesson 9-3)

GETTING READY FOR THE NEXT LESSON

PREREQUISITE SKILLS Evaluate each expression if $n = -3$, $n = 0$, and $n = 3$. (Lesson 1-6)

30. $n - 5$

31. $n + 2$

32. $2n$

33. $\frac{1}{3}n$

Problem-Solving Strategy
A Follow-Up of Lesson 9-5

What You'll Learn
Solve problems by writing an equation.

Write an Equation

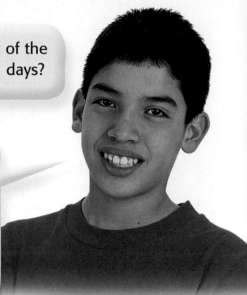

Mario, are you going to get your first draft of the 1,000-word English assignment done in 3 days?

Well, Ashley, at last count, I had 400 words. Let's see. If I subtract 400 from 1,000, I'll know the number of words I have left to write. Then if I divide by the 3 days left, I'll know the number of words I need to write each day.

Explore	We know the total number of words needed, how many have been written, and how many days are left.
Plan	We can write an equation.
	Let w = the words to be written each day.

3 days times w words a day ⟶ $3w$
plus 400 ⟶ $+\ 400$
must equal 1,000 words. ⟶ $=1{,}000$

Solve	$3w + 400 - 400 = 1{,}000 - 400$	Subtract 400 from each side.
	$3w = 600$	Simplify
	$\dfrac{3w}{3} = \dfrac{600}{3}$	Divide each side by 3.
	$w = 200$	

Mario needs to write 200 words each day.

Examine	Check the answer in the original situation. If Mario writes 200 words a day for 3 days, he will have written 600 words. Add the 400 words he has already written to 600 to get 1,000. The answer checks.

Analyze the Strategy

1. **Explain** how each equation represents the situation above.
   Equation A: $1{,}000 - 400 - 3w = 0$
   Equation B: $1{,}000 = 400 + 3w$

2. **Write** an equation to describe the following situation. There are 1,200 words in an assignment, 500 words are completed, and there are 4 days left to work.

Solve. Use the write an equation strategy.

3. **MONEY** Taylor thinks she was overcharged when she bought 8 CD's at $2 each and a CD player for $15 at a garage sale. She paid a total of $39. Write the equation that describes this problem and solve. Was she overcharged? Explain.

4. **NUMBER THEORY** A number is multiplied by 2. Then 7 is added to the product. After subtracting 3, the result is 0. Write and solve an equation for this problem.

Mixed Problem Solving

Solve. Use any strategy.

5. **SPORTS** Violetta, Brian, and Shanté play volleyball, soccer, and basketball. One of the girls is Brian's next-door neighbor. No person's sport begins with the same letter as his or her first name. Brian's neighbor plays volleyball. Which sport does each person play?

6. **THEATER** Ticket prices for a theater are shown in the table.

Ticket Prices	
Adult	$7.25
Student	$3.50
Child under 4	$1.75

The Stevens family needs 2 adult tickets, 3 student tickets, and 1 child's ticket. What is the total cost for the Stevens family to attend the play?

7. **ANIMALS** The table shows the weights of various animals. If there are 2,000 pounds in one ton, how many bobcats would it take to equal 2 tons?

Animal Weights	
Animal	**Weight (lb)**
zebra	600
anteater	100
bonobo	80
bobcat	20

Source: www.colszoo.org

8. **TRANSPORTATION** The sixth grade class is planning a field trip. 348 students and 18 teachers will be going on the field trip. If each bus holds 48 people, how many buses will they need?

9. Anoki is selling cotton candy at the school carnival. The machine holds enough for 16 cotton candy treats. If he needs to refill the machine every 30 minutes, how many cotton candy treats can he expect to sell in 3 hours?

10. **PATTERNS** Draw the next two figures in the pattern shown below.

11. **MONEY** Wesley wants to collect all 50 U.S. special edition quarters. Five quarters are released each year. He has already collected the first four years. Write an equation to find the number of years that Wesley still has to collect quarters to have all 50.

12. **STANDARDIZED TEST PRACTICE**
The perimeter of a rectangular garden is 72 feet. The length of the garden is 20 feet. Which equation *cannot* be used to find the width, w?

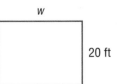

Ⓐ $72 = 2w + 2 \times 20$

Ⓑ $72 = 2(w + 20)$

Ⓒ $72 - 2 \times 20 = 2w$

Ⓓ $w = 72 - 2 \times 20$

Function Machines

What You'll LEARN

Illustrate functions using function machines.

A *function machine* takes a number called the *input* and performs one or more operations on it to produce a new value called the *output*.

Materials

• scissors
• tape

ACTIVITY *Work in small groups.*

Make a function machine for the rule $n - 4$.

STEP 1 Cut a sheet of paper in half lengthwise.

STEP 2 Cut four slits into one of the halves of paper as shown. The slits should be at least one inch wide.

STEP 3 Using the other half of the paper, cut two narrow strips. These strips should be able to slide through the slits you cut on the first sheet of paper.

STEP 4 On one of the narrow strips, write the numbers 10 through 6 as shown. On the other strip, write the numbers 6 through 2 as shown.

The numbers on both strips should align.

10	6
9	5
8	4
7	3
6	2

STEP 5 Place the strips into the slits so that the numbers 10 and 6 can be seen. Then tape the ends of the strips together at the top. When you pull the strips, they should move together.

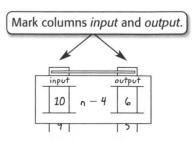

Mark columns *input* and *output*.

input		output
10	$n - 4$	6

STEP 6 Write the function rule $n - 4$ between the input and output as shown.

STEP 7 Use the function machine to find the output value for each input value. Copy and complete the function table showing the input and output.

Input	Output
10	6
9	■
8	■
7	■
6	■

Your Turn Make a function machine for each rule. Use the input values 0, 1, 2, and 3 for n. Record the input and output in a function table.

a. $n + 3$ b. $n + 5$ c. $n - 2$

d. $n - 3$ e. $n \times 2$ f. $n \times 3$

Writing Math

Work in small groups.

1. **Explain** what a function machine would do for the rule $n \times 4$.

2. Use the function machine at the right. Copy and complete the function table. Then write the function rule for the table.

input	output
10	5
6	3
4	2
2	1

Input	Output
10	5
8	■
6	■
4	■
2	■

3. Explain how a function machine would evaluate the rule $n \times 3 + 4$.

4. **Make** a function machine using the rule $n \times 3 + 4$. Use the numbers 1–5 as the input values. Record the input and output values in a function table.

5. **Create** your own function machine. Write pairs of inputs and outputs and have the other members of your group determine the rule.

6. **Tell** what the function rule is for each set of input and output values.

a.

Input	Output
3	−2
4	−1
5	0
6	1
7	2

b.

Input	Output
2	4
3	6
4	8
5	10
6	12

7. **Explain** why using a function machine is like finding a pattern.

Functions

What You'll LEARN

Complete function tables and find function rules.

NEW Vocabulary

function
function table
function rule

WHEN am I ever going to use this?

LIFE SCIENCE A brown bat can eat 600 mosquitoes an hour.

1. Write an expression to represent the number of mosquitoes a brown bat can eat in 2 hours.

2. Write an expression to represent the number of mosquitoes a brown bat can eat in 5 hours.

3. Write an expression to represent the number of mosquitoes a brown bat can eat in t hours.

The number of mosquitoes eaten by a bat is a **function** of the number of hours. The results can be organized in a **function table**.

Input	Function Rule	Output
Number of Hours (t)	$600t$	Mosquitoes Eaten
1	600(1)	600
2	600(2)	1,200
3	600(3)	1,800

The **function rule** describes the relationship between each input and output.

EXAMPLE Complete a Function Table

1 Complete the function table.

The function rule is $x + 4$. Add 4 to each input.

Input (x)	Output ($x + 4$)
-2	■
1	■
4	■

Input		Output
-2	$+ 4 \rightarrow$	2
1	$+ 4 \rightarrow$	5
4	$+ 4 \rightarrow$	8

Input (x)	Output ($x + 4$)
-2	2
1	5
4	8

Your Turn Copy and complete each function table.

a.
Input (x)	Output ($x - 2$)
-2	■
1	■
4	■

b.
Input (x)	Output ($2x$)
-1	■
0	■
3	■

EXAMPLE **Find the Rule for a Function Table**

2 Find the rule for the function table.

Input (x)	Output (■)
−3	−1
1	$\frac{1}{3}$
6	2

Study the relationship between each input and output.

Input   **Output**

$-3 \quad \times \frac{1}{3} \rightarrow \quad -1$

$1 \quad \times \frac{1}{3} \rightarrow \quad \frac{1}{3}$

$6 \quad \times \frac{1}{3} \rightarrow \quad 2$

The output is one-third of the input.

So, the function rule is $\frac{1}{3}x$, or $\frac{x}{3}$.

Your Turn Find the rule for each function table.

c.

Input (x)	Output (■)
−3	−12
1	4
4	16

d.

Input (x)	Output (■)
4	−1
8	3
10	5

REAL-LIFE MATH

How Does a Criminalist Use Math?

Criminalists can determine the height of a victim by measuring certain bones and using formulas to make predictions.

Online Research
For more information about a career as a criminalist, visit: msmath1.net/careers

EXAMPLE **Solve a Problem Using a Function**

3 **CRIMINOLOGY** A criminalist knows that an adult male's height, in centimeters, is about 72 centimeters more than 2.5 times the length of his tibia, t (shin bone). How tall is a man whose tibia is 30 centimeters?

First, determine the function rule.

Let t = length of tibia.

The function rule is $2.5t + 72$.

> 72 centimeters more than means to add 72.

Then, replace t in the rule $2.5t + 72$ with the length of the tibia, 30.

$2.5t + 72 = 2.5(30) + 72$ Replace t with 30.

$\qquad\qquad = 75 + 72$ Multiply 2.5 and 30.

$\qquad\qquad = 147$ Add 75 and 72.

The man is about 147 centimeters tall.

1. **Make** a function table for the function rule $4x$. Use inputs of $-4, -2, 0$, and 4.

2. **OPEN ENDED** Make a function table. Then write a function rule. Choose three input values and find the output values.

3. **FIND THE ERROR** Nicole and Olivia are finding the function rule when each output is 5 less than the input. Who is correct? Explain.

> Nicole
> Function rule: $5 - x$

> Olivia
> Function rule: $x - 5$

GUIDED PRACTICE

Copy and complete each function table.

4.

Input (x)	Output (x + 3)
−2	■
0	■
2	■

5.

Input (x)	Output (3x)
−3	■
0	■
6	■

Find the rule for each function table.

6.

x	■
0	−1
2	1
4	3

7.

x	■
−3	6
1	−2
4	−8

8. If the input values are $-3, 0$, and 6 and the corresponding outputs are $1, 4$, and 10, what is the function rule?

Practice and Applications

HOMEWORK HELP

For Exercises	See Examples
9–10, 19–20	1
11–18	2
21–22	3

Extra Practice
See pages 613, 632.

Copy and complete each function table.

9.

Input (x)	Output (x − 4)
−2	■
0	■
8	■

10.

Input (x)	Output ($\frac{1}{2}x$)
−6	■
0	■
3	■

Find the rule for each function table.

11.

x	■
−1	1
0	2
6	8

12.

x	■
−1	−6
1	−4
3	−2

13.

x	■
−1	−2
0	0
6	12

14.

x	■
−2	$-\frac{2}{5}$
0	0
10	2

Find the rule for each function table.

15.

x	■
−2	6
1	−3
3	−9

16.

x	■
−2	12
1	9
4	6

17.

x	■
0	−1.6
2	0.4
4	2.4

18.

x	■
−2	4
1	1
4	16

19. If a function rule is $2x + 2$, what is the output for an input of 3?

20. If a function rule is $5x - 3$, what is the output for -2?

MONEY MATTERS For Exercises 21 and 22, use the following information.
For a school project, Sarah and her friends made hair scrunchies to sell for $3 each and friendship bracelets to sell for $4 each.

HAIR SCRUNCHIES
$3 EACH

FRIENDSHIP BRACELETS
$4 EACH

21. Write a function rule to represent the total selling price of scrunchies (s) and bracelets (b).

22. What is the price of 10 scrunchies and 12 bracelets?

23. **MONEY** Suppose the estimated 223 million Americans who have jugs or bottles of coins around their homes put coins back into circulation at a rate of $10 a year. Make a function table showing the amount that would be recirculated in 1, 2, and 3 years.

24. **CRITICAL THINKING** Find the rule for the function table.

x	−2	−1	2	3
■	−2	0	6	8

Standardized Test Practice and Mixed Review

25. **MULTIPLE CHOICE** Find the rule for the function table shown.

Ⓐ $x \div 8$ Ⓑ $\frac{1}{8}x$ Ⓒ $8 - x$ Ⓓ $8 + x$

x	■
1	7
4	4
10	−2

26. **MULTIPLE CHOICE** The school store makes a profit of 5¢ for each pencil sold. Which expression best represents the profit on 25 pencils?

Ⓕ 0.05×25 Ⓖ 5×0.25 Ⓗ $25 \div 5$ Ⓘ $25 - 5$

27. **SHOPPING** Ping bought 3 T-shirts. His cost after using a $5-off total purchase coupon was $31. How much did each T-shirt cost? (Lesson 9-5)

Solve each equation. Use models if necessary. (Lesson 9-4)

28. $6x = 24$ 29. $7y = -42$ 30. $-12 = 5m$ 31. $4p = 11$

GETTING READY FOR THE NEXT LESSON

PREREQUISITE SKILL Graph each point on a coordinate plane. (Lesson 8-6)

32. $A(4, -2)$ 33. $B(3, 4)$ 34. $C(-5, 0)$ 35. $D(-1, -3)$

Graphing Functions

WHEN **am I ever going to use this?**

SAVINGS Suppose you put $2 a week in savings.

1. Copy and complete the table to find the amount you would save in 2, 3, and 6 weeks.

2. On grid paper, graph the ordered pairs (number, amount saved).

3. Describe how the points appear on the grid.

4. What happens to the amount saved as the number of weeks increases?

Savings		
Number of Weeks	Multiply by 2.	Amount Saved
1	2 × 1	$2
2		
3		
6		

The amount saved depends on the number of weeks. You can represent the function "multiply by 2" with an equation.

amount saved ⟶   ⟵ number of months

$$y = 2x$$

EXAMPLE **Graph a Function**

1 Make a function table for the rule $y = 3x$. Use input values of −2, 0, and 2. Then graph the function.

Step 1 Record the input and output in a function table. List the input and output as ordered pairs.

Input	Function Rule	Output	Ordered Pairs
(x)	(3x)	(y)	(x, y)
−2	3(−2)	−6	(−2, −6)
0	3(0)	0	(0, 0)
2	3(2)	6	(2, 6)

Step 2 Graph the ordered pairs on the coordinate plane.

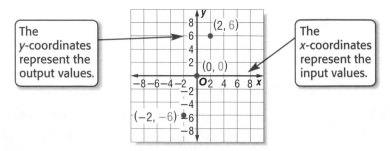

The y-coordinates represent the output values.

The x-coordinates represent the input values.

Line Graphs
The arrowheads indicate that the line extends in both directions.

Step 3 The points appear to lie on a line. Draw the line that contains these points. The line is the graph of $y = 3x$. For any point on this line, $y = 3x$.

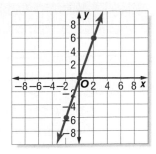

● Your Turn

a. Make a function table for the rule $y = x - 4$ using input values of 0, 2, and 4. Then graph the function.

EXAMPLE   **Make a Function Table for a Graph**

2 Make a function table for the graph. Then determine the function rule.

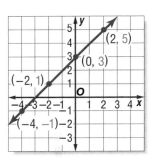

Use the ordered pairs to make a function table.

Input (x)	Output (y)	(x, y)
−4	−1	(−4, −1)
−2	1	(−2, 1)
0	3	(0, 3)
2	5	(2, 5)

Study the input and output. Look for a rule.

Input		**Output**
−4	+ 3	−1
−2	+ 3	1
0	+ 3	3
2	+ 3	5

3 is added to each input to get the output.

The function rule is $y = x + 3$.

● Your Turn

b. Make a function table for the graph. Then determine the function rule.

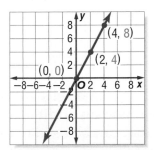

Skill and Concept Check

Writing Math
Exercise 1

1. **Explain** the difference between a function table and the graph of a function.

2. **OPEN ENDED** Draw the graph of a function that passes through the point (0, 0). Name three points on the graph.

GUIDED PRACTICE

Make a function table for each rule with the given input values. Then graph the function.

3. $y = x + 5; -2, 0, 2$

4. $y = \frac{x}{2}; -4, 0, 4$

5. Make a function table for the graph at the right. Then determine the function rule.

6. Make a function table for the rule $y = x - 5$ using 1, 3, and 6 as the input. Then graph the function.

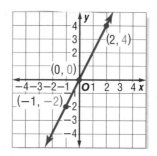

Practice and Applications

Make a function table for each rule with the given input values. Then graph the function.

7. $y = x - 2; 0, 2, 4$

8. $y = 2x; -1, 1, 2$

9. $y = 2n - 3; -3, 0, 4$

10. $y = 2n + 3; 2, \frac{1}{2}, 0$

11. $y = x + 4; -5, -2, 1$

12. $y = 2x + 4; -2, 1, 3$

13. $y = -4x; 2, 0, -2$

14. $y = \frac{1}{2}x + 1; -2, 0, 4$

HOMEWORK HELP

For Exercises	See Examples
7–14	1
15–17	2

Extra Practice
See pages 613, 632.

Make a function table for each graph. Then determine the function rule.

15.

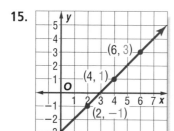

16.

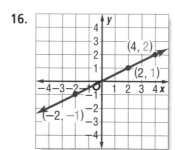

17.

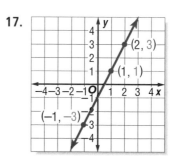

18. **CATALOGS** A catalog that sells gift wrap charges $3 for each roll of gift wrap ordered and an additional $1 for shipping of each roll. Write a function rule that can be used to find the cost, including shipping, of any number of rolls of gift wrap.

MONEY MATTERS For Exercises 19–21, use the following information.
Ben's summer job pays $50 a week, and he must pay $30 for a uniform.
Rachel earns $45 a week and does not need a uniform.

19. Write the function rule for each person's wages.

20. Graph each function on the same coordinate plane.

21. What does the intersection of the two graphs represent?

22. CRITICAL THINKING Determine the rule for the line that passes
through $A(-2, -1)$ and $B(3, 9)$.

EXTENDING THE LESSON
Some function rules result in a
curved line on the graph. These
are called *nonlinear functions*.

Example $y = x^2$

Input (x)	Output (x²)
−2	4
−1	1
0	0
1	1
2	4

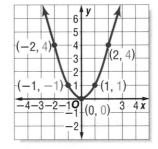

Make a function table for each rule with the given input values.
Then graph the function.

23. $y = n^3$; −2, −1, 0, 1, 2

24. $y = n^2 - 2$; −2, −1, 0, 1, 2

Standardized Test Practice and Mixed Review

25. MULTIPLE CHOICE Which is the output for the input −1 using the
rule $y = 2x - 3$?

    A −5   **B** −4   **C** −3   **D** −1

26. SHORT RESPONSE Find a function rule for the graph at
the right.

27. If input values are 3, 5, and 8 and the corresponding outputs
are 5, 7, and 10, what is the function rule? (Lesson 9-6)

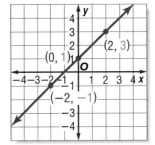

Solve each equation. Check your solution. (Lesson 9-5)

28. $4t - 1 = 11$ **29.** $7x - 10 = -38$ **30.** $-4 = 2y + 8$

WebQuest **Interdisciplinary Project**

Weather Watchers
It's time to complete your project. Use the data you have gathered about weather
patterns in your state to prepare a Web page or poster. Be sure to include two graphs
and a report with your project.
msmath1.net/webquest

Study Guide and Review

Vocabulary and Concept Check

Addition Property of
 Equality (p. 345)
Additive Identity (p. 334)
Associative Property (p. 334)
coefficient (p. 350)

Commutative Property (p. 334)
Distributive Property (p. 333)
function (p. 362)
function rule (p. 362)
function table (p. 362)

inverse operations (p. 339)
Multiplicative Identity (p. 334)
Subtraction Property of
 Equality (p. 340)
two-step equation (p. 355)

Choose the correct term or number to complete each sentence.

1. The (Commutative, Associative) Property states that the order in which numbers are added or multiplied does not change the sum or product.

2. To solve a multiplication equation, you can (divide, multiply) to undo the multiplication.

3. A(n) (function, output) describes a relationship between two quantities.

4. The equation $2b + 3 = 11$ is an example of a (one-step, two-step) equation.

5. A (function rule, coordinate system) describes the relationship between each input and output.

6. The Distributive Property states that when (multiplying, dividing) a number by a sum, multiply each number inside the parentheses by the number outside the parentheses.

Lesson-by-Lesson Exercises and Examples

9-1 Properties (pp. 333–336)

Rewrite each expression using the Distributive Property. Then evaluate.

7. $4(7 + 2)$
8. $(14 + 9)8$
9. $(3 \times 8) + (3 \times 12)$
10. $(9 \times 6) + (9 \times 13)$

Identify the property shown by each equation.

11. $14 + (11 + 7) = (14 + 11) + 7$
12. $(7 \times 4)3 = 3(7 \times 4)$
13. $12 + 15 + 28 = 12 + 28 + 15$
14. $(2 \times 28) \times 3 = 2 \times (28 \times 3)$

Example 1 Rewrite $4(2 + 9)$ using the Distributive Property. Then evaluate.

$$4(2 + 9) = 4(2) + 4(9) \quad \text{Distributive Property}$$
$$= 8 + 36 \quad \text{Multiply.}$$
$$= 44 \quad \text{Add.}$$

Example 2 Identify the property shown by $8 + (7 + 13) = (8 + 7) + 13$.

The grouping of the numbers to be added changes. This is the Associative Property of Addition.

 msmath1.net/vocabulary_review

9-2 Solving Addition Equations (pp. 339–342)

Solve each equation. Use models if necessary.

15. $c + 8 = 11$ 16. $x + 15 = 14$

17. $54 = m - 9$ 18. $-5 = -2 + x$

19. $w + 13 = -25$ 20. $17 + d = -2$

21. $23 = h + 11$ 22. $19 + r = 11$

23. **WEATHER** In the morning, the temperature was $-8°F$. By noon, the temperature had risen $14°$. What was the temperature at noon?

Example 3 Solve $x + 8 = 10$.

$$
\begin{array}{rl}
x + 8 = & 10 \\
\underline{-8 = -8} & \text{Subtract 8 from each side.} \\
x \quad = & 2 \quad \text{Simplify.}
\end{array}
$$

Example 4 Solve $y + 7 = 3$.

$$
\begin{array}{rl}
y + 7 = & 3 \\
\underline{-7 = -7} & \text{Subtract 7 from each side.} \\
y \quad = & -4 \quad \text{Simplify.}
\end{array}
$$

9-3 Solving Subtraction Equations (pp. 344–347)

Solve each equation. Use models if necessary.

24. $z - 7 = 11$ 25. $s - 9 = -12$

26. $14 = m - 5$ 27. $-4 = y - 9$

28. $h - 2 = -9$ 29. $-6 = g - 4$

30. $p - 22 = -7$ 31. $d - 3 = -14$

32. Find the value of c if $c - 9 = -3$.

33. If $d - 1.2 = 6$, what is the value of d?

Example 5 Solve $a - 5 = -3$.

$$
\begin{array}{rl}
a - 5 = & -3 \\
\underline{+5 = +5} & \text{Add 5 to each side.} \\
a \quad = & 2 \quad \text{Simplify.}
\end{array}
$$

Example 6 Solve $4 = m - 9$.

$$
\begin{array}{rl}
4 = & m - 9 \\
\underline{+9 = \quad +9} & \text{Add 9 to each side.} \\
13 = & m \quad \text{Simplify.}
\end{array}
$$

9-4 Solving Multiplication Equations (pp. 350–353)

Solve each equation. Use models if necessary.

34. $4b = 32$ 35. $5y = 60$

36. $-3m = 21$ 37. $-18 = -6c$

38. $7a = -35$ 39. $28 = -2d$

40. $-4x = 10$ 41. $-6y = -9$

42. **ALGEBRA** The product of a number and 8 is -56. What is the number?

Example 7 Solve $-6y = 24$.

$$
\begin{array}{rl}
-6y = 24 & \text{Write the equation.} \\
\dfrac{-6y}{-6} = \dfrac{24}{-6} & \text{Divide each side by } -6. \\
y = -4 & \text{Simplify.}
\end{array}
$$

9-5 Solving Two-Step Equations (pp. 355–357)

Solve each equation. Use models if necessary.

43. $3p - 4 = 8$

44. $2x + 5 = 3$

45. $8 + 6w = 50$

46. $5m + 6 = -9$

47. $6 = 3y - 12$

48. $-15 = 5 + 2t$

Example 8 Solve $4x - 9 = 15$.

$$4x - 9 = 15 \quad \text{Write the equation.}$$
$$\underline{+9 = +9} \quad \text{Add 9 to each side.}$$
$$4x = 24 \quad \text{Simplify.}$$
$$\frac{4x}{4} = \frac{24}{4} \quad \text{Divide each side by 4.}$$
$$x = 6 \quad \text{Simplify.}$$

9-6 Functions (pp. 362–365)

Copy and complete the function table.

49.

Input (x)	Output (x + 3)
−2	▧
1	▧
5	▧

Find the rule for each function table.

50.

x	▢
−2	2
1	5
4	8

51.

x	▢
−5	−11
0	−1
2	3

Example 9 Complete the function table.

Input (x)	Output (x − 4)
−5	▧
1	▧
3	▧

The function rule is $x - 4$. Subtract 4 from each input value.

Input		Output
−5	−4 →	−9
1	−4 →	−3
3	−4 →	−1

9-7 Graphing Functions (pp. 366–369)

Copy and complete each function table. Then graph the function.

52.

Input (x)	Output (x + 3)
−2	▧
0	▧
3	▧

53.

Input (x)	Output (−2x)
−3	▧
−1	▧
2	▧

54. Make a function table for the rule $y = x - 1$ using input values of −2, 0, and 2. Graph the function.

Example 10 Graph the function represented by the function table.

Input (x)	Output (2x + 1)
−1	−1
0	1
2	5

Graph the ordered pairs $(-1, -1)$, $(0, 1)$, and $(2, 5)$. Draw the line that contains the points.

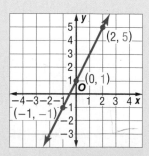

Practice Test

Vocabulary and Concepts

1. **Explain** the Commutative Property. Give an example using addition.

2. **Describe** the process used to solve a two-step equation.

3. **Explain** how to graph the function $y = 2x + 1$.

Skills and Applications

Identify the property shown by each equation.

4. $5 \times (3 \times 2) = (5 \times 3) \times 2$

5. $14 + 9 = 9 + 14$

Rewrite each expression using the Distributive Property. Then evaluate.

6. $2(12 + 5)$

7. $16(12) + 16(8)$

Solve each equation. Use models if necessary.

8. $-5 = x + 11$

9. $w + 17 = 29$

10. $m - 9 = 3$

11. $p - 5 = -1$

12. $-6d = 42$

13. $12 = c + (-2)$

14. $2b = -8$

15. $15 = 3n$

16. $g - 4 = -3$

17. $6x + 4 = 10$

18. $24 = 3y - 6$

19. $-5m = -30$

20. Copy and complete the function table.

Input (x)	Output (2x + 3)
−2	■
1	■
2	■

21. Find the rule for the function table.

x	■
−3	−1
0	2
1	3

Make a function table for each given rule and input values. Then graph the function.

22. $y = x - 4$; $-1, 2, 6$

23. $y = 3x$; $-2, 1, 4$

24. $y = -2x - 2$; $-3, 0, 1$

Standardized Test Practice

Ⓐ Ⓑ Ⓒ Ⓓ

25. **MULTIPLE CHOICE** Fresno, California, f, and Buffalo, New York, b, are 3 time zones apart. Use the function rule $f = b - 3$ to find the time in Buffalo when it is 3:30 P.M. in Fresno.

Ⓐ 4:30 P.M.   Ⓑ 6:30 P.M.   Ⓒ 9:30 P.M.   Ⓓ 12:30 P.M.

CHAPTER 9 Standardized Test Practice

PART 1 Multiple Choice

Record your answers on the answer sheet provided by your teacher or on a sheet of paper.

1. What is the sum of 27 and 59? (Prerequisite Skill, p. 589)

   Ⓐ 76   Ⓑ 86
   Ⓒ 96   Ⓓ 906

2. Juan earned $37.00 baby-sitting last week. If he had not baby-sat on Friday, about how much money would he have earned? (Lesson 3-4)

Money Earned Baby-Sitting	
Day	Amount Earned
Monday	$4.50
Wednesday	$4.50
Friday	$12.75
Saturday	$15.25

   Ⓕ about $13   Ⓖ about $24
   Ⓗ about $25   Ⓘ about $37

3. Which of the following is the least common multiple of 12 and 8? (Lesson 5-4)

   Ⓐ 24   Ⓑ 48
   Ⓒ 72   Ⓓ 96

4. Dion has 60 baseball cards. He gave away $\frac{3}{4}$ of them to Amy. How many did he give to Amy? (Lesson 7-2)

   Ⓕ 15   Ⓖ 20
   Ⓗ 30   Ⓘ 45

5. Which of the following represents 20 feet above sea level? (Lesson 8-1)

   Ⓐ −20 ft   Ⓑ −2 ft
   Ⓒ +2 ft   Ⓓ +20 ft

> **TEST-TAKING TIP**
>
> **Questions 7 and 8** On multiple choice test items involving solving equations, you can replace the variable in the equation with the values given in each answer choice. The answer choice that results in a true statement is the correct answer.

6. Which of the following shows another way to write 6(2 + 8)? (Lesson 9-1)

   Ⓕ 6 − 2 − 8
   Ⓖ 6 + 2 + 8
   Ⓗ 6 × 2 + 6 × 8
   Ⓘ 12 × 48

7. What is the value of b in the equation $22 + b = 34$? (Lesson 9-2)

   Ⓐ 6   Ⓑ 12
   Ⓒ 20   Ⓓ 56

8. After giving 16 comic books to her friends, Carmen had 64 comic books left. She used the equation $x − 16 = 64$ to figure out how many comic books she started with. What is the value of x in the equation? (Lesson 9-3)

   Ⓕ 4   Ⓖ 48
   Ⓗ 76   Ⓘ 80

9. What is the function rule that relates the input and output values in the function table? (Lesson 9-6)

n	▢
0	1
1	3
2	5
3	7
4	9

   Ⓐ $n + 1$
   Ⓑ $n − 1$
   Ⓒ $2n + 1$
   Ⓓ $2n − 1$

PART 2 Short Response/Grid In

Record your answers on the sheet provided by your teacher or on a sheet of paper.

10. If Jim rounds the weight of $3\frac{7}{8}$ pounds of green beans to the nearest pound to estimate the price, what weight will he use? (Lesson 6-1)

11. Sakowski Tailors are sewing band uniforms. They need $5\frac{1}{4}$ yards of fabric for each uniform. How many yards of fabric are needed for 12 uniforms? (Lesson 7-3)

12. What rule was used to create the following pattern? (Lesson 7-6)

56, 48, 40, 32, ?

13. The table shows the lowest extreme temperatures for four U.S. cities. Order the temperatures from least to greatest. (Lesson 8-1)

Lowest Extreme Temperatures	
City	Temp. (°F)
Anchorage	−34
Chicago	−27
Los Angeles	28
Duluth	−39

14. A football team lost 8 yards on their first play. If they gained 9 yards on the next play, how many total yards did they advance? (Lesson 8-2)

15. Find the value of m that makes $-32 \div m = -8$ true. (Lesson 8-5)

16. What ordered pair names point P on the coordinate grid? (Lesson 8-6)

17. Gloria had 24 coins in her collection. At a yard sale, Gloria bought a tin filled with coins. She now has 39 coins in her collection. Use $24 + y = 39$ to find the number of coins she added to her collection. (Lesson 9-2)

18. What is the value of m if $m - 5 = -7$? (Lesson 9-3)

19. What output value completes the following function table? (Lesson 9-6)

Input	Output
1	4
2	7
3	10
4	13
5	■

PART 3 Extended Response

Record your answers on a sheet of paper. Show your work.

20. Three friends went to the skateboard arena. The admission was $3.50 each. Two people had to rent boards. The total cost for the three to skateboard was $15.50. What was the cost to rent a skateboard? Explain how you found the solution. (Lesson 9-5)

21. The values of a function are shown below. (Lessons 9-6 and 9-7)

x	y
0	−2
1	3
2	8
3	13

a. Graph the function on a coordinate plane.

b. Identify the corresponding y-values for $x = 4$ and $x = 5$.

c. What is the function rule?

UNIT 5
Ratio and Proportion

Chapter 10

Ratio, Proportion, and Percent

Chapter 11

Probability

In Unit 3, you learned how fractions and decimals are related. In this unit, you will learn how these numbers are also related to ratios, proportions, and percents, and how they can be used to describe real-life probabilities.

TAKE ME OUT TO THE BALLGAME

Baseball, one of America's favorite pastimes, is overflowing with mathematics. The National Baseball Statisticians Organization has asked you to step up to the plate! They need your help to analyze several seasons of baseball data. You'll also be asked to create a scale drawing of a professional baseball field. The game is about to begin. Let's see if you can hit a homerun!

Log on to msmath1.net/webquest to begin your WebQuest.

CHAPTER 10

Ratio, Proportion, and Percent

"What do insects have to do with math?"

Most insects are very small. A drawing or photograph of an insect often shows the insect much larger than it is in real life. **For example, this photograph shows a praying mantis about three times as large as an actual praying mantis.**

You will find the actual dimensions of certain insects in Lesson 10-3.

GETTING STARTED

Take this quiz to see whether you are ready to begin Chapter 10. Refer to the lesson number in parentheses if you need more review.

▶ Vocabulary Review

Choose the correct number to complete each sentence.

1. To write 0.28 as a fraction, write the decimal as a fraction using (100, 1,000) as the denominator. (Lesson 5-6)

2. The fraction $\frac{5}{8}$ is equivalent to (0.875, 0.625). (Lesson 5-7)

▶ Prerequisite Skills

Multiply. (Lesson 4-2)

3. 0.28×25 4. 364×0.88

5. 154×0.18 6. 0.03×16

7. 0.5×220 8. 0.25×75

Draw a model to represent each fraction.
(Lesson 5-5)

9. $\frac{2}{4}$ 10. $\frac{1}{6}$

11. $\frac{3}{5}$ 12. $\frac{2}{3}$

Write each fraction as a decimal.
(Lesson 5-7)

13. $\frac{3}{8}$ 14. $\frac{46}{100}$

15. $\frac{7}{10}$ 16. $\frac{1}{5}$

Multiply. (Lesson 7-2)

17. $\frac{1}{4} \times 360$ 18. $\frac{3}{4} \times 96$

19. $\frac{2}{5} \times 125$ 20. $\frac{7}{9} \times 27$

FOLDABLES™
Study Organizer

Ratio, Proportion, and Percent Make this Foldable to help you organize your notes. Begin with a piece of graph paper.

STEP 1 Fold
Fold one sheet of grid paper in thirds lengthwise.

STEP 2 Fold and Cut
Unfold lengthwise and fold one-fourth down widthwise. Cut to make three tabs as shown.

STEP 3 Unfold and Label
With the tabs unfolded, label the paper as shown.

Definitions & Notes	Definitions & Notes	Definitions & Notes
Examples	Examples	Examples

STEP 4 Refold and Label
Refold the tabs and label as shown.

Ratio	Proportion	Percent
Examples	Examples	Examples

Reading and Writing As you read and study the chapter, write definitions, notes, and examples for ratios, proportions, and percents under the tabs.

What You'll LEARN

Express ratios and rates in fraction form.

NEW Vocabulary

ratio
equivalent ratios
rate
unit rate

WHEN am I ever going to use this?

CLOTHES The table shows how many socks of each color are in a drawer.

Socks	
Color	Number
Black	6
White	12
Navy	2

1. Write a sentence that compares the number of navy socks to the number of white socks. Use the word *less* in your sentence.

2. Write a sentence that compares the number of black socks to the number of white socks. Use the word *half* in your sentence.

3. Write a sentence comparing the number of white socks to the total number of socks. Use a fraction in your sentence.

There are many ways to compare numbers. A **ratio** is a comparison of two numbers by division. If there are 6 black socks and a total of 20 socks, then the ratio comparing the black socks to the total socks can be written as follows.

$$\frac{6}{20} \qquad 6 \text{ to } 20 \qquad 6 \text{ out of } 20 \qquad 6:20$$

A common way to express a ratio is as a fraction in simplest form.

The GCF of 6 and 20 is 2. → $\frac{6}{20} \overset{\div 2}{\underset{\div 2}{=}} \frac{3}{10}$ ← The simplest form of $\frac{6}{20}$ is $\frac{3}{10}$.

EXAMPLE Write a Ratio in Simplest Form

① **SPORTS** Write the ratio that compares the number of footballs to the number of tennis balls.

$$\begin{array}{l} \text{footballs} \rightarrow \\ \text{tennis balls} \rightarrow \end{array} \quad \frac{4}{6} \overset{\div 2}{\underset{\div 2}{=}} \frac{2}{3} \quad \leftarrow \boxed{\text{The GCF of 4 and 6 is 2.}}$$

STUDY TIP

Look Back To review **simplifying fractions**, see Lesson 5-2.

The ratio of footballs to tennis balls is $\frac{2}{3}$, 2 to 3, or 2:3.

For every 2 footballs, there are 3 tennis balls.

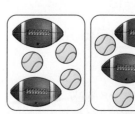

The two ratios in Example 1 are **equivalent ratios** since $\frac{4}{6} = \frac{2}{3}$.

Use Ratios to Compare Parts of a Whole

2 **FOOD** Write the ratio that compares the number of pretzels to the total number of snacks.

$$\text{pretzels} \rightarrow \atop \text{snacks} \rightarrow \quad \frac{4}{12} = \frac{1}{3}$$

÷4 ÷4

> The GCF of 4 and 12 is 4.

The ratio of pretzels to the total number of snacks is $\frac{1}{3}$, 1 to 3, or 1:3. For every one pretzel, there are three total snacks.

Your Turn Write each ratio as a fraction in simplest form.
a. 3 drums to 18 trumpets
b. 8 gerbils to 36 pets

A **rate** is a ratio of two measurements having different kinds of units. Two examples are shown below.

> Dollars and pounds are different kinds of units.

> Miles and hours are different kinds of units.

$12 for 3 pounds   60 miles in 3 hours

When a rate is simplified so that it has a denominator of 1, it is called a **unit rate**. An example of a unit rate is $3 per pound, which means $3 per 1 pound.

Find Unit Rate

3 **BIRDS** Use the information at the left to find how many miles a roadrunner can run in one hour.

$$\frac{54 \text{ miles}}{4 \text{ hours}} = \frac{13.5 \text{ miles}}{1 \text{ hour}}$$

÷4 ÷4

Divide the numerator and the denominator by 4 to get a denominator of 1.

So, a roadrunner can run about 13.5 miles in one hour.

REAL-LIFE MATH

BIRDS The roadrunner is the state bird of New Mexico. Roadrunners prefer running to flying. It would take 4 hours for a roadrunner to run about 54 miles.

Source: www.50states.com

Skill and Concept Check

Writing Math
Exercises 3 & 4

1. **Write** the ratio 6 geese out of 15 birds in three different ways.

2. **OPEN ENDED** Explain the difference between a rate and a unit rate. Give an example of each.

3. **FIND THE ERROR** Brian and Marta are writing the rate $56 in 4 weeks as a unit rate. Who is correct? Explain.

> Brian
> $$\frac{\$56}{4 \text{ weeks}} = \frac{\$14}{1 \text{ week}}$$

> Marta
> $$\frac{\$56}{4 \text{ weeks}} = \frac{\$28}{2 \text{ weeks}}$$

4. **NUMBER SENSE** The ratio of videocassettes to digital videodiscs is 1 to 4. Explain the meaning of this ratio.

GUIDED PRACTICE

Write each ratio as a fraction in simplest form.

5. 6 wins to 8 losses

6. 15 pens to 45 pencils

7. 9 salmon out of 21 fish

8. 4 roses out of 24 flowers

Write each ratio as a unit rate.

9. $9 for 3 cases of soda

10. 25 meters in 2 seconds

11. **MONEY** Two different packages of batteries are shown. Determine which is less expensive per battery, the 4-pack or the 8-pack. Explain.

4-pack **$3.60**

8-pack **$6.80**

Practice and Applications

Write each ratio as a fraction in simplest form.

12. 14 dimes to 24 nickels

13. 15 rubies to 25 emeralds

14. 16 pigs to 10 cows

15. 8 circles to 22 squares

16. 6 mustangs out of 21 horses

17. 4 cellular phones out of 18 phones

18. 10 girls out of 24 students

19. 32 apples out of 72 pieces of fruit

Write each ratio as a unit rate.

20. 180 words in 3 minutes

21. $36 for 4 tickets

22. $1.50 for 3 candy bars

23. $1.44 for a dozen eggs

HOMEWORK HELP	
For Exercises	See Examples
12–19, 26, 27	1, 2
20–23, 28	3

Extra Practice
See pages 613, 633.

24. **SHOPPING** Luke purchased a 16-ounce bag of potato chips for $2.56 and a 32-ounce bag of tortilla chips for $3.52. Which of these snack foods is less expensive per ounce? Explain.

25. **SCHOOL** Draw a picture showing 4 pencils and a number of pens in which the ratio of pencils to pens is 2:3.

HOCKEY For Exercises 26 and 27, use the graphic at the right. Write each ratio in simplest form.

26. What ratio compares the appearances of the Rangers to the appearances of the Red Wings?

27. What ratio compares the appearances of the Maple Leafs to the appearances of the Bruins?

28. **DINOSAURS** A pterodactyl could fly 75 miles in three hours. At this rate, how far could a pterodactyl travel in 1 hour?

29. **CRITICAL THINKING** If 9 out of 24 students received below a 75% on the test, what ratio of students received a 75% or above?

USA TODAY Snapshots®

Skating for Lord Stanley's Cup
Teams with the most appearances in the NHL Stanley Cup finals since 1927[1]:

Montreal Canadiens — 29
Detroit Red Wings — 22
Toronto Maple Leafs — 19
Boston Bruins — 17
New York Rangers — 10

1 – National Hockey League assumed control of Stanley Cup competition after 1926

Source: NHL   By Ellen J. Horrow and Sam Ward, USA TODAY

Standardized Test Practice and Mixed Review

30. **MULTIPLE CHOICE** Dr. Rodriguez drove 384.2 miles on 17 gallons of gasoline. At this rate, how many miles could he drive on 1 gallon?

    Ⓐ 22.5 mi   Ⓑ 22.6 mi   Ⓒ 126 mi   Ⓓ none of the above

31. **SHORT RESPONSE** Find the ratio of the number of vowels in the word *Mississippi* to the number of consonants as a fraction in simplest form.

32. Make a function table for the rule $y = -2x$. Use input values of -1, 0, and 1. Then graph the function. (Lesson 9-7)

Find the rule for each function table. (Lesson 9-6)

33.

x	■
0	-2
1	-1
2	0

34.

x	■
-2	-1
0	1
3	4

35.

x	■
-3	0
-1	2
2	5

GETTING READY FOR THE NEXT LESSON

PREREQUISITE SKILL Multiply. (Page 590)

36. 6×15 37. 5×9 38. 12×3 39. 8×12

Ratios and Tangrams

What You'll LEARN

Explore ratios and the relationship between ratio and area.

Materials

• 2 sheets of patty paper
• scissors

INVESTIGATE *Work with a partner.*

A tangram is a puzzle that is made by cutting a square into seven geometric figures. The puzzle can be formed into many different figures.

In this lab, you will use a tangram to explore ratios and the relationship between ratio and area.

STEP 1 Begin with one sheet of patty paper. Fold the top left corner to the bottom right corner. Unfold and cut along the fold so that two large triangles are formed.

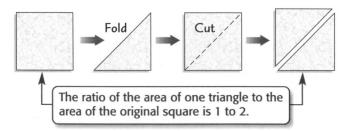

The ratio of the area of one triangle to the area of the original square is 1 to 2.

STEP 2 Use one of the cut triangles. Fold the bottom left corner to the bottom right corner. Unfold and cut along the fold. Label the triangles A and B.

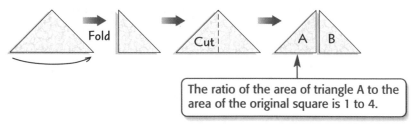

The ratio of the area of triangle A to the area of the original square is 1 to 4.

STEP 3 Use the other large triangle from step 1. Fold the bottom left corner to the bottom right corner. Make a crease and unfold. Next, fold the top down along the crease as shown. Make a crease and cut along the second crease line. Cut out the small triangle and label it C.

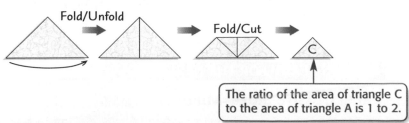

The ratio of the area of triangle C to the area of triangle A is 1 to 2.

STEP 4 Use the remaining piece. Fold it in half from left to right. Cut along the fold. Using the left figure, fold the bottom left corner to the bottom right corner. Cut along the fold and label the triangle D and the square E.

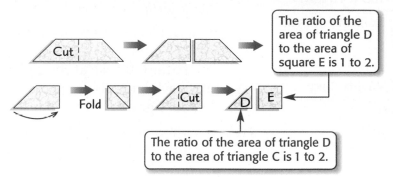

The ratio of the area of triangle D to the area of square E is 1 to 2.

The ratio of the area of triangle D to the area of triangle C is 1 to 2.

STEP 5 Use the remaining piece. Fold the bottom left corner to the top right corner. Cut along the fold. Label the triangle F and the other figure G.

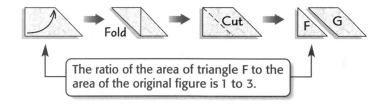

The ratio of the area of triangle F to the area of the original figure is 1 to 3.

Writing Math

Work with a partner.

1. Suppose the area of triangle B is 1 square unit. Find the area of each triangle below.

   a. triangle C   **b.** triangle F

2. Explain how the area of each of these triangles compares to the area of triangle B.

3. **Explain** why the ratio of the area of triangle C to the original large square is 1 to 8.

4. **Tell** why the area of square E is equal to the area of figure G.

5. **Find** the ratio of the area of triangle F to the original large square. Explain your reasoning.

6. Complete the table. Write the fraction that compares the area of each figure to the original square. What do you notice about the denominators?

Figure	A	B	C	D	E	F	G
Fractional Part of the Large Square							

Solving Proportions

HANDS-ON Mini Lab

Materials
• pattern blocks

Work with a partner.

Pattern blocks can be used to explore ratios that are equivalent. The pattern blocks at the right show how each large figure is made using smaller figures.

1. Complete each ratio so that the ratios comparing the areas are equivalent.

   a.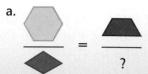

   b.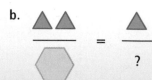

2. How did you find which figure made the ratios equivalent?

3. Suppose a green block equals 2, a blue block equals 4, a yellow block equals 6, and a red block equals 3. Write a pair of equivalent ratios.

4. What relationship exists in these equivalent ratios?

The ratios $\frac{4}{6}$ and $\frac{2}{3}$ are equivalent. That is, $\frac{4}{6} = \frac{2}{3}$. The equation $\frac{4}{6} = \frac{2}{3}$ is an example of a **proportion**.

Key Concept — Proportion

Words	A proportion is an equation stating that two ratios are equivalent.

Symbols

Arithmetic

$$\overset{\times 3}{\underset{\times 3}{\frac{2}{5} = \frac{6}{15}}}$$

Algebra

$$\frac{a}{b} = \frac{c}{d}, b \neq 0, d \neq 0$$

For two ratios to form a proportion, their **cross products** must be equal.

2 × 9 is one cross product. → $\frac{2}{3} \bowtie \frac{6}{9}$ ← 3 × 6 is the other cross product.

$$2(9) \overset{?}{=} 3(6)$$

$$18 = 18 \quad \longleftarrow \text{The cross products are equal.}$$

Key Concept — Property of Proportions

Words	The cross products of a proportion are equal.
Symbols	

Arithmetic	Algebra
If $\frac{2}{5} = \frac{6}{15}$, then $2 \times 15 = 5 \times 6$.	If $\frac{a}{b} = \frac{c}{d}$, then $ad = bc$.

When one value in a proportion is unknown, you can use cross products to *solve the proportion*.

STUDY TIP

Mental Math In some cases, you can solve a proportion mentally by using equivalent fractions. Consider the proportion $\frac{3}{4} = \frac{x}{16}$. Since $4 \times 4 = 16$ and $3 \times 4 = 12$, $x = 12$.

EXAMPLES — Solve a Proportion

Solve each proportion.

1 $\frac{5}{7} = \frac{25}{m}$

$5 \times m = 7 \times 25$ Cross products

$5m = 175$ Multiply.

$\frac{5m}{5} = \frac{175}{5}$ Divide each side by 5.

$m = 35$

The solution is 35.

2 $\frac{1.2}{1.5} = \frac{y}{5}$

$1.2 \times 5 = 1.5 \times y$ Cross products

$6 = 1.5y$ Multiply.

$\frac{6}{1.5} = \frac{1.5y}{1.5}$ Divide each side by 1.5.

$4 = y$

The solution is 4.

Your Turn Solve each proportion.

a. $\frac{5}{9} = \frac{z}{54}$ b. $\frac{5}{8} = \frac{40}{x}$ c. $\frac{k}{7} = \frac{18}{6}$

Proportions can be used to solve real-life problems.

REAL-LIFE CAREERS

How Does a Dentist Use Math?
Dentists use math when determining the amount of material needed to fill a cavity in a patient's tooth.

Online Research
For information about a career as a dentist, visit:
msmath1.net/careers

EXAMPLE — Use a Proportion to Solve a Problem

3 **TOOTHPASTE** Out of the 32 students in a health class, 24 prefer using gel toothpaste. Based on these results, how many of the 500 students in the school can be expected to prefer using gel toothpaste?

Write and solve a proportion. Let s represent the number of students who can be expected to prefer gel toothpaste.

prefer gel toothpaste → $\frac{24}{32} = \frac{s}{500}$ ← prefer gel toothpaste

total students in class →   ← total students in school

$24 \times 500 = 32 \times s$ Cross products

$12{,}000 = 32s$ Multiply.

$\frac{12{,}000}{32} = \frac{32s}{32}$ Divide.

$375 = s$

So, 375 students can be expected to prefer gel toothpaste.

Skill and Concept Check

Writing Math
Exercises 1–3

1. **Determine** whether each pair of ratios form a proportion. Explain your reasoning.

   a. $\dfrac{1}{8}, \dfrac{8}{64}$
   b. $\dfrac{7}{12}, \dfrac{8}{15}$
   c. $\dfrac{0.7}{0.9}, \dfrac{2.1}{2.7}$

2. **OPEN ENDED** Write a proportion with $\dfrac{7}{8}$ as one of the ratios.

3. **Which One Doesn't Belong?** Identify the ratio that does not form a proportion with the others. Explain your reasoning.

   $\dfrac{8}{12}$ $\dfrac{40}{60}$ $\dfrac{36}{44}$ $\dfrac{24}{36}$

GUIDED PRACTICE

Solve each proportion.

4. $\dfrac{5}{4} = \dfrac{a}{36}$
5. $\dfrac{3}{4} = \dfrac{x}{20}$
6. $\dfrac{w}{1.8} = \dfrac{3.5}{1.4}$

7. **SCHOOL** At West Boulevard Middle School, the teacher to student ratio is 3 to 78. If there are 468 students enrolled at the school, how many teachers are there at the school?

Practice and Applications

Solve each proportion.

8. $\dfrac{2}{5} = \dfrac{w}{15}$
9. $\dfrac{3}{4} = \dfrac{z}{28}$
10. $\dfrac{7}{d} = \dfrac{35}{10}$
11. $\dfrac{4}{x} = \dfrac{16}{28}$

12. $\dfrac{p}{3} = \dfrac{25}{15}$
13. $\dfrac{h}{8} = \dfrac{6}{16}$
14. $\dfrac{6}{7} = \dfrac{18}{c}$
15. $\dfrac{21}{35} = \dfrac{3}{r}$

16. $\dfrac{1.4}{2.6} = \dfrac{4.2}{n}$
17. $\dfrac{g}{4.7} = \dfrac{0.6}{9.4}$
18. $\dfrac{1.8}{b} = \dfrac{9}{2.5}$
19. $\dfrac{1.6}{6.4} = \dfrac{k}{1.6}$

HOMEWORK HELP

For Exercises	See Examples
8–21	1, 2
22–24, 26–28	3

Extra Practice
See pages 614, 633.

20. What is the solution of $\dfrac{1}{3} = \dfrac{x}{14}$? Round to the nearest tenth.

21. Find the solution of $\dfrac{m}{2} = \dfrac{5}{12}$ to the nearest tenth.

22. **MONEY** Suppose you buy 2 CDs for $21.99. How many CDs can you buy for $65.97?

SURVEYS For Exercises 23 and 24, use the table at the right. It shows which physical education class activities are favored by a group of students.

23. Write a proportion that could be used to find the number of students out of 300 that can be expected to pick sit-ups as their favorite physical education activity.

24. How many of the students can be expected to pick sit-ups as their favorite physical education class activity?

Favorite Physical Education Class Activity

Activity	Number of Responses
pull-ups	2
running	7
push-ups	3
sit-ups	8

PARENTS For Exercises 25–27, use the graphic that shows what grade parents gave themselves for their involvement in their children's education.

25. If $6\% = \dfrac{6}{100}$, what fraction of the parents gave themselves a B?

26. Suppose 500 parents were surveyed. Write a proportion that could be used to find how many of them gave themselves a B.

27. How many of the 500 parents gave themselves a B?

28. **PRIZES** A soda company is having a promotion. Every 3 out of 72 cases of soda contains a $5 movie rental certificate. If there are 384 cases of soda on display in a store, how many of the cases can be expected to contain a $5 movie rental certificate?

29. **CRITICAL THINKING** Suppose 24 out of 180 people said they like hiking, and 5 out of every 12 hikers buy Turf-Tuff hiking boots. In a group of 270 people, how many would you expect to have Turf-Tuff hiking boots?

USA TODAY

USA TODAY Snapshots®

Parents make the grade
The majority of parents give themselves A's or B's for involvement in their children's education. Parents assess their performance:

A (Superior) **38%**
B (Above Average) ✓ **42%**
C (Average) **17%**
D (Below Average) **2%**
F (Failing) **1%**

Source: Opinion Research Corp.

By In-Sung Yoo and Adrienne Lewis, USA TODAY

Standardized Test Practice and Mixed Review

Ⓐ Ⓑ Ⓒ Ⓓ

30. **MULTIPLE CHOICE** If you work 22 hours a week and earn $139.70, how much money do you earn per hour?

   Ⓐ $6.50   Ⓑ $6.35   Ⓒ $6.05   Ⓓ $5.85

31. **SHORT RESPONSE** If an airplane travels 438 miles per hour, how many miles will it travel in 5 hours?

Express each ratio as a unit rate. (Lesson 10-1)

32. 56 wins in 8 years   33. $12 for 5 hot dogs

Copy and complete each function table. Then graph the function.
(Lesson 9-7)

34.
Input	Output ($n - 3$)
−2	■
0	■
2	■

35.
Input	Output ($3n$)
−2	■
0	■
2	■

GETTING READY FOR THE NEXT LESSON

PREREQUISITE SKILL Multiply or divide. (Pages 590 and 591, Lessons 4-2 and 4-3)

36. 9×3 37. 1.5×4 38. $56 \div 4$ 39. $161.5 \div 19$

 msmath1.net/self_check_quiz

Spreadsheet Investigation

A Follow-Up of Lesson 10-2

What You'll LEARN

Use a spreadsheet to solve problems involving proportions.

Solving Proportions

Spreadsheets can be used to help solve proportion problems.

ACTIVITY

Your class is going to make peanut butter cocoa cookies for a school party. The ingredients needed to make enough cookies for 16 people are shown. Find how much of each ingredient is needed to make enough cookies for the school party.

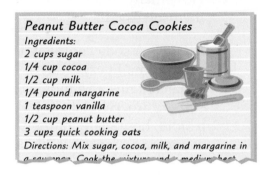

Peanut Butter Cocoa Cookies
Ingredients:
2 cups sugar
1/4 cup cocoa
1/2 cup milk
1/4 pound margarine
1 teaspoon vanilla
1/2 cup peanut butter
3 cups quick cooking oats
Directions: Mix sugar, cocoa, milk, and margarine in a saucepan. Cook the mixture over medium heat.

Set up a spreadsheet like the one shown to find the amount of ingredients needed to serve a given number of people.

Cell B1 is where you enter how many people will be served.

One recipe yields enough cookies for 16 people.

Peanut Butter Cocoa Cookies

	A	B	C	D
1	People to Serve			
2	Batches Needed	=B1/16		
3	Ingredient	Recipe	Amount	Unit
4	Sugar	2	=B4*B2	Cups
5	Cocoa Mix	0.25		Cups
6	Milk	0.5		Cups
7	Margarine	0.25		Pound
8	Vanilla	1		Teaspoon
9	Peanut Butter	0.5		Cups
10	Quick Cook Oats	3		Cups

Sheet1 She

The spreadsheet will calculate the amount of each ingredient you must have to make the number of cookies needed.

EXERCISES

1. Explain the formula in B2.
2. What does the formula in C4 represent?
3. What formulas should be entered in cells C5 through C10?
4. How does the spreadsheet use proportions?
5. Adjust your spreadsheet to find the amount of ingredients needed for 128 students.

10-3 Scale Drawings and Models

What You'll LEARN

Use scale drawings and models to find actual measurements.

NEW Vocabulary

scale drawing
scale model
scale

WHEN am I ever going to use this?

MAPS A map of a portion of Tennessee is shown. On the map, one inch equals 14 miles.

1. Explain how you would use a ruler to find the number of miles between any two cities on the map.

2. Use the method you described in Exercise 1 to find the actual distance between Haletown and Jasper.

3. What is the actual distance between Kimball and Signal Mountain?

A map is an example of a scale drawing. **Scale drawings** and **scale models** are used to represent objects that are too large or too small to be drawn or built at actual size.

The **scale** gives the ratio that compares the measurements on the drawing or model to the measurements of the real object. The measurements on a drawing or model are proportional to measurements on the actual object.

EXAMPLE Find Actual Measurements

1 **INSECTS** A scale model of a firefly has a scale of 1 inch = 0.125 inch. If the length of the firefly on the model is 3 inches, what is the actual length of the firefly?

Let x represent the actual length.

Scale Model   **Firefly**

model length → $\dfrac{1}{0.125} = \dfrac{3}{x}$ ← model length
actual length →   ← actual length

$$1 \times x = 0.125 \times 3 \quad \text{Find the cross products.}$$

$$x = 0.375 \quad \text{Multiply.}$$

The actual length of the firefly is 0.375 inch.

---

2 **STATES** On a map of Arizona, the distance between Meadview and Willow Beach is 14 inches. If the scale on the map is 2 inches = 5 miles, what is the actual distance between Meadview and Willow Beach?

Let d represent the actual distance.

	Map Scale		Actual Distance

map distance → $\dfrac{2}{5} = \dfrac{14}{d}$ ← map distance
actual distance → $\phantom{\dfrac{2}{5}}$ ← actual distance

$$2 \times d = 5 \times 14 \qquad \text{Find the cross products.}$$
$$2d = 70 \qquad \text{Multiply.}$$
$$\frac{2d}{2} = \frac{70}{2} \qquad \text{Divide.}$$
$$d = 35$$

The distance between Meadview and Willow Beach is 35 miles.

Skill and Concept Check

1. **Describe** the scale given in a scale drawing or a scale model.

2. **OPEN ENDED** Give an example of an object that is often shown as a scale model.

3. **FIND THE ERROR** Greg and Jeff are finding the actual distance between Franklin and Ohltown on a map. The scale is 1 inch = 12 miles, and the distance between the cities on the map is 3 inches. Who is correct? Explain.

Greg
$\dfrac{1}{12} = \dfrac{3}{x}$

Jeff
$\dfrac{1}{12} = \dfrac{x}{3}$

GUIDED PRACTICE

ARCHITECTURE For Exercises 4–7, use the following information.
On a set of blueprints, the scale is 2 inches = 3 feet. Find the actual length of each object on the drawing.

	Object	Drawing Length
4.	porch	4 inches
5.	window	3 inches

	Object	Drawing Length
6.	garage door	15 inches
7.	chimney	0.5 inch

8. **TREES** A model of a tree has a height of 4 inches. If the scale of the tree is 1 inch = 3 feet, what is the actual height of the tree?

9. **HOUSES** A scale model of a house has a scale of 1 inch = 2.5 feet. If the width of the house on the model is 12 inches, what is the actual width of the house?

Practice and Applications

HOMEWORK HELP

For Exercises	See Examples
10–14	1, 2

Extra Practice
See pages 614, 633.

BICYCLES On a scale model of a bicycle, the scale is 1 inch = 0.5 foot. Find the actual measurements.

	Object	Model Measurement
10.	diameter of the wheel	4.5 inches
11.	height of the bicycle	7 inches

INSECTS On a scale drawing of a praying mantis, the scale is 1 inch = $\frac{3}{4}$ inch. Find the actual measurements.

	Praying Mantis	Model Measurement
12.	height	$2\frac{1}{2}$ inches
13.	body length	$1\frac{3}{4}$ inches

14. **HISTORY** A model of the Titanic has a length of 2.5 feet. If the scale of the ship is 1 foot = 350 feet, what is the actual length of the Titanic?

15. **CRITICAL THINKING** Some toys that replicate actual vehicles have scales of 1:10, 1:18, or 1:64. For a model representing a motorcycle, which scale would be best to use? Explain.

Standardized Test Practice and Mixed Review

16. **MULTIPLE CHOICE** A drawing of a paperclip has a scale of 1 inch = $\frac{1}{8}$ inch. Find the actual length of the paperclip if the length on the drawing is 10 inches.

   Ⓐ 2 in.   Ⓑ $1\frac{1}{2}$ in.   Ⓒ $1\frac{1}{4}$ in.   Ⓓ $\frac{1}{2}$ in.

17. **MULTIPLE CHOICE** A drawing of a room measures 8 inches by 10 inches. If the scale is 1 inch = 5 feet, find the dimensions of the room.

   Ⓕ 40 ft by 50 ft   Ⓖ 35 ft by 45 ft
   Ⓗ 20 ft by 25 ft   Ⓘ 4 ft by 5 ft

18. Solve $\frac{7.3}{h} = \frac{14.6}{10.8}$. (Lesson 10-2)

Express each ratio as a fraction in simplest form. (Lesson 10-1)

19. 2 out of 18 games played   20. 180 out of 365 days worked

GETTING READY FOR THE NEXT LESSON

PREREQUISITE SKILL Model each fraction. (Lesson 5-3)

21. $\frac{1}{2}$ 22. $\frac{1}{4}$ 23. $\frac{3}{4}$ 24. $\frac{3}{5}$ 25. $\frac{2}{3}$

10-3b 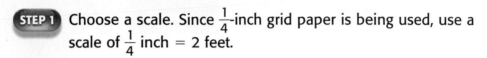 HANDS-ON LAB   *A Follow-Up of Lesson 10-3*

What You'll LEARN

Construct scale drawings.

Materials

• grid paper

Construct Scale Drawings

INVESTIGATE *Work with a partner.*

Jordan's bedroom measures 16 feet long and 12 feet wide. A scale drawing of the room can be drawn so that it is proportional to the actual room. In this lab, you will construct a scale drawing of Jordan's room.

STEP 1 Choose a scale. Since $\frac{1}{4}$-inch grid paper is being used, use a scale of $\frac{1}{4}$ inch = 2 feet.

STEP 2 Find the length and width of the room on the scale drawing. The scale tells us that each unit represents 2 feet. Since the room is 16 feet long, divide 16 by 2. Since the room is 12 feet wide, divide 12 by 2.

$$16 \div 2 = 8 \qquad 12 \div 2 = 6$$

STEP 3 Construct the scale drawing. On the drawing, the length of the room is 8 units and the width is 6 units.

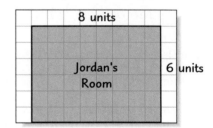

Your Turn

a. A rectangular flower bed is 4 feet wide and 14 feet long. Make a scale drawing of the flower bed that has a scale of $\frac{1}{4}$ inch = 2 feet.

b. A playground has dimensions 150 feet wide and 75 feet long. Make a scale drawing of the playground that has a scale of $\frac{1}{4}$ inch = 10 feet.

Writing Math

1. **Explain** how the scale is used to determine the dimensions of the object on the scale drawing.

2. **Describe** $\frac{1}{2}$-inch grid paper.

3. Suppose you were making a scale drawing of a football field. What size grid paper would you use? What would be an appropriate scale?

10-4 Modeling Percents

What You'll LEARN

Use models to illustrate the meaning of percent.

NEW Vocabulary

percent

MATH Symbols

% percent

WHEN am I ever going to use this?

CANDY Kimi asked 100 students in the cafeteria to tell which lollipop flavor was their favorite, cherry, grape, orange, or lime. The results are shown in the bar graph at the right.

1. What ratio compares the number of students who prefer grape flavored lollipops to the total number of students?

2. What decimal represents this ratio?

3. Draw a decimal model to represent this ratio.

Favorite Lollipop Flavors

(Bar graph: Number of Students vs Flavor. Cherry = 32, Grape = 45, Orange = 18, Lime = 5)

Ratios like 32 out of 100, 45 out of 100, 18 out of 100, or 5 out of 100, can be written as percents. A **percent** (%) is a ratio that compares a number to 100.

> ### Key Concept   Percent
>
> **Words**   A percent is a ratio that compares a number to 100.
>
> **Symbols**   75% = 75 out of 100

In Lesson 3-1, you learned that a 10 × 10 grid can be used to represent *hundredths*. Since the word percent means *out of one hundred*, you can also use a 10 × 10 grid to model percents.

STUDY TIP

Percent To model 100%, shade all of the squares since 100% means 100 out of 100.

EXAMPLE   Model a Percent

1 Model 18%.

18% means 18 out of 100.

So, shade 18 of the 100 squares.

Your Turn Model each percent.

a. 75%   b. 8%   c. 42%

You can use what you know about decimal models and percents to identify the percent of a model that is shaded.

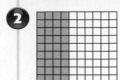

Identify each percent that is modeled.

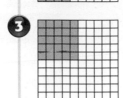 There are 40 out of 100 squares shaded.
So, the model shows 40%.

 There are 25 out of 100 squares shaded.
So, the model shows 25%.

 Your Turn Identify each percent modeled.

d.   e.   f.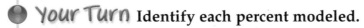

Skill and Concept Check

Writing Math
Exercises 1 & 3

1. **Explain** what it means if you have 50% of a pizza.

2. **OPEN ENDED** Draw a model that shows 23%.

3. **NUMBER SENSE** Santino has 100 marbles, and he gives 43% of them to Michael. Would it be reasonable to say that Santino gave Michael less than 50 marbles? Explain?

GUIDED PRACTICE

Model each percent.

4. 85%   5. 43%   6. 4%

Identify each percent that is modeled.

7. 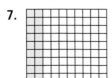   8. 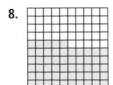   9.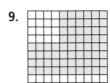

10. **MUSIC** Of the 100 CDs in a CD case, 67% are pop music and 33% are country. For which type of CDs are there more in the case? Use a model in your explanation.

Practice and Applications

HOMEWORK HELP

For Exercises	See Examples
11–16, 23	1
17–22	2

Extra Practice
See pages 614, 633.

Model each percent.

11. 15%   **12.** 65%   **13.** 48%   **14.** 39%   **15.** 9%   **16.** 3%

Identify each percent that is modeled.

17.   **18.**   **19.**

20.   **21.**   **22.**

23. SNOWBOARDING At a popular ski resort, 35% of all people who buy tickets are snowboarders. Make a model to show 35%.

24. Use a model to show which percent is greater, 27% or 38%.

25. CRITICAL THINKING The size of a photograph is increased 200%. Model 200%. What does an increase of 200% mean?

Standardized Test Practice and Mixed Review

For Exercises 26 and 27, use the table at the right.

26. MULTIPLE CHOICE How much time do most 13-year olds spend studying?

    Ⓐ do not study at all   Ⓑ less than 1 h

    Ⓒ 1–2 h   Ⓓ more than 2 h

Nightly Study Time for 13-year olds	
Time	**Percent**
Do not study	24%
Less than 1 hour	37%
1–2 hours	26%
More than 2 hours	8%

Source: National Center for Education Statistics

27. SHORT RESPONSE Which study time has the least percent of students?

28. GEOGRAPHY On a map, 1 inch = 20 miles. If the distance on the map between two cities is $2\frac{3}{4}$ inches, what is the actual distance? (Lesson 10-3)

Solve each proportion. (Lesson 10-2)

29. $\frac{2}{5} = \frac{x}{15}$ **30.** $\frac{x}{10} = \frac{18}{30}$ **31.** $\frac{2.5}{8} = \frac{10}{x}$

GETTING READY FOR THE NEXT LESSON

PREREQUISITE SKILL Write each fraction in simplest form. (Lesson 5-2)

32. $\frac{26}{100}$ **33.** $\frac{54}{100}$ **34.** $\frac{10}{100}$ **35.** $\frac{75}{100}$

 msmath1.net/self_check_quiz

Vocabulary and Concepts

1. **Define** *ratio*. (Lesson 10-1)

2. **State** the property of proportions. (Lesson 10-2)

Skills and Applications

Write each ratio as a fraction in simplest form. (Lesson 10-1)

3. 12 boys out of 20 students

4. 15 cookies to 40 brownies

Write each ratio as a unit rate. (Lesson 10-1)

5. 171 miles in 3 hours

6. $15 for 3 pounds

Solve each proportion. (Lesson 10-2)

7. $\dfrac{x}{6} = \dfrac{12}{18}$

8. $\dfrac{8}{20} = \dfrac{30}{x}$

9. $\dfrac{3}{d} = \dfrac{9}{4.8}$

10. $\dfrac{2.4}{7.2} = \dfrac{x}{3.6}$

11. **HEALTH** Suppose 27 out of 50 people living in one neighborhood of a community exercise regularly. How many people in a similar community of 2,600 people can be expected to exercise regularly? (Lesson 10-2)

ANIMALS A model of an African elephant has a scale of 1 inch = 2 feet. Find the actual dimensions of the elephant. (Lesson 10-3)

	Feature	Model Length
12.	trunk	4 inches
13.	shoulder height	7 inches
14.	ear	2 inches
15.	tusk	5 inches

Identify each percent modeled. (Lesson 10-4)

16.

17.

18.

Standardized Test Practice

19. **GRID IN** A team made four of 10 attempted goals. Which ratio compares the goals made to the goals attempted? (Lesson 10-1)

20. **SHORT RESPONSE** Use a model to explain which is less, 25% or 20%. (Lesson 10-4)

The Game Zone

A Place To Practice Your Math Skills

Fishin' for Ratios

● GET READY!

Players: two or three
Materials: scissors, 18 index cards

● GET SET!

- Cut all index cards in half.
- Write the ratios shown on half of the cards.
- Write a ratio equivalent to each of these ratios on the remaining cards.
- Two cards with equivalent ratios are considered matching cards.

$\frac{1}{2}$	$\frac{1}{4}$	$\frac{2}{3}$	$\frac{3}{4}$	$\frac{5}{8}$	$\frac{1}{3}$
$\frac{2}{5}$	$\frac{3}{7}$	$\frac{1}{5}$	$\frac{4}{5}$	$\frac{3}{5}$	$\frac{7}{8}$
$\frac{5}{7}$	$\frac{5}{9}$	$\frac{1}{8}$	$\frac{3}{8}$	$\frac{2}{7}$	$\frac{2}{9}$

● GO!

- Shuffle the cards. Then deal 7 cards to each player. Place the remaining cards facedown in a pile. Players set aside any pairs of matching cards that they were dealt.
- The first player asks for a matching card. If a match is made, then the player sets aside the match, and it is the next player's turn. If no match is made, then the player picks up the top card from the pile. If a match is made, then the match is set aside, and it is the next player's turn. If no match is made, then it is the next player's turn.
- **Who Wins?** After all of the cards have been drawn or when a player has no more cards, the player with the most matches wins.

What You'll LEARN

Express percents as fractions and vice versa.

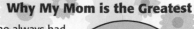

WHEN am I ever going to use this?

SURVEYS A group of adults were asked to give a reason why they honor their mom.

1. What was the second most popular reason?

2. What percent represents this section of the graph?

3. Based on the meaning of 22%, make a conjecture as to how you would write this percent as a fraction.

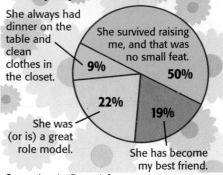

Why My Mom is the Greatest

She always had dinner on the table and clean clothes in the closet.

She survived raising me, and that was no small feat. **50%**

9%

22%

19%

She was (or is) a great role model.

She has become my best friend.

Source: Impulse Research Corp.

All percents can be written as fractions in simplest form.

> ### Key Concept
> **Percent to Fraction**
>
> To write a percent as a fraction, write the percent as a fraction with a denominator of 100. Then simplify.

EXAMPLES Write a Percent as a Fraction

Write each percent as a fraction in simplest form.

1 **50%**

50% means *50 out of 100.*

$50\% = \dfrac{50}{100}$ Write the percent as a fraction with a denominator of 100.

$= \dfrac{\overset{1}{\cancel{50}}}{\underset{2}{\cancel{100}}}$ or $\dfrac{1}{2}$ Simplify. Divide the numerator and the denominator by the GCF, 50.

STUDY TiP

Percents A percent can be greater than 100%. Since percent means *hundredths*, or *per 100*, a percent like 150% means 150 hundredths, or 150 per 100.

2 **125%**

125% means *125 for every 100.*

$125\% = \dfrac{125}{100}$

$= 1\dfrac{\overset{1}{\cancel{25}}}{\underset{4}{\cancel{100}}}$ or $1\dfrac{1}{4}$

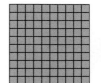

Your Turn Write each percent as a fraction in simplest form.

a. 10%   b. 97%   c. 135%

EXAMPLE **Write a Percent as a Fraction**

3 PATRIOTISM Use the table at the right. What fraction of those surveyed are extremely proud to be American?

The table shows that 65% of adults are extremely proud to be an American.

$65\% = \dfrac{65}{100}$ Write the percent as a fraction with a denominator of 100.

$ = \dfrac{13}{20}$ Simplify.

So, $\dfrac{13}{20}$ of those surveyed are extremely proud to be American.

Proud To Be An American	
Answer	**Percent**
no opinion	1%
a little/not at all	3%
moderately	6%
very	25%
extremely	65%

Source: Gallup Poll

Fractions can be written as percents. To write a fraction as a percent, write a proportion and solve it.

EXAMPLES **Write a Fraction as a Percent**

4 Write $\dfrac{9}{10}$ as a percent.

$\dfrac{9}{10} = \dfrac{n}{100}$ Set up a proportion.

$9 \times 100 = 10 \times n$ Write the cross products.

$900 = 10n$ Multiply.

$\dfrac{900}{10} = \dfrac{10n}{10}$ Divide each side by 10.

$90 = n$ Simplify.

So, $\dfrac{9}{10}$ is equivalent to 90%.

5 Write $\dfrac{7}{5}$ as a percent.

$\dfrac{7}{5} = \dfrac{c}{100}$ Set up a proportion.

$5 \times c = 7 \times 100$ Write the cross products.

$5c = 700$ Multiply.

$\dfrac{5c}{5} = \dfrac{700}{5}$ Divide each side by 5.

$c = 140$ Simplify.

So, $\dfrac{7}{5}$ is equivalent to 140%.

STUDY TIP

Percents
Remember that a percent is a number compared to 100. So, one ratio in the proportion is the fraction. The other ratio is an unknown number compared to 100.

Your Turn Write each fraction as a percent.

d. $\dfrac{3}{5}$ e. $\dfrac{1}{4}$ f. $\dfrac{1}{5}$

Skill and Concept Check

Writing Math
Exercises 1 & 2

1. **Explain** how to write any percent as a fraction.

2. **Which One Doesn't Belong?** Identify the number that does not have the same value as the other three. Explain your reasoning.

| 25% | $\frac{2}{8}$ | $\frac{7}{25}$ | $\frac{25}{100}$ |

3. **NUMBER SENSE** List three fractions that are less than 75%.

GUIDED PRACTICE

Write each percent as a fraction in simplest form.

4. 15%   5. 80%   6. 180%

Write each fraction as a percent.

7. $\frac{1}{4}$ 8. $\frac{2}{5}$ 9. $\frac{9}{4}$

10. **SOCCER** During the 2002 regular season, the Atlanta Beat women's soccer team won about 52% of their games. What fraction of their games did they win?

Practice and Applications

Write each percent as a fraction in simplest form.

11. 14%   12. 47%   13. 2%
14. 20%   15. 185%   16. 280%

Write each fraction as a percent.

17. $\frac{7}{10}$ 18. $\frac{7}{20}$ 19. $\frac{5}{4}$ 20. $\frac{7}{4}$
21. $\frac{1}{100}$ 22. $\frac{5}{100}$ 23. $\frac{3}{8}$ 24. $\frac{5}{6}$

25. **MONEY** What percent of a dollar is a nickel?

26. **MONEY** What percent of a dollar is a penny?

27. Write *ninety-eight percent* as a fraction in simplest form.

28. How is *sixty-four hundredths* written as a percent?

BASKETBALL For Exercises 29 and 30, use the table at the right.

29. What percent of the baskets did Kendra make?

30. What fraction of the baskets did Kendra miss?

HOMEWORK HELP

For Exercises	See Examples
11–16, 27	1, 2
30–32	3
17–26, 28–29	4, 5

Extra Practice
See pages 615, 633.

Kendra's Basketball Chart	
Baskets Made	Baskets Missed
ЖЖ ЖЖ ЖЖ III	ЖЖ II

SURVEY For Exercises 31–33, use the graph that shows how pressured parents feel about making sure their children have the things that other children have.

31. What fraction of the parents do not feel pressured? Write the fraction in simplest form.

32. What fraction of the parents feel not very pressured? Write the fraction in simplest form.

33. Write a sentence describing what fraction of the parents surveyed feel very pressured.

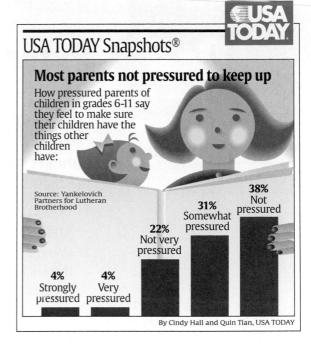

USA TODAY Snapshots®

Most parents not pressured to keep up

How pressured parents of children in grades 6-11 say they feel to make sure their children have the things other children have:

Source: Yankelovich Partners for Lutheran Brotherhood

38% Not pressured

31% Somewhat pressured

22% Not very pressured

4% Strongly pressured

4% Very pressured

By Cindy Hall and Quin Tian, USA TODAY

34. **CRITICAL THINKING** The table shows what fraction of the daily chores a father assigned to his son and daughters. If the remaining chores are for the father to complete, what percent of chores was left for him? Round to the nearest whole percent.

Person	son	daughter	daughter
Fraction	$\frac{1}{2}$	$\frac{1}{3}$	$\frac{1}{7}$

Standardized Test Practice and Mixed Review

35. **MULTIPLE CHOICE** Four-fifths of the sixth-grade students have siblings. What percent of these students do *not* have siblings?

　Ⓐ 15%　　　　Ⓑ 20%　　　　Ⓒ 25%　　　　Ⓓ 80%

36. **MULTIPLE CHOICE** Suppose 75% of teenagers use their home computers for homework. What fraction of teenagers is this?

　Ⓕ $\frac{3}{4}$　　　　Ⓖ $\frac{7}{10}$　　　　Ⓗ $\frac{3}{5}$　　　　Ⓘ $\frac{1}{4}$

Model each percent. (Lesson 10-4)

37. 32%　　　　　　　38. 65%　　　　　　　39. 135%

40. **ROLLER COASTERS** On a model of a roller coaster, the scale is 1 inch = 2 feet. If the width of the track on the model is 2.5 inches, what is the actual width? (Lesson 10-3)

GETTING READY FOR THE NEXT LESSON

PREREQUISITE SKILL Write each fraction as a decimal. (Lesson 5-7)

41. $\frac{65}{100}$　　　　42. $\frac{1}{8}$　　　　43. $\frac{0.5}{100}$　　　　44. $\frac{1}{5}$

Percents and Decimals

What You'll LEARN

Express percents as decimals and vice versa.

WHEN am I ever going to use this?

BUDGETS The graph shows the Balint's monthly budget.

1. What percent does the circle graph represent?

2. What fraction represents the section of the graph labeled rent?

3. Write the fraction from Exercise 2 as a decimal.

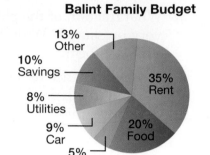

Balint Family Budget

13% Other
10% Savings
8% Utilities
9% Car
5% Clothes
35% Rent
20% Food

Percents can be written as decimals.

> **Key Concept**   **Percent as Decimal**
>
> To write a percent as a decimal, rewrite the percent as a fraction with a denominator of 100. Then write the fraction as a decimal.

EXAMPLES Write a Percent as a Decimal

Write each percent as a decimal.

1 **56%**

$$56\% = \frac{56}{100}$$ Rewrite the percent as a fraction with a denominator of 100.

$$= 0.56$$ Write the fraction as a decimal.

2 **120%**

$$120\% = \frac{120}{100}$$ Rewrite the percent as a fraction with a denominator of 100.

$$= 1.2$$ Write the fraction as a decimal.

3 **0.3%**

0.3% means three-tenths of one percent.

$$0.3\% = \frac{0.3}{100}$$ Rewrite the percent as a fraction with a denominator of 100.

$$= \frac{0.3}{100} \times \frac{10}{10}$$ Multiply by $\frac{10}{10}$ to eliminate the decimal in the numerator.

$$= \frac{3}{1,000} \text{ or } 0.003$$ Write the fraction as a decimal.

STUDY TiP

Mental Math To write a percent as a decimal, you can use a shortcut. Move the decimal point two places to the left, which is the same as dividing by 100.

Your Turn Write each percent as a decimal.

    a. 32%   **b.** 190%   **c.** 0.6%

You can also write a decimal as a percent.

> ## Key Concept   **Decimal as Percent**
>
> To write a decimal as a percent, write the decimal as a fraction whose denominator is 100. Then write the fraction as a percent.

EXAMPLES   **Write a Decimal as a Percent**

Write each decimal as a percent.

④ 0.38

$0.38 = \dfrac{38}{100}$ Write the decimal as a fraction.

$= 38\%$ Write the fraction as a percent.

⑤ 0.189

$0.189 = \dfrac{189}{1,000}$ Write the decimal as a fraction.

$= \dfrac{189 \div 10}{1,000 \div 10}$ Divide the numerator and the denominator by 10 to get a denominator of 100.

$= \dfrac{18.9}{100}$ or 18.9%   Write the fraction as a percent.

 Your Turn Write each decimal as a percent.

    d. 0.47   **e.** 0.235   **f.** 1.75

STUDY TIP

Mental Math To write a decimal as a percent, you can use this shortcut. Move the decimal point two places to the right, which is the same as multiplying by 100.

Skill and Concept Check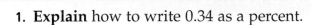

Writing Math
Exercises 1 & 2

1. **Explain** how to write 0.34 as a percent.

2. **Which One Doesn't Belong?** Identify the decimal that cannot be written as a percent greater than 1. Explain your reasoning.

0.4	0.048	0.0048	0.484

GUIDED PRACTICE

Write each percent as a decimal.

  3. 27%   **4.** 15%   **5.** 0.9%   **6.** 115%

Write each decimal as a percent.

  7. 0.32   **8.** 0.15   **9.** 0.125   **10.** 0.291

11. PASTA According to the *American Pasta Report,* 12% of Americans say that lasagna is their favorite pasta. What decimal is equivalent to 12%?

Practice and Applications

HOMEWORK HELP

For Exercises	See Examples
12–19, 29, 30, 31	1, 2, 3
20–27, 28	4, 5

Extra Practice
See pages 615, 633.

Express each percent as a decimal.

12. 2% **13.** 6% **14.** 17% **15.** 35%

16. 0.7% **17.** 0.3% **18.** 125% **19.** 104%

Express each decimal as a percent.

20. 0.5 **21.** 0.4 **22.** 0.22 **23.** 0.99

24. 0.175 **25.** 0.355 **26.** 0.106 **27.** 0.287

28. How is seventy-two thousandths written as a percent?

29. Write four and two tenths percent as a decimal.

30. **LIFE SCIENCE** About 95% of all species of fish have skeletons made of bone. Write 95% as a decimal.

31. **TAXES** The sales tax in Allen County is 5%. Write 5% as a decimal.

 Data Update Use the Internet or another source to find the sales tax for your state. Visit: **msmath1.net/data_update** to learn more.

Replace each ● with <, >, or = to make a true sentence.

32. 25% ● 0.20 **33.** 0.46 ● 46% **34.** 2.3 ● 23%

CRITICAL THINKING

35. Order 23.4%, 2.34, 0.0234, and 20.34% from least to greatest.

36. Order $2\frac{1}{4}$, 0.6, 2.75, 40%, and $\frac{7}{5}$ from greatest to least.

37. Graph $\frac{2}{5}$, 1, 0.5, 30%, −1, 2.0%, on a number line.

Standardized Test Practice and Mixed Review

38. **MULTIPLE CHOICE** Which percent is greater than 0.5?
　Ⓐ 56%　　Ⓑ 49%　　Ⓒ 45%　　Ⓓ 44%

39. **SHORT RESPONSE** The sales tax on the baseball cap Tionna is buying is 8.75%. Write the percent as a decimal.

Write each percent as a fraction in simplest form. (Lesson 10-5)

40. 24% **41.** 38% **42.** 125% **43.** 35%

44. 36 out of 100 is what percent? (Lesson 10-4)

GETTING READY FOR THE NEXT LESSON

PREREQUISITE SKILL Multiply. (Lesson 7-2)

45. $\frac{1}{5} \times 200$　　**46.** $\frac{1}{2} \times 1{,}500$　　**47.** $\frac{3}{5} \times 35$　　**48.** $\frac{3}{4} \times 32$

msmath1.net/self_check_quiz

Percent of a Number

At a department store, a backpack is on sale for 30% off the original price. If the original price of the backpack is $50, how much will you save?

In this situation, you know the percent. You need to find what part of the original price you will save. To find the percent of a number by using a model, follow these steps:

- Draw a percent model that represents the situation.
- Use the percent model to find the percent of the number.

ACTIVITY *Work with a partner.*

1 Use a model to find 30% of $50.

STEP 1 Draw a rectangle as shown on grid paper. Since percent is a ratio that compares a number to 100, label the units on the right from 0% to 100% as shown.

STEP 2 Since $50 represents the original price, mark equal units from $0 to $50 on the left side of the model as shown.

STEP 3 Draw a line from 30% on the right side to the left side of the model as shown.

The model shows that 30% of $50 is $15. So, you will save $15.

Your Turn

Draw a model to find the percent of each number.

a. 20% of 120   **b.** 60% of 70   **c.** 90% of 400

Suppose a bicycle is on sale for 35% off the original price. How much will you save if the original price of the bicycle is $180?

ACTIVITY *Work with a partner.*

2 Use a model to find 35% of $180.

STEP 1 Draw a rectangle as shown on grid paper. Label the units on the right from 0% to 100% to represent the percents as shown.

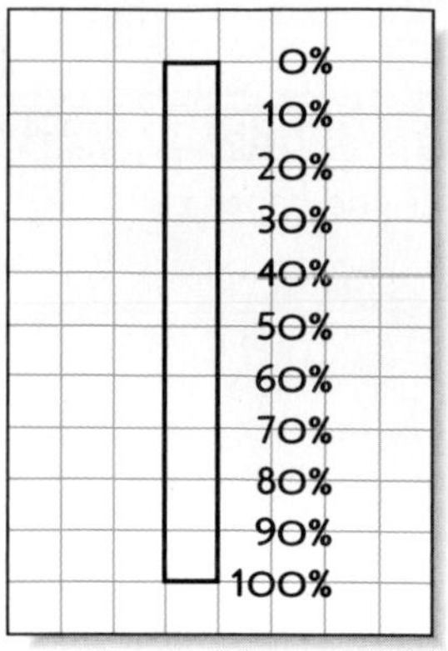

0%
10%
20%
30%
40%
50%
60%
70%
80%
90%
100%

STEP 2 The original price is $180. So, mark equal units from $0 to $180 on the left side of the model as shown.

STEP 3 Draw a line from 35% on the right side to the left side of the model.

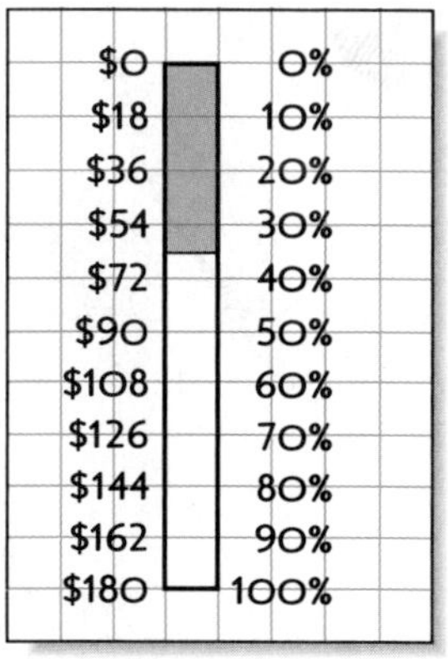

The model shows that 35% of $180 is halfway between $54 and $72, or $63. So, you will save $63.

$0	0%
$18	10%
$36	20%
$54	30%
$72	40%
$90	50%
$108	60%
$126	70%
$144	80%
$162	90%
$180	100%

Your Turn Draw a model to find the percent of each number. If it is not possible to find an exact answer from the model, estimate.

d. 25% of 140   **e.** 7% of 50   **f.** 0.5% of 20

Writing Math

1. **Explain** how to determine the units that get labeled on the left side of the percent model.

2. **Write** a sentence explaining how you can find 7% of 50.

3. **Explain** how knowing 10% of a number will help you find the percent of a number when the percent is a multiple of 10%.

4. **Explain** how knowing 10% of a number can help you determine whether a percent of a number is a reasonable amount.

Percent of a Number

What You'll LEARN

Find the percent of a number.

WHEN am I ever going to use this?

SAFETY A local police department wrote a report on how fast over the speed limit cars were traveling in a school zone. The results are shown in the graph.

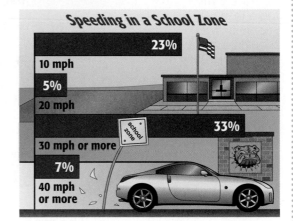

Speeding in a School Zone

10 mph — 23%
20 mph — 5%
30 mph or more — 33%
40 mph or more — 7%

1. What percent of the cars were traveling 20 miles per hour over the speed limit?

2. Write a multiplication sentence that involves a percent that could be used to find the number of cars out of 300 that were traveling 20 miles an hour over the speed limit.

To find the percent of a number such as 23% of 300, 33% of 300, or 7% of 300, you can use one of the following methods.

• Write the percent as a fraction and then multiply, or
• Write the percent as a decimal and then multiply.

EXAMPLE Find the Percent of a Number

 Find 5% of 300.

To find 5% of 300, you can use either method.

Method 1 Write the percent as a fraction.	**Method 2** Write the percent as a decimal.
$5\% = \frac{5}{100}$ or $\frac{1}{20}$	$5\% = \frac{5}{100}$ or 0.05
$\frac{1}{20}$ of $300 = \frac{1}{20} \times 300$ or 15	0.05 of $300 = 0.05 \times 300$ or 15

So, 5% of 300 is 15. Use a model to check the answer.

The model confirms that 5% of 300 is 15.

STUDY TIP

Percent of a Number
A calculator can also be used to find the percent of a number. For example, to find 5% of 300, push
5 [2nd] [%] [×] 300 [ENTER =] .
The result is 15.

Find the Percent of a Number

2 Find 120% of 75.

Method 1 Write the percent as a fraction.	**Method 2** Write the percent as a decimal.

Method 1 Write the percent as a fraction.

$120\% = \frac{120}{100}$ or $1\frac{1}{5}$

$1\frac{1}{5}$ of $75 = 1\frac{1}{5} \times 75$

$= \frac{6}{5} \times 75$

$= \frac{6}{5} \times \frac{75}{1}$ or 90

Method 2 Write the percent as a decimal.

$120\% = \frac{120}{100}$ or 1.2

1.2 of $75 = 1.2 \times 75$ or 90

So, 120% of 75 is 90. Use a model to check the answer.

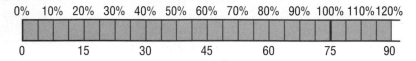

The model confirms that 120% of 75 is 90.

3 **STATISTICS** The graphic shows that 12.2% of college students majoring in medicine say they couldn't leave home for college without their stuffed animals. If a college has 350 students majoring in medicine, how many can be expected to have stuffed animals in their dorm room?

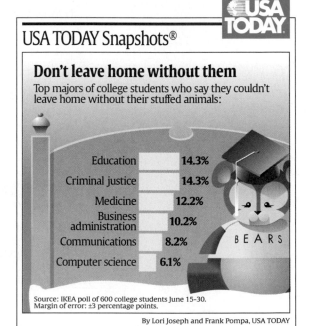

To find 12.2% of 350, write the percent as a decimal. Then use a calculator to multiply.

$12.2\% = \frac{12.2}{100}$ or 0.122   Divide 12.2 by 100 to get 0.122.

0.122 of $350 = 0.122 \times 350 = 42.7$ Use a calculator.

So, about 43 students can be expected to have stuffed animals in their dorm room.

Your Turn Find the percent of each number.

a. 55% of 160   **b.** 140% of 125   **c.** 0.3% of 500

Skill and Concept Check

1. **Explain** how to find 40% of 65 by changing the percent to a decimal.

2. **OPEN ENDED** Write a problem in which the percent of the number results in a number greater than the number itself.

3. **FIND THE ERROR** Gary and Belinda are finding 120% of 60. Who is correct? Explain your reasoning.

Gary
120% of 60 = $1\frac{1}{5} \times 60$
= 72

Belinda
120% of 60 = 12.0 × 60
= 720

GUIDED PRACTICE

Find the percent of each number.

4. 30% of 90   5. 50% of 78   6. 4% of 65

7. 7% of 7   8. 150% of 38   9. 0.4% of 20

10. **MONEY** A skateboard is on sale for 85% of the regular price. If it is regularly priced at $40, how much is the sale price?

Practice and Applications

Find the percent of each number.

11. 15% of 60   12. 12% of 800   13. 75% of 120

14. 25% of 80   15. 2% of 25   16. 4% of 9

17. 7% of 85   18. 3% of 156   19. 150% of 90

20. 125% of 60   21. 0.5% of 85   22. 0.3% of 95

23. What is 78% of 265?   24. Find 24% of 549.

HOMEWORK HELP

For Exercises	See Examples
11–24	1, 2, 3
25–32	3

Extra Practice
See pages 615, 633.

25. **BOOKS** Chad and Alisa donated 30% of their book collection to a local children's hospital. If they had 180 books, how many did they donate to the hospital?

26. **FOOTBALL** The Mooney High School football team won 75% of their football games. If they played 12 games, how many did they win?

SCHOOL For Exercises 27–29, use the diagram at the right that shows Sarah's and Morgan's test scores.

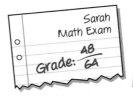

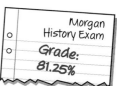

Sarah
Math Exam
Grade: $\frac{48}{64}$

Morgan
History Exam
Grade: 81.25%

27. What percent of the questions did Sarah score correctly?

28. What percent did Sarah score incorrectly?

29. If there were 64 questions on the test, how many did Morgan answer correctly?

SAFETY For Exercises 30–32, use the graph that shows what percent of the 50 states require motorcycle riders to wear a helmet.

Motorcycle Safety

Helmets not required 6%

Helmets required for riders under 18 54%

Helmets required for all riders 40%

Source: U.S. Department of Transportation

30. How many states require helmets for all riders?

31. How many states require helmets for riders under 18?

32. How many states do not require a helmet?

33. **MULTI STEP** Suppose you buy a sweater and a pair of jeans. The total of the two items before tax is $65.82. If sales tax is 6%, how much money will you need for the total cost of the items, including tax?

CRITICAL THINKING Solve each problem.

34. What percent of 70 is 14?

35. What percent of 240 is 84?

36. 45 is 15% of what number?

37. 21 is 30% of what number?

EXTENDING THE LESSON Simple interest is the amount of money paid or earned for the use of money. $I = prt$ is a formula that can be used to find the simple interest. I is the interest, p is the principal, r is the rate, and t is the time. Suppose you place $750 in a savings account that pays 2.9% interest for one year.

$$I = 750 \times 0.029 \times 1 \qquad \text{You will earn \$21.75 in one year.}$$

Find the interest earned on $550 for each rate for one year.

38. 0.3%

39. 12%

40. 19.5%

Standardized Test Practice and Mixed Review

41. **MULTIPLE CHOICE** At Langley High School, 19% of the 2,200 students walk to school. How many students walk to school?

   (A) 400   (B) 418   (C) 428   (D) 476

42. **MULTIPLE CHOICE** Which number is 124% of 260?

   (F) 3.224   (G) 32.24   (H) 322.4   (I) 3,224

43. Write 1.35 as a percent. (Lesson 10-6)

Write each percent as a fraction in simplest form. (Lesson 10-5)

44. 30%

45. 28%

46. 145%

47. 85%

GETTING READY FOR THE NEXT LESSON

PREREQUISITE SKILL Multiply. (Lesson 7-2)

48. $\frac{1}{2} \times 150$

49. $\frac{2}{5} \times 25$

50. $\frac{3}{4} \times 48$

51. $\frac{2}{3} \times 21$

Problem-Solving Strategy
A Preview of Lesson 10-8

What You'll Learn
Solve problems by solving a simpler problem.

Solve a Simpler Problem

Hey Yutaka, a total of 350 students voted on whether a tiger or a dolphin should be the new school's mascot. I heard that 70% of the students voted for the tiger.

Well Justin, I'm glad the tiger won! I wonder how many students voted for the tiger. We could find 70% of 350. But, I know a way to **solve a simpler problem** using mental math.

Explore	We know the number of students who voted and that 70% of the students voted for the tiger. We need to find the number of students who voted for the tiger.
Plan	Solve a simpler problem by finding 10% of 350 and then use the result to find 70% of 350.
Solve	10% of 350 = 35 Since there are seven 10%s in 70%, multiply 35 by 7. 35 × 7 = 245 So, 245 students voted for the tiger.
Examine	Since 70% of 350 is 245, the answer is correct.

Analyze the Strategy

1. **Explain** when you would use the solve a simpler problem strategy.

2. **Explain** why the students found it simpler to work with 10%.

3. **Think** of another way the students could have solved the problem.

4. **Write** a problem than can be solved by working a simpler problem. Then write the steps you would take to find the solution.

Solve. Use the solve a simpler problem strategy.

5. **SCHOOL** Refer to the example on page 413. If 30% of the students voted for the dolphin as a school mascot, how many of the 350 students voted for the dolphin?

6. **GEOGRAPHY** The total area of Minnesota is 86,939 square miles. Of that, about 90% is land area. About how much of Minnesota is not land area?

Mixed Problem Solving

Solve. Use any strategy.

7. **MONEY** A total of 32 students are going on a field trip. Each student must pay $4.75 for travel and $5.50 for dining. About how much money should the teacher collect in all from the students?

8. **VENN DIAGRAMS** The Venn diagram shows information about the members in Jacob's scout troop.

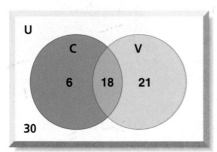

U = all members in the troop
C = members with a camping badge
V = members with a volunteer badge

How many more members have a badge than do not have a badge?

9. **MONEY** Kip wants to leave a 15% tip on a $38.79 restaurant bill. About how much money should he leave for the tip?

10. **SCIENCE** Sound travels through air at a speed of 1,129 feet per second. At this rate, how far will sound travel in 1 minute?

11. **TRAVEL** Mr. Ishikawa left Houston at 3:00 P.M. and arrived in Dallas at 8:00 P.M., driving a distance of approximately 240 miles. During his trip, he took a one-hour dinner break. What was Mr. Ishikawa's average speed?

12. **PATTERNS** Find the area of the sixth figure in the pattern shown.

13. **SALES** A sales manager reported to his sales team that sales increased 34.7% over last month's sales total of $98,700. About how much did the team sell this month?

14. **SCHOOL** Jewel's math scores for her last four tests were 94, 87, 90, and 89. What score does she need on the next test to average a score of 91?

15. **STANDARDIZED TEST PRACTICE**
The circle graph shows the results of a favorite juice survey. Which percents best describe the data?

Favorite Juice

	Apple	Grape	Orange	Mixed Fruit
A	25%	30%	15%	60%
B	32%	18%	21%	29%
C	10%	35%	10%	45%
D	45%	15%	35%	5%

You will use the solve a simpler problem strategy in the next lesson.

Estimating with Percents

What You'll LEARN

Estimate the percent of a number.

WHEN am I ever going to use this?

SHOPPING A store is having a back-to-school sale. All school supplies are on sale.

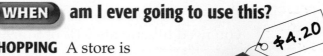

$4.20

$1.99

1. What would be the cost of the notebook at 10% off?

2. What would be the cost of the pencils at 25% off? Round to the nearest cent.

3. Explain how you might estimate the cost of the notebook at 10% off and the cost of the pencils at 25% off.

Sometimes when finding the percent of a number, an exact answer is not needed. So, you can estimate. The table below shows some commonly used percents and their fraction equivalents.

Key Concept			Percent-Fraction Equivalents	
$20\% = \frac{1}{5}$	$50\% = \frac{1}{2}$	$80\% = \frac{4}{5}$	$25\% = \frac{1}{4}$	$33\frac{1}{3}\% = \frac{1}{3}$
$30\% = \frac{3}{10}$	$60\% = \frac{3}{5}$	$90\% = \frac{9}{10}$	$75\% = \frac{3}{4}$	$66\frac{2}{3}\% = \frac{2}{3}$
$40\% = \frac{2}{5}$	$70\% = \frac{7}{10}$	$100\% = 1$		

EXAMPLES Estimate the Percent of a Number

Estimate each percent.

1 **52% of 298**

52% is close to 50% or $\frac{1}{2}$.

Round 298 to 300.

$\frac{1}{2}$ of 300 is 150.

So, 52% of 298 is about 150.

2 **60% of 27**

60% is $\frac{3}{5}$.

Round 27 to 25 since it is divisible by 5.

$$\frac{3}{5} \times 25 = \frac{3}{\cancel{5}_1} \times \frac{\cancel{25}^5}{1}$$

$$= 15$$

So, 60% of 27 is about 15.

Your Turn Estimate each percent.

a. 48% of $76   b. 18% of 42   c. 25% of 41

3 **MONEY** A DVD that originally costs $15.99 is on sale for 50% off. If you have $9, would you have enough money to buy the DVD?

To determine whether you have enough money to buy the DVD, you need to estimate 50% of $15.99.

$$50\% \times \$15.99 \rightarrow \frac{1}{2} \times \$16 \text{ or } \$8$$

Since $8 is less than $9, you should have enough money.

Estimation can be used to find what percent of a figure is shaded.

EXAMPLE **Estimate the Percent of a Figure**

Standardized Test Practice

4 **MULTIPLE-CHOICE TEST ITEM** Which of the following is a reasonable percent for the percent of the figure that is shaded?

  A 25%   **B** 40%
  C 60%   **D** 80%

Test-Taking Tip
The Princeton Review

When taking a multiple-choice test, eliminate the choices you know to be incorrect. The percent of the model shaded is clearly greater than 50%. So, eliminate choices A and B.

Read the Test Item

You need to find what percent of the circles are shaded.

Solve the Test Item

13 out of 15 circles are shaded.

$\frac{13}{15}$ is about $\frac{12}{15}$ or $\frac{4}{5}$

$\frac{4}{5} = 80\%$

So, about 80% of the figure is shaded. The answer is D.

Skill and Concept Check

Writing Math

Exercise 2

1. **List** three commonly used percent-fraction equivalents.

2. **OPEN ENDED** Write about a real-life situation when you would need to estimate the percent of a number.

GUIDED PRACTICE

Estimate each percent.

3. 38% of $50   4. 59% of 16   5. 75% of 33

6. **TIPS** Abigail wants to give a 20% tip to a taxi driver. If the fare is $23.78, what would be a reasonable amount to tip the driver?

Practice and Applications

HOMEWORK HELP

For Exercises	See Examples
7–18	1, 2
19–20, 24	3
21–23	4

Extra Practice
See pages 616, 633.

Estimate each percent.

7. 21% of 96

8. 42% of 16

9. 79% of 82

10. 74% of 45

11. 26% of 125

12. 89% of 195

13. 31% of 157

14. 77% of 238

15. 69% of 203

16. 33% of 92

17. 67% of 296

18. 99% of 350

19. TIPS Dakota and Emma want to give a 20% tip for a food bill of $64.58. About how much should they leave for the tip?

20. BANKING Louisa deposited 25% of the money she earned baby-sitting into her savings account. If she earned $237.50, about how much did she deposit into her savings account?

Estimate the percent that is shaded in each figure.

21.

22.

23.

24. GEOGRAPHY The Atlantic coast has 2,069 miles of coastline. Of that, about 28% is located in Florida. About how many miles of coastline does Florida have?

25. MULTI STEP If you answered 9 out of 25 problems incorrectly on a test, about what percent of answers were correct? Explain.

26. CRITICAL THINKING Order the percents 40% of 50, 50% of 50, and $\frac{1}{2}$% of 50 from least to greatest.

Standardized Test Practice and Mixed Review

27. MULTIPLE CHOICE Refer to the graph at the right. If 3,608 people were surveyed, which expression could be used to estimate the number of people that are influenced by a friend or relative when buying a CD?

 Ⓐ $\frac{1}{8} \times 3,600$ Ⓑ $\frac{1}{5} \times 3,600$

 Ⓒ $\frac{1}{4} \times 3,600$ Ⓓ $\frac{1}{6} \times 3,600$

Who Influences People Age 16–40 to Buy CDs

45% Heard on Radio
15% Heard from Friend/Relative
10% Heard in Store

Source: Edison Media Research

28. SHORT RESPONSE Estimate 35% of 95.

29. Find 20% of 129. (Lesson 10-7)

Express each decimal as a percent. (Lesson 10-6)

30. 0.31   **31.** 0.05   **32.** 0.113   **33.** 0.861

CHAPTER 10 Study Guide and Review

Vocabulary and Concept Check

cross products (p. 386)	rate (p. 381)	scale model (p. 391)
equivalent ratios (p. 381)	ratio (p. 380)	unit rate (p. 381)
percent (%) (p. 395)	scale (p. 391)	
proportion (p. 386)	scale drawing (p. 391)	

State whether each sentence is *true* or *false*. If *false*, replace the underlined word or number to make a true sentence.

1. A ratio is a comparison of two numbers by <u>multiplication</u>.
2. A <u>rate</u> is a ratio of two measurements that have different units.
3. Three tickets for $7.50 expressed as a rate is <u>$1.50</u> per ticket.
4. A <u>percent</u> is an equation that shows that two ratios are equivalent.
5. The model shown at the right represents <u>85%</u>.
6. The cross products of a proportion are <u>equal</u>.
7. A <u>scale drawing</u> shows an object exactly as it looks, but it is generally larger or smaller.
8. A percent is a ratio that compares a number to <u>10</u>.
9. The decimal 0.346 can be expressed as <u>3.46%</u>.

Lesson-by-Lesson Exercises and Examples

10-1 Ratios (pp. 380–383)

Write each ratio as a fraction in simplest form.

10. 12 blue marbles out of 20 marbles
11. 9 goldfish out of 36 fish
12. 15 carnations out of 40 flowers
13. 18 boys out of 21 students

Write each ratio as a unit rate.

14. 3 inches of rain in 6 hours
15. 189 pounds of garbage in 12 weeks
16. $24 for 4 tickets
17. 78 candy bars in 3 packages

Example 1 Write the ratio 30 sixth graders out of 45 students as a fraction in simplest form.

$$\frac{30}{45} = \frac{2}{3} \quad \text{The GCF of 30 and 45 is 15.}$$

(÷ 15)

Example 2 Write the ratio 150 miles in 4 hours as a unit rate.

$$\frac{150 \text{ miles}}{4 \text{ hours}} = \frac{37.5 \text{ miles}}{1 \text{ hour}}$$

(÷ 4)

Divide the numerator and the denominator by 4 to get the denominator of 1.

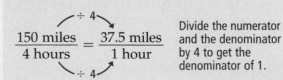 msmath1.net/vocabulary_review

10-2 Solving Proportions (pp. 386–389)

Solve each proportion.

18. $\dfrac{7}{11} = \dfrac{m}{33}$

19. $\dfrac{12}{20} = \dfrac{15}{k}$

20. $\dfrac{g}{20} = \dfrac{9}{12}$

21. $\dfrac{25}{h} = \dfrac{10}{12}$

22. **SCHOOL** At Rio Middle School, the teacher to student ratio is 3 to 42. If there are 504 students enrolled at the school, how many teachers are there at the school?

Example 3 Solve the proportion $\dfrac{9}{12} = \dfrac{g}{8}$.

$9(8) = 12g$ Cross products

$72 = 12g$ Multiply.

$\dfrac{72}{12} = \dfrac{12g}{12}$ Divide each side by 12.

$6 = g$ The solution is 6.

10-3 Scale Drawings and Models (pp. 391–393)

On a scale model of a fire truck, the scale is 2 inches = 5 feet. Find the actual measurements.

	Truck	Model
23.	length	12 inches
24.	width	4 inches
25.	height	7.2 inches

26. **BUILDINGS** On an architectural drawing, the height of a building is $15\frac{3}{4}$ inches. If the scale on the drawing is $\frac{1}{2}$ inch = 1 foot, find the height of the actual building.

Example 4 On a scale drawing of a room, the scale is 1 inch = 2 feet. What is the actual length of the room?

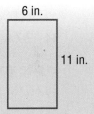

6 in.

11 in.

Write a proportion.

$\begin{array}{l} \text{drawing width} \rightarrow \\ \text{actual width} \rightarrow \end{array} \dfrac{1 \text{ in.}}{2 \text{ ft}} = \dfrac{11 \text{ in.}}{x \text{ ft}} \begin{array}{l} \leftarrow \text{drawing width} \\ \leftarrow \text{actual width} \end{array}$

$1 \cdot x = 2 \cdot 11$ Find cross products.

$1x = 22$ Simplify.

$x = 22$ Multiply.

The actual length of the room is 22 feet.

10-4 Modeling Percents (pp. 395–397)

Model each percent.

27. 20%

28. 75%

29. 5%

30. 50%

31. Tell what percent is modeled in the figure shown.

Example 5 Model 55%.

55% means 55 out of 100. So, shade 55 of the 100 squares.

10-5 **Percents and Fractions** (pp. 400–403)

Write each percent as a fraction in simplest form.

32. 3%   33. 18%

34. 48%   35. 120%

Write each fraction as a percent.

36. $\dfrac{3}{5}$ 37. $\dfrac{7}{8}$

38. $\dfrac{8}{5}$ 39. $\dfrac{3}{100}$

Example 6   Write 24% as a fraction in simplest form.

$24\% = \dfrac{24}{100}$ Express the percent as a fraction with a denominator of 100.

$= \dfrac{\overset{6}{\cancel{24}}}{\underset{25}{\cancel{100}}}$ Simplify. Divide numerator and denominator by the GCF, 4.

$= \dfrac{6}{25}$

10-6 **Percents and Decimals** (pp. 404–406)

Write each percent as a decimal.

40. 2.2%   41. 38%

42. 140%   43. 66%

44. 90%   45. 55%

Write each decimal as a percent.

46. 0.003   47. 1.3

48. 0.65   49. 0.591

50. 1.75   51. 0.73

Example 7   Write 46% as a decimal.

$46\% = \dfrac{46}{100}$ Rewrite the percent as a fraction with a denominator of 100.

$= 0.46$ Write the fraction as a decimal.

Example 8   Write 0.85 as a percent.

$0.85 = \dfrac{85}{100}$ Write the decimal as a fraction.

$= 85\%$ Write the fraction as a percent.

10-7 **Percent of a Number** (pp. 409–412)

Find the percent of each number.

52. 40% of 150   53. 5% of 340

54. 18% of 90   55. 8% of 130

56. 170% of 30   57. 125% of 120

Example 9   Find 42% of 90.

42% of $90 = 0.42 \times 90$ Change the percent to a decimal.

$= 37.8$ Multiply.

10-8 **Estimating with Percents** (pp. 415–417)

Estimate each percent.

58. 40% of 78   59. 73% of 20

60. 25% of 122   61. 19% of 99

62. 48% of 48   63. 41% of 243

64. **SCHOOL** Jenna answered 8 out of 35 questions incorrectly on a test. About what percent of the answers did she answer correctly?

Example 10   Estimate 33% of 60.

33% is close to $33\dfrac{1}{3}\%$ or $\dfrac{1}{3}$.

$\dfrac{1}{3} \times 60 = \dfrac{1}{\underset{1}{\cancel{3}}} \times \dfrac{\overset{20}{\cancel{60}}}{1}$ Rewrite 60 as a fraction with a denominator of 1.

$= 20$ Simplify.

So, 33% of 60 is about 20.

Vocabulary and Concepts

1. **Draw** a model that shows 90%.

2. **Explain** how to change a percent to a fraction.

Skills and Applications

Write each ratio as a fraction in simplest form.

3. 12 red blocks out of 20 blocks

4. 24 chips out of 144 chips

5. **BIRDS** If a hummingbird flaps its wings 250 times in 5 seconds, how many times does a hummingbird flap its wings each second?

Solve each proportion.

6. $\dfrac{4}{6} = \dfrac{x}{15}$

7. $\dfrac{10}{p} = \dfrac{2.5}{8}$

8. $\dfrac{n}{1.3} = \dfrac{6}{5.2}$

9. **GEOGRAPHY** On a map of Texas, the scale is 1 inch = 30 miles. Find the actual distance between Dallas and Houston if the distance between these cities on the map is 8 inches.

Write each percent as a decimal and as a fraction in simplest form.

10. 42%

11. 20%

12. 4%

13. 110%

14. Write $\dfrac{2}{5}$ as a percent.

15. Write 0.8% as a decimal.

Express each decimal as a percent.

16. 0.3

17. 0.87

18. 0.149

19. **MONEY** Ian used 35% of his allowance to buy a book. If Ian received $20 for his allowance, how much did he use to buy the book?

20. Find 60% of 35.

21. What is 2% of 50?

Estimate each percent.

22. 9.5% of 51

23. 49% of 26

24. 308% of 9

Standardized Test Practice

25. **MULTIPLE CHOICE** In which model is about 25% of the figure shaded?

PART 1 Multiple Choice

Record your answers on the answer sheet provided by your teacher or on a sheet of paper.

1. Use the table to find the total weight of one jar of jam, one package of cookies, and one box of crackers. (Lesson 3-5)

Gourmet Food Catalog	
Item	**Weight (oz)**
jam	6.06
cookies	18.73
crackers	12.12

- Ⓐ 26.81 oz
- Ⓑ 36.00 oz
- Ⓒ 36.91 oz
- Ⓓ 37.45 oz

2. The box shown originally contained 24 bottles of juice. What fraction represents the number of juice bottles that remain? (Lesson 5-2)

- Ⓕ $\frac{5}{24}$
- Ⓖ $\frac{1}{4}$
- Ⓗ $\frac{1}{2}$
- Ⓘ $\frac{6}{13}$

3. At a party, the boys ate $\frac{1}{3}$ of a pizza. The girls ate $\frac{1}{4}$ of another pizza. What fraction of a whole pizza did they eat altogether? (Lesson 6-4)

- Ⓐ $\frac{1}{12}$
- Ⓑ $\frac{2}{7}$
- Ⓒ $\frac{7}{12}$
- Ⓓ $\frac{5}{6}$

4. There are $3\frac{3}{4}$ pies to be shared equally among 5 people. How much of a pie will each person get? (Lesson 7-5)

- Ⓕ $\frac{1}{5}$
- Ⓖ $\frac{1}{3}$
- Ⓗ $\frac{1}{2}$
- Ⓘ $\frac{3}{4}$

TEST-TAKING TIP

Question 6 When setting up a proportion, make sure the numerators and the denominators in each ratio have the same units, respectively.

5. Which ratio compares the number of apples to the total number of pieces of fruit? (Lesson 10-1)

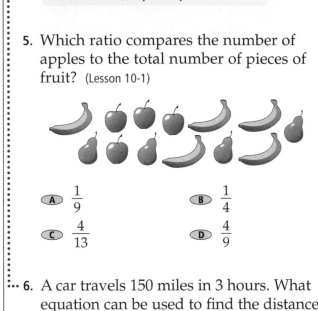

- Ⓐ $\frac{1}{9}$
- Ⓑ $\frac{1}{4}$
- Ⓒ $\frac{4}{13}$
- Ⓓ $\frac{4}{9}$

6. A car travels 150 miles in 3 hours. What equation can be used to find the distance the car will travel in 10 hours? (Lesson 10-2)

- Ⓕ $\frac{3}{150} = \frac{d}{10}$
- Ⓖ $\frac{3}{d} = \frac{150}{10}$
- Ⓗ $\frac{3}{150} = \frac{10}{d}$
- Ⓘ $\frac{150}{3} = \frac{10}{d}$

7. Which figures have more than 25% of their area shaded? (Lesson 10-4)

Figure 1   Figure 2

Figure 3   Figure 4

- Ⓐ 1 and 2
- Ⓑ 1 and 4
- Ⓒ 2 and 3
- Ⓓ 3 and 4

PART 2   Short Response/Grid In

Record your answers on the answer sheet provided by your teacher or on a sheet of paper.

8. What is the quotient of 315 divided by 5? (Prerequisite Skill, p. 591)

9. What are the next 3 numbers in the pattern 960, 480, 240, 120, …? (Lesson 1-1)

10. What is the total area of the figure shown? (Lesson 1-8)

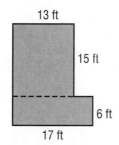

11. The stem-and-leaf plot shows the cost of different pairs of jeans. How many of the jeans cost more than $34? (Lesson 2-5)

Stem	Leaf
2	5 6 7 9
3	0 0 4 5 8
4	0 0 0 0 2

$3|8 = \$38$

12. Nina buys a sports magazine that costs $3.95 for a monthly issue. How much will it cost her if she buys one magazine each month for a year? (Lesson 4-1)

13. Write the mixed number modeled below in simplest form. (Lesson 5-3)

14. Evaluate $a - b$ if $a = \frac{2}{5}$ and $b = \frac{1}{4}$. (Lesson 6-4)

15. What value of m satisfies the equation $m + 16 = 40$? (Lesson 9-2)

16. What is the value of y in $3y + 24 = 30$? (Lesson 9-5)

17. What is the function rule for the x- and y-values shown? (Lesson 9-6)

x	y
0	−3
1	−1
2	1
3	3
4	5

18. In a survey, 12 out of 15 adults preferred a certain brand of chewing gum. How many adults would prefer that particular brand if 100 adults were surveyed? (Lesson 10-2)

19. What is 25% written as a fraction? (Lesson 10-5)

20. Melissa bought a sweatshirt that originally cost $30. If the sweatshirt was on sale for 25% off, what was the discount? (Lesson 10-7)

PART 3   Extended Response

Record your answers on a sheet of paper. Show your work.

21. Dante made a scale model of a tree. The actual tree is 32 feet tall, and the height of the model he made is 2 feet. (Lessons 10-2 and 10-3)

2 ft

  a. Write a proportion that Dante could use to find the actual height that one foot on the drawing represents.

  b. How many actual feet does one foot on the model represent?

  c. Suppose a branch on the actual tree is 4 feet long. How long would this branch be on the model of the tree?

CHAPTER 11 Probability

"What does soccer have to do with math?"

Suppose a soccer player scored 6 goals in her last 15 attempts.
You can use this ratio to find the probability of the soccer player scoring on her next attempt. In addition, you can use the probability to predict the number of goals she will score in her next 50 attempts.

You will solve problems about sports such as soccer, football, and golf in Lessons 11-3, 11-4, and 11-5.

GETTING STARTED

Take this quiz to see whether you are ready to begin Chapter 11. Refer to the lesson or page number in parentheses if you need more review.

▶ ## Vocabulary Review

Complete each sentence.

1. A fraction is in __?__ form when the GCF of the numerator and denominator is one. (Lesson 5-2)

2. A __?__ is a comparison of two numbers by division. (Lesson 10-1)

▶ ## Prerequisite Skills

Write each fraction in simplest form.
(Lesson 5-2)

3. $\dfrac{15}{40}$

4. $\dfrac{7}{63}$

5. $\dfrac{21}{30}$

6. $\dfrac{8}{28}$

7. $\dfrac{9}{21}$

8. $\dfrac{9}{45}$

Multiply. Write in simplest form.
(Lesson 7-2)

9. $\dfrac{5}{6} \times \dfrac{3}{10}$

10. $\dfrac{1}{8} \times \dfrac{7}{9}$

11. $\dfrac{11}{12} \times \dfrac{4}{5}$

12. $\dfrac{2}{3} \times \dfrac{9}{14}$

13. $\dfrac{2}{5} \times \dfrac{3}{4}$

14. $\dfrac{4}{7} \times \dfrac{2}{3}$

Solve each proportion. (Lesson 10-2)

15. $\dfrac{a}{6} = \dfrac{5}{15}$

16. $\dfrac{b}{9} = \dfrac{56}{84}$

17. $\dfrac{8}{y} = \dfrac{28}{42}$

18. $\dfrac{4}{c} = \dfrac{12}{30}$

19. $\dfrac{33}{44} = \dfrac{x}{12}$

20. $\dfrac{3}{8} = \dfrac{d}{24}$

FOLDABLES™
Study Organizer

Probability Make this Foldable to help you organize information about probability. Begin with one sheet of notebook paper.

STEP 1 **Fold**
Fold the paper lengthwise to the holes.

STEP 2 **Unfold and Cut**
Unfold the paper and cut five equal tabs as shown.

STEP 3 **Label**
Label lesson numbers and titles as shown.

Reading and Writing As you read and study each lesson, write notes and examples under the appropriate tab.

Simulations

Materials

- 3 two-colored counters
- cups
- spinner

A **simulation** is a way of acting out a problem situation. Simulations can be used to find probability, which is the chance something will happen. When you find a probability by doing an experiment, you are finding **experimental probability**.

To explore experimental probability using a simulation, you can use these steps.

- Choose the most appropriate manipulative to aid in simulating the problem. Choose among counters, number cubes, coins, or spinners.

- Act out the problem for many trials and record the results to find an experimental probability.

ACTIVITY   *Work with a partner.*

1   **Use cups and counters to explore the experimental probability that at least two of three children in a family are girls.**

STEP 1 Place three counters in a cup and toss them onto your desk.

STEP 2 Count the number of red counters. This represents the number of boys. The number of yellow counters represents the number of girls.

STEP 3 Record the results in a table like the one shown.

Trial	Outcome		
1	B	B	G
2	B	G	G
3			
⋮			
50			

STEP 4 Repeat Steps 1–3 for 50 trials.

Suppose 23 of the 50 trials have at least two girls. The experimental probability that at least two of the three children in a family are girls is $\frac{23}{50}$ or 0.46.

Your Turn

a. Describe a simulation to explore the experimental probability that two of five children in a family are boys. Then conduct your experiment. What is the experimental probability?

Spinners can also be used in simulations.

ACTIVITY *Work with a partner.*

2 The probability of the Hornets beating the Jets is 0.5.
The probability of the Hornets beating the Flashes is 0.25.
Find the experimental probability that the Hornets beat
both the Jets and the Flashes.

STEP 1 A probability of 0.5 is equal to $\frac{1}{2}$.
This means that the Hornets should
win 1 out of 2 games. Make a spinner
as shown. Label one section "win"
and the other section "lose".

STEP 2 A probability of 0.25 is equal to $\frac{1}{4}$.
This means that the Hornets should
win 1 out of 4 games. Make a spinner
as shown. Label one section "win"
and the other sections "lose".

STEP 3 Spin each spinner and
record the results in a table
like the one shown at the
right. Repeat for 100 trials.

	Outcome	
Trial	Hornets and Jets	Hornets and Flashes
1	L	W
2	W	W
3		
⋮		
100		

STEP 4 Use the results of the
trials to write the ratio
$\frac{\text{beat both teams}}{100}$. The ratio
represents the experimental
probability that the Hornets beat both teams.

● **Your Turn**

b. The probability of rain on Monday is 0.75, and the probability
of rain on Tuesday is 0.4. Describe a simulation you could use
to explore the probability of rain on both days. Conduct your
simulation to find the experimental probability of rain on
both days.

Writing Math

1. **Explain** experimental probability.

2. How is a simulation used to find the experimental probability
of an event?

Theoretical Probability

What You'll LEARN

Find and interpret the theoretical probability of an event.

NEW Vocabulary

outcomes
event
theoretical probability
complementary events

REVIEW Vocabulary

ratio: comparison of two numbers by division **(Lesson 10-1)**

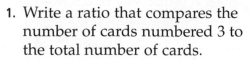 **am I ever going to use this?**

GAMES Drew and Morgan are playing cards. Morgan needs to draw a 3 in order to make a match and win the game. The cards shown are shuffled and placed facedown on the table.

1. Write a ratio that compares the number of cards numbered 3 to the total number of cards.

2. What percent of the cards are numbered 3?

3. Does Morgan have a good chance of winning? Explain.

4. What would happen to her chances of winning if cards 1, 4, 7, 9, and 10 were added to the cards shown?

5. What happens to her chances if only cards 3 and 8 are facedown on the table?

It is equally likely to select any one of the five cards. The player hopes to select a card numbered 3. The five cards represent the possible **outcomes**. The specific outcome the player is looking for is an **event**, or favorable outcome.

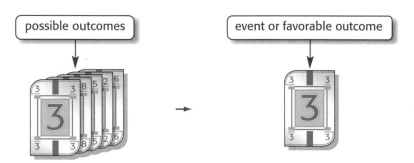

Theoretical probability is the chance that some event will occur. It is based on known characteristics or facts. You can use a ratio to find probability.

Key Concept   **Theoretical Probability**

Words   The theoretical probability of an event is a ratio that compares the number of favorable outcomes to the number of possible outcomes.

Symbols $P(\text{event}) = \dfrac{\text{number of favorable outcomes}}{\text{number of possible outcomes}}$

The probability that an event will occur is a number from 0 to 1, including 0 and 1. The closer a probability is to 1, the more likely it is to happen.

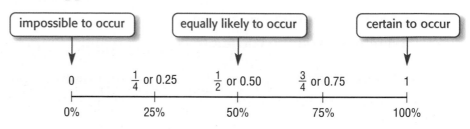

| impossible to occur | | equally likely to occur | | certain to occur |

$$0 \qquad \frac{1}{4} \text{ or } 0.25 \qquad \frac{1}{2} \text{ or } 0.50 \qquad \frac{3}{4} \text{ or } 0.75 \qquad 1$$

0%   25%   50%   75%   100%

EXAMPLES Find Probability

There are eight equally likely outcomes on the spinner.

1 **Find the probability of spinning red.**

$$P(\text{red}) = \frac{\text{number of favorable outcomes}}{\text{number of possible outcomes}}$$

$$= \frac{1}{8}$$

The probability of landing on red is $\frac{1}{8}$, 0.125, or 12.5%.

2 **Find the probability of spinning blue or yellow.**

$$P(\text{blue or yellow}) = \frac{\text{number of favorable outcomes}}{\text{number of possible outcomes}}$$

$$= \frac{2}{8} \text{ or } \frac{1}{4} \quad \text{Simplify.}$$

The probability of landing on blue or yellow is $\frac{1}{4}$, 0.25, or 25%.

Complementary events are two events in which either one or the other must happen, but they cannot happen at the same time. An example is a coin landing on heads or not landing on heads. The sum of the probabilities of complementary events is 1.

EXAMPLE Use Probability to Solve a Problem

3 **WEATHER** The morning newspaper reported a 20% chance of snow. What is the probability that it will not snow?

The two events are complementary. So, the sum of the probabilities is 1.

$$P(\text{snow}) + P(\text{not snowing}) = 1$$

$$0.2 + P(\text{not snowing}) = 1 \quad \text{Replace } P(\text{snow}) \text{ with 0.2.}$$

$$\underline{-0.2 \qquad\qquad\qquad\qquad = -0.2} \quad \text{Subtract 0.2 from each side.}$$

$$P(\text{not snowing}) = 0.8$$

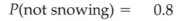

Skill and Concept Check

1. **OPEN ENDED** Give an example of a situation in which the probability of an event occurring is 0.

2. **Explain** what you can conclude about an event if its probability is 1.

3. **FIND THE ERROR** Laura and Lourdes are finding the probability of rolling a 5 on a number cube. Who is correct? Explain your reasoning.

Laura
Favorable: 5
Possible: 1, 2, 3, 4, 5, 6
$P(5) = \frac{1}{6}$

Lourdes
Favorable: 5
Unfavorable: 1, 2, 3, 4, 6
$P(5) = \frac{1}{5}$

GUIDED PRACTICE

A counter is randomly chosen. Find each probability. Write each answer as a fraction, a decimal, and a percent.

4. $P(4)$
5. $P(2 \text{ or } 5)$
6. $P(\text{less than } 8)$
7. $P(\text{greater than } 4)$
8. $P(\text{prime})$
9. $P(\text{not } 6)$

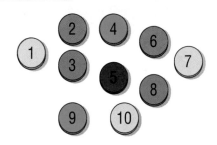

10. **SCHOOL** The probability of guessing the answer to a true-false question correctly is 50%. What is the probability of guessing the answer incorrectly?

Practice and Applications

The spinner shown is spun once. Find each probability. Write each answer as a fraction, a decimal, and a percent.

11. $P(Z)$
12. $P(U)$
13. $P(A \text{ or } M)$
14. $P(C, D \text{ or } A)$
15. $P(\text{vowel})$
16. $P(\text{consonant})$

HOMEWORK HELP

For Exercises	See Examples
11–25	1, 2
26	3

Extra Practice
See pages 616, 634.

A number cube is rolled. Find the probability of each event. Write each answer as a fraction, a decimal, and a percent.

17. $P(3)$
18. $P(4 \text{ or } 6)$
19. $P(\text{greater than } 4)$
20. $P(\text{less than } 1)$
21. $P(\text{even})$
22. $P(\text{odd})$
23. $P(\text{multiple of } 2)$
24. $P(\text{not } 3 \text{ and not } 4)$
25. $P(\text{not less than } 2)$

WEATHER For Exercises 26 and 27, use the following information.
A morning radio announcer reports that the chance of rain today is 85%.

26. What is the probability that it will not rain?

27. Should you carry an umbrella? Explain.

One marble is selected without looking from the bag shown. Write a sentence explaining how likely it is for each event to happen.

28. P(green)

29. P(yellow)

30. P(purple)

31. P(blue)

32. SCHOOL There are 160 girls and 96 boys enrolled at Grant Middle School. The school newspaper is randomly selecting a student to be interviewed. Find the probability of selecting a girl. Write as a fraction, a decimal, and a percent.

CRITICAL THINKING For Exercises 33 and 34, use the following information.

A spinner for a board game has more than three equal sections, and the probability of the spinner stopping on blue is 0.5.

33. Draw two possible spinners for the game.

34. Explain why each spinner drawn makes sense.

EXTENDING THE LESSON

Another way to describe the chance of an event occurring is with odds. The *odds* in favor of an event is the ratio that compares the number of ways the event can occur to the ways that the event *cannot* occur.

ways to occur ⎯⎯⎯⎯ ⎯⎯⎯⎯ ways to not occur

odds of rolling a 3 or a 4 on a number cube → 2 : 4 or 1 : 2

Find the odds of each outcome if a number cube is rolled.

35. a 2, 3, 5, or 6

36. a number less than 3

37. an odd number

Standardized Test Practice and Mixed Review

38. MULTIPLE CHOICE A number cube is rolled. What is the probability of rolling a composite number?

(A) $\frac{1}{6}$ (B) $\frac{1}{3}$ (C) $\frac{1}{2}$ (D) $\frac{2}{3}$

39. SHORT RESPONSE The probability of spinning a 5 on the spinner is $\frac{2}{3}$. What number is missing from the spinner?

40. Estimate 31% of 15. (Lesson 10-8)

Find the percent of each number. (Lesson 10-7)

41. 32% of 148

42. 6% of 25

GETTING READY FOR THE NEXT LESSON

BASIC SKILL List all outcomes for each situation.

43. tossing a coin

44. rolling a number cube

45. selecting a month of the year

46. choosing a color of the American flag

Experimental and Theoretical Probability

INVESTIGATE *Work with a partner.*

In this lab, you will investigate the relationship between experimental probability and theoretical probability.

STEP 1 The table shows all of the possible outcomes when you roll two number cubes. The highlighted outcomes are doubles.

	1	2	3	4	5	6
1	(1, 1)	(1, 2)	(1, 3)	(1, 4)	(1, 5)	(1, 6)
2	(2, 1)	(2, 2)	(2, 3)	(2, 4)	(2, 5)	(2, 6)
3	(3, 1)	(3, 2)	(3, 3)	(3, 4)	(3, 5)	(3, 6)
4	(4, 1)	(4, 2)	(4, 3)	(4, 4)	(4, 5)	(4, 6)
5	(5, 1)	(5, 2)	(5, 3)	(5, 4)	(5, 5)	(5, 6)
6	(6, 1)	(6, 2)	(6, 3)	(6, 4)	(6, 5)	(6, 6)

Find the theoretical probability of rolling doubles.

STEP 2 Copy the table shown. Then roll a pair of number cubes and record the results in the table. Write D for doubles and N for not doubles. Repeat for 30 trials.

Trials	Outcome
1	N
2	D
3	
⋮	
30	

Writing Math

1. Find the experimental probability of rolling doubles for the 30 trials. How does the experimental probability compare to the theoretical probability? Explain any differences.

2. Compare the results of your experiment with the results of the other groups in your class. Why do you think experimental probabilities usually vary when an experiment is repeated?

3. Find the experimental probability for the entire class's trials. How does the experimental probability compare to the theoretical probability?

4. **Explain** why the experimental probability obtained in Exercise 3 may be closer in value to the theoretical probability than the experimental probability in Exercise 1.

Outcomes

What You'll LEARN

Find outcomes using lists, tree diagrams, and combinations.

NEW Vocabulary

sample space
tree diagram

WHEN **am I ever going to use this?**

MOVIES A movie theater's concession stand sign is shown.

1. List the possible ways to choose a soft drink, a popcorn, and a candy.

2. How many different ways are possible?

The set of all possible outcomes is called the **sample space**. The list you made above is the sample space of choices at the concession stand. The sample space for rolling a number cube and spinning the spinner are listed below.

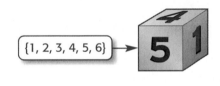 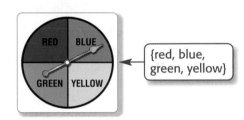

A tree diagram can also be used to show a sample space. When you make a tree diagram, you have an organized list of outcomes. A **tree diagram** is a diagram that shows all possible outcomes of an event.

EXAMPLE **Find a Sample Space**

1 **How many outfits are possible from a choice of jean or khaki shorts and a choice of a yellow, white, or blue shirt?**

Use a tree diagram. List each choice for shorts. Then pair each choice for shorts with each choice for a shirt.

READING Math

Outcomes The outcome JY means jean shorts and a yellow shirt.

Shorts	Shirt	Outcome
jean (J)	yellow (Y)	JY
	white (W)	JW
	blue (B)	JB
khaki (K)	yellow (Y)	KY
	white (W)	KW
	blue (B)	KB

There are six possible outfits.

When you know the number of outcomes, you can easily find the probability that an event will occur.

EXAMPLE Use a Tree Diagram to Find Probability

2 Lexi spins two spinners. What is the probability of spinning red on the first spinner and blue on the second spinner?

Spinner 1   Spinner 2

Use a tree diagram to find all of the possible outcomes.

Spinner 1	Spinner 2	Outcome
red (R)	red (R)	RR
	blue (B)	RB
blue (B)	red (R)	BR
	blue (B)	BB

One outcome has red and then blue. There are four possible outcomes. So, P(red, blue) = $\frac{1}{4}$, 0.25, or 25%.

Your Turn

a. Use a tree diagram to find how many different words can be made using the words *quick, slow,* and *sad* and the suffixes *-ness, -er,* and *-ly*.

b. A penny is tossed, and a number cube is rolled. Use a tree diagram to find the probability of getting heads and a 5.

You can also make a list of possible outcomes.

EXAMPLE Use a List to Find Sample Space

STUDY TIP

Combinations
Example 3 is an example of a *combination*. A combination is an arrangement, or listing, of objects in which order is *not* important.

3 PETS The names of three bulldog puppies are shown. In how many different ways can a person choose two of the three puppies?

Puppies' Names
Alex
Bailey
Chester

List all of the ways two puppies can be chosen.

AB   AC   BA   BC   CA   CB

From the list count only the different arrangements. In this case, AB is the same as BA.

AB   AC   BC

There are 3 ways to choose two of the puppies.

Your Turn

c. How many different ways can a person choose three out of four puppies?

Skill and Concept Check

1. **Define** *sample space* in your own words.

2. **OPEN ENDED** Give an example of a situation that has 8 outcomes.

GUIDED PRACTICE

Draw a tree diagram to show the sample space for each situation. Then tell how many outcomes are possible.

3. hamburger or cheeseburger with a soft drink, water, or juice

4. small, medium, large, or extra-large shirt in blue, white, or black

For Exercises 5 and 6, use the spinners shown. Each spinner is spun once.

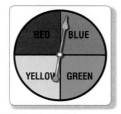
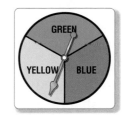

5. How many outcomes are possible?

6. What is *P*(blue, green) in that order?

Practice and Applications

Draw a tree diagram to show the sample space for each situation. Then tell how many outcomes are possible.

7. apple, peach, or cherry pie with milk, juice, or tea

8. nylon, leather, or mesh backpack in red, blue, gold, or black

9. roll two number cubes

10. toss a dime, quarter, and penny

HOMEWORK HELP

For Exercises	See Examples
7–10	1
11–18	2
19–20	3

Extra Practice
See pages 616, 634.

For Exercises 11–13, a coin is tossed, and a letter is chosen from the bag shown.

11. How many outcomes are possible?

12. Find *P*(heads, E).

13. What is the probability of tails and P, Z, or M?

SCHOOL **For Exercises 14 and 15, use the following information.**
A science quiz has one multiple-choice question with answer choices A, B, and C, and two true/false questions.

14. Draw a tree diagram that shows all of the ways a student can answer the questions.

15. Find the probability of answering all three questions correctly by guessing.

For Exercises 16–18, a coin is tossed, the spinner shown is spun, and a number cube is rolled.

16. How many outcomes are possible?

17. What is *P*(heads, purple, 5)?

18. Find *P*(tails, orange, less than 4).

 msmath1.net/self_check_quiz

19. GAMES How many ways can 2 video games be chosen from 5 video games?

20. BOOKS How many ways can 3 books be selected from 5 books?

CRITICAL THINKING For Exercises 21 and 22, use the following information.
One of the bags shown is selected without looking, and then one marble is selected from the bag without looking.

21. Draw a tree diagram showing all of the outcomes.

22. Is each outcome equally likely? Explain.

EXTENDING THE LESSON
A permutation is another way to show a sample space. A *permutation* is an arrangement or listing where order is important.

Example
How many ways can Katie, Jose, and Tara finish a race?
List the ways in a table as shown. There are 6 ways three people can finish a race.

1st Place	2nd Place	3rd Place
Katie	Jose	Tara
Katie	Tara	Jose
Jose	Katie	Tara
Jose	Tara	Katie
Tara	Jose	Katie
Tara	Katie	Jose

Order is important because Katie, Jose, Tara is not the same as Jose, Tara, Katie.

Make a list to find the number of outcomes for each situation.

23. How many ways can Maria, Ryan, Liana, and Kurtis serve as president, vice-president, secretary, and treasurer?

24. How many ways can an editor and a reporter be chosen for their school paper from Jenna, Rico, Maralan, Ron, and Venessa?

Standardized Test Practice and Mixed Review

25. MULTIPLE CHOICE The menu for Wedgewood Pizza is shown. How many different one-topping pizzas can a customer order?

    Ⓐ 4   Ⓑ 8   Ⓒ 16   Ⓓ 20

Wedgewood Pizza	
Size	**Toppings**
8-inch	cheese
10-inch	pepperoni
12-inch	sausage
14-inch	mushroom

26. SHORT RESPONSE How many different ways can you choose 2 of 4 different donuts?

A bag contains 5 red marbles, 6 green marbles, and 4 blue marbles. One marble is selected at random. Find each probability. (Lesson 11-1)

27. P(green)   **28.** P(green or blue)   **29.** P(red or blue)

30. Estimate 84% of 24. (Lesson 10-8)

GETTING READY FOR THE NEXT LESSON

PREREQUISITE SKILL Solve each proportion. (Lesson 8-2)

31. $\dfrac{k}{9} = \dfrac{10}{45}$ **32.** $\dfrac{18}{72} = \dfrac{m}{24}$ **33.** $\dfrac{5}{c} = \dfrac{30}{96}$ **34.** $\dfrac{15}{35} = \dfrac{3}{d}$

Bias

What You'll LEARN

Determine whether a group is biased.

INVESTIGATE *Work in three large groups.*

Have you ever heard a listener win a contest on the radio? The disc jockey usually asks the listener "What's your favorite radio station?" When asked this question, the winner is usually **biased** because he or she favors the station that is awarding them a prize. In this lab, you will investigate bias.

STEP 1 Each group chooses one of the questions below.

> **Question 1:** Should the amount of time between classes be lengthened since classrooms are far away?
>
> **Question 2:** Our teacher will throw a pizza party if she is teacher of the year. Who is your favorite teacher?
>
> **Question 3**: Many students in our school buy lunch. What is your favorite school lunch?

STEP 2 Each member of the group answers the question by writing his or her response on a sheet of paper.

STEP 3 Collect and then record the responses in a table.

Writing Math

Work with a partner.

1. **Compare** the responses of your group to the responses of the other groups. Which questions may result in bias?

2. **Describe** the ways in which the wording of these questions may have influenced your answers.

3. **Tell** how these questions can be rewritten so they do not result in answers that are biased.

Tell whether each of the following survey locations might result in bias.

Type of Survey	Survey Location
4. favorite hobby	model train store
5. favorite season	public library
6. favorite TV show	skating rink
7. favorite food	Mexican restaurant

NEW Vocabulary

survey
population
sample
random

HANDS-ON Mini Lab

Work in three large groups.

In this activity, you will make a prediction about the number of left-handed or right-handed students in your school.

STEP 1 Have one student in each group copy the table shown.

Left- or Right-Handed?	
Trait	**Students**
left-handed	
right-handed	

STEP 2 Count the number of left-handed students and right-handed students in your group. Record the results.

STEP 3 Predict the number of left-handed and right-handed students in your school.

STEP 4 Combine your results with the other groups in your class. Make a class prediction.

1. When working in a group, how did your group predict the number of left-handed and right-handed students in your school?

2. Compare your group's prediction with the class prediction. Which do you think is more accurate? Explain.

A **survey** is a method of collecting information. The group being studied is the **population**. Sometimes, the population is very large. To save time and money, part of the group, called a **sample**, is surveyed.

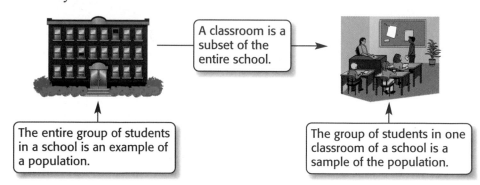

A classroom is a subset of the entire school.

The entire group of students in a school is an example of a population.

The group of students in one classroom of a school is a sample of the population.

STUDY TIP

Surveys A survey consists of questions that require a response. The most common types of surveys are personal interviews, telephone surveys, or mailed questionnaire surveys.

A good sample is:

• selected at **random**, or without preference,

• representative of the population, and

• large enough to provide accurate data.

EXAMPLE **Determine a Good Sample**

1 Every ninth person entering a grocery store one day is asked to state whether they support the proposed school tax for their school district. Determine whether the sample is a good sample.

- Asking every ninth person ensures a random survey.
- The sample should be representative of the larger population; that is, every person living in the school's district.
- The sample is large enough to provide accurate information.

So, this sample is a good sample.

Your Turn

a. One hundred people eating at an Italian restaurant are surveyed to name their favorite type of restaurant. Is this a good sample? Explain.

You can use the results of a survey to predict or estimate the actions of a larger group.

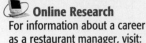

REAL-LIFE MATH

How Does a Restaurant Manager Use Math?
A restaurant manager uses math when he or she estimates food consumption, places orders with suppliers, and schedules the delivery of fresh food and beverages.

Online Research
For information about a career as a restaurant manager, visit: msmath1.net/careers

EXAMPLES **Make Predictions Using Proportions**

PIZZA Lorenzo asked every tenth student who walked into school to name their favorite pizza topping.

Favorite Pizza Topping	
Topping	**Students**
pepperoni	18
cheese	9
sausage	3
mushroom	2

2 What is the probability that a student will prefer pepperoni pizza?

$$P(\text{pepperoni}) = \frac{\text{number of students that like pepperoni}}{\text{number of students surveyed}}$$

$$= \frac{18}{32}$$

So, P(pepperoni) is $\frac{18}{32}$, 0.5625, or about 56%.

3 There are 384 students at the school Lorenzo attends. Predict how many students prefer pepperoni pizza.

STUDY TIP

Look Back To review solving **proportions**, see Lesson 10-2.

Use a proportion.   Let s = students who prefer pepperoni.

$\frac{18}{32} = \frac{s}{384}$ Write the proportion.

$18 \times 384 = 32 \times s$ Write the cross products.

$6{,}912 = 32s$ Multiply.

$\frac{6{,}912}{32} = \frac{32s}{32}$ Divide each side by 32.

$216 = s$

Of the 384 students, about 216 will prefer pepperoni pizza.

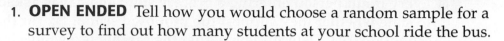

Skill and Concept Check

Writing
Math
Exercises 1 & 2

1. **OPEN ENDED** Tell how you would choose a random sample for a survey to find out how many students at your school ride the bus.

2. **FIND THE ERROR** Raheem and Elena are deciding on the best place to conduct a favorite movie star survey. Who is correct? Explain.

> Raheem
> Ask people standing in a line waiting to see a movie.

> Elena
> Ask people shopping at a local shopping center.

GUIDED PRACTICE

Determine whether the following sample is a good sample. Explain.

3. Every third student entering the school cafeteria is asked to name his or her favorite lunch food.

FOOD For Exercises 4 and 5, use the following information and the table shown.

Every tenth person entering a concert is asked to name his or her favorite milk shake flavor.

4. Find the probability that any person attending the concert prefers chocolate milk shakes.

5. Predict how many people out of 620 would prefer chocolate milk shakes.

Favorite Milk Shake	
Milk Shake	People
vanilla	30
chocolate	15
strawberry	10
mint	5

Practice and Applications

Determine whether each sample is a good sample. Explain.

6. Fifty children at a park are asked whether they like to play indoors or outdoors.

7. Every twentieth student leaving school is asked where the school should hold the year-end outing.

SOCCER For Exercises 8 and 9, use the following information.
In soccer, Isabelle scored 4 goals in her last 10 attempts.

8. Find the probability of Isabelle scoring a goal on her next attempt.

9. Suppose Isabelle attempts to score 20 goals. About how many goals will she make?

MUSIC For Exercises 10–13, use the table at the right to predict the number of students out of 450 that would prefer each type of music.

10. rock
11. alternative
12. country
13. pop

HOMEWORK HELP

For Exercises	See Examples
6–7	1
8	2
9–15	3

Extra Practice
See pages 617, 634.

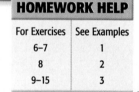

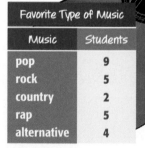

Favorite Type of Music	
Music	Students
pop	9
rock	5
country	2
rap	5
alternative	4

VOLUNTEERING For Exercises 14 and 15, use the graph at the right.

14. There were about 300,000 kids aged 10–14 living in Colorado in 2000. Predict the number of kids that volunteered.

15. In 2000, there were about 75,000 kids aged 10–14 living in Hawaii. Make a prediction as to the number of kids in this age group that did not volunteer.

 Data Update Find an estimate for the number of kids aged 10–14 currently living in your state. Predict the number of these kids that volunteer. Visit **msmath1.net/data_update** to learn more.

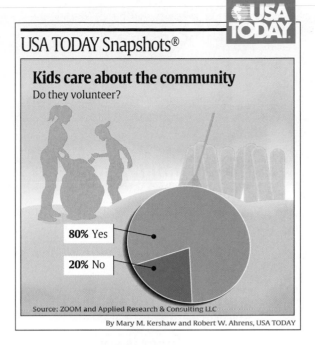

USA TODAY Snapshots®

Kids care about the community
Do they volunteer?

80% Yes

20% No

Source: ZOOM and Applied Research & Consulting LLC

By Mary M. Kershaw and Robert W. Ahrens, USA TODAY

16. **CRITICAL THINKING** Use the spinner shown at the right. If the spinner is spun 400 times, about how many times will the spinner stop on red or yellow?

Standardized Test Practice and Mixed Review

17. **MULTIPLE CHOICE** Students in a classroom were asked to name their favorite pet. The results are shown. If there are 248 students in the school, how many will prefer cats?

  Ⓐ 82   Ⓑ 74   Ⓒ 62   Ⓓ 25

Pet	Students
dog	12
cat	9
hamster	4
turtle	4
rabbit	2
guinea pig	5

18. **MULTIPLE CHOICE** Which sample is *not* a good sample?

  Ⓕ survey: favorite flower; location: mall

  Ⓖ survey: favorite hobby; location: model train store

  Ⓗ survey: favorite holiday; location: fast-food restaurant

  Ⓘ survey: favorite sport; location: library

19. How many ways can a person choose 3 videos from a stack of 6 videos? (Lesson 11-2)

Sarah randomly turns to a page in a 12-month calendar. Find each probability. (Lesson 11-1)

20. $P(\text{April or May})$ 21. $P(\text{not June})$ 22. $P(\text{begins with a J})$

GETTING READY FOR THE NEXT LESSON

PREREQUISITE SKILL Write each fraction in simplest form. (Lesson 5-2)

23. $\frac{4}{16}$ 24. $\frac{12}{36}$ 25. $\frac{8}{20}$ 26. $\frac{9}{24}$

 msmath1.net/self_check_quiz

Vocabulary and Concepts

1. **Explain** how a ratio is used to find the probability of an event. (Lesson 11-1)

2. In your own words, **describe** *sample*. (Lesson 11-3)

Skills and Applications

The spinner is spun once. Find each probability. Write each answer as a fraction, a decimal, and a percent. (Lesson 11-1)

3. $P(\text{red})$

4. $P(\text{blue or red})$

5. $P(\text{orange})$

6. $P(\text{not yellow})$

7. Draw a tree diagram to show the sample space for a choice of chicken, beef, or fish and rice or potatoes for dinner. (Lesson 11-2)

8. **PETS** How many different ways can a person choose 2 hamsters from a group of 4 hamsters? (Lesson 11-2)

9. How many outcomes are possible if a coin is tossed once and a number cube is rolled once? (Lesson 11-2)

10. Every tenth person leaving a grocery store is asked his or her favorite color. Determine whether the sample is a good sample. Explain. (Lesson 11-3)

Standardized Test Practice

11. **MULTIPLE CHOICE** There is a 25% chance that tomorrow's baseball game will be cancelled due to bad weather. What is the probability that the baseball game will *not* be cancelled? (Lesson 11-1)

    Ⓐ 85%   Ⓑ 75%

    Ⓒ 65%   Ⓓ 55%

12. **SHORT RESPONSE** Use the results shown to predict the number of people out of 300 that would prefer vanilla ice cream. (Lesson 11-3)

Favorite Ice Cream	
Flavor	**Amount**
vanilla	9
chocolate	8
strawberry	3

The GameZone

A Place To Practice Your Math Skills

Match 'Em Up

● **GET READY!**

Players: two
Materials: 16 index cards

● **GET SET!**

• Make a set of 8 probability cards using the numbers below.

$$\frac{1}{6} \quad \frac{1}{2} \quad 0 \quad \frac{1}{3} \quad \frac{5}{6} \quad \frac{2}{3} \quad 1 \quad \frac{1}{2}$$

• Make a set of 8 event cards describing the results of rolling a number cube.

6	3, 4, or 5	7	less than 6
composite	1 or prime	odd	greater than 0

● **GO!**

• Mix up each set of cards and place the cards facedown as shown.

• Player 1 turns over an event card and a probability card. If the number corresponds to the probability that the event occurs, the player removes the cards, a point is scored, and the player turns over two more cards.

Event Cards

☐ ☐ ☐ ☐ ☐ ☐ ☐ ☐

Probability Cards

☐ ☐ ☐ ☐ ☐ ☐ ☐ ☐

• If the fraction does not correspond to the event, the player turns the cards facedown, no points are scored, and it is the next player's turn.

• Take turns until all cards are matched.

• **Who Wins?** The player with more points wins.

What You'll LEARN

Find probability using area models.

HHANDS-ON **Mini Lab**

Materials
- centimeter grid paper
- rice
- colored pencils

Work with a partner.

Let's investigate the relationship between area and probability.

- Copy the squares on grid paper and shade them as shown.

- Randomly drop 50 pieces of rice onto the squares, from about 8 inches above.

- Record the number of pieces of rice that land in the blue region.

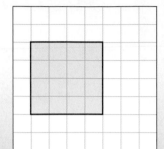

1. Find P(a piece of rice lands in blue region).

2. How does this ratio compare to the ratio $\dfrac{\text{area of blue region}}{\text{area of target}}$? Explain.

With a very large sample, the experimental probability should be close to the theoretical probability. The probability can be expressed as the ratio of the areas.

Key Concept	Probability and Area
Words	The probability of landing in a specific region of a target is the ratio of the area of the specific region to the area of the target.
Symbols	$P(\text{specific region}) = \dfrac{\text{area of specific region}}{\text{area of the target}}$

EXAMPLE **Find Probability Using Area Models**

1 Find the probability that a randomly thrown dart will land in the shaded region of the dartboard.

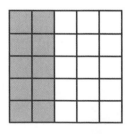

$$P(\text{shaded region}) = \frac{\text{area of shaded region}}{\text{area of the target}}$$
$$= \frac{10}{25} \text{ or } \frac{2}{5}$$

So, the probability is $\frac{2}{5}$, 0.4, or 40%.

2 **CHEERLEADING** A cheerleading squad plans to throw T-shirts into the stands using a sling shot. Find the probability that a T-shirt will land in the upper deck of the stands. Assume it is equally likely for a shirt to land anywhere in the stands.

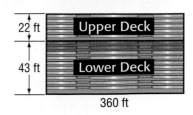

$$P(\text{upper deck}) = \frac{\text{area of upper deck}}{\text{area of the stands}}$$

Area of Upper Deck

$\ell \times w = 360 \times 22$
$\qquad = 7{,}920 \text{ sq ft}$

Area of Stands

$\ell \times w = 360 \times 65$
$\qquad = 23{,}400 \text{ sq ft}$

The upper and the lower deck make up the stands.

$$P(\text{upper deck}) = \frac{\text{area of upper deck}}{\text{area of the stands}}$$

$$= \frac{7{,}920}{23{,}400}$$

$$\approx \frac{1}{3}$$

So, the probability that a T-shirt will land in the upper deck of the stands is about $\frac{1}{3}$, $0.33\overline{3}$, or $33.\overline{3}\%$.

3 **CHEERLEADING** Predict how many times a T-shirt will land in the upper deck of the stands if the cheerleaders throw 15 T-shirts.

Write a proportion that compares the number of T-shirts landing in the upper deck to the number of T-shirts thrown. Let $n =$ the number of T-shirts landing in the upper deck.

$\dfrac{n}{15} = \dfrac{1}{3}$ ← T-shirts landing in upper deck
            ← T-shirts thrown

$n \times 3 = 15 \times 1$ Write the cross products.

$\qquad 3n = 15$ Multiply.

$\qquad \dfrac{3n}{3} = \dfrac{15}{3}$ Divide each side by 3.

$\qquad n = 5$

About 5 T-shirts will land in the upper deck.

Your Turn

a. Refer to the diagram in Example 2. Find the probability that a T-shirt will land in the lower deck of the stands.

b. Predict the number of T-shirts that will land in the lower deck if 36 T-shirts are thrown.

c. Predict the number of T-shirts that will land in the lower deck if 90 T-shirts are thrown.

STUDY TIP

Look Back To review **area of rectangles**, see Lesson 1-8.

REAL-LIFE MATH

CHEERLEADING Eighty percent of the schools in the U.S. have cheerleading squads. The most popular sport for cheerleading is football.

Source: www.about.com

Skill and Concept Check

Writing Math

Exercise 2

1. **OPEN ENDED** Draw a dartboard in which the probability of a dart landing in the shaded area is 60%.

2. **Draw** a model in which the probability of a dart landing in the shaded region is 25%. Then explain how to change the model so that the probability of a dart landing in the shaded region is 50%.

GUIDED PRACTICE

Find the probability that a randomly thrown dart will land in the shaded region of each dartboard.

3.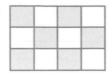

4.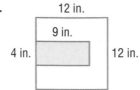

5. If a dart is randomly thrown 640 times at the dartboard in Exercise 4, about how many times will the dart land in the shaded region?

Practice and Applications

Find the probability that a randomly thrown dart will land in the shaded region of each dartboard.

HOMEWORK HELP

For Exercises	See Examples
6–11	1
12–17	2, 3

Extra Practice
See pages 617, 634.

6.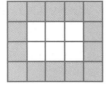

7.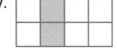

8.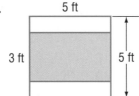

9.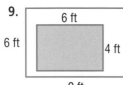

10.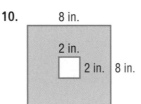

11.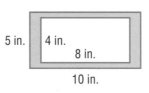

GAMES For Exercises 12–14, use the following information and the dartboard shown.

To win a prize, a dart must land on an even number. It is equally likely that the dart will land anywhere on the dartboard.

12. What is the probability of winning a prize?

13. Predict how many times a dart will land on an even number if it is thrown 9 times.

14. **MULTI STEP** To play the game, it costs $2 for 3 darts. About how much money would you have to spend in order to win 5 prizes?

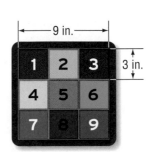

15. If a dart is randomly thrown 320 times at the dartboard in Exercise 10, about how many times will the dart land in the shaded region?

GOLF For Exercises 16 and 17, use the following information and the diagram at the right that shows part of one hole on a golf course.
Suppose a golfer tees off and it is equally likely that the ball lands somewhere in the area of the course shown.

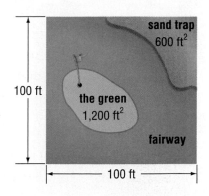

16. What is the probability that the ball lands in the sand trap?

17. If the golfer tees off from this hole 50 times, how many times can he expect the ball to land somewhere on the green?

18. CRITICAL THINKING Predict the number of darts that will land in each region of the dartboard shown if a dart is randomly thrown 200 times.

Standardized Test Practice and Mixed Review

19. MULTIPLE CHOICE Find the probability that a randomly thrown dart will land in the shaded region of the dartboard shown.

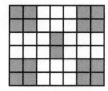

- **A** $\frac{2}{7}$
- **B** $\frac{2}{5}$
- **C** $\frac{3}{7}$
- **D** $\frac{3}{4}$

20. MULTIPLE CHOICE If 100 darts are thrown at the dartboard above, about how many would you expect to land in the shaded region?

- **F** 40
- **G** 43
- **H** 45
- **I** 46

21. Every tenth person entering the main entrance of a Yankees' baseball game is asked to name their favorite baseball player. Determine whether this is a good sample. Explain. (Lesson 11-3)

22. SCHOOL How many ways can a teacher choose 3 vocabulary words to put on a quiz from 4 different vocabulary words? (Lesson 11-2)

Find the percent of each number. (Lesson 10-7)

23. 30% of 60   **24.** 8% of 12   **25.** 110% of 150

GETTING READY FOR THE NEXT LESSON

PREREQUISITE SKILL Multiply. Write in simplest form. (Lesson 7-2)

26. $\frac{1}{5} \times \frac{3}{4}$ **27.** $\frac{2}{3} \times \frac{3}{7}$ **28.** $\frac{8}{15} \times \frac{3}{4}$ **29.** $\frac{4}{7} \times \frac{14}{15}$

What You'll Learn
Solve problems by making a table.

Make a Table

Hey Mario, my cat is going to have kittens! I'm hoping for three kittens, so I can keep one and give two to my cousins.

If your cat does have three kittens, what's the probability that she will have one female and two males? Let's **use a table** to find out.

	Explore	Each kitten will be either a male or a female. We need to find the probability that one kitten is female and the other two kittens are male.
	Plan	Make a table to list all of the possibilities to find the probability.

1st Kitten		2nd Kitten		3rd Kitten		Outcomes
F	M	F	M	F	M	
	X		X		X	M M M
	X		X	X		M M F
	X	X			X	M F M
	X	X		X		M F F
X			X		X	F M M
X			X	X		F M F
X		X			X	F F M
X		X		X		F F F

Solve

There are 3 ways a female, male, and male can be born. There are 8 possible outcomes. So, the probability is $\frac{3}{8}$, or 0.375, or 37.5%.

Examine If you draw a tree diagram to find the possible outcomes, you will find that the answer is correct.

Analyze the Strategy

1. **Explain** when to use the make a table strategy to solve a problem.
2. **Tell** the advantages of organizing information in a table. What are the disadvantages?
3. **Write** a problem that can be solved using the make a table strategy.

Apply the Strategy

Solve. Use the make a table strategy.

4. **CHILDREN** Find the probability that in a family with three children there is one boy and two girls.

5. **TESTS** A list of test scores is shown.

68	77	99	86	73	75	100
86	70	97	93	80	91	72
85	98	79	77	65	89	71

   How many more students scored 71 to 80 than 91 to 100?

6. **SCHOOL** The birth months of the students in Miss Miller's geography class are shown below. Make a frequency table of the data. How many more students were born in June than in August?

Birth Months		
June	July	April
March	July	June
October	May	August
June	April	October
May	October	April
September	December	January

Mixed Problem Solving

Solve. Use any strategy.

7. **MONEY** Julissa purchased her school uniforms for $135. This was the price after a 15% discount. What was the original price of her uniforms?

GEOMETRY For Exercises 8 and 9, use the square shown at the right.

7 ft

8. Find the length of each side of the square.

9. What is the perimeter of the square?

10. **GAMES** Sara tosses a beanbag onto an alphabet board. It is equally likely that the bag will land on any letter. Find the probability that the beanbag will land on one of the letters in her name.

11. **SCHOOL** Of the 150 students at Lincoln Middle School, 55 are in the orchestra, 75 are in marching band, and 25 are in both orchestra and marching band. How many students are in neither orchestra nor marching band?

12. **MONEY** Tetuso bought a clock radio for $9 less than the regular price. If he paid $32, what was the regular price?

13. **ROLLER COASTERS** The list shows how many roller coasters 20 kids rode at an amusement park.

5	10	0	12	8	7	2	6	4	1
0	6	3	11	5	9	13	8	14	3

   Make a frequency table to find how many more kids ride roller coasters 5 to 9 times than 10 to 14 times.

14. **MONEY** Luke collected $2 from each student to buy a gift for their teacher. If 27 people contributed, how much money was collected?

15. **STANDARDIZED TEST PRACTICE** Fabric that costs $6.59 per yard is on sale for 20% off per yard. Abigail needs to purchase $5\frac{3}{8}$ yards of the fabric. Which expression shows the amount of change c she should receive from a $50 bill?

    A $c = 50 - \left(6.59 + 0.20 + 5\frac{3}{8}\right)$

    B $c = 50 - (6.59)(0.80) - 5\frac{3}{8}$

    C $c = 50 - (6.59)(0.80)\left(5\frac{3}{8}\right)$

    D $c = 50 + (6.59)(0.80)\left(5\frac{3}{8}\right)$

Probability of Independent Events

What You'll LEARN

Find the probability of independent events.

NEW Vocabulary

independent events

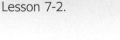

STUDY TIP

Look Back To review **multiplying fractions**, see Lesson 7-2.

HANDS-ON Mini Lab

Materials
- number cube
- bag
- 5 marbles

Work with a partner.

Make a tree diagram that shows the sample space for rolling a number cube and choosing a marble from the bag.

1. How many outcomes are in the sample space?
2. What is the probability of rolling a 5 on the number cube?
3. Find the probability of selecting a yellow marble.
4. Use the tree diagram to find $P(5 \text{ and yellow})$.
5. Describe the relationship between $P(5)$, $P(\text{yellow})$, and $P(5 \text{ and yellow})$.

In the Mini Lab, the outcome of rolling the number cube does not affect the outcome of choosing a marble. Two or more events in which the outcome of one event does not affect the outcome of the other event are **independent events**.

Key Concept — Probability of Independent Events

The probability of two independent events is found by multiplying the probability of the first event by the probability of the second event.

EXAMPLE — Find Probability of Independent Events

1 A coin is tossed, and the spinner shown is spun. Find the probability of tossing heads and spinning a 3.

$P(\text{heads}) = \frac{1}{2}$ $P(3) = \frac{1}{4}$

$P(\text{heads and 3}) = \frac{1}{2} \times \frac{1}{4}$ or $\frac{1}{8}$

So, the probability is $\frac{1}{8}$, 0.125, or 12.5%.

Your Turn Find the probability of each event.

a. $P(\text{tails and even})$

b. $P(\text{heads and less than 4})$

Standardized Test Practice

EXAMPLE Find Probability of Independent Events

2 **GRID-IN TEST ITEM** Amanda placed 2 red marbles and 6 yellow marbles into a bag. She selected 1 marble without looking, replaced it, and then selected a second marble. Find the probability that each marble selected was *not* yellow.

Read the Test Item To find the probability, find P(not yellow and not yellow).

Solve the Test Item

First marble: P(not yellow) $= \frac{2}{8}$ or $\frac{1}{4}$

Second marble: P(not yellow) $= \frac{2}{8}$ or $\frac{1}{4}$

So, P(not yellow and not yellow) is $\frac{1}{4} \times \frac{1}{4}$ or $\frac{1}{16}$.

Fill in the Grid

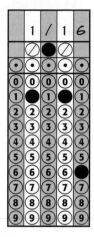

Test-Taking Tip
The Princeton Review

Read the Test Item
When reading the question, always look for more than one way to solve the problem. For example, the probability question in Example 2 can also be solved by finding P(red and red).

Skill and Concept Check

Writing Math
Exercises 1 & 3

1. **Explain** how to find the probability of two independent events.

2. **OPEN ENDED** Give an example of two independent events.

3. **Which One Doesn't Belong?** Identify the situation that is *not* a pair of independent events. Explain your reasoning.

| tossing two coins | choosing one marble from a bag of marbles without replacing it | rolling two number cubes | spinning two spinners |

GUIDED PRACTICE

One marble is selected from the bag shown without looking, and the spinner shown is spun. Find the probability of each event.

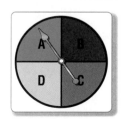

4. P(D and blue)

5. P(B and yellow)

6. P(consonant and green)

7. **PETS** What is the probability that in a litter of 2 puppies the first puppy born is female and the second puppy born is male?

8. **PENS** A desk drawer contains 8 blue pens and 4 red pens. Suppose one pen is selected without looking, replaced, and another pen is selected. What is the probability that the first pen selected is blue and the second pen is red?

Practice and Applications

HOMEWORK HELP

For Exercises	See Examples
9–15	1
17–26	2

Extra Practice
See pages 617, 634.

A letter card is chosen, and a number cube is rolled. Find the probability of each event.

9. P(A and 5)

10. P(K and even)

11. P(J and less than 3)

12. P(vowel and 7)

13. P(consonant and odd)

14. P(M or T and greater than 3)

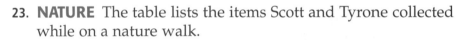

SPORTS **For Exercises 15 and 16, use the following information.**
The probability that the Jets beat the Zips is 0.4. The probability that the Jets beat the Cats is 0.6.

15. What is the probability that the Jets beat both the Zips and the Cats?

16. Explain whether the Jets beating both the Zips and the Cats is less likely or more likely to happen than the Jets beating either the Zips or the Cats.

One marble is chosen from the bag shown without looking, replaced, and another marble chosen. Find each probability.

17. P(green and red)

18. P(green and orange)

19. P(purple and red)

20. P(orange and blue)

21. P(green and purple)

22. P(purple and purple)

23. **NATURE** The table lists the items Scott and Tyrone collected while on a nature walk.

Name	Stones	Pinecone	Acorn
Scott	12	12	16
Tyrone	6	8	10

Each person reaches into his bag and randomly selects an object. Find the probability that Scott chooses an acorn and Tyrone chooses a stone.

24. **SCHOOL** A quiz has one true/false question and one multiple-choice question with possible answer choices A, B, C, and D. If you guess each answer, what is the probability of answering both questions correctly?

CANDY **For Exercises 25 and 26, use the graph at the right. It shows the colors of candy that come in a bag of chocolate candies. Suppose one candy is chosen without looking, replaced, and another candy chosen.**

25. Find P(red and not green).

26. What is P(blue or green and not yellow)?

Colors in Each Bag of Chocolate

30% 20% 20% 10% 10% 10%

27. **RESEARCH** Use the Internet or another source to find one example of probability being used in everyday life. Write a probability question and then solve your question.

28. Give a counterexample for the following statement.
If the probability of event A is less than $\frac{3}{4}$ and the probability of event B is less than $\frac{1}{2}$, then the probability of A and B is less than $\frac{1}{4}$.

29. **CRITICAL THINKING** Two spinners are spun. The probability of landing on red on both spinners is 0.2. The probability of landing on red on the second spinner is 0.5. What is the probability of landing on red on the first spinner?

30. **MULTIPLE CHOICE** A coin is tossed and a card is chosen from a set of ten cards labeled 1–10. Find P(tails, prime).

  (A) $\frac{1}{10}$ (B) $\frac{1}{5}$ (C) $\frac{3}{10}$ (D) $\frac{2}{5}$

31. **MULTIPLE CHOICE** A yellow and a blue number cube are rolled. Find the probability of rolling a 6 on the yellow number cube and a number less than 5 on the blue number cube.

  (F) $\frac{1}{18}$ (G) $\frac{1}{9}$ (H) $\frac{1}{6}$ (I) $\frac{2}{3}$

32. It is equally likely that a thrown dart will land anywhere on the dartboard shown. Find the probability of a randomly thrown dart landing in the shaded region.
(Lesson 11-4)

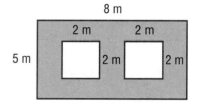

33. **SOCCER** Ian scores 6 goals in every 10 attempts. About how many goals will he score if he attempts 15 goals?
(Lesson 11-3)

Subtract. Write in simplest form. (Lesson 6-5)

34. $5\frac{1}{6} - 3\frac{8}{9}$ 35. $8\frac{2}{3} - 3\frac{6}{7}$ 36. $6\frac{3}{5} - 4\frac{2}{3}$ 37. $9\frac{2}{3} - 5\frac{4}{5}$

WebQuest **Interdisciplinary Project**

Take Me Out to the Ballgame

It's time to complete your project. Use the scale drawing you've created and the data you have gathered about your baseball teams to prepare a Web page or poster. Be sure to include a spreadsheet with your project.

msmath1.net/webquest

Vocabulary and Concept Check

complementary events (p. 429)	population (p. 438)	survey (p. 438)
event (p. 428)	random (p. 438)	theoretical probability (p. 428)
independent events (p. 450)	sample (p. 438)	tree diagram (p. 433)
outcomes (p. 428)	sample space (p. 433)	

Choose the letter of the term that best matches each phrase.

1. the ratio of the number of ways an event can occur to the number of possible outcomes

2. a specific outcome or type of outcome

3. when one event occurring does not affect another event

4. events that happen by chance

5. a diagram used to show all of the possible outcomes

6. the set of all possible outcomes

7. a randomly selected group chosen for the purpose of collecting data

a. tree diagram
b. outcomes
c. random events
d. event
e. sample
f. probability
g. independent events
h. sample space

Lesson-by-Lesson Exercises and Examples

 Theoretical Probability (pp. 428–431)

One coin shown is chosen without looking. Find each probability. Write each answer as a fraction, a decimal, and a percent.

8. P(nickel)   9. P(not dime)

10. P(quarter or penny)

11. P(nickel or dime)

A number cube is rolled. Find each probability. Write each answer as a fraction, a decimal, and a percent.

12. P(5)   13. P(less than 4)

14. P(odd)   15. P(at least 5)

Example 1
The spinner shown is spun once. Find the probability of spinning blue.

There are six equally likely outcomes on the spinner. One of the six is blue.

P(blue)

$= \dfrac{\text{number of ways to spin blue}}{\text{total number of possible outcomes}}$

$= \dfrac{1}{6}$

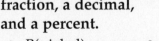 msmath1.net/vocabulary_review

Outcomes (pp. 433–436)

Draw a tree diagram to show the sample space for each situation. Then tell how many outcomes are possible.

16. a choice of black or blue jeans in tapered, straight, or baggy style

17. a choice of soup or salad with beef, chicken, fish, or pasta

18. a choice of going to a basketball game, an amusement park, or a concert on a Friday or a Saturday

A coin is tossed, and a number cube is rolled.

19. How many outcomes are possible?
20. Find P(tails, 2).

21. **MARBLES** How many ways can 3 marbles be selected from a bag of 6 different marbles?

Example 2 Suppose you have a choice of a sugar cone (S) or a waffle cone (W) and blueberry (B), mint (M), or peach (P) ice cream?

a. **How many ice cream cones are possible?**

Cone	Ice cream	Outcome
S	B	SB
	M	SM
	P	SP
W	B	WB
	M	WM
	P	WP

There are 6 possible ice cream cones.

b. **If you choose at random, find the probability of selecting a sugar cone with blueberry ice cream or a waffle cone with mint ice cream.**

P(blueberry/sugar or mint/waffle)
$= \frac{2}{6}$ or $\frac{1}{3}$

Making Predictions (pp. 438–441)

SCHOOL For Exercises 22 and 23, use the following information.

Out of 40 students, 14 are interested in publishing a school newspaper.

22. What is the probability that a student at this school would be interested in publishing a school newspaper?

23. If there are 420 students, how many would you expect to be interested in publishing a school newspaper?

24. Twenty residents of Florida are asked whether they prefer warm or cold weather. Determine whether the sample is a good sample. Explain.

Example 3 If 12 out of 50 people surveyed prefer to watch TV after 11 P.M., how many people out of 1,000 would prefer to watch TV after 11 P.M.?

Let p represent the number of people who would prefer to watch TV after 11 P.M.

$\frac{12}{50} = \frac{p}{1,000}$ Use a proportion.

$12 \times 1,000 = 50 \times p$ Find the cross products.

$12,000 = 50p$ Multiply.

$\frac{12,000}{50} = \frac{50p}{50}$ Divide.

$240 = p$ Simplify.

Of the 1,000 people, 240 would prefer to watch TV after 11 P.M.

11-4 Probability and Area (pp. 444–447)

Find the probability that a randomly thrown dart will land in the shaded region of each dartboard.

25.

26.

27.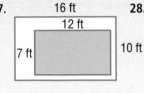
16 ft
12 ft
7 ft

28.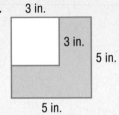
3 in.
3 in.
5 in.
5 in.

10 ft

29. Suppose you threw a dart 150 times at the dartboard in Exercise 28. How many times would you expect it to land in the shaded region?

Example 4 The figure shown represents a dartboard. Find the probability that a randomly thrown dart lands in the shaded region.

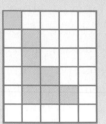

P(shaded region)

$$= \frac{\text{area of shaded region}}{\text{area of target}}$$

$$= \frac{8}{30} \text{ or } \frac{4}{15}$$

So, the probability is $\frac{4}{15}$, $1.26\overline{6}$, or $26.\overline{6}\%$.

11-5 Probability of Independent Events (pp. 450–453)

A coin is tossed and a number cube is rolled. Find the probability of each event.

30. P(heads and 4)
31. P(tails and even)
32. P(heads and 5 or 6)
33. P(tails and 2, 3, or 4)
34. P(heads and prime)

35. **EARTH SCIENCE** The probability of rain on Saturday is 0.6. The probability of rain on Sunday is 0.3. What is the probability that it will rain on both days?

Example 5 Two number cubes are rolled.

a. Find P(odd and 4).

$$P(\text{odd}) = \frac{1}{2} \qquad P(4) = \frac{1}{6}$$

$$P(\text{odd and 4}) = \frac{1}{2} \times \frac{1}{6} \text{ or } \frac{1}{12}$$

So, the probability is $\frac{1}{12}$, $0.08\overline{3}$, or $8.\overline{3}\%$.

b. What is P(6 and less than 5)?

$$P(6) = \frac{1}{6} \qquad P(\text{less than 5}) = \frac{2}{3}$$

$$P(\text{6 and less than 5}) = \frac{1}{6} \times \frac{2}{3} \text{ or } \frac{1}{9}$$

So, the probability is $\frac{1}{9}$, $0.11\overline{1}$, or $11.\overline{1}\%$.

Practice Test

Vocabulary and Concepts

1. **List** three characteristics of a good sample.

2. **Define** *independent events*.

Skills and Applications

A set of 20 cards is numbered 1–20. One card is chosen without looking. Find each probability. Write as a fraction, a decimal, and a percent.

3. $P(8)$ 4. $P(3 \text{ or } 10)$ 5. $P(\text{prime})$ 6. $P(\text{odd})$

7. **PENS** How many ways can 2 pens be chosen from 4 pens?

MUSIC For Exercises 8 and 9, use the table at the right and the following information.
Alonso asked every fourth sixth grade student who walked into a school dance to name their favorite sport.

8. Find the probability a student prefers football.

9. If there are 375 students in the sixth grade, how many can be expected to prefer football?

Favorite Sport	
Sport	**Students**
football	52
soccer	22
baseball	16
hockey	10

Find the probability that a randomly thrown dart will land in the shaded region of each dartboard.

10.
11.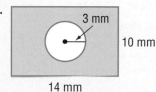
12.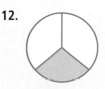

A coin is tossed and the spinner shown is spun. Find the probability of each event.

13. $P(\text{tails and blue})$ 14. $P(\text{heads and not green})$

15. $P(\text{tails and pink, yellow, or green})$

Standardized Test Practice

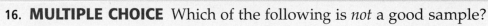

16. **MULTIPLE CHOICE** Which of the following is *not* a good sample?

  Ⓐ survey: best movie; location: home improvement store

  Ⓑ survey: worst song; location: park

  Ⓒ survey: favorite dessert; location: ice cream shop

  Ⓓ survey: least favorite school subject; location: school play

PART 1 Multiple Choice

Record your answers on the answer sheet provided by your teacher or on a sheet of paper.

1. Find 200×2. (Prerequisite Skill, p. 590)
   - A 4
   - B 40
   - C 100
   - D 400

2. Which list orders 34.1, 33.9, 33.8, 34.2, 34.9 from least to greatest? (Lesson 3-2)
   - F 34.1, 33.9, 34.2, 33.8, 34.9
   - G 33.8, 33.9, 34.1, 34.2, 34.9
   - H 34.1, 34.9, 34.2, 33.9, 33.8
   - I 33.9, 33.8, 34.1, 34.2, 34.9

3. To the nearest tenth, how many times greater is the cost of a pancake breakfast at Le Café than at Pancakes & More? (Lesson 4-4)

Pancake Breakfast	
Restaurant	Cost
Skyview Chalet	$3.60
Delightful Diner	$4.10
Pancakes & More	$2.20
Le Café	$6.40

   - A 2.9
   - B 3.1
   - C 3.4
   - D 4.2

4. What is 0.40 written as a fraction? (Lesson 5-6)
   - F $\frac{4}{100}$
   - G $\frac{6}{30}$
   - H $\frac{2}{5}$
   - I $\frac{40}{10}$

5. What is the value of $\frac{7}{12} \times \frac{4}{15}$? (Lesson 7-2)
   - A $\frac{1}{150}$
   - B $\frac{7}{45}$
   - C $\frac{7}{30}$
   - D $\frac{11}{27}$

6. A coin is dropped into a river from 6 meters above the water's surface. If the coin falls to a depth of 4 meters, what is the total distance that the coin falls? (Lesson 8-2)
   - F -6 m
   - G -4 m
   - H 4 m
   - I 10 m

7. Suppose 60% of kids like video games. Predict the number of kids that prefer video games out of a group of 200 kids. (Lesson 10-8)
   - A 40
   - B 60
   - C 120
   - D 180

8. What letter should be placed on the spinner so that the probability of landing on that letter would be $\frac{1}{2}$? (Lesson 11-1)

   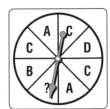

   - F A
   - G B
   - H C
   - I D

9. How many different jeans and shirt combinations can be made with blue jeans and black jeans, and a white shirt, a blue shirt, and a yellow shirt? (Lesson 11-2)
   - A 6
   - B 12
   - C 16
   - D 20

10. One marble is selected from each jar without looking. Find the probability that both marbles are black. (Lesson 11-5)

   - F $\frac{1}{5}$
   - G $\frac{6}{15}$
   - H $\frac{2}{3}$
   - I $\frac{14}{17}$

TEST-TAKING TIP

Question 9 Drawing a tree diagram may help you find the answer.

PART 2 Short Response/Grid In

Record your answers on the answer sheet provided by your teacher or on a sheet of paper.

11. Refer to the table below that shows the length of time spent at the zoo.

Family	Time (h)
Martinez	2
Kowalkski	3
Bridi	1
Munemo	4
Smith	2
Elstein	5

What is the mode of the data? (Lesson 2-7)

12. Shaquille is making trail mix. He uses the ingredients below.

Ingredient	Amount (lb)
peanuts	$1\frac{1}{4}$
raisins	$\frac{1}{6}$
dried fruit	$\frac{2}{3}$
almonds	$\frac{1}{4}$

How many pounds of trail mix did he make? (Lesson 6-4)

13. Margaretta notices that 23 out of 31 birds landed in a large oak tree in her backyard and the remaining birds landed in a small pine tree in her backyard. What percent of birds landed in the oak tree? (Lesson 10-7)

14. The probability of a coin landing on heads is $\frac{1}{2}$. The 1 in the numerator stands for the number of ways that a coin can land heads up. What does the 2 stand for? (Lesson 11-1)

15. If a dart is thrown at the dartboard shown, what is the probability that it will hit region A? (Lesson 11-4)

B	B	C
C	B	C
A	B	C

16. What is the probability that a dart will land within the large square but outside the small square? (Lesson 11-4)

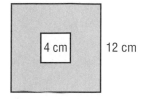

4 cm   12 cm

17. The probability that it will rain on Saturday is 65%. The probability that it will rain on Sunday is 40%. What is the probability that it will rain on both days? (Lesson 11-5)

PART 3 Extended Response

Record your answers on a sheet of paper. Show your work.

18. Michelle can select her lunch from the following menu. (Lesson 11-2)

Sandwich	Drink	Fruit
cheese	apple juice	apple
ham	milk	banana
tuna	orange juice	pear

a. What are the possible combinations of one sandwich, one drink, and one piece of fruit Michelle can choose? Show these combinations in a tree diagram.

b. If the tuna sandwich were removed from the menu, how many fewer lunch choices would Michelle have?

c. If an orange is added to the original menu, how many lunch combinations will there be?

UNIT 6
Measurement and Geometry

In this unit, you will solve problems involving customary and metric measures, angle measure, area, and volume.

ROAD TRIP

Let's hit the road! Come join us on a cross-country trip to see the nation. In preparation, you'll need a map to figure out how far you're traveling. You're also going to need to load up your car with all the necessary travel essentials. Don't overdo it though, there's only so much room in there. Put on your geometry thinking cap and let's get packing!

Log on to msmath1.net/webquest to begin your WebQuest.

CHAPTER 12 Measurement

"What do pumpkins have to do with math?"

A pumpkin pie recipe calls for 15 ounces of pumpkin. *About* how many pies can be made with eight pounds of pumpkin? To estimate, you need to change pounds to ounces. Being able to convert measures of length, weight, and capacity is a useful skill for solving many real-life problems.

You will solve problems by converting customary units in Lessons 12-1 and 12-2.

GETTING STARTED

Take this quiz to see whether you are ready to begin Chapter 12. Refer to the lesson number in parentheses if you need more review.

▶ Vocabulary Review

State whether each sentence is *true* or *false*. If *false*, replace the underlined word to make a true sentence.

1. When <u>adding</u> decimals, you must align the decimal point. (Lesson 3-5)

2. When a decimal between 0 and 1 is <u>divided</u> by a whole number greater than one, the result is a greater number. (Lesson 4-1)

▶ Prerequisite Skills

Add. (Lesson 3-5)

3. $8.73 + 11.96$ 4. $54.26 + 21.85$

5. $3.04 + 9.92$ 6. $76.38 + 44.15$

7. $7.9 + 8.62$ 8. $15.37 + 9.325$

Subtract. (Lesson 3-5)

9. $17.46 - 3.29$ 10. $68.05 - 24.38$

11. $9.85 - 2.74$ 12. $73.91 - 50.68$

13. $8.4 - 3.26$ 14. $27 - 8.62$

Multiply. (Lesson 4-1)

15. 3.8×100 16. 5.264×10

17. 6.75×10 18. $8.9 \times 1,000$

19. 7.18×100 20. 24.9×100

Divide. (Lesson 4-3)

21. $9.8 \div 100$ 22. $12.25 \div 10$

23. $4.5 \div 10$ 24. $26.97 \div 100$

25. $7.3 \div 100$ 26. $16.4 \div 1,000$

Measurement Make this Foldable to help you organize your notes on metric and customary units. Begin with a sheet of 11" by 17" paper.

STEP 1 Fold
Fold the paper in half along the length. Then fold in thirds along the width.

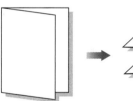

STEP 2 Unfold and Cut
Open and cut along the two top folds to make three strips. Cut off the first strip.

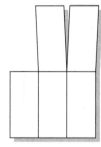

STEP 3 Refold
Refold the two top strips down and fold the entire booklet in thirds along the length.

STEP 4 Unfold and Label
Unfold and draw lines along the folds. Label as shown.

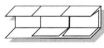

Reading and Writing As you read and study the chapter, write notes and definitions on the tabs. Write examples and computations under the tabs.

Area and Perimeter

What You'll LEARN

Explore changes in area and perimeter of rectangles.

Materials

• centimeter grid paper

INVESTIGATE *Work in groups of three.*

If you increase the side lengths of a rectangle, how are the area and the perimeter affected? In this lab, you will investigate relationships between the areas and perimeters of original figures and those of the newly created figures.

STEP 1 On centimeter grid paper, draw and label a rectangle with a length of 6 centimeters and a width of 2 centimeters.

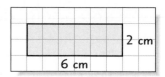

STEP 2 Find the area and perimeter of this original rectangle. Then record the information in a table like the one shown.

Rectangle	Length (cm)	Width (cm)	Area (sq cm)	Perimeter (cm)
original	6	2		
A	12	4		
B	18	6		
C	24	8		

STEP 3 Repeat Steps 1 and 2 for rectangles A, B, and C, whose dimensions are shown in the table.

Writing Math

1. Describe how the dimensions of rectangles A, B, and C are different from the original rectangle.

2. Describe how the area of the original rectangle changed when the length and width were both doubled.

3. Describe how the perimeter of the original rectangle changed when the length and width were both doubled.

4. Describe how the area and the perimeter of the original rectangle changed when the length and width were both tripled.

5. **Draw** a rectangle whose length and width are half those of the original rectangle. Describe how the area and perimeter changes.

6. Suppose the perimeter of a rectangle is 15 centimeters. **Make a conjecture** about the perimeter of the rectangle if the length and the width are both doubled.

Look Back You can review **area** and **perimeter of rectangles** in Lessons 1-8 and 4-5, respectively.

12-1

Length in the Customary System

What You'll LEARN

Change units of length and measure length in the customary system.

NEW Vocabulary

inch
foot
yard
mile

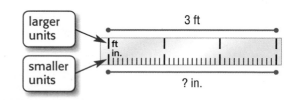

HANDS-ON Mini Lab

Work with a partner.

Materials
• string
• scissors
• yardstick
• tape measure

STEP 1 Using string, measure and cut the lengths of your arm and your shoe.

STEP 2 Use the strings to find the classroom length in *arms* and classroom width in *shoes*. Record the nonstandard measures.

Measure	Nonstandard	Standard
classroom length	—— arms	—— yards
classroom width	—— shoes	—— feet

STEP 3 Use a yardstick or tape measure to find the length in yards and width in feet. Record the standard measures.

1. Compare your nonstandard and standard measures with the measures of other groups. Are they similar? Explain.

2. Explain the advantages and the disadvantages of using nonstandard measurement and standard measurement.

The most commonly used customary units of length are shown below.

Key Concept	Customary Units Of Length
Unit	**Model**
1 **inch** (in.)	width of a quarter
1 **foot** (ft) = 12 in.	length of a large adult foot
1 **yard** (yd) = 3 ft	length from nose to fingertip
1 **mile** (mi) = 1,760 yd	10 city blocks

To change from larger units of length to smaller units, multiply.

STUDY TIP

Measurement
When changing from larger units to smaller units, there will be a greater number of smaller units than larger units.

EXAMPLE Change Larger Units to Smaller Units

1 3 ft = __?__ in.

Since 1 foot = 12 inches, multiply by 12.

$3 \times 12 = 36$

So, 3 feet = 36 inches.

larger units → smaller units

3 ft
? in.

To change from smaller units to larger units, divide.

Measurement
When changing from smaller units to larger units, there will be fewer larger units than smaller units.

EXAMPLE **Change Smaller Units to Larger Units**

2 21 ft = __?__ yd

Since 3 feet = 1 yard, divide by 3.

$21 \div 3 = 7$

So, 21 feet = 7 yards.

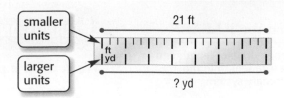

smaller units

larger units

21 ft

ft
yd

? yd

There will be fewer larger units than smaller units.

Your Turn Complete.

a. 5 ft = __?__ in.   **b.** 3 yd = __?__ ft   **c.** 2,640 yd = __?__ mi

Rulers are usually separated into eighths of an inch.

$\frac{1}{8}$ inch

in.   1   2   3

The longest marks on a ruler represent an inch, the next smaller marks represent $\frac{1}{2}$ inch, and so on.

EXAMPLE **Draw a Line Segment**

3 Draw a line segment measuring $2\frac{3}{8}$ inches.

Draw a line segment from 0 to $2\frac{3}{8}$.

in.   1   2   3

EXAMPLE **Measure Length**

4 **KEYS** Measure the length of the key to the nearest half, fourth, or eighth inch.

in.   1   2

The key is between $1\frac{3}{4}$ inches and $1\frac{7}{8}$ inches. It is closer to $1\frac{3}{4}$ inches.

The length of the key is about $1\frac{3}{4}$ inches.

Skill and Concept Check

Writing Math
Exercise 1

1. **Describe** how you would change 12 feet to yards.

2. **OPEN ENDED** Draw a segment that is between $1\frac{1}{2}$ inches and $2\frac{1}{4}$ inches long. State the measure of the segment to the nearest fourth inch and eighth inch.

GUIDED PRACTICE

Complete.

3. 4 yd = __?__ ft   4. 72 in. = __?__ yd   5. 4 mi = __?__ yd

Draw a line segment of each length.

6. $1\frac{1}{4}$ in.   7. $\frac{5}{8}$ in.

Measure the length of each line segment or object to the nearest half, fourth, or eighth inch.

8. •————————•   9.

10. Which is greater: $2\frac{1}{2}$ yards or 8 feet? Explain.

11. **IDENTIFICATION** Measure the length and width of a student ID card or driver's license to the nearest eighth inch.

Practice and Applications

Complete.

12. 5 yd = __?__ in.   13. 6 yd = __?__ ft   14. $6\frac{1}{2}$ ft = __?__ in.

15. 3 mi = __?__ ft   16. 48 in. = __?__ ft   17. 10 ft = __?__ yd

HOMEWORK HELP

For Exercises	See Examples
12–17	1, 2
18–21	3
22–27	4

Extra Practice
See pages 618, 635.

Draw a line segment of each length.

18. $2\frac{1}{2}$ in.   19. $3\frac{1}{4}$ in.   20. $\frac{3}{4}$ in.   21. $1\frac{3}{8}$ in.

Measure the length of each line segment or object to the nearest half, fourth, or eighth inch.

22.   23. 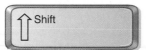   24. •————————•

25. •——•   26.   27.

28. Which is greater: $1\frac{1}{3}$ yards or 45 inches? Explain.

29. Which is greater: 32 inches or $2\frac{1}{2}$ feet? Explain.

30. TOOLS Measure the width of the bolt. What size wrench would tighten it: $\frac{1}{2}$ inch, $\frac{5}{8}$ inch, or $\frac{3}{4}$ inch?

31. MONEY Measure the length of a dollar bill to the nearest sixteenth inch.

32. RESEARCH Estimate the height of an adult giraffe. Then use the Internet or another source to find the actual height.

For Exercises 33–35, estimate the length of each object. Then measure to find the actual length.

33. the length of your bedroom to the nearest yard

34. the width of a computer mouse to the nearest half inch

35. the height of your dresser to the nearest foot

36. CRITICAL THINKING How many sixteenth inches are in a foot? How many half inches are in a yard?

EXTENDING THE LESSON The symbol for foot is ' and the symbol for inches is ". You can add and subtract lengths with different units.

```
                    7'  8"    ◄── Add the inches.
  Add the feet. ──►  + 3'  6" ◄──
                    10'14"      Since 14" = 1'2", the sum is 10' + 1'2" or 11'2".
```

Add or subtract.

37. 15'7" + 1'5"

38. 6'9" − 4'3"

39. 2 yd 2 ft + 4 yd 2 ft

Standardized Test Practice and Mixed Review

40. MULTIPLE CHOICE Choose the greatest measurement.

Ⓐ 53 in.   Ⓑ $4\frac{5}{7}$ ft   Ⓒ $\frac{1}{8}$ ft   Ⓓ $\frac{1}{8}$ yd

41. SHORT RESPONSE The length of a football field is 100 yards. How many inches is this?

42. A coin is tossed, and the spinner is spun. Find P(tails and 4). (Lesson 11-5)

43. Find the probability that a randomly thrown dart will land in the shaded region. (Lesson 11-4)

GETTING READY FOR THE NEXT LESSON

PREREQUISITE SKILL Multiply or divide. (Pages 590, 591)

44. 4×8 **45.** 16×5 **46.** $5,000 \div 2,000$ **47.** $400 \div 8$

12-1b Spreadsheet Investigation

A Follow-Up of Lesson 12-1

Area and Perimeter

What You'll LEARN

Use a spreadsheet to compare areas of rectangles with the same perimeter.

A computer spreadsheet is a useful tool for comparing different rectangular areas that have the same perimeter.

ACTIVITY

Suppose 24 sections of fencing, each one foot long, are to be used to enclose a rectangular vegetable garden. What are the dimensions of the garden with the largest possible area?

If w represents the width of the garden, then $12 - w$ represents the length.

Set up the spreadsheet as shown. The possible widths are listed in column A. The spreadsheet calculates the lengths in column B, the perimeters in column C, and the areas in column D.

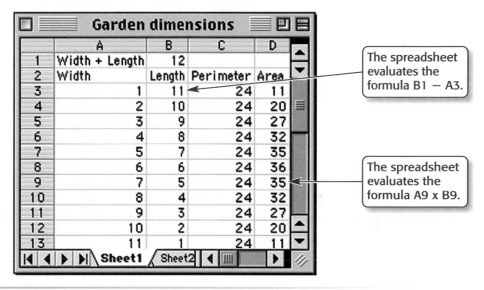

The spreadsheet evaluates the formula B1 − A3.

The spreadsheet evaluates the formula A9 x B9.

EXERCISES

1. **Explain** why the formula for the length is $12 - w$ instead of $24 - w$.

2. Which garden size results in the largest area?

3. Which cell should you modify to find the largest area that you can enclose with 40 sections of fencing, each 1 foot long?

4. Use the spreadsheet to find the dimensions of the largest area you can enclose with 40, 48, and 60 feet of fencing.

5. **Make a conjecture** about the shape of a rectangular garden when the area is the largest.

Capacity and Weight in the Customary System

What You'll LEARN

Change units of capacity and weight in the customary system.

NEW Vocabulary

fluid ounce
cup
pint
quart
gallon
ounce
pound
ton

Link to READING

Everyday Meaning of Capacity: the maximum amount that can be contained, as in a theater filled to capacity

HANDS-ON Mini Lab

Materials
- gallon container
- quart container
- pint container
- water

Work in groups of 4 or 5.

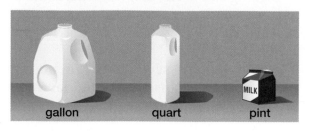
gallon   quart   pint

STEP 1 Fill the pint container with water. Then pour the water into the quart container. Repeat until the quart container is full. Record the number of pints needed to fill the quart.

STEP 2 Fill the quart container with water. Then pour the water into the gallon container. Repeat until the gallon container is full. Record the number of quarts needed to fill the gallon.

Complete.

1. 1 quart = __?__ pints
2. 2 quarts = __?__ pints
3. 1 gallon = __?__ quarts
4. 1 gallon = __?__ pints

5. What fractional part of 1 gallon would fit in 1 pint?

6. How many gallons are equal to 12 quarts? Explain.

The most commonly used customary units of capacity are shown.

Key Concept	Customary Units Of Capacity
Unit	**Model**
1 **fluid ounce** (fl oz)	2 tablespoons of water
1 **cup** (c) = 8 fl oz	coffee cup
1 **pint** (pt) = 2 c	small ice cream container
1 **quart** (qt) = 2 pt	large measuring cup
1 **gallon** (gal) = 4 qt	large plastic jug of milk

As with units of length, to change from larger units to smaller units, multiply. To change from smaller units to larger units, divide.

EXAMPLES Change Units of Capacity

Complete.

1 $3 \text{ qt} = \underline{\quad?\quad} \text{ pt}$

$3 \times 2 = 6$

So, 3 quarts = 6 pints.

THINK 1 quart = 2 pints

Multiply to change a larger unit to a smaller unit.

2 $64 \text{ fl oz} = \underline{\quad?\quad} \text{ pt}$

$64 \div 8 = 8$

$8 \div 2 = 4$

So, 64 fluid ounces = 4 pints.

THINK 8 fluid ounces = 1 cup and 2 cups = 1 pint. You need to divide twice.

Divide to change fluid ounces to cups. So 64 fl oz = 8 c.

Divide to change cups to pints.

The most commonly used customary units of weight are shown below.

Key Concept	Customary Units Of Weight
Unit	**Model**
1 **ounce** (oz)	pencil
1 **pound** (lb) = 16 oz	package of notebook paper
1 **ton** (T) = 2,000 lb	small passenger car

EXAMPLES Change Units of Weight

3 **TRUCKS** A truck weighs 7,000 pounds. How many tons is this?

$7,000 \text{ lb} = \underline{\quad?\quad} \text{ T}$ THINK 2,000 pounds = 1 ton

$7,000 \div 2,000 = 3\frac{1}{2}$ Divide to change pounds to tons.

So, 7,000 pounds = $3\frac{1}{2}$ tons.

4 **PARTIES** How many 4-ounce party favors can be made with 5 pounds of mixed nuts?

First, find the total number of ounces in 5 pounds.

$5 \times 16 = 80$ Multiply by 16 to change pounds to ounces.

Next, find how many sets of 4 ounces are in 80 ounces.

$80 \text{ oz} \div 4 \text{ oz} = 20$

So, 20 party favors can be made with 5 pounds of mixed nuts.

Your Turn Complete.

a. $4 \text{ pt} = \underline{\quad?\quad} \text{ c}$ **b.** $32 \text{ fl oz} = \underline{\quad?\quad} \text{ c}$ **c.** $40 \text{ oz} = \underline{\quad?\quad} \text{ lb}$

Skill and Concept Check

Writing Math Exercises 2 & 3

1. **State** the operation that you would use to change pints to quarts.

2. **Explain** whether 1 cup of sand and 1 cup of cotton balls would have the same capacity, the same weight, both, or neither.

3. **OPEN ENDED** Without looking at the labels, estimate the weight or capacity of three packaged food items in your kitchen. Then compare your estimate to the actual weight or capacity.

GUIDED PRACTICE

Complete.

4. 7 pt = __?__ c

5. 24 qt = __?__ gal

6. 16 pt = __?__ gal

7. 5 c = __?__ fl oz

8. 10,000 lb = __?__ T

9. $3\frac{1}{2}$ lb = __?__ oz

10. **OCEAN** Giant clams can weigh as much as $\frac{1}{4}$ ton. How many pounds is this?

11. **LIFE SCIENCE** Owen estimates that the finches eat 8 ounces of birdseed a day at his feeder. If he buys a 10-pound bag of birdseed, about how many days will it last?

Practice and Applications

Complete.

12. 5 qt = __?__ pt

13. 8 gal = __?__ qt

14. 24 fl oz = __?__ c

15. 32 qt = __?__ gal

16. $6\frac{1}{2}$ pt = __?__ c

17. 13 qt = __?__ gal

18. 9 gal = __?__ pt

19. 24 fl oz = __?__ pt

20. 1,500 lb = __?__ T

21. 112 oz = __?__ lb

22. 84 oz = __?__ lb

23. $4\frac{1}{2}$ T = __?__ lb

HOMEWORK HELP

For Exercises	See Examples
12–23	1, 2
24–25, 31–34	3, 4

Extra Practice
See pages 618, 635.

24. How many pounds are in 30 tons?

25. How many gallons equal 8 cups?

26. Which is less: 14 cups or 5 pints? Explain.

27. Which is greater: $3\frac{1}{2}$ pints or 60 fluid ounces? Explain.

Choose the better estimate for each measure.

28. the amount of milk in a bowl of cereal: 1 cup or 1 quart

29. the amount of cough syrup in one dosage: 2 fluid ounces or 1 pint

30. the weight of a bag of groceries: 3 ounces or 3 pounds

31. Estimate how many cups of soda are in a 12-ounce can. Then find the actual amount.

32. Estimate the number of pints in a bottle of laundry detergent. Then find the exact number.

33. CHOCOLATE Refer to the graphic at the right. How many tons of chocolate candy did Americans consume in 2000?

34. BAKING A pumpkin pie recipe calls for 15 ounces of pumpkin. *About* how many pies can be made with 8 pounds of pumpkin?

35. WRITE A PROBLEM Write a problem that can be solved by converting customary units of capacity or weight.

36. MULTI STEP Peni has 12 quart jars and 24 pint jars to fill with strawberry jam. If her recipe makes 5 gallons of jam, will she have enough jars? Explain.

MULTI STEP For Exercises 37 and 38, refer to the information below.
During the Ironman Triathlon World Championships, about 25,000 cookies and 250,000 cups of water are given away. Each cup contains 8 fluid ounces, and each cookie weighs 2.5 ounces.

37. About how many gallons of water are given away?

38. About how many pounds of cookies are given away?

39. CRITICAL THINKING What number can you divide by to change 375 fluid ounces directly to quarts?

USA TODAY

USA TODAY Snapshots®

America is sweet on chocolate
Of the 6.5 billion pounds of candy eaten by Americans last year, more than half was chocolate. Candy consumed in 2000 in billions of pounds:

Chocolate — 3.3
Non-chocolate — 2.7
Gum — 0.5

Source: Candy USA

By William Risser and Adrienne Lewis, USA TODAY

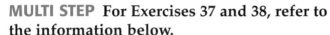

Standardized Test Practice and Mixed Review

40. MULTIPLE CHOICE About how many glasses of water, each containing 12 fluid ounces, would it take to fill a gallon jug?

ⓐ 10   ⓑ 9   ⓒ 8   ⓓ 7

41. MULTIPLE CHOICE A can of green beans weighs 13 ounces. How many pounds does a case of 24 cans weigh?

ⓕ 1.5 lb   ⓖ 15 lb   ⓗ 19.5 lb   ⓘ 312 lb

42. MEASUREMENT Measure the width of your pencil to the nearest eighth inch. (Lesson 12-1)

A coin is tossed, and a number cube is rolled. Find the probability of each event. (Lesson 11-5)

43. P(heads and 1)   **44.** P(tails and 5 or 6)   **45.** P(heads or tails and odd)

GETTING READY FOR THE NEXT LESSON

PREREQUISITE SKILL Estimate each measure. (Lesson 12-1)

46. the width of a quarter   **47.** the width of a doorway

The Metric System

What You'll LEARN

Measure in metric units.

Materials

• tape measure

INVESTIGATE *Work as a class.*

The basic unit of length in the metric system is the *meter*. All other metric units of length are defined in terms of the meter.

The most commonly used metric units of length are shown in the table.

Metric Unit	Symbol	Meaning
millimeter	mm	thousandth
centimeter	cm	hundredth
meter	m	one
kilometer	km	thousand

A metric ruler or tape measure is easy to read. The ruler below is labeled using *centimeters*.

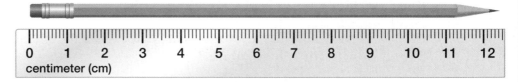

The pencil below is about 12.4 centimeters long.

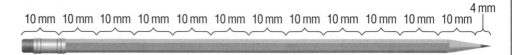

To read *millimeters*, count each individual unit or mark on the metric ruler.

There are ten millimeter marks for each centimeter mark. The pencil is about 124 millimeters long.

$$124 \text{ mm} = 12.4 \text{ cm}$$

There are 100 centimeters in one meter. Since there are 10 millimeters in one centimeter, there are 10×100 or 1,000 millimeters in one meter. The pencil is $\frac{124}{1,000}$ of a meter or 0.124 meter long.

$$124 \text{ mm} = 12.4 \text{ cm}$$
$$12.4 \text{ cm} = 0.124 \text{ m}$$

ACTIVITY   *Work with a partner.*

Use metric units of length to measure various items.

STEP 1 Copy the table.

Object	Measure		
	mm	cm	m
length of pencil			
length of sheet of paper			
length of your hand			
width of your little finger			
length of table or desk			
length of chalkboard eraser			
width of door			
height of door			
distance from doorknob to the floor			
length of classroom			

STEP 2 Use a metric ruler or tape measure to measure the objects listed in the table. Complete the table.

Writing Math

1. **Tell** which unit of measure is most appropriate for each item. How did you decide which unit is most appropriate?

2. **Examine** the pattern between the numbers in each column. What relationship do the numbers have to each other?

3. **Select** three objects around your classroom that would be best measured in meters, three objects that would be best measured in centimeters, and three objects that would be best measured in millimeters. Explain your choices.

4. **Write** the name of a common object that you think has a length that corresponds to each length. Explain your choices.

   a. 5 centimeters   b. 3 meters   c. 1 meter   d. 75 centimeters

Measure the sides of each rectangle in centimeters. Then find its perimeter and area.

5.

6.

7.

12-3 Length in the Metric System

What You'll LEARN

Use metric units of length.

NEW Vocabulary

meter
metric system
millimeter
centimeter
kilometer

Link to READING

Everyday meaning of
mill-: one thousand, as
a millennium is one
thousand years

WHEN am I ever going to use this?

SCIENCE The table shows the deepest points in several oceans.

1. What unit of measure is used?

2. What is the depth of the deepest point?

3. Use the Internet or another source to find the meaning of meter. Then write a sentence explaining how a meter compares to a yard.

Deepest Ocean Points

Ocean	Point	Depth (m)
Pacific	Mariana Trench	10,924
Atlantic	Puerto Rico Trench	8,648
Indian	Java Trench	7,125

Source: www.geography.about.com

A **meter** is the basic unit of length in the metric system. The **metric system** is a decimal system of weights and measures. The most commonly used metric units of length are shown below.

Key Concept — Metric Units of Length

Unit	Model	Benchmark
1 **millimeter** (mm)	thickness of a dime	25 mm ≈ 1 inch
1 **centimeter** (cm)	half the width of a penny	2.5 cm ≈ 1 inch
1 meter (m)	width of a doorway	1 m ≈ 1.1 yard
1 **kilometer** (km)	six city blocks	1.6 km ≈ 1 mile

The segment at the right is 2.5 centimeters or 25 millimeters long. This is about 1 inch in customary units.

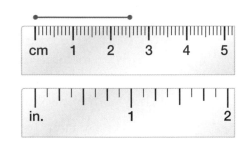

EXAMPLE Use Metric Units of Length

1 **Write the metric unit of length that you would use to measure the width of a paper clip.**

Compare the width of a paper clip to the models described above.

The width of a paper clip is greater than the thickness of a dime, but less than half the width of a penny. So, the millimeter is an appropriate unit of measure.

Use Metric Units of Length

Write the metric unit of length that you would use to measure each of the following.

2 **height of a desk**

Since the height of a desk is close to the width of a doorway, the meter is an appropriate unit of measure.

3 **distance across Indiana**

Since the distance across Indiana is much greater than 6 city blocks, this is measured in kilometers.

INDIANA

4 **width of a floppy disk**

Since the width of a floppy disk is greater than half the width of a penny and much less than the width of a doorway, the centimeter is an appropriate unit of measure.

Your Turn **Write the metric unit of length that you would use to measure each of the following.**

a. thickness of a nickel **b.** height of a cereal box

Standardized Test Practice

EXAMPLE **Estimate Length in Metric Units**

5 **MULTIPLE-CHOICE TEST ITEM** Which is the best estimate for the measurement of the line segment?

●————————————————●

Ⓐ 0.5 km   Ⓑ 0.5 m   Ⓒ 6.0 cm   Ⓓ 6.0 mm

Read the Test Item

You need to determine the best estimate for the measure of the segment.

Solve the Test Item

The segment is much smaller than 0.5 kilometer, or 3 city blocks, so choice A can be eliminated. The segment is also quite a bit smaller than 0.5 meter, or half the width of a doorway, so choice B can also be eliminated.

Two centimeters is about the width of a penny, so 6.0 centimeters is about the width of 3 pennies. This estimate is the most reasonable. The answer is C.

Check The segment is much longer than 6 times the thickness of a dime, so choice D is not correct.

Test-Taking Tip
The Princeton Review

Eliminate Choices
Eliminate any answer choices with units that are unreasonably small or large for the given measure.

Skill and Concept Check

1. **Name** the four most commonly used metric units of length and describe an object having each length. Use objects that are different from those given in the lesson.

2. **OPEN ENDED** Give two examples of items that can be measured with a meterstick and two examples of items that cannot reasonably be measured with a meterstick.

3. **Which One Doesn't Belong?** Identify the measure that does not have the same characteristic as the other three. Explain your reasoning.

| 4.3 km | 35 mm | 23 yd | 9.5 cm |

GUIDED PRACTICE

Write the metric unit of length that you would use to measure each of the following.

4. thickness of a calculator

5. distance from home to school

6. height of a tree

7. width of a computer screen

8. About how many centimeters is the thickness of a textbook?

Practice and Applications

Write the metric unit of length that you would use to measure each of the following.

9. thickness of a note pad

10. thickness of a watchband

11. length of a trombone

12. width of a dollar bill

13. length of a bracelet

14. length of the Mississippi River

15. distance from Knoxville, Tennessee, to Asheville, North Carolina

16. distance from home plate to first base on a baseball field

HOMEWORK HELP

For Exercises	See Examples
9–16	1–4
25–27	5

Extra Practice
See pages 618, 635.

Measure each line segment or side of each figure in centimeters and millimeters.

17. •————————————•   18. •———•

19.    20.

21. Which customary unit of length is comparable to a meter?

22. Is a mile or a foot closer in length to a kilometer?

23. Which is greater: 15 millimeters or 3 centimeters? Explain.

24. Which is less: 3 feet or 1 meter? Explain.

Estimate the length of each of the following. Then measure to find the actual length.

25. CD case   26. chalkboard   27. eraser on end of pencil

28. **MAPS** Use a centimeter ruler to find the distance between Houston and Jacinto City, Texas, on the map at the right.

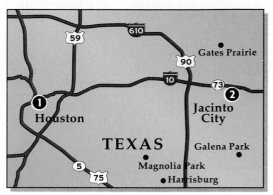

29. **FENCES** If you were to build a fence around a garden, would you need to be accurate to the nearest kilometer, to the nearest meter, or to the nearest centimeter? Explain.

30. **GEOMETRY** On centimeter grid paper, draw and label a square that has a perimeter of 12 centimeters.

31. **CRITICAL THINKING** Order 4.8 mm, 4.8 m, 4.8 cm, 0.48 m, and 0.048 km from greatest to least measurement.

Standardized Test Practice and Mixed Review

32. **MULTIPLE CHOICE** Which is the *best* estimate for the height of a flagpole?

  Ⓐ 10 km   Ⓑ 10 m   Ⓒ 10 cm   Ⓓ 10 mm

33. **MULTIPLE CHOICE** What is the length of the pencil to the nearest centimeter?

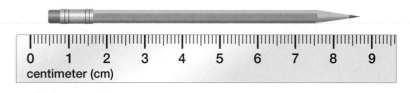

  Ⓕ 9 cm   Ⓖ 8 cm   Ⓗ 7 cm   Ⓘ 6 cm

34. **PAINTING** Painters used 170 gallons of white topcoat to paint the famous Hollywood sign. How many quarts is this? (Lesson 12-2)

Complete. (Lesson 12-1)

35. 4 ft = __?__ in.   36. __?__ ft = $\frac{1}{2}$ mi   37. 144 in. = __?__ yd

GETTING READY FOR THE NEXT LESSON

PREREQUISITE SKILL Name an item sold in a grocery store that is measured using each type of unit. (Lesson 12-2)

38. milliliter   39. gram   40. liter

A Follow-Up of Lesson 12-3

Significant Digits

All measurements are approximations. The *precision* of a measurement is the exactness to which a measurement is made. It depends on the smallest unit on the measuring tool.

The smallest unit on the ruler below is 1 centimeter.

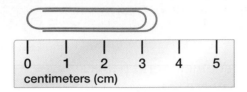

The length is *about* 3 centimeters.

The smallest unit on the ruler below is 1 millimeter, or 0.1 centimeter.

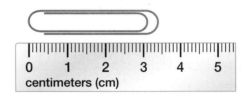

The length is *about* 3.4 centimeters.

Both paper clips have the same length, but the second measurement, 3.4 centimeters, is more precise than the first.

Another method of measuring that is more precise is to use significant digits. *Significant digits* include all of the digits of a measurement that you know for sure, plus one estimated digit. Consider the length of the pen shown below.

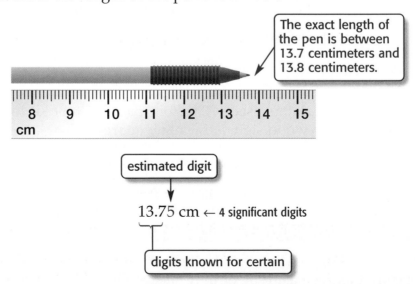

The exact length of the pen is between 13.7 centimeters and 13.8 centimeters.

estimated digit

13.75 cm ← 4 significant digits

digits known for certain

So, using significant digits, the length of the pen is 13.75 centimeters.

ACTIVITY   *Work in groups of three.*

STEP 1   Choose a measuring tool to measure the height of a desk. Find the measure using the smallest unit on your measuring tool and record your measure in a table like the one shown below.

Object	Measure	
	Using Smallest Unit	**Using Significant Digits**
height of desk		
length of calculator		
length of a pencil		
height of door		
length of classroom		
width of classroom		

STEP 2   Determine the height of the desk using significant digits. Record the measure in your table.

STEP 3   Repeat Steps 1 and 2 for all the objects listed in the table.

Writing Math

Identify the smallest unit of each measuring tool.

1. cm   1   2   3   4   2. in.   1   2

3. **Explain** how you used significant digits to find the measures.

4. **Choose** the most precise unit of measurement: inches, feet, or yards. Explain.

5. When adding measurements, the sum should have the same precision as the *least precise* measurement. Find the perimeter of the rectangle at the right using significant digits.

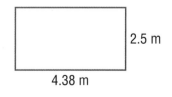
2.5 m

4.38 m

6. **Describe** a real-life situation in which a family member or neighbor used measurement precision. Then describe a real-life situation in which a less precise or estimated measure is sufficient.

7. Find the area of your classroom using the precision unit measures from your table and using the significant digit measures. Which area is more precise? Explain.

Vocabulary and Concepts

1. **List** four units of length in the customary system. (Lesson 12-1)

2. **Describe** the metric system. (Lesson 12-3)

Skills and Applications

Complete. (Lesson 12-1)

3. 10,560 ft = __?__ mi   4. __?__ in. = 2 yd   5. 18 ft = __?__ yd

Draw a line segment of each length. (Lesson 12-1)

6. $2\frac{1}{2}$ in.   7. $\frac{3}{4}$ in.

Measure the length of each line segment or object to the nearest half, fourth, or eighth inch. (Lesson 12-1)

8. •————————•   9.    10.

Complete. (Lesson 12-2)

11. 22 pt = __?__ qt   12. __?__ qt = 14 gal   13. 36 oz = __?__ lb

14. __?__ fl oz = 5 c   15. 9 pt = __?__ c   16. __?__ gal = 48 pt

17. **ICE CREAM** A container of ice cream contains 20 half-cup servings. How many quarts is this? (Lesson 12-2)

18. Write the metric unit of length that you would use to measure the thickness of a pencil. (Lesson 12-3)

Standardized Test Practice

Ⓐ Ⓑ Ⓒ Ⓓ

19. **MULTIPLE CHOICE** How much punch is made with 1 pint of ginger ale, 1 cup of orange juice, and 3 cups of pineapple juice?

   Ⓐ 1 pint   Ⓑ 2 pints
   Ⓒ 3 pints   Ⓓ 5 pints

20. **SHORT RESPONSE** State the measure of the line segment in centimeters and millimeters. (Lesson 12-3)

The Game Zone

A Place To Practice Your Math Skills

Math Skill

Using Metric Measures

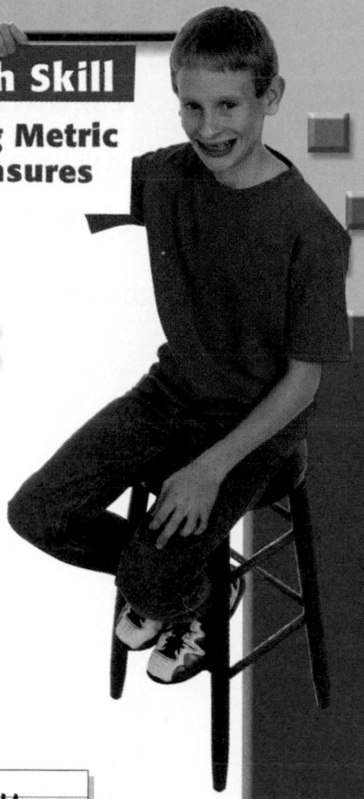

Mystery Measurements

GET READY!

Players: two
Materials: 2 metersticks, 12 index cards, scissors

GET SET!

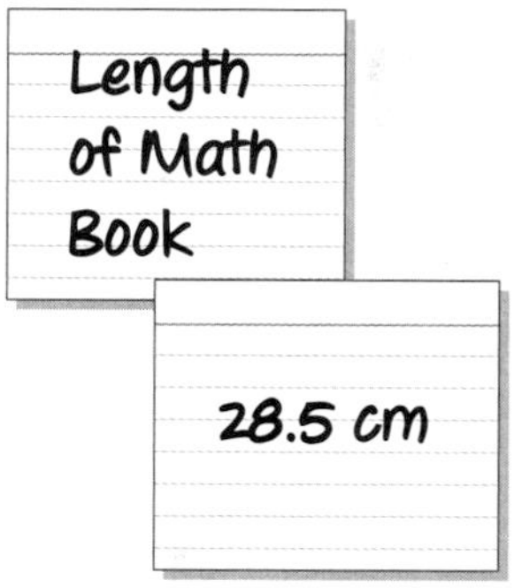
Length of Math Book
28.5 cm

- Working alone, each player secretly measures the length, width, or depth of six objects in the classroom and records them on a piece of paper. The measures may be in millimeters, centimeters, or meters. This will serve as the answer sheet at the end of the game.
- Each player takes 6 index cards and cuts them in half, making 12 cards.
- For each object, the measurement is recorded on one card. A description of what was measured is recorded on another card. Make sure each measurement is different.

GO!

- Each player shuffles his or her cards.
- Keeping the cards facedown, players exchange cards.
- At the same time, players turn over all the cards given to them.
- Each player attempts to match each object with its measure.
- **Who Wins?** The person with more correct matches after 5 minutes is the winner.

Mass and Capacity in the Metric System

What You'll LEARN

Use metric units of mass and capacity.

NEW Vocabulary

milligram
gram
kilogram
milliliter
liter

HANDS-ON Mini Lab

Materials
- balance
- small paper clips
- roll of breath mints
- 2 pencils

Work in groups of 2 or 3.

STEP 1 Place the roll of mints on one side of the balance and the paper clips on the other side until the scale balances. How many paper clips were used?

STEP 2 Read the label on the mints to find its mass in grams. Record how many grams are in the roll.

STEP 3 Find the number of paper clips needed to balance 2 pencils of the same size.

1. How does the number of paper clips needed to balance the roll of breath mints compare to the mass of the roll in grams?
2. Estimate the mass of one paper clip.
3. How many paper clips were needed to balance 2 pencils?
4. What is the mass of 1 pencil in grams?

The most commonly used metric units of mass are shown below.

Key Concept		Metric Units of Mass
Unit	**Model**	**Benchmark**
1 **milligram** (mg)	grain of salt	1 mg ≈ 0.00004 oz
1 **gram** (g)	small paper clip	1 g ≈ 0.04 oz
1 **kilogram** (kg)	six medium apples	1 kg ≈ 2 lb

EXAMPLES Use Metric Units of Mass

Write the metric unit of mass that you would use to measure each of the following. Then estimate the mass.

1 sheet of notebook paper

A sheet of paper has a mass greater than a small paper clip, but less than six apples. So, the gram is the appropriate unit.

Estimate A sheet of paper is a little heavier than a paper clip.

One estimate for the mass of a sheet of paper is about 6 grams.

STUDY TIP

Mass The mass of an object is the amount of material it contains.

2 **bag of potatoes**

A bag of potatoes has a mass greater than six apples. So, the kilogram is the appropriate unit.

Estimate   A bag of potatoes is several times heavier than six apples.

One estimate for the mass of a bag of potatoes is about 2 or 3 kilograms.

Your Turn Write the metric unit of mass that you would use to measure each of the following. Then estimate the mass.

a. tennis ball   b. horse   c. aspirin

STUDY TIP

Capacity Capacity refers to the amount of liquid that can be held in a container.

The most commonly used metric units of capacity are shown below.

Key Concept		Metric Units of Capacity
Unit	**Model**	**Benchmark**
1 **milliliter** (mL)	eyedropper	1 mL ≈ 0.03 fl oz
1 **liter** (L)	small pitcher	1 L ≈ 1 qt

EXAMPLES   **Use Metric Units of Capacity**

Write the metric unit of capacity that you would use to measure each of the following. Then estimate the capacity.

3 **large fishbowl**

A large fishbowl has a capacity greater than a small pitcher. So, the liter is the appropriate unit.

Estimate   A fishbowl will hold about 2 small pitchers of water.

One estimate for the capacity of a fishbowl is about 2 liters.

4 **glass of milk**

A glass of milk is greater than an eyedropper and less than a small pitcher. So, the milliliter is the appropriate unit.

Estimate   There are 1,000 milliliters in a liter. A small pitcher can fill about 4 glasses.

One estimate for the capacity of a glass of milk is about 1,000 ÷ 4 or 250 milliliters.

Your Turn Write the metric unit of capacity that you would use to measure each of the following. Then estimate the capacity.

d. bathtub   e. 10 drops of food coloring

REAL-LIFE MATH

GOLDFISH Most pet goldfish range in length from 2.5 to 10 cm. However, in the wild, they may be up to 40 cm long.

Source: www.factmonster.com

EXAMPLE   Comparing Metric Units

5 ANIMALS The table shows part of the recommended daily diet for a gibbon. Does a gibbon eat more or less than one kilogram of carrots, bananas, and celery each day?

Find the total amount of carrots, bananas, and celery each day.

carrots	147 grams
bananas	270 grams
celery	+ 210 grams
total	627 grams

Daily Diet for a Gibbon	
Food	**Amount (g)**
lettuce	380
oranges	270
spinach	150
sweet potato	143
bananas	270
carrots	147
celery	210
green beans	210

One kilogram is equal to 1,000 grams. Since 627 is less than 1,000, a gibbon eats less than one kilogram of carrots, bananas, and celery each day.

Skill and Concept Check

Writing Math
Exercise 2

1. **OPEN ENDED** Name an item found at home that has a capacity of about one liter.

2. **NUMBER SENSE** The mass of a dime is recorded as 4. What metric unit was used to measure the mass? Explain.

GUIDED PRACTICE

Write the metric unit of mass or capacity that you would use to measure each of the following. Then estimate the mass or capacity.

3. nickel

4. bucket of water

5. laptop computer

6. juice in a lemon

7. light bulb

8. can of paint

FOOD For Exercises 9–12, use the list of ingredients for one dark chocolate cake at the right.

9. Is the total amount of sugar, chocolate, butter, and flour more or less than one kilogram?

10. Write the quantities of ingredients needed for two cakes.

11. Is the total amount of sugar, chocolate, butter, and flour for two cakes more or less than one kilogram? Explain.

12. Explain why people in the United States may have trouble using this recipe.

Dark Chocolate Cake
6 medium eggs
175 grams sugar
280 grams chocolate
100 grams butter
100 grams flour

Practice and Applications

Write the metric unit of mass or capacity that you would use to measure each of the following. Then estimate the mass or capacity.

13. candy bar
14. grape
15. large watermelon
16. cow
17. large bowl of punch
18. cooler of lemonade
19. canary
20. shoe
21. grain of sugar
22. postage stamp
23. raindrop
24. ink in a ballpoint pen

HOMEWORK HELP

For Exercises	See Examples
13–24	1–4
25–26, 28	5

Extra Practice
See pages 619, 635.

25. **SHOPPING** Your favorite cereal comes in a 1.7-kilogram box or a 39-gram box. Which box is larger? Explain.

26. **SHOPPING** Liquid soap comes in 1.89 liter containers and 221 milliliter containers. Which container is smaller? Explain.

27. **FOOD** Estimate the capacity of a can of cola in metric units. Determine the actual capacity of a can of cola. Compare your estimate with the actual capacity.

28. **MULTI STEP** The doctor told you to take 1,250 milligrams of aspirin for your sprained ankle. According to the bottle at the right, how many tablets should you take?

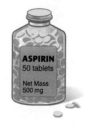

29. **CRITICAL THINKING** If you filled a 150-milliliter beaker with salt, would its mass be 150 milligrams? Explain.

Standardized Test Practice and Mixed Review

30. **MULTIPLE CHOICE** Approximately what is the mass of a bag of flour?
    (A) 2 kg
    (B) 2 mL
    (C) 2 g
    (D) 2 L

31. **SHORT RESPONSE** What metric unit would you use to measure the capacity of a tablespoon of water?

Write the metric unit of length that you would use to measure each of the following. (Lesson 12-3)

32. length of a hand
33. thickness of a folder

34. **MEASUREMENT** How many ounces are in $2\frac{1}{2}$ pounds? (Lesson 12-2)

GETTING READY FOR THE NEXT LESSON

PREREQUISITE SKILL Multiply or divide. (Lessons 4-1 and 4-3)

35. 2.5×10
36. $2.5 \times 1,000$
37. $2,500 \div 100$
38. $2.5 \div 10$

What You'll Learn
Solve problems using benchmarks.

Use Benchmarks

To make the punch for the party, we need to add one-half liter of juice concentrate to 3 liters of water. But we don't have any metric measuring containers.

I have a clean 2-liter cola bottle. Let's use that as a **benchmark**.

Explore	We need to measure 0.5 liter of concentrate. We have a 2-liter cola bottle.
Plan	We can take the 2-liter bottle and use a marker to visually divide it into four approximately equal sections. Each section will be about 0.5 liter.
Solve	Mark the 2-liter bottle into four sections. Pour the concentrate into the bottle until it reaches the first mark on the bottle.
Examine	Since 4 halves equal 2 wholes, a fourth of the bottle should equal 0.5 liter.

Analyze the Strategy

1. A benchmark is a measurement by which other items can be measured. **Explain** why the 2-liter bottle is a good benchmark to use for measuring the 0.5 liter of concentrate.

2. **Describe** how the students could measure the 3 liters of water for the punch.

3. **Write** a plan to estimate the length of a bracelet in centimeters without using a metric ruler.

Solve. Use the benchmark strategy.

4. **INTERIOR DESIGN** Lucas wants to put a border around his room. He needs to know the approximate length and width of the room in meters. He has some string, and he knows that the distance from the doorknob to the floor is about one meter. Describe a way Lucas could estimate the distances in meters.

5. **PROBABILITY** The students in Mrs. Lightfoot's math class want to determine the probability that a person picked at random from the class is taller than 200 centimeters. They know that the doorway is 3 meters high. Describe a way the students can determine who is taller than 200 centimeters.

Mixed Problem Solving

Solve. Use any strategy.

6. **GEOMETRY** Find the area of the figure.

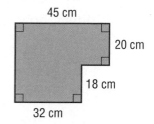

7. **MEASUREMENT** Garcia estimated that he takes 3 steps every 2 meters. How many steps will Garcia take for a distance of 150 meters?

8. **MULTI STEP** The Grayson Middle School softball team won three times as many games as they have lost this season. If they lost 5 games how many games did they play this season?

9. **GEOMETRY** Look at the pattern. What is the perimeter of the next figure in the pattern?

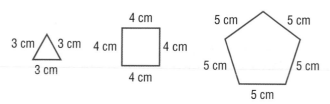

10. **MONEY** Antwon purchased a portable MP3 player for $129.98, including tax. How much change should he receive from $150?

11. **NUMBER SENSE** A number multiplied by itself is 676. What is the number?

12. **MEASUREMENT** What is the missing measurement in the pattern?

$$\ldots, \underline{\quad ? \quad}, \frac{1}{4} \text{ in.}, \frac{1}{8} \text{ in.}, \frac{1}{16} \text{ in.}, \ldots$$

13. **CRAFTS** Melissa has a piece of ribbon measuring $8\frac{3}{4}$ yards. How many pieces of ribbon each measuring $1\frac{3}{4}$ yards can be cut from the large piece of ribbon?

14. **BUSINESS** The North Shore Fish Market reported the following sales each day during the first half of April. Which is greater, the mean or the median sales during this time?

15. **STANDARDIZED TEST PRACTICE**
Katie has three books in her backpack. Which amount best describes the mass of the three books and Katie's backpack?

Ⓐ 6 g   Ⓑ 60 g   Ⓒ 6 kg   Ⓓ 60 kg

12-5 **Changing Metric Units**

What You'll LEARN

Change units within the metric system.

Link to READING

Everyday Meaning of
Cent: one one-hundredth of a dollar, as in 53 cents

HANDS-ON **Mini Lab**

Materials
- empty soda can
- empty 2-liter soda bottle
- water
- funnel

Work with a partner.

In the following activity, you will find how many milliliters are in a liter.

STEP 1 Fill an empty soda can with water. Then pour the contents into an empty 2-liter soda bottle, using a funnel.

STEP 2 Repeat Step 1 until the 2-liter bottle is full. Record how many cans it took to fill the bottle.

1. How many cans did it take to fill the 2-liter bottle?
2. If the capacity of a soda can is 355 milliliters, how would you find the number of milliliters in the 2-liter bottle?
3. How many milliliters are in 2 liters?
4. Based on this information, how could you find the number of milliliters in one liter?
5. How many milliliters are in one liter?

To change from one unit to another within the metric system, you either multiply or divide by powers of 10. The chart below shows the relationship between the units in the metric system and the powers of 10.

1,000	100	10	1	0.1	0.01	0.001
thousands	hundreds	tens	ones	tenths	hundredths	thousandths
kilo	hecto	deka	basic unit	deci	centi	milli

Each place value is 10 times the place value to its right.

To change from a larger unit to a smaller unit, you need to multiply. To change from a smaller unit to a larger unit, you need to divide.

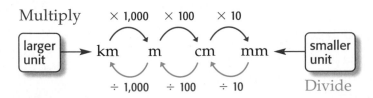

Multiply × 1,000 × 100 × 10

larger unit → km m cm mm ← smaller unit

÷ 1,000 ÷ 100 ÷ 10 Divide

EXAMPLES Change Metric Units

Complete.

1 135 g = __?__ kg

Since 1,000 grams = 1 kilogram, divide by 1,000.

$135 \div 1{,}000 = 0.135$

So, 135 g = 0.135 kg.

Check Since a kilogram is a larger unit than a gram, the number of kilograms should be less than the number of grams. The answer seems reasonable.

2 __?__ mm = 26.8 cm

Since 1 centimeter = 10 millimeters, multiply by 10.

$26.8 \times 10 = 268$

So, 268 mm = 26.8 cm.

Check Since a millimeter is a smaller unit than a centimeter, the number of millimeters should be greater than the number of centimeters. The answer seems reasonable.

Your Turn Complete.

a. 513 mL = __?__ L   **b.** 5 cm = __?__ mm   **c.** __?__ mg = 8.2 g

Sometimes you will need to change units to solve real-life problems.

EXAMPLE Change Units to Solve a Problem

3 **TRIATHLONS** Use the table at the right to determine the total number of meters in the San Diego International Triathlon.

San Diego International Triathlon		
Swim	**Bike**	**Run**
1 km	30 km	10 km

Words	Total equals sum of swim, bike, and run times 1,000.
Variable	$m = (s + b + r) \times 1{,}000$
Equation	$m = (1 + 30 + 10) \times 1{,}000$

$m = (1 + 30 + 10) \times 1{,}000$ Write the equation.

$m = 41 \times 1{,}000$ First, find the sum in the parentheses.

$m = 41{,}000$ Multiply.

There are 41,000 meters in the San Diego International Triathlon.

Skill and Concept Check

Writing Math
Exercises 1 & 3

1. **Explain** how to change liters to milliliters.

2. **OPEN ENDED** Write an equation involving metric units where you would divide by 100 to solve the equation.

3. **FIND THE ERROR** Julia and Trina are changing 590 centimeters to meters. Who is correct? Explain.

> Julia
> 590 × 100 = 59,000 m

> Trina
> 590 ÷ 100 = 5.9 m

GUIDED PRACTICE

Complete.

4. __?__ m = 75 mm
5. 205 mg = __?__ g
6. __?__ mm = 3.8 cm
7. 85 mm = __?__ cm
8. 0.95 g = __?__ mg
9. 0.05 L = __?__ mL

10. **SCOOTERS** A scooter goes up to 20 kilometers per hour. How many meters per hour can it travel?

Practice and Applications

Complete.

11. __?__ L = 95 mL
12. __?__ g = 1,900 mg
13. __?__ cm = 52 mm
14. 13 cm = __?__ mm
15. __?__ mg = 8.2 g
16. __?__ mL = 23.8 L
17. 0.4 L = __?__ mL
18. 4 m = __?__ cm
19. 354 cm = __?__ m
20. __?__ m = 35.6 cm
21. 500 mg = __?__ kg
22. 250 cm = __?__ km

HOMEWORK HELP	
For Exercises	See Examples
11–24, 29	1–2
25, 26	3
Extra Practice See pages 619, 635.	

23. How many centimeters are in 0.82 meter?

24. Change 7 milligrams to grams.

25. **EARTHQUAKES** Earthquakes originate between 5 and 700 kilometers below the surface of Earth. Write these distances in meters.

 Data Update What is the depth of the latest earthquake? Visit msmath1.net/data_update to learn more.

BIRDS For Exercises 26 and 27, use the table at the right.

26. Write the length and width of a robin's egg in millimeters.

27. Which bird has the larger eggs?

Size of Bird Eggs		
Bird	Length	Width
Robin	1.9 cm	1.5 cm
Turtledove	31 mm	23 mm

Source: *Animals as Our Companions*

FOOD For Exercises 28–31, use the table at the right.

Pancake Syrup	
Size	Price
350 mL	$2.80
1.06 L	$5.60

28. Find the cost to the nearest tenth of a cent for one milliliter of pancake syrup from the 350-milliliter bottle.

29. Change 1.06 liters to milliliters.

30. Find the cost to the nearest tenth of a cent for one milliliter of pancake syrup from the 1.06-liter bottle.

31. Which bottle of pancake syrup is a better buy?

32. **CRITICAL THINKING** A milliliter of water at 4°C has a mass of 1 gram. What is the mass of 1 liter of water at 4°C?

EXTENDING THE LESSON You can use proportions to change units of measure.

Example   250 mm = $\underline{\ ?\ }$ m

$$\begin{array}{cc} \text{meters} \to \\ \text{millimeters} \to \end{array} \quad \frac{1\text{ m}}{1{,}000\text{ mm}} = \frac{x\text{ m}}{250\text{ mm}} \quad \begin{array}{c} \leftarrow \text{meters} \\ \leftarrow \text{millimeters} \end{array}$$

$$1 \times 250 = 1{,}000 \times x \quad \text{Cross products}$$

$$250 = 1{,}000x \quad \text{Multiply.}$$

$$\frac{250}{1{,}000} = \frac{1{,}000x}{1{,}000} \quad \text{Divide.}$$

$$0.25 = x \quad \text{Simplify.}$$

So, 250 millimeters = 0.25 meter.

Write a proportion to solve each problem. Then complete.

33. 0.005 km = $\underline{\ ?\ }$ m   34. $\underline{\ ?\ }$ mL = 0.45 L   35. 263 g = $\underline{\ ?\ }$ kg

Standardized Test Practice and Mixed Review

36. **MULTIPLE CHOICE** Chicago's Buckingham Fountain contains 133 jets that spray approximately 52,990 liters of water into the air every minute. How many milliliters of water is this?

Ⓐ 52.99 mL   Ⓑ 5,299,000 mL

Ⓒ 52,990,000 mL   Ⓓ 529,900,000,000 mL

37. **GRID IN** Write 25.9 kilograms in grams.

38. **FOOD** Which is the better estimate for the capacity of a glass of milk, 360 liters or 360 milliliters? (Lesson 12-4)

Find the length of each line segment to the nearest tenth of a centimeter. (Lesson 12-3)

39. •————————•   40. •————————————•

GETTING READY FOR THE NEXT LESSON

PREREQUISITE SKILL Add or subtract. (Lesson 3-5)

41. 3.26 + 4.86   42. 9.32 − 4.78   43. 27.48 + 78.92   44. 7.18 − 2.31

Measures of Time

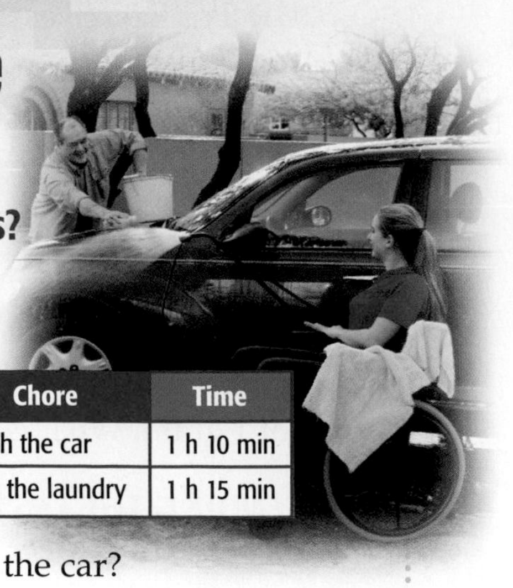

What You'll LEARN

Add and subtract measures of time.

NEW Vocabulary

second
minute
hour

WHEN am I ever going to use this?

MONEY MATTERS Bethany wants to buy a new coat. Her grandfather promises to pay her for each hour she spends doing extra chores. The table shows the amount of time spent on extra chores.

Chore	Time
wash the car	1 h 10 min
fold the laundry	1 h 15 min

1. How long did Bethany take to wash the car?
2. How long did Bethany take to fold laundry?
3. What is the sum of the minutes?
4. What is the sum of the hours?
5. How long did it take to wash the car and fold the laundry?

The most commonly used units of time are shown below.

Key Concept	Units of Time
Unit	**Model**
1 **second** (s)	time needed to say 1,001
1 **minute** (min) = 60 seconds	time for 2 average TV commercials
1 **hour** (h) = 60 minutes	time for 2 weekly TV sitcoms

To add or subtract measures of time, use the following steps.

Step 1 Add or subtract the seconds.

Step 2 Add or subtract the minutes.

Rename if necessary in each step.

Step 3 Add or subtract the hours.

EXAMPLE Add Units of Time

1 Find the sum of 4 h 20 min 45 s and 2 h 50 min 10 s.

Estimate 4 h 20 min 45 s is about 4 h, and 2 h 50 min 10 s is about 3 h.
4 h + 3 h = 7 h

$$\begin{array}{r} 4 \text{ h } 20 \text{ min } 45 \text{ s} \\ + 2 \text{ h } 50 \text{ min } 10 \text{ s} \\ \hline 6 \text{ h } 70 \text{ min } 55 \text{ s} \end{array}$$ Add seconds first, then minutes, and finally the hours.

70 minutes equals 1 hour 10 minutes.

So, the sum is 7 h 10 min 55 s.   Compare the answer to the estimate.

EXAMPLE Subtract Units of Time

② **OLYMPICS** The table shows the winners of the Women's Marathon in the 1996 and 2000 Summer Olympics. How much faster was Takahashi's time than Roba's time?

Olympic Year	Runner	Time
1996	Fatuma Roba	2 h 26 min 5 s
2000	Naoko Takahashi	2 h 23 min 10 s

Source: *The World Almanac*

Estimate Since the numbers are so close, use the minutes.

26 min − 23 min = 3 min

$$\begin{array}{r} 2 \text{ h } 26 \text{ min } \ 5 \text{ s} \\ - \ 2 \text{ h } 23 \text{ min } 10 \text{ s} \end{array}$$ Since you cannot subtract 10 seconds from 5 seconds, you must rename 26 minutes 5 seconds as 25 minutes 65 seconds.

$$\begin{array}{r} 2 \text{ h } 26 \text{ min } \ 5 \text{ s} \\ - \ 2 \text{ h } 23 \text{ min } 10 \text{ s} \end{array} \rightarrow \begin{array}{r} 2 \text{ h } 25 \text{ min } 65 \text{ s} \\ - \ 2 \text{ h } 23 \text{ min } 10 \text{ s} \\ \hline 0 \text{ h } \ \ 2 \text{ min } 55 \text{ s} \end{array}$$

Takahashi's time was 2 minutes 55 seconds faster than Roba's time. Compared to the estimate, the answer seems reasonable.

Your Turn Add or subtract.

a. $\quad \begin{array}{r} 5 \text{ h } 55 \text{ min} \\ + \ 6 \text{ h } 17 \text{ min} \end{array}$

b. $\quad \begin{array}{r} 8 \text{ h } 25 \text{ min} \\ - \ 3 \text{ h } 30 \text{ min} \end{array}$

c. $\quad \begin{array}{r} 5 \text{ h } 15 \text{ min } 10 \text{ s} \\ - \ 2 \text{ h } 30 \text{ min } 45 \text{ s} \end{array}$

Sometimes you need to determine the *elapsed time*, which is how much time has passed from beginning to end.

EXAMPLE Elapsed Time

③ **TRAVEL** A flight leaves Boston at 11:35 A.M. and arrives in Miami at 2:48 P.M. How long is the flight?

You need to find how much time has elapsed.

11:35 A.M. to 12:00 noon is 25 minutes.

12:00 noon to 2:48 P.M. is 2 hours 48 minutes.

The length of the flight is 25 minutes + 2 hours 48 minutes or 2 hours 73 minutes. Now rename 73 minutes as 1 hour 13 minutes.

2 h + 1 h 13 min = 3 h 13 min.

The length of the flight is 3 hours 13 minutes.

Skill and Concept Check

Writing Math
Exercise 3

1. **Write** the number of hours in 330 minutes.

2. **OPEN ENDED** Name a starting time in the morning and an ending time in the afternoon where the elapsed time is 3 hours 45 minutes.

3. **Which One Doesn't Belong?** Identify the time that is not the same as the others. Explain your reasoning.

| 2 h 36 min 16 s | 1 h 96 min 16 s | 2 h 35 min 76 s | 1 h 36 min 76 s |

GUIDED PRACTICE

Add or subtract.

4. 4 h 23 min 45 s
    + 6 h 52 min 20 s

5. 17 min 15 s
    − 9 min 24 s

6. 8 h   35 s
    + 7 h 29 min 54 s

Find each elapsed time.

7. 8:25 A.M. to 11:50 A.M.

8. 10:15 A.M. to 3:45 P.M.

9. **TELEVISION** Use the information at the right. How much more time did the average household spend watching television in 2000 than in 1950?

Average Household Time Spent Watching Television	
Year	Time
1950	4 h 35 min
2000	7 h 29 min

Source: Nielsen Media Research

Practice and Applications

Add or subtract.

10. 15 min 45 s
     + 20 min 10 s

11. 35 min 25 s
     + 17 min 30 s

12. 6 h 48 min 28 s
     − 2 h 29 min 14 s

13. 2 h 57 min 42 s
     − 1 h 23 min 19 s

14. 12 h 25 min
     + 19 h 53 min

15. 9 h 35 min
     + 2 h 59 min

16. 17 min 15 s
     − 9 min 24 s

17. 25 min 17 s
     − 12 min 38 s

18. 12 h   45 s
     + 8 h 45 min 16 s

19. 5 h   28 s
     + 3 h 8 min 40 s

20. 12 h   3 s
     − 3 h 14 min 6 s

21. 5 h
     − 1 h 15 min 12 s

HOMEWORK HELP

For Exercises	See Examples
10–21	1, 2
22–25	3

Extra Practice
See pages 619, 635.

Find each elapsed time.

22. 5:18 P.M. to 9:36 P.M.

23. 8:05 A.M. to 11:45 A.M.

24. 5:30 P.M. to 3:10 A.M.

25. 8:30 A.M. to 10:40 P.M.

26. A stopwatch is *always, sometimes,* or *never* a good way to measure the length of a movie. Explain.

496   Chapter 12 Measurement

27. **MULTI STEP** The three acts of a play are 28 minutes, 20 minutes, and 14 minutes long. There are 15-minute intermissions between acts. If the play starts at 7:30 P.M., when will it end?

28. **SCHOOL** Estimate the amount of time you need to get dressed for school. Then time yourself. How does your estimate compare with the actual time?

29. **CRITICAL THINKING** A flight from New York City to Los Angeles takes 6 hours. If the plane leaves New York City at 9:15 A.M., at what time will it arrive in Los Angeles? (*Hint*: Remember that there is a time difference.)

EXTENDING THE LESSON
You can use two formulas to change units of temperature.

To change Celsius C to Fahrenheit F, use the formula $F = \frac{9}{5}C + 32$.

Example Change 20°C to Fahrenheit.

$F = \frac{9}{5}C + 32$ Original formula

$F = \frac{9}{5}(20) + 32$ $C = 20$

$F = 36 + 32$ Simplify.

$F = 68$ 20°C equals 68°F.

To change Fahrenheit F to Celsius C, use the formula $C = \frac{5}{9}(F - 32)$.

Example Change 50°F to Celsius.

$C = \frac{5}{9}(F - 32)$ Original formula

$C = \frac{5}{9}(50 - 32)$ $F = 50$

$C = \frac{5}{9}(18)$ Simplify.

$C = 10$ 50°F equals 10°C.

Complete.

30. $15°C = \underline{\ ?\ }°F$

31. $0°C = \underline{\ ?\ }°F$

32. $59°F = \underline{\ ?\ }°C$

33. $104°F = \underline{\ ?\ }°C$

Standardized Test Practice and Mixed Review

34. **MULTIPLE CHOICE** The table shows the times for three flights leaving from three different airports in the Washington, D.C., area and traveling to Detroit. Which airport has the shortest travel time to Detroit?

Airport	Departure Time	Arrival Time
WDI	7:31 A.M.	10:05 A.M.
DCA	7:15 A.M.	9:43 A.M.
BWI	7:23 A.M.	9:53 A.M.

  Ⓐ WDI   Ⓑ DCA

  Ⓒ BWI   Ⓓ All three flights take the same length of time.

35. **SHORT RESPONSE** Suppose Mr. James puts a meat loaf in the oven at 11:49 A.M. It needs to bake for 1 hour and 33 minutes. What time should he take the meat loaf out of the oven?

Complete. (Lesson 12-5)

36. $\underline{\ ?\ }$ L = 450 mL

37. 6.5 m = $\underline{\ ?\ }$ cm

38. 8,800 g = $\underline{\ ?\ }$ kg

39. **MEASUREMENT** To measure the water in a washing machine, which metric unit of capacity would you use? (Lesson 12-4)

Vocabulary and Concept Check

centimeter (p. 476)	kilometer (p. 476)	minute (p. 494)
cup (p. 470)	liter (p. 485)	ounce (p. 471)
fluid ounce (p. 470)	meter (p. 476)	pint (p. 470)
foot (p. 465)	metric system (p. 476)	pound (p. 471)
gallon (p. 470)	mile (p. 465)	quart (p. 470)
gram (p. 484)	milligram (p. 484)	second (p. 494)
hour (p. 494)	milliliter (p. 485)	ton (p. 471)
inch (p. 465)	millimeter (p. 476)	yard (p. 465)
kilogram (p. 484)		

Choose the correct term or number to complete each sentence.

1. A centimeter equals (one tenth, one hundredth) of a meter.
2. You should (multiply, divide) to change from larger to smaller units.
3. One paper clip has a mass of about one (gram, kilogram).
4. One cup is equal to (8, 16) fluid ounces.
5. To convert from kilograms to grams, multiply by (100, 1,000).
6. The basic unit of capacity in the metric system is the (liter, gram).
7. To convert from 15 yards to feet, you should (multiply, divide) by 3.
8. You should (multiply, divide) to change from ounces to pounds.
9. One centimeter is (longer, shorter) than 1 millimeter.

Lesson-by-Lesson Exercises and Examples

 Length in the Customary System (pp. 465–468)

Complete.

10. $2 \text{ mi} = \underline{} \text{ ft}$ 11. $\underline{} \text{ in.} = 5 \text{ ft}$

12. $9 \text{ yd} = \underline{} \text{ ft}$ 13. $72 \text{ in.} = \underline{} \text{ yd}$

14. $\underline{} \text{ in.} = 3 \text{ mi}$ 15. $\underline{} \text{ yd} = 180 \text{ in.}$

Draw a line segment of each length.

16. $2\frac{7}{8}$ in.   17. $1\frac{1}{2}$ in.

18. $1\frac{5}{8}$ in.   19. $3\frac{1}{4}$ in.

Example 1   Complete $36 \text{ ft} = \underline{} \text{ yd}$.

$36 \div 3 = 12$ Since 1 yard equals 3 feet, divide by 3.

So, $36 \text{ ft} = 12 \text{ yd}$.

Example 2   Draw a line segment measuring $1\frac{3}{8}$ inches.

Draw a line segment from 0 to $1\frac{3}{8}$.

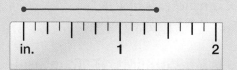

msmath1.net/vocabulary_review

12-2 **Capacity and Weight in the Customary System** (pp. 470–473)

Complete.

20. 3 pt = __?__ fl oz 21. __?__ qt = 44 c
22. 5 T = __?__ lb   23. 2.75 gal = __?__ pt
24. __?__ lb = 12 oz 25. __?__ pt = 8 qt
26. 64 fl oz = __?__ c 27. 3 gal = __?__ qt

28. **FOOD** Lauren bought 9 gallons of apple cider for the school party. How many 1-cup servings will she be able to serve?

Example 3 Complete 5 qt = __?__ pt.

THINK   2 pints are in 1 quart.

$5 \times 2 = 10$ Multiply to change a larger unit to a smaller unit.

So, 5 quarts = 10 pints.

12-3 **Length in the Metric System** (pp. 476–479)

Write the metric unit of length that you would use to measure each of the following.

29. height of your school
30. the length of the state of Florida
31. thickness of slice of bread
32. distance across school gym
33. length of your arm
34. length of a paper clip

Example 4 Write the metric unit of length that you would use to measure the height of a slide on the school playground.

Compare the slide with an item in the table on page 476. The height of a slide is larger than half the width of a penny and smaller than six city blocks. So, you would use the meter to measure the slide.

12-4 **Mass and Capacity in the Metric System** (pp. 484–487)

Write the metric unit of mass or capacity that you would use to measure each of the following. Then estimate the mass or capacity.

35. a candy apple
36. a pitcher of lemonade
37. a snowflake
38. an automobile
39. a puppy
40. a can of soda

41. **SHOPPING** Your favorite juice comes in 1.5 liter containers and 355 milliliter containers. Which container has less juice?

Example 5 Write the metric unit of mass that you would use to measure a cell phone. Then estimate the mass.

The mass of a cell phone is greater than a paper clip, but less than 6 apples. So, the gram is the appropriate unit.

Estimate There are 1,000 grams in a kilogram. A cell phone is much heavier than a paper clip, but not nearly as heavy as six apples.

One estimate for the mass of a cell phone is about 250 grams.

Changing Metric Units (pp. 490–493)

Complete.

42. 300 mL = __?__ L

43. __?__ g = 1 mg

44. __?__ m = 0.75 km

45. 5.02 kg = __?__ g

46. 345 cm = __?__ m

47. __?__ m = 23.6 mm

48. 5,200 m = __?__ km

49. 35 m = __?__ cm

50. How many centimeters are in 0.74 meter?

51. **PUNCH** Sabrina mixes 6.3 liters of punch. How many milliliters is this?

52. **DISTANCE** Sam's house is 0.6 kilometer from Jose's house. How many meters is this?

Example 6 Complete 9.2 g = __?__ mg.

To change from grams to milligrams, multiply by 1,000.

$9.2 \times 1,000 = 9,200$

So, 9.2 g = 9,200 mg.

Example 7 Complete 523 mm = __?__ cm.

To change from millimeters to centimeters, divide by 10 since 10 millimeters = 1 centimeter.

$523 \div 10 = 52.3$

So, 523 mm = 52.3 cm.

Measures of Time (pp. 494–497)

Add or subtract.

53. 5 h 20 min
    + 2 h 16 min

54. 7 h 45 min
    − 4 h 32 min

55. 9 h 7 min
    − 8 h 7 min 8 s

56. 2 h 35 min
    + 6 h 41 min

57. Find the sum of 7 h 20 min and 2 h 48 min 10 s.

58. What is the sum of 3 h 35 min 40 s and 6 h 50 min 40 s?

59. **MUSIC** Latisha's piano lesson started at 4:45 P.M. and ended at 5:30 P.M. How long was her lesson?

60. **TRAVEL** Aaron flew from Tampa, Florida, to New York City. His plane left Tampa at 6:34 A.M., and the flight took 3 hours 55 minutes. What time did he arrive in New York City?

Example 8 Find the sum of 3 h 50 min and 2 h 15 min.

   3 h 50 min
 + 2 h 15 min
   5 h 65 min

Rename 65 min as 1 h 5 min.

5 h + 1 h 5 min = 6 h 5 min

Example 9 Find the difference of 5 h 10 min and 2 h 29 min.

   5 h 10 min
 − 2 h 29 min

You cannot subtract 29 min. from 10 min, so rename 5 h 10 min as 4 h 70 min.

   4 h 70 min
 − 2 h 29 min
   2 h 41 min

Vocabulary and Concepts

1. **List** three commonly used customary units of weight.

2. **State** the meaning of the prefixes kilo-, centi-, and milli-.

3. **Describe** the operation necessary to convert metric units from a prefix of centi- to a prefix of milli-.

Skills and Applications

Complete.

4. 48 in. = __?__ ft

5. 2 yd = __?__ in.

6. __?__ pt = 6 qt

7. __?__ fl oz = 3 c

8. 48 c = __?__ gal

9. __?__ yd = 8 mi

10. 328 mL = __?__ L

11. __?__ mm = 0.7 cm

12. 150 g = __?__ kg

13. __?__ km = 57 m

14. 10,000 mg = __?__ g

15. 7.1 L = __?__ mL

16. Draw a line segment that is $4\frac{3}{4}$ inches long.

17. **ICE CREAM** A baseball team orders 5 gallons of ice cream for its end-of-season party. How many cups of ice cream is this?

Write the metric unit of length that you would use to measure each of the following.

18. length of a skateboard

19. height of a giraffe

Write the metric unit of mass or capacity that you would use to measure each of the following. Then estimate the mass or capacity.

20. five $1 bills

21. a bucket of water

Add or subtract.

22. 7 h 20 min
    + 3 h 18 min

23. 19 min 30 s
    − 12 min 40 s

24. 7 h 20 min
    + 2 h 48 min 10 s

Standardized Test Practice

25. **MULTIPLE CHOICE** Determine which container of milk is the best buy by finding the price to the nearest cent for one pint of milk.

    Ⓐ 1 pt
    Ⓑ 1 qt
    Ⓒ 0.5 gal
    Ⓓ 1 gal

Milk	
Size	Price
0.5 gal	$1.09
1 qt	$1.29
1 pt	$0.75
1 gal	$2.25

Standardized Test Practice

PART 1 Multiple Choice

Record your answers on the answer sheet provided by your teacher or on a sheet of paper.

1. The expression $10d$ converts the number of U.S. dollars to the approximate number of Mexican pesos. If d is the number of U.S. dollars, how many Mexican pesos can you get for $17? (Lesson 1-6)

   Ⓐ 17 pesos   Ⓑ 27 pesos

   Ⓒ 107 pesos   Ⓓ 170 pesos

2. Estimate the sum of three pizzas costing $9.75, $10.20, and $12.66. (Lesson 3-2)

   Ⓕ $33   Ⓖ $32

   Ⓗ $31   Ⓘ $30

3. What is the circumference of the circle? (Lesson 4-6)

   6 cm

   Ⓐ 9.14 cm

   Ⓑ 18.84 cm

   Ⓒ 37.68 cm

   Ⓓ 50.09 cm

4. Which addition sentence describes the model below? (Lesson 8-2)

   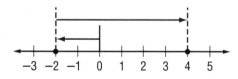

   Ⓕ $6 + (-2) = -4$ Ⓖ $-6 + 4 = -2$

   Ⓗ $-2 + 6 = 4$ Ⓘ $-6 + (-4) = 2$

5. Abigail had a collection of 36 CDs. She sold some of them at a yard sale and had 12 left. Use the equation $36 - s = 12$ to determine s, the number of CDs Abigail sold. (Lesson 9-3)

   Ⓐ 3   Ⓑ 12   Ⓒ 18   Ⓓ 24

6. Identify the model at the right as a percent. (Lesson 10-4)

   Ⓕ 30%   Ⓖ 40%

   Ⓗ 50%   Ⓘ 60%

7. Cristóbal paid $25.88 for shoes that were reduced by 25%. What was the original price? (Lesson 10-7)

   Ⓐ $29.99   Ⓑ $30.95

   Ⓒ $34.50   Ⓓ $44.95

8. Kylie measured four pieces of wood as shown. Which piece was the longest? (Lesson 12-1)

Wood Piece	Measurement
1	2 ft 5 in.
2	36 in.
3	2 ft 3 in.
4	31 in.

   Ⓕ 1   Ⓖ 2   Ⓗ 3   Ⓘ 4

9. How many cups are in 2 gallons? (Lesson 12-2)

   Ⓐ 4 c   Ⓑ 8 c   Ⓒ 16 c   Ⓓ 32 c

10. How many milliliters are equivalent to 3 liters? (Lesson 12-5)

    Ⓕ 30 mL   Ⓖ 300 mL

    Ⓗ 3,000 mL   Ⓘ 30,000 mL

11. It took Pablo 1 hour 40 minutes and 15 seconds to complete a walkathon. It took Set-Su 1 hour 50 minutes and 9 seconds to complete the walkathon. How much longer did Set-Su take to complete the walkathon? (Lesson 12-6)

    Ⓐ 9 min 51 s   Ⓑ 9 min 54 s

    Ⓒ 10 min 6 s   Ⓓ 10 min 54 s

PART 2 Short Response/Grid In

Record your answers on the answer sheet provided by your teacher or on a sheet of paper.

12. What number completes the factor tree? (Lesson 1-3)

$$42$$
$$7 \quad \times \quad 6$$
$$7 \times 2 \times \ ?$$

13. Samuel, Alexis, Madeline, and Malik each threw one dart at a dartboard. The player closest to the bull's eye wins. Who won? (Lesson 5-5)

Player	Distance to Bull's Eye
Samuel	$\frac{7}{16}$ in.
Alexis	$\frac{5}{8}$ in.
Madeline	$\frac{1}{4}$ in.
Malik	$\frac{1}{2}$ in.

14. Find the next number in the sequence.

$12, 10\frac{3}{4}, 9\frac{1}{2}, 8\frac{1}{4}, \dots$ (Lesson 7-6)

For Questions 15–17, use the function table at the right.

15. Find a rule for the function table. (Lesson 9-6)

16. Find the output for an input of -3. (Lesson 9-6)

17. Graph the function. (Lesson 9-7)

x	y
0	0
1	3
2	6
3	9

18. What color should the missing shaded region on the spinner be labeled in order to make the probability of landing on that color $\frac{1}{4}$? (Lesson 11-1)

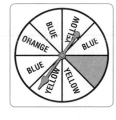

19. Stadiums that host the World Cup must have soccer fields that are 75 yards wide. How many feet is this? (Lesson 12-1)

20. How many grams are equivalent to 34.7 kilograms? (Lesson 12-5)

PART 3 Extended Response

Record your answers on a sheet of paper. Show your work.

21. Victor tutors younger students at the community center on Saturdays. It takes Victor 20 minutes to get ready and leave the house. The walk to the bus stop is 10 minutes long, and the bus ride to the community center is 40 minutes long. Finally, Victor walks 5 minutes from the bus stop to the community center. (Lesson 12-6)

  a. If Victor is scheduled to arrive at 3:00 P.M., what is the latest time at which he could start getting ready to leave?

  b. Explain the strategy you used to answer part a.

  c. Victor has to be home by 7:00 P.M. If it takes him 5 minutes longer to get home, at what time should he leave the community center?

  d. How long does Victor spend tutoring on Saturdays?

"What do road signs have to do with math?"

Many road signs are geometric in shape. In fact, they are often shaped like squares, rectangles, and triangles. You can use geometric properties to classify the shapes of road signs, determine the types of angles found in road signs, and list the similarities and differences among their shapes.

You will solve problems about road signs in Lessons 13-1 and 13-4.

GETTING STARTED

Take this quiz to see whether you are ready to begin Chapter 13. Refer to the lesson number in parentheses if you need more review.

▶ Vocabulary Review

Choose the correct term to complete each sentence.

1. A rectangle with sides of equal length is a (square, rhombus).
   (Lesson 1-8)

2. The expression $2\ell + 2w$ can be used to find the (area, perimeter) of a rectangle. (Lesson 4-5)

▶ Prerequisite Skills

Identify which figure *cannot* be folded so that one half matches the other half.

3. A B C

4. A B C

Tell whether each pair of figures has the same size and shape.

5.    6.

Find the length of each line segment in centimeters. (Lesson 12-3)

7. •————————•

8. •——————————————•

Angles and Polygons
Make this Foldable to help you organize information about angles and polygons. Begin with six half-sheets of notebook paper.

STEP 1 Fold and Cut
Fold a half-sheet of paper in half lengthwise. Then cut a 1" tab along the left edge through one thickness.

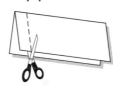

STEP 2 Glue and Label
Glue the 1" tab down. Write the word *Geometry* on this tab. Then write the lesson and title on the front tab.

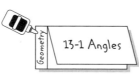

13-1 Angles

STEP 3 Label
Write *Definitions* and *Examples* under the tab.

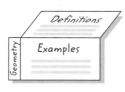

Definitions
Examples

STEP 4 Repeat and Staple
For each lesson, repeat Steps 1–3 using the remaining sheets of paper. Staple together to form a booklet.

13-1 Angles

Reading and Writing As you read and study each lesson, write definitions and examples of geometry terms from the lesson.

Angles

What You'll LEARN

Classify and measure angles.

NEW Vocabulary

angle
side
vertex
degree
right angle
acute angle
obtuse angle
straight angle
complementary
supplementary

MATH Symbols

$m\angle A$ measure of angle A

WHEN am I ever going to use this?

GARDENING The circle graph shows what Mai-Lin planted in her garden this spring.

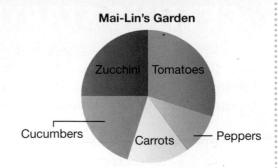

Mai-Lin's Garden

1. Mai-Lin planted the most of which food? Explain how you came to this conclusion.

2. Of which did she plant the least?

3. The percents 30%, 25%, 20%, 15%, and 10% correspond to the sections in the graph. Explain how you would match each percent with its corresponding section.

Each section of the circle graph above shows an angle. **Angles** have two **sides** that share a common endpoint called the **vertex** of the angle.

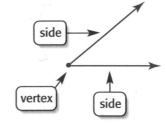

The most common unit of measure for angles is the **degree**. A circle can be separated into 360 equal-sized parts. Each part would make up a one-degree (1°) angle.

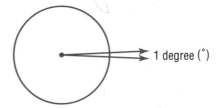

1 degree (°)

Angles can be classified according to their measure.

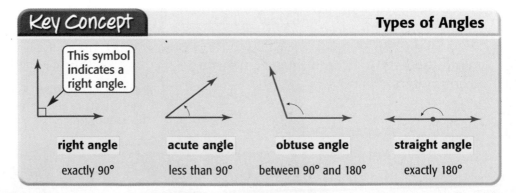

Key Concept   **Types of Angles**

This symbol indicates a right angle.

right angle	acute angle	obtuse angle	straight angle
exactly 90°	less than 90°	between 90° and 180°	exactly 180°

Use a protractor to find the measure of each angle. Then classify each angle as *acute, obtuse, right,* or *straight.*

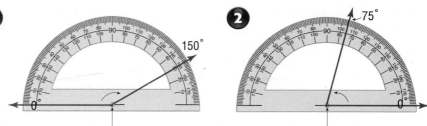

①

150°

②

75°

Align the center of the protractor with the vertex of the angle.

The angle measures 150°.
It is an obtuse angle.

The angle measures 75°.
It is an acute angle.

Your Turn Use a protractor to find the measure of each angle. Then classify each angle as *acute, obtuse, right,* or *straight.*

a.

b. ◄────────────►

Some pairs of angles are **complementary** or **supplementary**.

Key Concept   **Pairs of Angles**

Words Two angles whose sum is 90° are complementary angles.	**Models**

$m\angle 1 = 30°, m\angle 2 = 60°,$
$m\angle 1 + m\angle 2 = 90°$

Words Two angles whose sum is 180° are supplementary angles.	**Models**

$m\angle 1 = 120°, m\angle 2 = 60°,$
$m\angle 1 + m\angle 2 = 180°$

EXAMPLE   **Find Missing Angle Measures**

③ ALGEBRA Angles M and N are supplementary. If $m\angle M = 85°$, what is the measure of $\angle N$?

$m\angle M + m\angle N =$	$180°$	Supplementary angles
$85° + m\angle N =$	$180°$	Replace $m\angle M$ with 85°.
$-85°$ $=$	$-85°$	Subtract 85° from each side.
$m\angle N =$	$95°$	

So, $m\angle N = 95°$.   Since 95° + 85° = 180°, the answer is correct.

Skill and Concept Check

1. **OPEN ENDED** Draw an obtuse angle.

2. **FIND THE ERROR** Maria and David are measuring angles. Who is correct? Explain.

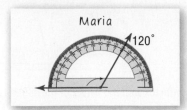
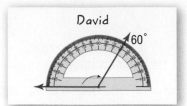

GUIDED PRACTICE

Use a protractor to find the measure of each angle. Then classify each angle as *acute*, *obtuse*, *right*, or *straight*.

3.

4.

5.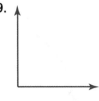

6. **ALGEBRA** Angles *G* and *H* are complementary. Find $m\angle H$ if $m\angle G = 47°$.

Practice and Applications

Use a protractor to find the measure of each angle. Then classify each angle as *acute*, *obtuse*, *right*, or *straight*.

7.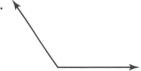

8.

9.

10.

11.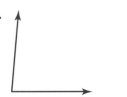

12.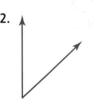

HOMEWORK HELP	
For Exercises	See Examples
7–14, 17	1, 2
15–16	3
Extra Practice See pages 620, 636.	

SIGNS Determine what types of angles are found in each road sign.

13.

14.

15. **ALGEBRA** If $m\angle A = 127°$ and $\angle A$ and $\angle B$ are supplementary, what is $m\angle B$?

16. **ALGEBRA** Angles *J* and *K* are complementary. Find $m\angle J$ if $m\angle K = 58°$.

READING For Exercises 17 and 18, use the graphic shown at the right.

17. Find the approximate measure of each angle formed by the sections of the circle graph.

18. Find the sum of the measures of the angles of the circle graph.

19. **CRITICAL THINKING** How would you change the grade of the hill so that it is not so steep?

USA TODAY Snapshots®

Families read to kindergarteners
Number of times each week family members read books to kindergartners in the class of 1998-99:

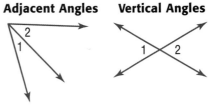

Source: National Center for Education Statistics

By Hilary Wasson and Julie Snider, USA TODAY

EXTENDING THE LESSON *Adjacent angles* are angles that share a common side and have the same vertex. *Vertical angles* are nonadjacent angles formed by a pair of lines that intersect.

20. Draw a different pair of angles that are adjacent and a pair of angles that are vertical angles.

Adjacent Angles   Vertical Angles

Determine whether ∠1 and ∠2 are *adjacent angles* or *vertical angles*.

21.

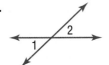

22.

23.

24.

Standardized Test Practice and Mixed Review

25. **SHORT RESPONSE** Give an example of two angle measures in which the angles are supplementary.

26. **MULTIPLE CHOICE** Which term best describes an 89.9° angle?

   A acute   **B** right   **C** straight   **D** obtuse

27. Find the sum of 13 h 45 min and 27 h 50 min. (Lesson 12-6)

Complete. (Lesson 12-5)

28. 310 mm = __?__ cm   29. 0.25 km = __?__ m   30. __?__ g = 895 mg

GETTING READY FOR THE NEXT LESSON

BASIC SKILL Use a ruler to draw a diagram that shows how the hands on a clock appear at each time.

31. 9:00   32. 12:10   33. 3:45   34. 2:30

 msmath1.net/self_check_quiz

Using Angle Measures

What You'll LEARN

Draw angles and estimate measures of angles.

HANDS-ON Mini Lab

Materials
• paper plate
• ruler
• scissors
• protractor

Work with a partner.

To estimate the measure of an angle, use angles that measure 45°, 90°, and 180°.

STEP 1 Fold a paper plate in half to find the center of the plate.

STEP 2 Cut wedges as shown. Then measure and label each angle.

1. Use the wedges to estimate the measure of each angle shown.

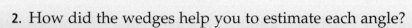

2. How did the wedges help you to estimate each angle?

3. Explain how the 90° and 45° wedges can be used to estimate the angle at the right. What is a reasonable estimate for the angle?

4. How would you estimate the measure of any angle without using the wedges?

To estimate the measure of an angle, compare it to an angle whose measure you know.

EXAMPLE Estimate Angle Measures

1. **Estimate the measure of the angle shown.**

   Compare the given angle to an angle whose measure you know. The angle is a little less than a 90° angle. So, a reasonable estimate is about 80°.

Your Turn Estimate the measure of each angle.

a.

b.

A protractor and a *straightedge*, or ruler, can be used to draw angles.

EXAMPLE **Draw an Angle**

2 Draw a 74° angle.

STUDY TIP

Checking Reasonableness
You can check whether you have used the correct scale by comparing your angle with an estimate of its size.

Step 1 Draw one side of the angle. Then mark the vertex and draw an arrow.

Step 2 Place the center point of the protractor on the vertex. Align the mark labeled 0 on the protractor with the line. Find 74° on the correct scale and make a pencil mark.

Step 3 Use a straightedge to draw the side that connects the vertex and the pencil mark.

Your Turn Use a protractor and a straightedge to draw angles having the following measurements.

c. 68°   d. 105°   e. 85°

Skill and Concept Check

Writing Math
Exercises 1 & 3

1. **Explain** how you would draw an angle measuring 65°.

2. **OPEN ENDED** Draw an angle whose measure is about 45°.

3. **Which One Doesn't Belong?** Identify the angle that does not measure about 45°. Explain your reasoning.

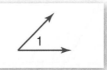   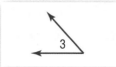   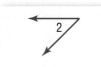   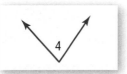

GUIDED PRACTICE

Estimate the measure of each angle.

4.

5.

Use a protractor and a straightedge to draw angles having the following measurements.

6. 25°   7. 140°

Estimate the measure of each angle.

8.

9.

10.

11.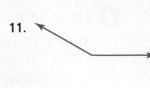

HOMEWORK HELP

For Exercises	See Examples
8–11, 20, 21	1
12–19, 22	2

Extra Practice
See pages 620, 636.

Use a protractor and a straightedge to draw angles having the following measurements.

12. 75°   **13.** 50°   **14.** 45°   **15.** 20°

16. 115°   **17.** 175°   **18.** 133°   **19.** 79°

20. TIME Is the measure of the angle formed by the hands on a clock at 3:20 P.M. greater than, less than, or about equal to 90°? Explain.

21. RESEARCH Use the Internet or another source to find a photo of a humpback whale. Draw an example of the angle formed by the two tail fins. Then give a reasonable estimate for the measure of this angle.

22. SCIENCE Most globes show that Earth's axis inclines 23.5° from vertical. Use the data below to draw diagrams that show the inclination of axis of each planet listed.

23.5°

Planet	Uranus	Neptune	Pluto	Venus
Inclination of Axis	97.9°	29.6°	122°	177.3°

23. CRITICAL THINKING Describe how the corner of a textbook can be used to estimate the measure of an angle.

24. MULTIPLE CHOICE Estimate the measure of the angle shown.

  Ⓐ 60°   Ⓑ 120°   Ⓒ 90°   Ⓓ 150°

25. SHORT RESPONSE Draw an angle having a measure of 66°.

26. ALGEBRA Angles W and U are supplementary. If $m\angle W = 56°$, what is the measure of $\angle U$? (Lesson 13-1)

Find each elapsed time. (Lesson 12-6)

27. 4:25 P.M. to 10:42 P.M.   **28.** 9:30 P.M. to 2:10 A.M.

GETTING READY FOR THE NEXT LESSON

PREREQUISITE SKILL Find the measure of each line segment in centimeters. What is the length of half of each segment? (Lesson 12-3)

29. •———• 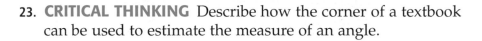   **30.** •——•   **31.** •————•

Construct Congruent Segments and Angles

A **line segment** is a straight path between two endpoints. To indicate line segment JK, write $\overline{JK}$. Line segments that have the same length are called **congruent segments**. In this lab, you will construct congruent segments using a **compass**.

ACTIVITY *Work with a partner.*

1 **STEP 1** Draw $\overline{JK}$. Then use a straightedge to draw a line segment longer than $\overline{JK}$. Label it $\overline{LM}$.

STEP 2 Place the compass at J and adjust the compass setting so you can place the pencil tip on K. The compass setting equals the length of $\overline{JK}$.

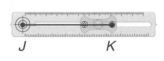

STEP 3 Using this setting, place the compass tip at L. Draw an arc to intersect $\overline{LM}$. Label the intersection P. $\overline{LP}$ is congruent to $\overline{JK}$.

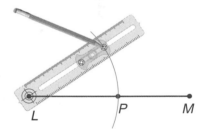

Your Turn Trace each segment. Then construct a segment congruent to it.

a. X —— Y b. R —— S c. M —————— N

Writing Math

1. Explain, in your own words, how to construct a line segment that is congruent to a given line segment.

2. Find the measure of $\overline{JK}$ above. How does this compare to the measure of $\overline{LP}$?

3. Suppose the length of $\overline{JK}$ is 26 centimeters. If $\overline{JK}$ is separated into two congruent parts, what will be the length of each part? Explain.

The angle in step 1 shown below can be named in two ways, ∠JKM and ∠MKJ. The vertex is always the middle letter.

You can also construct congruent angles with a compass.

ACTIVITY *Work with a partner.*

2

STEP 1 Draw ∠JKM. Then use a straightedge to draw $\overrightarrow{ST}$.

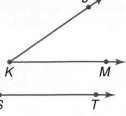

STEP 2 Place the tip of the compass at *K*. Draw an arc to intersect both sides of ∠JKM. Label the points of intersection *X* and *Y*.

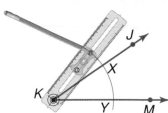

STEP 3 Using this setting, place the compass at point *S*. Draw an arc to intersect $\overrightarrow{ST}$. Label the intersection *W*.

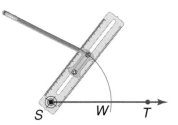

STEP 4 Place the point of the compass on *Y*. Adjust so that the pencil tip is on *X*.

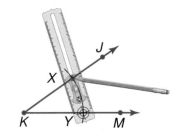

STEP 5 Using this setting, place the compass at *W*. Draw an arc to intersect the arc in Step 3. Label the intersection *U*. Draw $\overrightarrow{SU}$. ∠JKM is congruent to ∠UST.

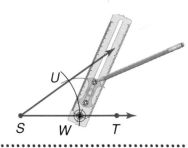

Writing Math

4. Explain the relationship between ∠JKM and ∠UST.

5. Explain how to construct an angle that is congruent to a 65° angle.

What You'll LEARN

Bisect line segments and angles.

NEW Vocabulary

bisect
congruent
perpendicular

MATH Symbols

$\cong$ is congruent to
$\overline{AB}$ segment AB

WHEN **am I ever going to use this?**

ENGLISH The table lists a few words that contain the prefix *bi-*.

1. Use the Internet or a dictionary to find the meaning of each word.
2. What do the meanings have in common?
3. What does the prefix *bi-* mean?
4. **Make a conjecture** about what it means to bisect something.

Words that Contain the Prefix *bi-*
bicycle
bimonthly
bilingual
biathlon
binocular

To **bisect** something means to separate it into two equal parts. You can use a straightedge and a compass to bisect a line segment.

EXAMPLE **Bisect a Line Segment**

1 **Use a straightedge and a compass to bisect $\overline{AB}$.**

Step 1 Draw $\overline{AB}$.

Step 2 Place the compass at point A. Using a setting greater than one half the length of $\overline{AB}$, draw two arcs as shown.

Step 3 Using the same setting, place the compass at point B. Draw an arc above and below as shown.

Step 4 Use a straightedge to align the intersections. Draw a segment that intersects $\overline{AB}$. Label the intersection M.

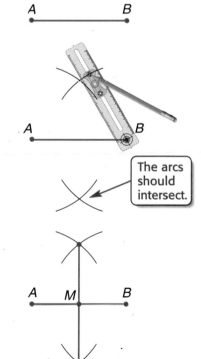

The arcs should intersect.

The vertical line segment bisects $\overline{AB}$ at M. The segments $\overline{AM}$ and $\overline{MB}$ are **congruent**. This can be written as $\overline{AM} \cong \overline{MB}$. This means that the measure of $\overline{AM}$ is equal to the measure of $\overline{MB}$. The line segments are also **perpendicular**. That is, they meet at right angles.

A compass and straightedge can also be used to bisect an angle.

EXAMPLE **Bisect an Angle**

2 Use a straightedge and a compass to bisect ∠MNP.

Step 1 Draw ∠MNP.

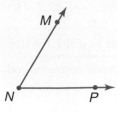

Step 2 Place the compass at point N and draw an arc that intersects both sides of the angle. Label the points of intersection X and Y.

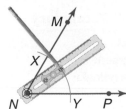

Step 3 With the compass at point X, draw an arc as shown.

Step 4 Using the same setting, place the compass point at Y and draw another arc as shown. Label the intersection Z.

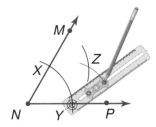

Step 5 Use a straightedge to draw $\overrightarrow{NZ}$.

$\overrightarrow{NZ}$ bisects ∠MNP. Therefore, ∠MNZ and ∠ZNP are congruent. This can be written as ∠MNZ ≅ ∠ZNP.

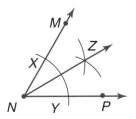

Your Turn

a. Draw a line segment measuring 6 centimeters. Then use a straightedge and compass to bisect the segment.

b. Draw a 120° angle. Then use a straightedge and a compass to bisect the angle.

Skill and Concept Check

Writing Math
Exercise 1

1. **Describe** the result of bisecting an angle.

2. **OPEN ENDED** Draw a pair of congruent segments.

GUIDED PRACTICE

Draw each line segment or angle having the given measurement. Then use a straightedge and a compass to bisect the line segment or angle.

3. 3 cm 　　　　　　　　　　4. 85°

msmath1.net/extra_examples

Practice and Applications

HOMEWORK HELP

For Exercises	See Examples
5–6, 9	1
7–8, 10	2

Extra Practice
See pages 620, 636.

Draw each line segment or angle having the given measurement. Then use a straightedge and a compass to bisect the line segment or angle.

5. 2 cm
6. $1\frac{1}{2}$ in.
7. 60°
8. 135°

9. Draw and then bisect a $1\frac{3}{4}$-inch segment. Find the measure of each segment.

10. Draw a 129° angle. Then bisect it. What is the measure of each angle?

For Exercises 11–14, refer to segment $\overline{AF}$. Identify the the point that bisects each given line segment.

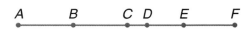

11. $\overline{AC}$
12. $\overline{AF}$
13. $\overline{CF}$
14. $\overline{BF}$

FANS For Exercises 15–17, refer to the fan at the right.

15. Name the side of the fan that appears to bisect the section represented by $\angle XVY$.

16. Name two other angles and their bisectors.

17. Use a protractor to verify your answers to Exercises 15 and 16. Are the answers reasonable? Explain.

18. **CRITICAL THINKING** Explain how to construct a line passing through a given point and perpendicular to a given line using a compass and a straightedge.

Standardized Test Practice and Mixed Review

19. **MULTIPLE CHOICE** In the figure shown, which ray appears to bisect $\angle DBC$?

   Ⓐ $\overrightarrow{BD}$ Ⓑ $\overrightarrow{BA}$ Ⓒ $\overrightarrow{BC}$ Ⓓ $\overrightarrow{BE}$

20. **MULTIPLE CHOICE** An angle that measures 37° is bisected. What is the measure of each angle?

   Ⓕ 17°   Ⓖ 17.5°   Ⓗ 18.5°   Ⓘ 19°

Use a protractor and a straightedge to draw angles having the following measurements. (Lesson 13-2)

21. 75°
22. 25°
23. 110°

24. **ALGEBRA** Angles G and H are supplementary angles. If $m\angle G = 115°$, find $m\angle H$. (Lesson 13-1)

GETTING READY FOR THE NEXT LESSON

BASIC SKILL Draw an example of each figure.

25. rectangle
26. parallelogram
27. triangle

Vocabulary and Concepts

1. **Explain** the difference between acute angles and obtuse angles.
   (Lesson 13-1)

2. **Define** *bisect*. (Lesson 13-3)

Skills and Applications

Use a protractor to find the measure of each angle. Then classify each angle as *acute*, *obtuse*, *right*, or *straight*. (Lesson 13-1)

3.

4.

5. **ALGEBRA** If $m\angle A = 108°$ and $\angle A$ and $\angle B$ are supplementary, find $m\angle B$. (Lesson 13-1)

Use a protractor to draw angles having the following measurements.
(Lesson 13-2)

6. $35°$ 7. $110°$ 8. $80°$

Draw each line segment or angle having the given measurement. Then use a straightedge and a compass to bisect the line segment or angle. (Lesson 13-3)

9. $120°$ 10. $30°$ 11. 2.5 in.   12. 3 cm

Standardized Test Practice

13. **MULTIPLE CHOICE** Which angle measures between $45°$ and $90°$?
    (Lesson 13-2)

14. **SHORT RESPONSE** A snowflake is shown. Name an angle and its bisector. (Lesson 13-3)

    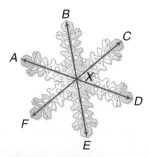

The GameZone

A Place To Practice Your Math Skills

Wild Angles

● **GET READY!**

Players: two
Materials: 21 index cards, spinner

5°	5°	10°	10°	15°	15°
20°	20°	25°	25°	30°	30°
35°	35°	40°	40°	45°	45°
50°	50°	55°	55°	60°	60°
65°	65°	70°	70°	75°	75°
80°	80°	85°	85°	90°	90°
Wild	Wild	Wild	Wild	Wild	Wild

● **GET SET!**

- Cut the index cards in half.
- Label the cards and spinner as shown.

● **GO!**

- Shuffle the cards and then deal five cards to each player. Place the remaining cards facedown in a pile.

- A player spins the spinner.

- Using two cards, the player forms a pair whose sum results in the type of angle spun. A wild card represents any angle measure. Each pair is worth 2 points.

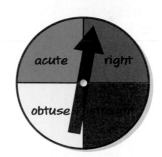

- If a pair cannot be formed, the player discards one card and selects another from the facedown pile. If a pair is formed, the player sets aside the two cards and gets 2 points. Then it is the other player's turn. If no pair is formed, it is the other player's turn.

- **Who Wins?** The first player to reach 20 points wins.

What You'll Learn
Solve problems by drawing a diagram.

Draw a Diagram

Hey Shawn, the science club is going to plant cacti in the school courtyard. The courtyard is 46 feet by 60 feet, and each planting bed will be 6 feet across.

Yes, Margarita, I heard that. The planting beds will be square, and they will be 8 feet apart and 6 feet away from any walls. How many beds can we make?

Explore	We know all the dimensions. We need to find how many beds will fit inside of the courtyard.
Plan	Let's draw a diagram to find how many planting beds will fit in the courtyard.

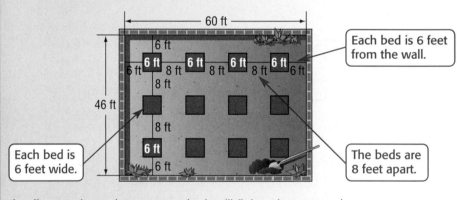

Each bed is 6 feet from the wall.

Each bed is 6 feet wide.

The beds are 8 feet apart.

Solve	The diagram shows that 12 cactus beds will fit into the courtyard.
Examine	Make sure the dimensions meet the requirements. The distance across is 46 feet, and the distance down is 60 feet. So the answer is correct.

Analyze the Strategy

1. **Explain** why you think the students chose to draw a diagram to solve the problem.

2. **Determine** the number of cactus beds that could be planted if the courtyard measured 74 feet by 88 feet.

3. **Write** a problem that can be solved by making a diagram.

Solve. Use the draw a diagram strategy.

4. **TRAVEL** Jasmine lives in Glacier and works in Alpine. There is no direct route from Glacier to Alpine, so Jasmine drives through either Elm or Perth. How many different ways can she drive to work?

5. **DECORATING** For the Spring Dance, there are 5 columns arranged in the shape of a pentagon. Large streamers are hung from each column to every other column. How many streamers are there in all?

Mixed Problem Solving

Solve. Use any strategy.

6. **CAMPING** Robin bought a tent for camping. Each of the four sides of the tent needs three stakes to secure it properly to the ground. How many stakes are needed?

BASKETBALL For Exercises 7 and 8, refer to the table.

Game	Tally	Three-Point Shots
1	JHT II	7
2	JHT	5
3	JHT IIII	9
4	IIII	4
5	JHT I	6

7. What is the mean number of three-point shots made by the team for games 1–5?

8. What is the median number of three-point shots made for games 1–5?

9. **GEOMETRY** A kite has two pairs of congruent sides. If two sides are 56 centimeters and 34 centimeters, what is the perimeter of the kite?

10. **FOOD** Jesse works at the local sandwich shop. There are 4 different kinds of bread and 6 different kinds of meat to choose from. How many different sandwiches could be made using one kind of bread and two different kinds of meat?

11. **WEATHER** What is the difference between the hottest and coldest temperatures in the world?

Hottest Temp.	Coldest Temp.
134°F	−128°F

12. **PATTERNS** A number is doubled and then 9 is subtracted. If the result is 15, what was the original number?

13. **STANDARDIZED TEST PRACTICE** Refer to the Venn diagram.

Ecology Club and Honor Society

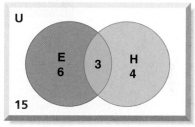

U = all students in the class
E = ecology club   H = honor society

Which statement is *not* true?

Ⓐ There are more kids in the honor society than the ecology club.

Ⓑ One fourth of the class is in the honor society.

Ⓒ There are only two more students in the ecology club than the honor society.

Ⓓ 15 students are not in either club.

You will use the draw a diagram strategy in the next lesson.

Two-Dimensional Figures

NEW Vocabulary

polygon
triangle
quadrilateral
pentagon
hexagon
heptagon
octagon
regular polygon
scalene triangle
isosceles triangle
equilateral triangle
rectangle
square
parallelogram
rhombus

 HANDS-ON Mini Lab

Materials
- six flexible straws
- ruler
- protractor

Work with a partner.

STEP 1 Using four flexible straws, insert an end of one straw into the end of another straw as shown.

STEP 2 Form a square.

1. What is true about the angles and sides of a square?
2. Using two more straws, what changes need to be made to the square to form a rectangle that is not a square?
3. How are rectangles and squares alike? How do they differ?
4. Push on one vertex of the rectangle so it is no longer a rectangle. What is true about the opposite sides?

In geometry, flat figures such as squares or rectangles are *two-dimensional* figures. A **polygon** is a simple, closed, two-dimensional figure formed by three or more sides.

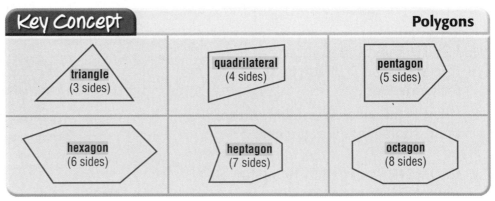

Key Concept Polygons

| triangle (3 sides) | quadrilateral (4 sides) | pentagon (5 sides) |
| hexagon (6 sides) | heptagon (7 sides) | octagon (8 sides) |

When all of the sides of a polygon are congruent and all of the angles are congruent, the polygon is a **regular polygon**.

READING Math

Congruent Markings
The red tick marks show congruent sides. The red arcs show congruent angles.

EXAMPLE Identify a Polygon

1 **Identify the polygon. Then tell if it is a regular polygon.**

The polygon has 5 sides. So, it is a pentagon. Since the sides and angles are congruent, it is a regular polygon.

Certain triangles and quadrilaterals have special names.

Figure	Characteristics
scalene triangle	• No sides congruent.
isosceles triangle	• At least two sides congruent.
equilateral triangle	• All sides congruent. • All angles congruent.
rectangle	• Opposite sides congruent. • All angles are right angles. • Opposite sides parallel.
square	• All sides congruent. • All angles are right angles. • Opposite sides parallel.
parallelogram	• Opposite sides congruent. • Opposite sides parallel. • Opposite angles congruent.
rhombus	• All sides congruent. • Opposite sides parallel. • Opposite angles congruent.

EXAMPLE Analyze Two-Dimensional Figures

② FLAGS Many aircraft display the American flag in the shape of a parallelogram to show motion. Identify and describe the similarities and the differences between a rectangle and a parallelogram.

These shapes are alike because they both have four sides, opposite sides parallel, and opposite sides congruent.

They are different because a rectangle has four right angles and a parallelogram does not necessarily have four right angles.

Skill and Concept Check

Writing Math
Exercise 2

1. **Draw** an example of each polygon listed. Mark any congruent sides, congruent angles, and right angles.

   a. hexagon   **b.** regular octagon   **c.** parallelogram

   d. triangle   **e.** equilateral triangle   **f.** rectangle

2. **OPEN ENDED** Describe two different real-life items that are shaped like a polygon.

GUIDED PRACTICE

Identify each polygon. Then tell if it is a regular polygon.

3.   4.   5.

6. **SIGNS** Identify and then describe the similarities and the differences between the shapes of the road signs shown.

Practice and Applications

Identify each polygon. Then tell if it is a regular polygon.

7.   8.   9.

10.   11.   12.

HOMEWORK HELP

For Exercises	See Examples
7–12, 20	1
17	2

Extra Practice
See pages 621, 636.

13. Draw a quadrilateral that is not a parallelogram.

14. Draw a triangle with only two equal sides. Identify this triangle.

15. Draw a scalene triangle having angle measures 55°, 40°, and 85°.

16. **BIRD HOUSES** The front of a bird house is shaped like a regular pentagon. If the perimeter of the front is 40 inches, how long is each side?

17. Describe the similarities and the differences between a square and a rhombus.

Give a counterexample for each statement.

18. All parallelograms are rectangles.   19. All quadrilaterals are parallelograms.

20. GEOGRAPHY Name the polygon formed by the boundaries of each state shown at the right.

COLORADO   UTAH   NEW MEXICO

21. ALGEBRA The sum of the measures of a regular octagon is 1,080°. Write and solve an equation to find the measure of one of the angles.

Tell whether each statement is *sometimes*, *always*, or *never* true.

22. Parallelograms are squares.

23. A rhombus is a square.

24. A rectangle is a parallelogram.

25. A square is a rhombus.

26. CRITICAL THINKING Explain how to construct the following using a compass and a straightedge.

    a. an equilateral triangle   **b.** an isosceles triangle

EXTENDING THE LESSON The sum of the angles of a quadrilateral is 360°.

Find each missing measure.

27.

28.

29.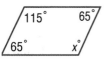

30. Draw a quadrilateral whose angles measure 90°, 70°, 120°, and 80°.

Standardized Test Practice and Mixed Review

31. MULTIPLE CHOICE Which polygon is *not* a regular polygon?

  A   **B**   **C**   **D**

32. MULTIPLE CHOICE Which is *not* a characteristic of a rectangle?

  F All sides are congruent.   **G** All angles are right angles.

  H Opposite sides are parallel.   **I** All angles are congruent.

33. Refer to the angles at the right. Identify the ray that bisects ∠JKM.
(Lesson 13-3)

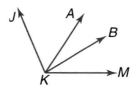

Use a protractor to draw angles having the following measurements. Then classify each angle as *acute*, *right*, *obtuse*, or *straight*. (Lesson 13-2)

34. 35°   **35.** 100°   **36.** 180°

GETTING READY FOR THE NEXT LESSON

BASIC SKILL Identify which figure *cannot* be folded so that one half matches the other half.

37.

38.

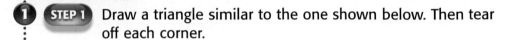

HANDS-ON LAB

A Follow-Up of Lesson 13-4

What You'll LEARN

Explore, classify, and draw triangles and quadrilaterals.

Materials

- notebook paper
- scissors
- protractor
- dot paper
- colored pencils

Triangles and Quadrilaterals

There are many different properties and characteristics of triangles and quadrilaterals. In this lab, you will explore these properties and characteristics. Triangle means three angles. Let's first explore how the three angles of a triangle are related.

ACTIVITY *Work with a partner.*

1 **STEP 1** Draw a triangle similar to the one shown below. Then tear off each corner.

STEP 2 Rearrange the torn pieces as shown.

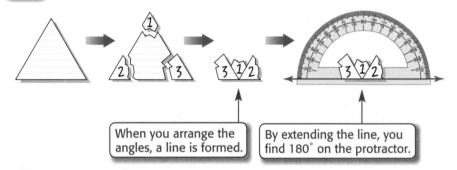

When you arrange the angles, a line is formed.

By extending the line, you find 180° on the protractor.

STEP 3 Repeat Steps 1 and 2 with a different triangle.

Therefore, the sum of the measures of the angles of a triangle is 180°.

Triangles can be classified according to their angles.

ACTIVITY *Work with a partner.*

2 **STEP 1** Draw the triangle shown at the right on dot paper. Then cut it out.

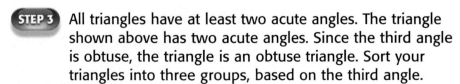

STEP 2 Draw nine more different triangles on dot paper. Then cut out each triangle.

STEP 3 All triangles have at least two acute angles. The triangle shown above has two acute angles. Since the third angle is obtuse, the triangle is an obtuse triangle. Sort your triangles into three groups, based on the third angle.

STEP 4 Name the groups *acute*, *right*, and *obtuse*.

Your Turn

Find the missing angle measure for each triangle shown. Then classify each triangle as *acute*, *right*, or *obtuse*.

a.

b.
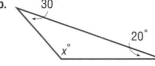

c.

In Activity 2, you classified triangles by their angles. Now you will classify quadrilaterals by their *sides* and *angles*.

ACTIVITY · *Work with a partner.*

3

STEP 1 Draw the two quadrilaterals shown on dot paper. Then cut them out.

STEP 2 Draw nine more different quadrilaterals on dot paper. Then cut out each quadrilateral.

STEP 3 The first quadrilateral shown can be classified as a quadrilateral with four right angles. Sort your quadrilaterals into three groups, based on any characteristic. Write a description of the quadrilaterals in each group.

Writing Math

1. If a triangle has angles with measures 45°, 35°, and 100°, what type of triangle is it? Explain.

2. Is the statement *All rectangles are parallelograms, but not all parallelograms are rectangles* true or false? Explain.

3. Tell why a triangle must always have at least two acute angles. Include drawings in your explanation.

4. Two different quadrilaterals each have four congruent sides. However, one has four 90° angles, and the other has no 90° angles. Draw the figures and compare them using the given characteristics.

Lines of Symmetry

What You'll LEARN

Describe and define lines of symmetry.

NEW Vocabulary

line symmetry
line of symmetry
rotational symmetry

HANDS-ON Mini Lab

Materials
• tracing paper

Work with a partner.
A butterfly, a dragonfly, and a lobster have a common characteristic that relates to math.

STEP 1 Trace the outline of each figure.

STEP 2 Draw a line down the center of each figure.

1. Compare the left side of the figure to the right side.
2. Draw another figure that has the same characteristic as a butterfly, a dragonfly, and a lobster.

When two halves of a figure match, the figure is said to have **line symmetry**. The line that separates the figure into two matching halves is called a **line of symmetry**.

EXAMPLES Draw Lines of Symmetry

Draw all lines of symmetry for each figure.

1

This figure has 1 line of symmetry.

2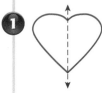

The letter J has no lines of symmetry.

3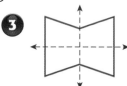

This hexagon has 2 lines of symmetry.

Your Turn Trace each figure. Then draw all lines of symmetry.

a.

b.

c.

Test-Taking Tip *The Princeton Review*

Taking the Test
If you are not permitted to write in the test booklet, copy the figure onto paper.

EXAMPLE **Identify Line Symmetry**

4 MULTIPLE-CHOICE TEST ITEM

The Navy signal flag for the number 5 is shown. How many lines of symmetry does this flag have?

ⓐ 2   ⓑ 4   ⓒ 8   ⓓ none

Read the Test Item You need to find all of the lines of symmetry for the flag.

Solve the Test Item Draw all lines of symmetry. It is a good idea to number each line so that you do not count a line twice.

There are 4 lines of symmetry.

The answer is B.

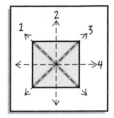

Some figures can be turned or rotated less than 360° about a fixed point so that the figure looks exactly as it did before being turned. These figures are said to have **rotational symmetry**.

EXAMPLES **Identify Rotational Symmetry**

Tell whether each figure has rotational symmetry.

5

0°   90°   180°   270°   360°

When the figure is rotated 180°, the figure looks as it did before it was rotated. So, the figure has rotational symmetry.

6

0°   90°   180°   270°   360°

The figure appears as it did before being rotated only after being rotated 360°. So, it does *not* have rotational symmetry.

Your Turn Tell whether each figure has rotational symmetry. Write *yes* or *no*.

d.

e.

Skill and Concept Check

Writing
Math
Exercises 1 & 3

1. **Describe** line symmetry and rotational symmetry.

2. **OPEN ENDED** Draw a figure that has rotational symmetry.

3. **FIND THE ERROR** Daniel and Jonas are finding the lines of symmetry for a regular pentagon. Who is correct? Explain.

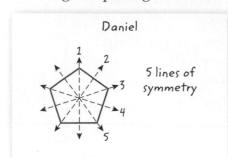

Daniel

5 lines of symmetry

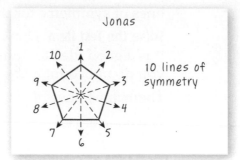

Jonas

10 lines of symmetry

GUIDED PRACTICE

Trace each figure. Then draw all lines of symmetry.

4.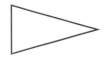

5.

6.

Tell whether each figure has rotational symmetry. Write *yes* or *no*.

7.

8.

9.

Practice and Applications

Trace each figure. Then draw all lines of symmetry.

HOMEWORK HELP	
For Exercises	See Examples
10–17, 25	1, 2, 3, 4
18–24	5–6
Extra Practice See pages 621, 636.	

10.

11.

12.

13.

14.

15.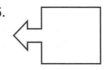

16. **MUSIC** How many lines of symmetry does a violin have?

17. **DECORATING** Find the number of lines of symmetry for a square picture frame.

Tell whether each figure has rotational symmetry. Write *yes* or *no*.

18.

19.

20.

21.

22.

23.

24. **SCIENCE** Does a four-leaf clover have rotational symmetry?

25. **FLAGS** The flag of Japan is shown. How many lines of symmetry does the flag have?

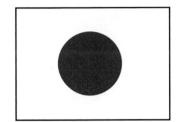

 Data Update What U.S. state flags have line symmetry? Visit msmath1.net/data_update to learn more.

26. **CRITICAL THINKING** Which figures below have both line and rotational symmetry?

a.   b.   c.   d.

Standardized Test Practice and Mixed Review

27. **MULTIPLE CHOICE** Which figure does *not* have line symmetry?

Ⓐ   Ⓑ   Ⓒ   Ⓓ

28. **SHORT RESPONSE** What capital letters of the alphabet have rotational symmetry?

29. **FOOD** Identify the shape of the front of a cereal box. Tell if it is a regular polygon. (Lesson 13-4)

30. Draw a 6-centimeter line segment. Then use a straightedge and a compass to bisect the line segment. (Lesson 13-3)

GETTING READY FOR THE NEXT LESSON

BASIC SKILL Tell whether each pair of figures have the same size and shape.

31.

32.

33.

Transformations

A **transformation** is a movement of a figure. The three types of transformations are a **translation** (slide), a **reflection** (flip), and a **rotation** (turn).

In a translation, a figure is slid horizontally, vertically, or both.

Materials

- grid paper
- pattern blocks
- geomirror
- colored pencils

ACTIVITY *Work with a partner.*

1 Perform a translation of a figure on a coordinate grid.

STEP 1 Trace a parallelogram-shaped pattern block onto the coordinate grid. Label the vertices *ABCD*.

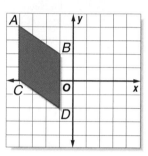

STEP 2 Slide the pattern block over 5 units to the right.

STEP 3 Trace the figure in its new position. Label the vertices *A'*, *B'*, *C'*, and *D'*.

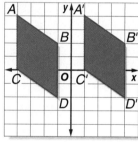

Parallelogram *A'B'C'D'* is the image of parallelogram *ABCD* after a translation 5 units right.

READING Math

Notation The notation *A'* is read *A prime*. This notation is used to name a point after a translation.

In a reflection, a figure is flipped over a line.

ACTIVITY *Work with a partner.*

2 Perform a reflection of a figure on a coordinate grid.

STEP 1 Trace a parallelogram-shaped pattern block as shown. Label the vertices *A*, *B*, *C*, and *D*.

STEP 2 Place a geomirror on the *y*-axis.

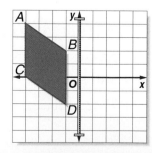

STEP 3 Trace the reflection of the parallelogram. Label the vertices *A'*, *B'*, *C'*, and *D'*.

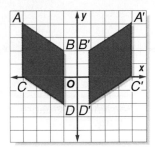

Parallelogram *A'B'C'D'* is the image of parallelogram *ABCD* reflected over the *y*-axis.

In a rotation, a figure is rotated about a point.

ACTIVITY *Work with a partner.*

3 **Perform a rotation of a figure on a coordinate grid.**

STEP 1 Trace a parallelogram-shaped pattern block onto the coordinate grid as shown. Label the vertices *A*, *B*, *C*, and *D*.

STEP 2 Rotate the figure 90° clockwise, using the origin as the point of rotation.

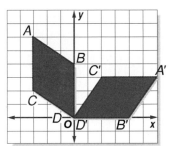

STEP 3 Trace the rotation of the figure. Label the vertices *A'*, *B'*, *C'*, and *D'*.

Parallelogram *A'B'C'D'* is the image of parallelogram *ABCD* rotated 90° clockwise about the origin.

Your Turn

Using the pattern block shown, perform each transformation described on a coordinate grid.

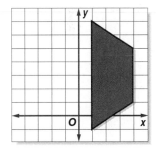

a. a translation 5 units left

b. a reflection across the *y*-axis

c. a 90° rotation counterclockwise

Writing Math

1. A square is transformed across the *y*-axis. How could this transformation be interpreted as a slide, a flip, and a turn?

2. A rectangle has vertices at (3, 2), (7, 2), (3, 8), and (7, 8). What happened to the figure if its transformed image has vertices at (8, 2), (12, 2), (8, 8), and (12, 8)?

Similar and Congruent Figures

13-6

What You'll LEARN

Determine congruence and similarity.

NEW Vocabulary

similar figures
congruent figures
corresponding parts

Link to READING

similar: nearly, but not exactly, the same or alike.

WHEN am I ever going to use this?

PATTERNS The triangle at the right is called *Sierpinski's triangle*. Notice how the pattern is made up of various equilateral triangles.

1. How many different-sized triangles are in the pattern?

2. Compare the size and shape of these triangles.

Figures that have the same shape but not necessarily the same size are called **similar figures**. Here are some examples.

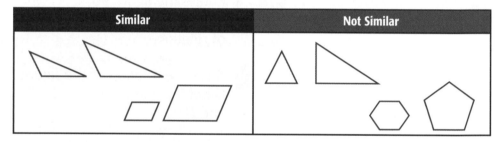

Figures that have the same size and shape are **congruent figures**. Consider the following.

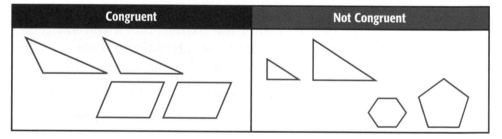

EXAMPLES Identify Similar and Congruent Figures

Tell whether each pair of figures is *similar, congruent,* or *neither.*

1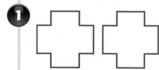

The figures have the same size and shape. They are congruent.

2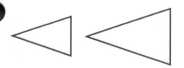

The figures have the same shape but not the same size. They are similar.

Your Turn Tell whether each pair of figures is *similar*, *congruent*, or *neither*.

a.  b.   c.

The parts of congruent figures that "match" are called **corresponding parts**.

EXAMPLES Apply Similarity and Congruence

WINDMILLS The arms of the windmill shown have congruent quadrilaterals.

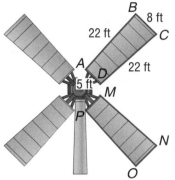

3 What side of quadrilateral *ABCD* corresponds to side $\overline{MP}$?

Side $\overline{AD}$ corresponds to side $\overline{MP}$.

4 What is the perimeter of quadrilateral *MNOP*?

The perimeter of quadrilateral *ABCD* is 5 + 22 + 8 + 22, or 57 feet. Since the quadrilaterals are congruent, they have the same size and shape. So, the perimeter of quadrilateral *MNOP* is 57 feet.

Skill and Concept Check

1. **Describe** similarity and congruence.

2. **Draw** two figures that are congruent and two figures that are not congruent.

3. **OPEN ENDED** Draw a pair of similar triangles and a pair of congruent quadrilaterals.

> **Writing Math**
> Exercise 1

GUIDED PRACTICE

Tell whether each pair of figures is *congruent*, *similar*, or *neither*.

4.    5. 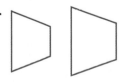   6.

For Exercises 7 and 8, use the figures shown at the right. Triangle *DEF* and △*XYZ* are congruent triangles.

7. What side of △*DEF* corresponds to side $\overline{XZ}$?

8. Find the measure of side $\overline{EF}$.

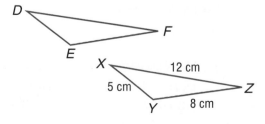

Practice and Applications

Tell whether each pair of figures is *congruent*, *similar*, or *neither*.

9.
10.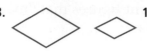
11.

12.
13.
14.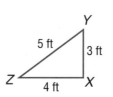

HOMEWORK HELP

For Exercises	See Examples
9–16	1, 2
17–20	3, 4

Extra Practice
See pages 621, 636.

15. **STATUES** Are a model of the Statue of Liberty and the actual Statue of Liberty similar figures? Explain.

16. Describe a transformation or a series of motions that will show that the two shapes shown on the coordinate system are congruent.

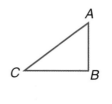

For Exercises 17–20, use the congruent triangles at the right.

17. What side of △*ABC* corresponds to side $\overline{YZ}$?
18. Name the side of △*XYZ* that corresponds to side $\overline{AB}$.
19. What is the measure of side $\overline{AC}$?
20. Find the perimeter of △*ABC*.

CRITICAL THINKING Tell whether each statement is *sometimes*, *always*, or *never* true. Explain your reasoning.

21. All rectangles are similar.
22. All squares are similar.

Standardized Test Practice and Mixed Review

23. **MULTIPLE CHOICE** Which polygons are congruent?

Ⓐ
Ⓑ
Ⓒ
Ⓓ

24. **SHORT RESPONSE** Draw two similar triangles in which the size of one is twice the size of the other.

Draw all lines of symmetry for each figure. (Lesson 13-5)

25. square
26. regular pentagon
27. equilateral triangle

28. **WINDOWS** A window is shaped like a regular hexagon. If the perimeter of the window is 54 inches, how long is each side?
(Lesson 13-4)

 msmath1.net/self_check_quiz

13-6b HANDS-ON LAB

A Follow-Up of Lesson 13-6

Tessellations

A pattern formed by repeating figures that fit together without gaps or overlaps is a **tessellation**. Tessellations are formed using slides, flips, or turns of congruent figures.

What You'll LEARN

Create tessellations using pattern blocks.

Materials
• pattern blocks

ACTIVITY *Work with a partner.*

STEP 1 Select the three pattern blocks shown.

STEP 2 Choose one of the blocks and trace it on your paper. Choose a second block that will fit next to the first without any gaps or overlaps and trace it.

STEP 3 Trace the third pattern block into the tessellation.

STEP 4 Continue the tessellation by expanding the pattern.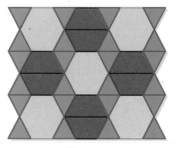

Your Turn

Create a tessellation using the pattern blocks shown.

a.  b.  c.

Writing Math

1. **Tell** if a tessellation can be created using a square and an equilateral triangle. Justify your answer with a drawing.

2. What is the sum of the measures of the angles where the vertices of the figures meet? Is this true for all tessellations?

3. Name two figures that cannot be used to create a tessellation. Use a drawing to justify your answer.

Vocabulary and Concept Check

acute angle (p. 506)	line of symmetry (p. 528)	rhombus (p. 523)
angle (p. 506)	line symmetry (p. 528)	right angle (p. 506)
bisect (p. 515)	obtuse angle (p. 506)	rotational symmetry (p. 529)
complementary (p. 507)	octagon (p. 522)	scalene triangle (p. 523)
congruent (p. 515)	parallelogram (p. 523)	side (p. 506)
congruent figures (p. 534)	pentagon (p. 522)	similar figures (p. 534)
corresponding parts (p. 535)	perpendicular (p. 515)	square (p. 523)
degree (p. 506)	polygon (p. 522)	straight angle (p. 506)
equilateral triangle (p. 523)	quadrilateral (p. 522)	supplementary (p. 507)
heptagon (p. 522)	rectangle (p. 523)	triangle (p. 522)
hexagon (p. 522)	regular polygon (p. 522)	vertex (p. 506)
isosceles triangle (p. 523)		

Choose the letter of the term that best matches each phrase.

1. a six-sided figure
2. a polygon with all sides and all angles congruent
3. a ruler or any object with a straight side
4. the point where two edges of a polygon intersect
5. the most common unit of measure for an angle
6. an angle whose measure is between 0° and 90°
7. an angle whose measure is between 90° and 180°

a. regular polygon
b. vertex
c. straightedge
d. acute angle
e. obtuse angle
f. degree
g. hexagon

Lesson-by-Lesson Exercises and Examples

13-1 **Angles** (pp. 506–509)

Use a protractor to find the measure of each angle. Then classify each angle as *acute, obtuse, right,* or *straight*.

8.   9.

10. **ALGEBRA** Angles M and N are supplementary. If $m\angle M = 68°$, find $m\angle N$.

Example 1 Use a protractor to find the measure of the angle. Then classify the angle as *acute, obtuse, right,* or *straight*.

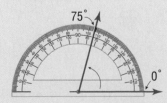

The angle measures 75°. Since it is less than 90°, it is an acute angle.

 msmath1.net/vocabulary_review

13-2 Using Angle Measures (pp. 510–512)

Use a protractor and a straightedge to draw angles having the following measurements.

11. 36°   12. 127°
13. 180°   14. 90°

Estimate the measure of each angle.

15. 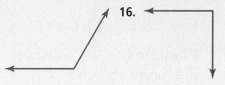   16.

Example 2 Use a protractor and a straightedge to draw a 47° angle.

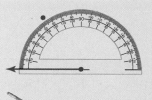

Draw one side of the angle. Align the center of the protractor and the 0° with the line. Find 47°. Make a mark.

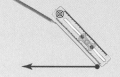

Draw the other side of the angle.

13-3 Bisectors (pp. 515–517)

Draw each line segment or angle having the given measure. Then use a straightedge and a compass to bisect the line segment or angle.

17. 2 cm   18. 2 in.
19. 100°   20. 65°

21. **TOOLS** A rake is shown. Identify the segment that bisects ∠CXD.

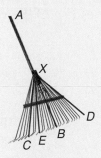

Example 3 Use a straightedge and a compass to bisect $\overline{AB}$.

Draw the segment.

Place the compass at A. Using a setting greater than one half the length of AB, draw two arcs as shown. Using the same setting, place the compass at B. Draw two arcs as shown.

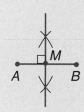

Use a straightedge to align the intersections. Draw a segment that intersects AB. Label the intersection.

13-4 Two-Dimensional Figures (pp. 522–525)

Identify each polygon. Then tell if it is a regular polygon.

22.   23.

Example 4 Identify the polygon shown. Then tell if the polygon is a regular polygon.

The polygon has eight sides. So, it is an octagon. Since the sides and angles are congruent, it is a regular polygon.

Trace each figure. Then draw all lines of symmetry.

24.

25.

26.

27.

Tell whether each figure has rotational symmetry. Write *yes* or *no*.

28.

29.

30.

31.

Example 5   Draw all lines of symmetry for the figure shown.

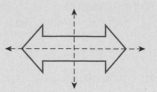

This figure has 2 lines of symmetry.

Example 6   Tell whether the figure shown has rotational symmetry.

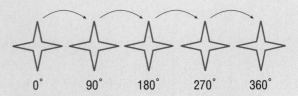

When the figure is rotated less than 360°, it looks as it did before it was rotated. So, the figure has rotational symmetry.

13-6   **Similar and Congruent Figures** (pp. 534–536)

The triangles shown are congruent.

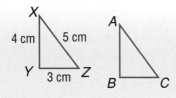

32. What side of △XYZ corresponds to side $\overline{AC}$?

33. Find the measure of $\overline{CB}$.

34. Tell whether the pair of figures shown at the right is *congruent, similar,* or *neither.*

Example 7   The figures shown are congruent. What side of quadrilateral *RSTU* corresponds to $\overline{MN}$?

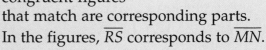

The parts of congruent figures that match are corresponding parts. In the figures, $\overline{RS}$ corresponds to $\overline{MN}$.

Vocabulary and Concepts

1. **Define** *polygon*.

2. **Explain** the difference between similar figure and congruent figures.

Skills and Applications

Use a protractor to find the measure of each angle. Then classify each angle as *acute*, *obtuse*, *right*, or *straight*.

3.

4.

5.

6. **ALGEBRA** Angles G and H are complementary angles. If $m\angle G = 37°$, find $m\angle H$.

Use a protractor and a straightedge to draw angles having the following measurements.

7. 25°

8. 135°

9. Draw a line segment that measures 4 centimeters. Then use a straightedge and a compass to bisect the line segment.

Identify each polygon. Then tell if it is a regular polygon.

10.

11.

12.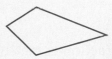

13. **SAFETY** The traffic sign shown warns motorists of a slow-moving vehicle. How many lines of symmetry does the sign have?

Tell whether each pair of figures is *congruent*, *similar*, or *neither*.

14.

15.

Standardized Test Practice

16. **SHORT RESPONSE** Triangle *DEF* and △*MNP* are congruent triangles. What is the perimeter of △*MNP*?

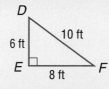
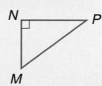

PART 1 Multiple Choice

Record your answers on the answer sheet provided by your teacher or on a sheet of paper.

1. Justin buys 5 balloons and a pack of gum. Theo buys 7 balloons. Which expression could be used to find the total amount they spent? (Lesson 4-1)

Item	Cost
Balloon	$0.10
Pack of gum	$0.25

- Ⓐ $0.10(5 + 7) + $0.25
- Ⓑ $0.10 + (5)(7) + $0.25
- Ⓒ $0.10(5) + $0.25(7)
- Ⓓ ($0.10 + $0.25)(5 + 7)

2. Find $6\frac{5}{6} - 2\frac{1}{3}$. (Lesson 6-5)

- Ⓕ $3\frac{5}{6}$
- Ⓖ $4\frac{5}{18}$
- Ⓗ $4\frac{1}{2}$
- Ⓘ $4\frac{2}{3}$

3. Which fraction comes next in the pattern below? (Lesson 7-6)

$$\frac{405}{243}, \frac{135}{81}, \frac{45}{27}, \frac{15}{9}, ?$$

- Ⓐ $\frac{5}{3}$
- Ⓑ $\frac{6}{3}$
- Ⓒ $\frac{8}{3}$
- Ⓓ $\frac{15}{4}$

4. What type of figure is formed if the points at $(-2, 2)$, $(2, 2)$, $(0, -2)$, $(-4, -2)$, and $(-2, 2)$ are connected in order? (Lesson 8-6)

- Ⓕ triangle
- Ⓖ pentagon
- Ⓗ trapezoid
- Ⓘ parallelogram

5. Which ratio compares the shaded part of the rectangle to the part that is not shaded? (Lesson 10-1)

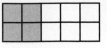

- Ⓐ $\frac{2}{5}$
- Ⓑ $\frac{4}{5}$
- Ⓒ $\frac{4}{10}$
- Ⓓ $\frac{2}{3}$

6. The table shows the contents of two bags containing red and blue gumballs. If one gumball is taken from each bag, find the probability that two red gumballs are taken. (Lesson 11-5)

- Ⓕ $\frac{1}{8}$
- Ⓖ $\frac{1}{4}$
- Ⓗ $\frac{3}{8}$
- Ⓘ $\frac{3}{5}$

Bag	Red	Blue
1st	2	2
2nd	1	3

7. Which line is a line of symmetry in the figure? (Lesson 13-5)

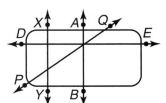

- Ⓐ $\overleftrightarrow{PQ}$
- Ⓑ $\overleftrightarrow{DE}$
- Ⓒ $\overleftrightarrow{AB}$
- Ⓓ $\overleftrightarrow{XY}$

8. Which triangle appears to be congruent to △NOP? (Lesson 13-6)

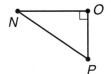

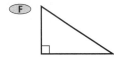

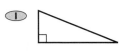

TEST-TAKING TIP

Question 7 To check that you have chosen the correct line of symmetry, pretend that the line is a fold in the page. The images on each side of the fold should be identical.

PART 2 Short Response/Grid In

Record your answers on the answer sheet provided by your teacher or on a piece of paper.

9. Write $\frac{75}{100}$ as a decimal. (Lesson 5-7)

10. What is the value of $-81 \div (-3)$? (Lesson 8-5)

11. Refer to the table. What is the function rule for these x- and y-values? (Lesson 9-6)

x	0	1	2	3	4
y	0	$\frac{1}{2}$	1	$1\frac{1}{2}$	2

12. A zookeeper wants to weigh a donkey. What is the most appropriate metric unit for the zookeeper to use when measuring the mass of the donkey? (Lesson 12-4)

13. What types of angles are in $\triangle ABC$? (Lesson 13-1)

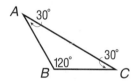

14. Draw a pair of angles that are complementary. (Lesson 13-1)

15. The shaded area below represents the students in the sixth grade who like jazz music. Is the measure of the angle of the shaded area closer to 30° or 60°? (Lesson 13-2)

Sixth Graders Who Like Jazz

16. What is the name for a 6-sided polygon? (Lesson 13-4)

17. Which two-dimensional shape makes up the surface of the box shown below? (Lesson 13-4)

18. How many lines of symmetry does the leaf have? (Lesson 13-5)

PART 3 Extended Response

Record your answers on a sheet of paper. Show your work.

19. Kelsey is designing a decorative border for her bedroom. The design will repeat and is made up of different geometric shapes.

a. The first figure she draws meets the following criteria.

> • a polygon
> • has exactly 2 lines of symmetry
> • not a quadrilateral

Draw a possible figure. (Lesson 13-5)

b. The next figure drawn meets the criteria listed below.

> • a polygon
> • no lines of symmetry
> • has an obtuse angle

Draw the possible figure. (Lesson 13-5)

c. The characteristics of the third figure drawn are listed below.

> • two similar polygons
> • one positioned inside the other

Draw the possible figure. (Lesson 13-6)

Geometry: Measuring Area and Volume

"What does architecture have to do with math?"

Two- and three-dimensional figures are often found in architecture. The Rock and Roll Hall of Fame in Cleveland, Ohio, contains two-dimensional figures such as triangles, rectangles, and parallelograms, and three-dimensional figures such as prisms, pyramids, and cylinders. The properties of geometric figures can be used to find the area and the volume of buildings.

You will solve a problem about architecture in Lesson 14-2.

GETTING STARTED

Take this quiz to see whether you are ready to begin Chapter 14. Refer to the lesson or page number in parentheses if you need more review.

▶ Vocabulary Review

Complete each sentence.

1. A(n) __?__ is a number expressed using exponents. (Lesson 1-4)

2. The number that is multiplied in a power is called the __?__. (Lesson 1-4)

3. __?__ is the distance around a circle. (Lesson 4-6)

▶ Prerequisite Skills

Evaluate each expression. (Lesson 1-4)

4. 8^2 5. $(1.2)^2$

6. $(0.5)^2$ 7. 11^2

8. 7^2 9. 10^2

Estimate each sum. (Lesson 3-4)

10. $17.6 + 8.41 + 3.2$

11. $20.9 + 4.25 + 9.1$

12. $2.7 + 6.9 + 13.8$

13. $15.67 + 11.8 + 7.3$

Multiply. (Lesson 7-2)

14. $\frac{1}{2} \times 6 \times 6$ 15. $\frac{1}{2} \times 5 \times 8$

16. $\frac{1}{2} \times 8 \times 3$ 17. $\frac{1}{2} \times 4 \times 7$

Multiply. (Page 590)

18. $2 \times 7 \times 5$ 19. $9 \times 6 \times 4$

20. $4 \times 11 \times 3$ 21. $10 \times 8 \times 2$

FOLDABLES™ Study Organizer

Area and Volume Make this Foldable to help you organize information about measuring area and volume.

STEP 1 Fold
Fold a sheet of 11" × 17" paper in thirds lengthwise.

STEP 2 Open and Fold
Fold a 2" tab along the short side. Then fold the rest into fifths.

STEP 3 Unfold and Label
Unfold and draw lines along the folds. Label as shown.

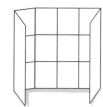

Reading and Writing As you read and study the chapter, write the formulas for area and volume, and list the characteristics of each geometric figure.

14-1 Area of Parallelograms

What You'll LEARN

Find the areas of parallelograms.

NEW Vocabulary

base
height

HANDS-ON Mini Lab

Materials
- grid paper
- scissors

Work with a partner.

You can explore how the areas of parallelograms and rectangles are related.

STEP 1 Draw and then cut out a rectangle as shown.

length (ℓ)

width (w)

STEP 2 Cut a triangle from one side of the rectangle and move it to the other side to form a parallelogram.

height (h)

base (b)

1. How does a parallelogram relate to a rectangle?

2. What part of the parallelogram corresponds to the length of the rectangle?

3. What part corresponds to the rectangle's width?

4. Write a formula for the area of a parallelogram.

In the Mini Lab, you showed that the area of a parallelogram is related to the area of a rectangle. To find the area of a parallelogram, multiply the measures of the base and the height.

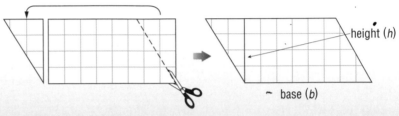

The shortest distance from the base to the opposite side is the **height** of the parallelogram.

height

The **base** of a parallelogram can be any one of its sides.

base

Key Concept

Area of a Parallelogram

Words The area *A* of a parallelogram is the product of any base *b* and its height *h*.

Model

b

h

Symbols $A = bh$

Find the area of each parallelogram.

1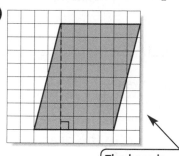

$A = bh$
$A = 6 \cdot 8$
$A = 48$

The base is 6 units, and the height is 8 units.

The area is 48 square units or 48 units^2.

2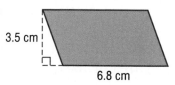

3.5 cm

6.8 cm

$A = bh$
$A = 6.8 \times 3.5$
$A = 23.8$

The area is 23.8 square centimeters or 23.8 cm^2.

READING Math

Area Measurement
An area measurement can be written using abbreviations and an exponent of 2.
For example:
square units = units^2
square inches = in^2
square feet = ft^2
square meters = m^2

Your Turn Find the area of each parallelogram. Round to the nearest tenth if necessary.

a.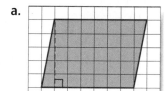

b.

16.2 m

3.7 m

Many real-life objects are parallelograms.

REAL-LIFE CAREERS

How Does an Architect Use Math?

Architects use geometry when they find the area of buildings.

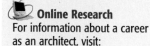**Online Research**
For information about a career as an architect, visit:
msmath1.net/careers

EXAMPLE **Use Area to Solve a Problem**

3 **ARCHITECTURE** An architect is designing a parallelogram-shaped lobby for a small office building. What is the area of the floor plan?

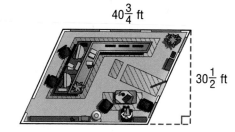

$40\frac{3}{4}$ ft

$30\frac{1}{2}$ ft

Since the floor plan of the lobby is a parallelogram, use the formula $A = bh$.

$A = bh$ Area of a parallelogram

$A = \left(40\frac{3}{4}\right)\left(30\frac{1}{2}\right)$ Replace b with $40\frac{3}{4}$ and h with $30\frac{1}{2}$.

$A = \left(\frac{163}{4}\right)\left(\frac{61}{2}\right)$ Estimate $40\frac{3}{4} \times 30\frac{1}{2} \rightarrow 40 \times 30 = 1,200$

$A = \frac{9,943}{8}$ or $1,242\frac{7}{8}$ Write the mixed numbers as improper fractions. Multiply. Then simplify.

The area of the lobby's floor plan is $1,242\frac{7}{8}$ square feet.

Notice that this is reasonable compared to the estimate of 1,200.

Writing Math

Exercise 1

1. **Explain** how the formula for the area of a parallelogram is related to the formula for the area of a rectangle.

2. **OPEN ENDED** Draw and label two different parallelograms each with an area of 16 square units.

GUIDED PRACTICE

Find the area of each parallelogram. Round to the nearest tenth if necessary.

3.

4.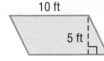
10 ft
5 ft

5.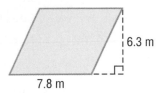
6.3 m
7.8 m

Practice and Applications

Find the area of each parallelogram. Round to the nearest tenth if necessary.

HOMEWORK HELP

For Exercises	See Examples
6–11	1, 2
12–13, 15, 18	3

Extra Practice
See pages 622, 637.

6.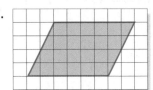

7.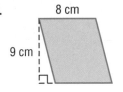

8.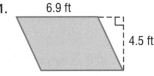
8 cm
9 cm

9.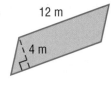
12 m
4 m

10.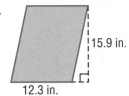
15.9 in.
12.3 in.

11.
6.9 ft
4.5 ft

12. What is the measure of the area of a parallelogram whose base is $8\frac{4}{5}$ inches and whose height is $6\frac{3}{8}$ inches?

13. Find the area of a parallelogram with base 6.75 meters and height 4.8 meters.

14. **ALGEBRA** If $x = 5$ and $y < x$, which parallelogram has the greatest area?

    a.   b.   c.
    a. y, x b. y, x c. x, x

15. What is a reasonable estimate for the area of a parallelogram with a base of $19\frac{3}{4}$ inches and a height of $15\frac{1}{8}$ inches?

16. **MEASUREMENT** How many square feet are in 4 square yards?

17. **MEASUREMENT** Find the number of square inches in 9 square feet.

18. **WEATHER** A local meteorologist alerted people of a thunderstorm warning for the region shown on the map. What is the area of the region that is under a thunderstorm warning?

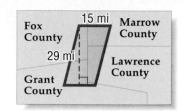

ERASERS For Exercises 19–20, use the eraser shown.

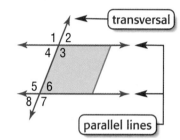

19. Write an equation to find the measure of the base of the side of the eraser.

20. Find the measure of the base of the side of the eraser.

21. **CRITICAL THINKING** The base and height of a parallelogram are doubled. How does the area change?

EXTENDING THE LESSON By extending the sides of a parallelogram, special angles are formed. Notice that the line intersecting a pair of parallel lines is called a *transversal*.

- interior angles: $\angle 3$, $\angle 4$, $\angle 5$, $\angle 6$
- exterior angles: $\angle 1$, $\angle 2$, $\angle 7$, $\angle 8$
- alternate interior angles: $\angle 3$ and $\angle 5$, $\angle 4$ and $\angle 6$
- alternate exterior angles: $\angle 1$ and $\angle 7$, $\angle 2$ and $\angle 8$
- corresponding angles: $\angle 1$ and $\angle 5$, $\angle 2$ and $\angle 6$, $\angle 3$ and $\angle 7$, $\angle 4$ and $\angle 8$

22. Give a definition for each type of angle listed above.

23. Describe the relationship between alternate interior angles.

24. Draw a parallelogram. Then extend its sides. Identify a pair of alternate interior angles and a pair of alternate exterior angles.

25. If $m\angle 3 = 110°$, what angles are congruent to $\angle 3$?

26. If $m\angle 6 = 65°$, find $m\angle 1$.

Standardized Test Practice and Mixed Review

27. **MULTIPLE CHOICE** Find the area of the parallelogram.

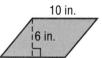

Ⓐ 72 in²   Ⓑ 60 in²   Ⓒ 30 in²   Ⓓ 16 in²

28. **SHORT RESPONSE** What is the height of a parallelogram if its area is 219.6 square meters and its base is 12 meters?

29. Draw a pair of similar quadrilaterals. (Lesson 13-6)

Trace each figure. Then draw all lines of symmetry. (Lesson 13-5)

30.

31.

32.

GETTING READY FOR THE NEXT LESSON

PREREQUISITE SKILL Multiply. (Lesson 7-2)

33. $\frac{1}{2} \cdot 8 \cdot 9$ 34. $\frac{1}{2} \cdot 12 \cdot 5$ 35. $\frac{1}{2} \cdot 25 \cdot 4$ 36. $\frac{1}{2} \cdot 48 \cdot 3$

Area of Triangles

In this lab, you will find the area of a triangle using the properties of parallelograms.

What You'll LEARN

Find the area of a triangle using the properties of parallelograms.

Materials
- grid paper
- colored pencils
- scissors

ACTIVITY *Work with a partner.*

STEP 1 Draw a triangle as shown. Label the height and the base.

STEP 2 Draw a dashed line segment that is 7 units high and parallel to the height of the triangle.

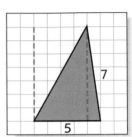

STEP 3 Draw a solid line segment that is 5 units long and parallel to the base. Draw another segment to form the parallelogram.

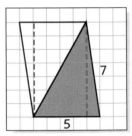

The area of the parallelogram is 5 × 7 or 35 square units.

The area of the triangle is half the area of the parallelogram. So, the area of the triangle is 35 ÷ 2 or 17.5 square units.

Your Turn

Draw the triangle shown on grid paper. Then draw a parallelogram and find the area of the triangle.

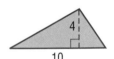

Writing Math

1. Suppose a parallelogram has an area of 84 square units with a height of 7 units. **Describe** a triangle related to this parallelogram, and find the triangle's area, base, and height.

2. **Draw** a parallelogram that is related to the triangle at the right. How could you use the drawing to find the area of the triangle?

3. **Write** a formula for the area of a triangle.

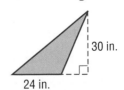

WHEN am I ever going to use this?

GAMES Tri-Ominos is a game played with triangular game pieces that are all the same size.

1. Compare the two triangles.

2. What figure is formed by the two triangles?

3. **Make a conjecture** about the relationship that exists between the area of one triangle and the area of the entire figure.

A parallelogram can be formed by two congruent triangles. Since congruent triangles have the same area, the area of a triangle is one half the area of the parallelogram.

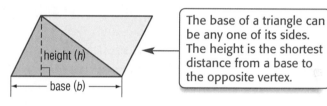

The base of a triangle can be any one of its sides. The height is the shortest distance from a base to the opposite vertex.

Key Concept — Area of a Triangle

Words	The area A of a triangle is one half the product of the base b and its height h.	Model
Symbols	$A = \frac{1}{2}bh$	

EXAMPLE Find the Area of a Triangle

1. **Find the area of the triangle.**

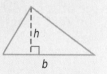

By counting, you find that the measure of the base is 6 units and the height is 4 units.

$A = \frac{1}{2}bh$ — Area of a triangle

$A = \frac{1}{2}(6)(4)$ — Replace b with 6 and h with 4.

$A = \frac{1}{2}(24)$ — Multiply. $6 \times 4 = 24$

$A = 12$ — The area of the triangle is 12 square units.

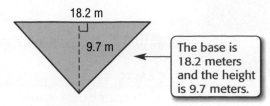

EXAMPLE Find the Area of a Triangle

2 Find the area of the triangle.

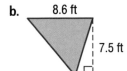

18.2 m

9.7 m

> The base is 18.2 meters and the height is 9.7 meters.

$A = \frac{1}{2}bh$ Area of a triangle

$A = \frac{1}{2}(18.2)(9.7)$ Replace *b* with 18.2 and *h* with 9.7.

0.5 ☒ 18.2 ☒ 9.7 ⊟ **88.27**   Use a calculator.

To the nearest tenth, the area of the triangle is 88.3 square meters.

Your Turn Find the area of each triangle. Round to the nearest tenth if necessary.

a.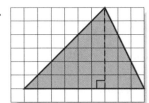

b.
8.6 ft

7.5 ft

Standardized········· Test Practice

EXAMPLE Use Area to Solve a Problem

3 MULTIPLE-CHOICE TEST ITEM

Which ratio compares the area of the shaded triangle to the area of the large square?

Ⓐ 1 to 4   Ⓑ 1 to 8

Ⓒ 1 to 16   Ⓓ 1 to 32

Read the Test Item

You need to find the ratio that compares the area of the triangle to the area of the large square.

Solve the Test Item

First find the area of the triangle and the area of the square.

Area of Triangle

$A = \frac{1}{2}bh$

$A = \frac{1}{2}(2)(1)$ or 1 unit^2

Area of Square

$A = s^2$

$A = 4^2$ or 16 units^2

Now find the ratio. Since $\dfrac{\text{area of triangle}}{\text{area of square}} = \dfrac{1\ \text{unit}^2}{16\ \text{units}^2}$, the ratio is 1 to 16. So, the answer is C.

Test-Taking Tip
The Princeton Review

Formulas
Most standardized tests list any geometry formulas you will need to solve problems. However, it is always a good idea to familiarize yourself with the formulas before taking the test.

Skill and Concept Check

Writing
Math
Exercise 2

1. **OPEN ENDED** Draw two different triangles each having an area of 24 square feet.

2. **FIND THE ERROR** Susana and D.J. are finding the area of the triangle. Who is correct? Explain.

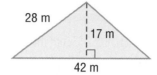

Susana
$A = \frac{1}{2}(28)(42)$
$A = 588 \text{ m}^2$

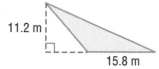

D.J.
$A = \frac{1}{2}(17)(42)$
$A = 357 \text{ m}^2$

28 m

17 m

42 m

GUIDED PRACTICE

Find the area of each triangle. Round to the nearest tenth if necessary.

3.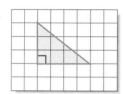

4.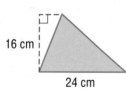

8 ft

12 ft

5.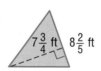

11.2 m

15.8 m

6. **SPORTS** The width of a triangular hang glider measures 9 feet, and the height of the wing is 6 feet. How much fabric was used for the wing of the glider?

Practice and Applications

Find the area of each triangle. Round to the nearest tenth if necessary.

HOMEWORK HELP

For Exercises	See Examples
7–14, 17–20	1, 2

Extra Practice
See pages 622, 637.

7.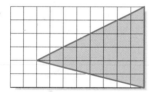

8.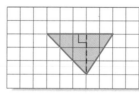

9.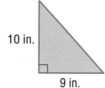

10 in.

9 in.

10.

16 cm

24 cm

11.

7.5 m

10.5 m

12.

$7\frac{3}{4}$ ft $8\frac{2}{5}$ ft

13. height: $4\frac{2}{3}$ in., base: $\frac{3}{4}$ in.

14. height: 7.5 cm, base: 5.6 cm

15. Which is larger, a triangle with an area of 25 square yards or a triangle with an area of 25 square meters?

16. Which is smaller, a triangle with an area of 1 square foot or a triangle with an area of 64 square inches?

17. **ARCHITECTURE** The main entrance of the Rock and Roll Hall of Fame is a triangle with a base of about 241 feet and a height of about 165 feet. Find the area of this triangle.

GEOGRAPHY For Exercises 18 and 19, use the diagram shown and the following information. The Bermuda Triangle is an imaginary triangle connecting Florida to the Bermuda Islands to Puerto Rico and back to Florida.

18. Estimate the area of the region enclosed by the Bermuda Triangle.

19. Find the actual area of the Bermuda Triangle.

20. **COLLEGE** Jack's dorm room is shaped like a triangle. The college brochure says it has an area of 304 square feet. The room is 15 feet long. Estimate the width of the room at its widest point.

21. Find the area and the perimeter of the figure at the right.

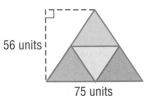

CRITICAL THINKING For Exercises 22–25, use the figure shown.

22. Find the area of the figure.

23. Find the measure of the base and height of the four smaller triangles.

24. What is the area of one small triangle?

25. Is your answer reasonable? Explain.

Standardized Test Practice and Mixed Review

26. **MULTIPLE CHOICE** In the diagram, the triangle on the left has an area of 3 square feet. What is the area of the figure on the right?

    (A) 8 ft^2 (B) 12 ft^2 (C) 18 ft^2 (D) 22 ft^2

Area = 3 ft^2 Area = ? ft^2

27. **MULTIPLE CHOICE** Find the area of the triangle.

    (F) 27 units^2 (G) 36 units^2
    (H) 40 units^2 (I) 54 units^2

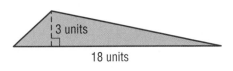

28. **GEOMETRY** Find the area of a parallelogram whose base is 20 millimeters and height is 16 millimeters. (Lesson 14-1)

Tell whether each pair of figures is *similar, congruent,* or *neither.* (Lesson 13-6)

29.   30.    31.

GETTING READY FOR THE NEXT LESSON

PREREQUISITE SKILL Evaluate each expression. (Lesson 1-4)

32. 9^2 33. 12^2 34. 0.6^2 35. 1.5^2

Area of Trapezoids

What You'll LEARN

Find the area of a trapezoid using the properties of triangles.

Materials

• grid paper

A trapezoid is a quadrilateral with one pair of opposite sides parallel. In this lab, you will explore how to find the area of a trapezoid using the formula for the area of a triangle.

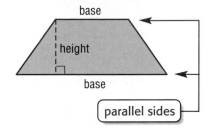

ACTIVITY

STEP 1 Draw a trapezoid. Separate it into two triangles as shown.

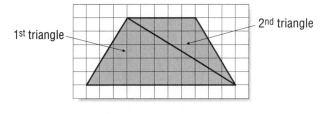

STEP 2 Draw and label the height and base of each triangle.

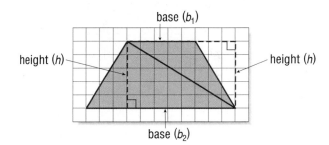

STEP 3 Write a formula for the area of the trapezoid.

area of trapezoid = area of first $\triangle$ + area of second $\triangle$

$$= \frac{1}{2}b_1h + \frac{1}{2}b_2h$$

$$= \frac{1}{2}h(b_1 + b_2) \quad \text{Distributive Property}$$

Your Turn

a. Find the area of the trapezoid above.

Writing Math

1. Explain why the area of a trapezoid is related to the area of a triangle.

2. Why can $A = \frac{1}{2}b_1h + \frac{1}{2}b_2h$ be written as $A = \frac{1}{2}h(b_1 + b_2)$?

3. Explain how you would separate any trapezoid into triangles to find its area.

Area of Circles

What You'll LEARN

Find the areas of circles.

REVIEW Vocabulary

circumference: the distance around a circle (Lesson 4-6)

MATH Symbols

π approximately 3.14

HANDS-ON Mini Lab

Materials
• paper plate
• scissors

Work with a partner.

You can use a paper plate to explore the area of circles.

STEP 1 Fold a paper plate into eighths.

STEP 2 Unfold the plate and cut along the creases.

STEP 3 Arrange the pieces to form the figure shown.

1. What shape does the figure look like?

2. What part of the circle represents the figure's height?

3. Relate the circle's circumference to the base of the figure.

4. How would you find the area of the figure?

A circle can be separated into parts as shown. The parts can then be arranged to form a figure that resembles a parallelogram.

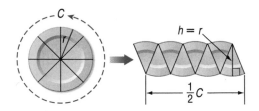

You can use the formula for the area of a parallelogram to find the formula for the area of a circle.

$A = bh$ Area of a parallelogram

$A = \left(\frac{1}{2}C\right)r$ The base is one half the circumference. The height is the radius.

$A = \frac{1}{2}(2\pi r)r$ Replace C with $2\pi r$, the formula for circumference.

$A = \pi \cdot r \cdot r$ Simplify. $\frac{1}{2} \cdot 2 = 1$

$A = \pi r^2$ Simplify. $r \cdot r = r^2$

Key Concept

Area of a Circle

Words   The area A of a circle is the product of π and the square of the radius r.

Model

Symbols $A = \pi r^2$

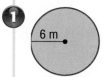
EXAMPLES **Find Areas of Circles**

Find the area of each circle to the nearest tenth. Use 3.14 for π.

1

6 m	$A = \pi r^2$	Area of a circle
	$A \approx 3.14 \times 6^2$	Replace π with 3.14 and r with 6. Estimate $3.14 \times 6^2 \to 3 \times 40 = 120$
	$A \approx 3.14 \times 36$	Evaluate 6^2.
	$A \approx 113.04$	Use a calculator.

The area is about 113.0 square meters.

2

8.6 ft

The diameter is 8.6 feet. So, the radius is $8.6 \div 2$ or 4.3 feet.

$A = \pi r^2$ Area of a circle

$A \approx 3.14 \times 4.3^2$ Replace π with 3.14 and r with 4.3. Estimate $3.14 \times 4.3^2 \to 3 \times 20 = 60$

$A \approx 3.14 \times 18.49$ Evaluate 4.3^2.

$A \approx 58.0586$ Use a calculator.

The area is about 58.1 square feet.

Your Turn **Find the area of each circle to the nearest tenth. Use 3.14 for π.**

a. 4 ft

b. 6.4 m

VOLCANOES Shield volcanoes are named for their broad and gently sloping shape that looks like a warrior's shield. In California and Oregon, many shield volcanoes have diameters of three or four miles and heights of 1,500 to 2,000 feet.

Source: U.S. Geological Survey

Many real-life objects are circular.

EXAMPLE **Use Area to Solve a Problem**

3 **VOLCANOES** The Belknap shield volcano is located in Oregon. This volcano is circular and has a diameter of 5 miles. About how much land does this volcano cover?

Use the area formula to find the area of the volcano.

$A = \pi r^2$ Area of a circle

$A \approx 3.14 \times 2.5^2$ Replace π with 3.14 and r with 2.5. Estimate $3.14 \times 2.5^2 \to 3 \times 6 = 18$

$A \approx 3.14 \times 6.25$ Evaluate 2.5^2.

$A \approx 19.625$ Use a calculator.

About 20 square miles of land is covered by the volcano.

1. **Explain** how to estimate the area of any circle.

2. **OPEN ENDED** Find a circular object in your classroom or home. Estimate and then find the actual area of the object.

3. **FIND THE ERROR** Whitney and Crystal are finding the circle's area. Who is correct? Explain.

12.5 units

> Whitney
> $A \approx 3.14 \times (12.5)^2$

> Crystal
> $A \approx 3.14 \times (6.25)^2$

GUIDED PRACTICE

Find the area of each circle to the nearest tenth. Use 3.14 for π.

4.
10 cm

5.
3.8 yd

6.
13 mi

7. **SCIENCE** An earthquake's epicenter is the point from which the shock waves radiate. What is the area of the region affected by an earthquake whose shock waves radiated 29 miles from its epicenter?

Practice and Applications

Find the area of each circle to the nearest tenth. Use 3.14 for π.

8.
3 ft

9.
8 cm

10.
18 in.

HOMEWORK HELP

For Exercises	See Examples
8–15	1, 2
16–18	3

Extra Practice
See pages 622, 637.

11.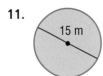
15 m

12.
5.5 km

13.
9.4 yd

14. What is the area of a circle whose radius is 7.75 meters?

15. Find the area of a circle with a diameter of $175\frac{3}{8}$ feet.

16. **WRESTLING** A wrestling mat is a square mat measuring 12 meters by 12 meters. Within the square, there is a circular ring whose radius is 4.5 meters. Find the area within the circle to the nearest tenth.

Data Update Find the area of each circle that appears on a regulation hockey rink. How do the areas of these circles compare to the one on a wrestling mat? Visit **msmath1.net/data_update** to learn more.

17. **SCHOOL** Suppose you are preparing a report on people's beliefs in space aliens. You redraw the circle graph shown at the right on the report cover. When redrawn, the graph has a diameter of 9.5 inches. Find the area of the section of the graph that represents the 20% section to the nearest tenth.

18. **TOOLS** A sprinkler that sprays water in a circular area can be adjusted to spray up to 30 feet. What is the maximum area of lawn that can be watered by the sprinkler?

19. **CRITICAL THINKING** Suppose you double the radius of a circle. How is the area affected?

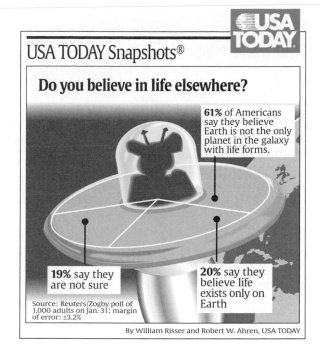

USA TODAY Snapshots®

Do you believe in life elsewhere?

61% of Americans say they believe Earth is not the only planet in the galaxy with life forms.

19% say they are not sure

20% say they believe life exists only on Earth

Source: Reuters/Zogby poll of 1,000 adults on Jan. 31; margin of error: ±3.2%

By William Risser and Robert W. Ahren, USA TODAY

EXTENDING THE LESSON The fraction $\frac{22}{7}$ can also be used for π.

Find the area of each circle. Use $\frac{22}{7}$ for π.

20.
7 ft

21.
28 cm

22.
$3\frac{1}{2}$ in.

Standardized Test Practice and Mixed Review

23. **SHORT RESPONSE** Find the area of a circular hot tub cover whose diameter measures 6.5 feet. Round to the nearest tenth.

24. **MULTIPLE CHOICE** Find the area of the shaded region of the figure shown. Use 3.14 for π.

    Ⓐ 53.38 cm²   Ⓑ 373.66 cm²

    Ⓒ 452.16 cm²   Ⓓ 530.66 cm²

5 cm — 12 cm

25. **GEOMETRY** What is the area of a triangle with a base 8 meters long and a height of 14 meters? (Lesson 14-2)

26. **GEOMETRY** Find the area of the parallelogram at the right. Round to the nearest tenth if necessary. (Lesson 14-1)

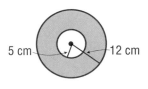
12 ft
7 ft

GETTING READY FOR THE NEXT LESSON

BASIC SKILL Sketch each object listed.

27. ice cream cone   28. shoe box   29. drinking straw

 msmath1.net/self_check_quiz

What You'll LEARN

Make circle graphs.

Materials

- colored pencils
- ruler
- compass
- protractor
- calculator

Making Circle Graphs

"How important is sunny weather in a vacation location?" The circle graph at the right shows how people responded to this question.

Sunshine While on Vacation

Not At All Important
Not Very Important
Important
Very Important
10%
15%
30%
45%

Source: Opinion Research Corp.

1. What percent of the people said that having sunshine while on vacation was not at all important?

2. What percent is represented by the whole circle graph? How many degrees are in the circle?

3. Explain when a circle graph is the best choice to display a set of data.

In this lab, you will learn to make circle graphs.

ACTIVITY *Work with a partner.*

A group of teenagers were asked to name their top priority for the school year. The results are shown at the right. Display the data in a circle graph.

Top Priorities for School Year	
Top Priority	**Percent**
Sports	12.5%
Good Grades	50%
Friends	25%
Boyfriend/Girlfriend	12.5%

STEP 1 Find the number of degrees for each percent. To do this, first write each percent as a decimal. Then multiply each decimal by 360, the total number of degrees in a circle graph.

Percent to Decimal	Multiply by 360
$12.5\% \rightarrow 0.125$	$0.125 \times 360 = 45$
$50\% \;\; \rightarrow 0.50$	$0.50 \;\; \times 360 = 180$
$25\% \;\; \rightarrow 0.25$	$0.25 \;\; \times 360 = 90$
$12.5\% \rightarrow 0.125$	$0.125 \times 360 = 45$

The sum should always be 360.

The results are the number of degrees in the corresponding sections of the circle graph.

STEP 2 Use a compass to draw a circle with at least a 1-inch radius. Draw the radius with the ruler.

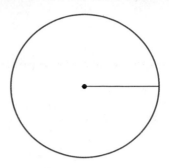

STEP 3 Use a protractor to draw an angle for the *Sports* section of the graph. Repeat Steps 1–3 for each category.

Top Priorities for School Year

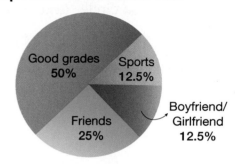

STEP 4 Shade each section of the graph. Then give the graph a title.

$50\% = \frac{1}{2}$ so 50% is $\frac{1}{2}$ of the graph.

$25\% = \frac{1}{4}$ so 25% is $\frac{1}{4}$ of the graph.

$12.5\% = \frac{1}{8}$ so 12.5% is $\frac{1}{8}$ of the graph.

Your Turn Display each set of data in a circle graph.

a.

Time Spent Playing Video Games	
Time (h)	Percent
0–1	35%
1–2	10%
2–3	25%
3 or more	30%

b.

Time Film Stays in Camera Before Being Developed	
Time (months)	Percent
0–6	45%
6–12	37.5%
13–18	12.5%
don't know	5%

Writing Math

1. **Compare** each circle graph to its corresponding table. Does the graph or table display the data more clearly? Explain.

2. **Examine** each data set you displayed. Explain how each set of data compares part to whole relationships.

3. **Give an example** of a data set that *cannot* be represented by a circle graph. What type of graph would you use to best represent the data set?

4. **Explain** how the area of a circle is related to making a circle graph.

Vocabulary and Concepts

1. **Write** in words the formula for the area of a parallelogram. (Lesson 14-1)
2. **Explain** the relationship between the radius and the diameter of a circle. (Lesson 14-3)

Skills and Applications

Find the area of each figure. Round to the nearest tenth if necessary.
(Lessons 14-1 and 14-2)

3.
6 ft
8 ft

4.
12 in.
15 in.

5.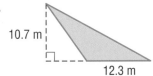
10.7 m
12.3 m

6. What is the measure of the area of a parallelogram whose base is $5\frac{2}{5}$ feet and whose height is $7\frac{1}{2}$ feet? (Lesson 14-1)

7. **BOATS** A sailboat has a triangular sail whose base is 10.5 feet and whose height is 30.75 feet. What is the area of the sail? (Lesson 14-2)

Find the area of each circle to the nearest tenth. Use 3.14 for π. (Lesson 14-3)

8.
2 in.

9.
4.5 cm

10.
8.4 yd

11. Find the area of a circle whose diameter is 10.5 inches. Round to the nearest tenth. (Lesson 14-3)

Standardized Test Practice

12. **MULTIPLE CHOICE** Which expression gives the area of the figure? (Lesson 14-1)

5 m 4 m
7 m

Ⓐ 5×4 Ⓑ 5×7
Ⓒ 4×7 Ⓓ $5 \times 4 \times 7$

13. **SHORT RESPONSE** A therapy pool is circular in shape. If the diameter is 9 meters, how much material is needed to make a cover for the pool? (Lesson 14-3)

The Game Zone

A Place To Practice Your Math Skills

Math Skill

Area of Circles

Times Up for Circles

- ● **GET READY!**

  Players: five
  Materials: poster board, compass, number cube, 1-minute timer

- ● **GET SET!**

  - Use a compass to draw the game board shown at the right.

  - Choose one player to be the official timekeeper and answer checker.

  - Divide into teams of two players.

- ● **GO!**

  - One player rolls a number cube onto the poster board.

  - The player's team member has one minute to find the area of the circle on which the number cube lands.

  - The answer checker checks the response and awards 5 points for a correct answer.

  - The other team takes its turn.

  - **Who Wins?** The team with the highest total score after five rounds wins.

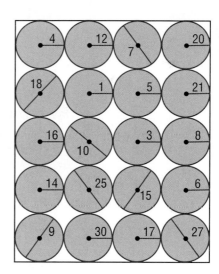

Three-Dimensional Figures

What You'll LEARN

Identify three-dimensional figures.

NEW Vocabulary

three-dimensional figure
face
edge
lateral face
vertex (vertices)
prism
base
pyramid
cone
cylinder
sphere
center

WHEN am I ever going to use this?

KITES A box kite and a delta kite are shown.

1. What shape does the delta kite resemble?

2. Name the shape that each side of the box kite resembles.

3. Describe how the shape of the box kite differs from the shape of the delta kite.

box kite

delta kite

Many common shapes are **three-dimensional figures**. That is, they have length, width, and depth (or height). Some terms associated with three-dimensional figures are face, edge, vertex, and lateral face.

A **face** is a flat surface.

The **edges** are the segments formed by intersecting faces.

The sides are called **lateral faces**.

The edges intersect at the **vertices**.

Two types of three-dimensional figures are prisms and pyramids.

Key Concept · Prisms and Pyramids

Prism	• Has at least three lateral faces that are rectangles. • The top and bottom faces are the **bases** and are parallel. • The shape of the base tells the name of the prism. Rectangular prism · Triangular prism · Square prism or cube
Pyramid	• Has at least three lateral faces that are triangles and only one base. • The base can be shaped like any closed figure with three or more sides. • The shape of the base tells the name of the pyramid. Triangular pyramid · Square pyramid

STUDY TIP

Three-Dimensional Figures In three-dimensional figures, dashed lines are used to indicate edges that are hidden from view.

Some three-dimensional figures have curved surfaces.

Key Concept		Cones, Cylinders, and Spheres
Cone	• Has only one base. • The base is a circle. • Has one vertex and no edges.	
Cylinder	• Has only two bases. • The bases are circles. • Has no vertices and no edges.	
Sphere	• All of the points on a sphere are the same distance from the **center**. • No faces, bases, edges, or vertices.	 center

EXAMPLES Identify Three-Dimensional Figures

Identify each figure.

1 One circular base, no edge, and no vertex.

The figure is a cone.

2 All of the faces are squares.

The figure is a cube.

Your Turn

a. Identify the figure shown at the right.

Skill and Concept Check

Writing Math
Exercise 2

1. **Determine** the number of vertices for each figure.

   a.   b.   c.   d.

2. **Explain** the difference between a two-dimensional and a three-dimensional figure.

GUIDED PRACTICE

Identify each figure.

3.   4.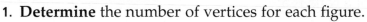

5. **FOOD** Draw the figure that represents a can of soup. Then identify the figure.

Practice and Applications

Identify each figure.

6.

7.

8.

9.

HOMEWORK HELP

For Exercises	See Examples
6–9	1, 2

Extra Practice
See pages 623, 637.

10. **SCHOOL** Draw the figure that represents a textbook. What is the name of this figure?

11. **SPORTS** Megaphones are used to intensify or direct the voice. Sketch the figure shown by a megaphone. Explain why it is not identified as a cone.

CRITICAL THINKING **For Exercises 12 and 13, draw figures to support your answer.**

12. What type of pyramid has exactly four faces?

13. What figure is formed if only the height of a cube is increased?

EXTENDING THE LESSON A *plane* is a flat surface that extends in all directions. The faces of a prism are parts of a plane. Two lines that are not in the same plane and do not intersect are *skew lines*.

14. Identify two other planes in the rectangular prism. Three vertices are needed to name a plane.

15. Name two other pairs of lines that are skew lines.

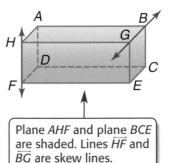

Plane *AHF* and plane *BCE* are shaded. Lines $\overleftrightarrow{HF}$ and $\overleftrightarrow{BG}$ are skew lines.

Standardized Test Practice and Mixed Review

16. **MULTIPLE CHOICE** The base of a cone is a __?__ .
    - Ⓐ triangle
    - Ⓑ circle
    - Ⓒ radius
    - Ⓓ rectangle

17. **MULTIPLE CHOICE** Identify the figure shown.
    - Ⓕ triangular pyramid
    - Ⓖ square pyramid
    - Ⓗ rectangular pyramid
    - Ⓘ triangular prism

Find the area of each circle described to the nearest tenth. Use 3.14 for π. (Lesson 14-3)

18. radius: 22 in.

19. diameter: 6 m

20. diameter: 4.6 ft

21. **GEOMETRY** What is the area of a triangle whose base is 52 feet and whose height is 38 feet? (Lesson 14-2)

GETTING READY FOR THE NEXT LESSON

PREREQUISITE SKILL **Multiply. Round to the nearest tenth.** (Lesson 4-2)

22. $8.2 \times 4.8 \times 2.1$

23. $5.9 \times 1.0 \times 7.3$

24. $1.0 \times 0.9 \times 1.3$

Three-Dimensional Figures

What You'll LEARN

Draw three-dimensional figures.

It is often helpful to draw a three-dimensional figure when trying to solve a problem.

Materials

• isometric dot paper
• ruler

ACTIVITY *Work with a partner.*

Use isometric dot paper to sketch a rectangular prism with length 4 units, height 2 units, and width 3 units.

STEP 1 Draw a parallelogram with sides 4 units and 3 units. This is the top of the prism.

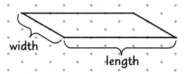

STEP 2 Start at one vertex. Draw a line passing through two dots. Repeat for the other three vertices. Draw the hidden edges as dashed lines.

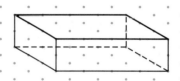

STEP 3 Connect the ends of the lines to complete the prism.

Writing Math

1. **Explain** which faces are the bases of the prism.

2. Use isometric dot paper to draw each figure.

   a. a cube with length, width, and height of 3 units

   b. a rectangular prism with length 4 units, width 2 units, and height 2 units

3. How would you draw a prism with a triangular base?

4. **Explain** why you think isometric dot paper is used to draw a three-dimensional object.

5. Suppose you need to draw a three-dimensional representation of a sphere. Do you think this method would work? Explain.

14-5a Problem-Solving Strategy

A Preview of Lesson 14-5

What You'll Learn

Solve problems by making a model.

Make a Model

Hey Jaime, one of our first jobs at the grocery store is to stack oranges in the shape of a square pyramid. The base of the pyramid should have 100 oranges and one orange needs to be on top.

We have 400 oranges, Patrick. Is that enough? Let's **make a model** to find out!

Explore	The oranges are to be stacked in the form of a square pyramid. There are to be 100 oranges on the base and one orange at the top. You need to know how many oranges are needed to make the pyramid.
Plan	Make a model to find the number of oranges needed to make the pyramid.
Solve	Use pennies to model the oranges. Begin with 100 pennies for the bottom layer. For each consecutive layer, one penny is placed where four pennies meet. Continue this pattern until one penny is on the top layer.

bottom layer	second layer	third layer	fourth layer
100	81	64	49

By continuing the pattern, you will find that $100 + 81 + 64 + 49 + 36 + 25 + 16 + 9 + 4 + 1$ or 385 oranges will be needed. So, we have enough.

Examine	Stack the pennies into a square pyramid with 100 pennies on the bottom and continue until one penny is on top. The result is 385.

Analyze the Strategy

1. **Tell** how making a model helped the students solve the problem.
2. **Write** a problem that can be solved by making a model.

Solve. Use the make a model strategy.

3. Cory is designing a stained glass window made of triangle pieces of glass. If the window frame is 3 feet by 4 feet and the height and base of the triangular pieces are 4 inches long, how many triangles are needed to fill the window?

4. **SALES** Karen is making a pyramid-shaped display of cereal boxes. The bottom layer of the pyramid has six boxes. If there is one less box in each layer and there are five layers in the pyramid, how many boxes will Karen need to make the display?

Mixed Problem Solving

Solve. Use any strategy.

5. **BOOKS** A bookstore arranges its best-seller books in the front window. In how many different ways can four best-seller books be arranged in a row?

6. **MONEY** Mrs. Rivas works in sales. Her base salary is $650 per week, and she makes a 5% commission on her sales. What is Mrs. Rivas' salary for four weeks if she has $8,000 in sales?

7. **PATTERNS** Draw the next figure.

8. **SCHOOL** The sixth grade class is planning a field trip. There are 575 students in the sixth grade. If each bus holds 48 people, about how many buses will they need?

9. **MONEY** How many hats can be purchased with $90 if the hats can only be bought in pairs?

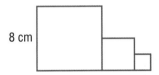

2 for $18.50

10. **FOOD** Robert bought 3 gallons of ice cream for a birthday party. If each serving size is about $\frac{1}{3}$ cup, how many servings will there be?

11. **GEOMETRY** The sides of each square in the figure are twice as long as the square on its immediate right. What is the perimeter of the entire figure?

8 cm

12. **GEOMETRY** A rectangular prism is made using exactly 8 cubes. Find the length, width, and height of the prism.

13. **STANDARDIZED TEST PRACTICE**
The graph below shows the number of parents that have participated in the booster organizations at Rancher Heights Middle School. If the trend continues, about how many parents can be expected to participate in the band booster organization in 2005?

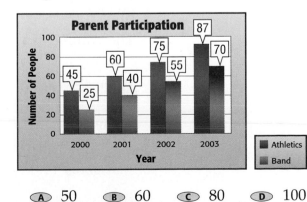

(A) 50   (B) 60   (C) 80   (D) 100

You will use models in the next lesson.

Volume of Rectangular Prisms

What You'll LEARN

Find the volume of rectangular prisms.

NEW Vocabulary

volume
cubic units

Work with a partner.

A rectangular prism and three different-sized groups of centimeter cubes are shown.

Group A   Group B   Group C

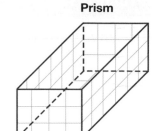

Prism

1. What are the dimensions of the prism?

2. Estimate how many of each group of cubes it will take to fill the prism. Assume that a group of cubes can be taken apart to fill the prism.

3. Use grid paper and tape to construct the prism. Then use centimeter cubes to find how many of each group of cubes it will take to fill the prism. Compare the results to your estimates.

4. Describe the relationship between the number of centimeter cubes that it takes to fill the prism and the product of the dimensions of the prism.

The amount of space inside a three-dimensional figure is the **volume** of the figure. Volume is measured in **cubic units**. This tells you the number of cubes of a given size it will take to fill the prism.

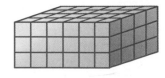

The volume of a rectangular prism is related to its dimensions.

Key Concept   **Volume of a Rectangular Prism**

Words   The volume V of a rectangular prism is the product of its length ℓ, width w, and height h.

Symbols $V = \ell w h$ **Model**

READING Math

Volume Measurement A volume measurement can be written using abbreviations and an exponent of 3. For example:
cubic units = units^3
cubic inches = in^3
cubic feet = ft^3
cubic meters = m^3

Another method you can use to find the volume of a rectangular prism is to multiply the area of the base (*B*) by the height (*h*).

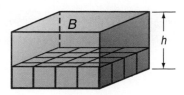

$V = Bh$ ← number of rows of cubes needed to fill the prism

area of the base, or the number of cubes needed to cover the base

EXAMPLE Find the Volume of a Rectangular Prism

1 Find the volume of the rectangular prism.

In the figure, $\ell = 12$ cm, $w = 10$ cm, and $h = 6$ cm.

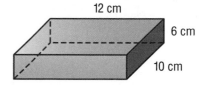

12 cm

6 cm

10 cm

Method 1 Use $V = \ell wh$.

$V = \ell wh$

$V = 12 \times 10 \times 6$

$V = 720$

The volume is 720 cm^3.

Method 2 Use $V = Bh$.

B, or the area of the base, is 10×12 or 120 square centimeters.

$V = Bh$

$V = 120 \times 6$

$V = 720$

The volume is 720 cm^3.

Your Turn Find the volume of each rectangular prism.

a.
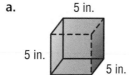
5 in.
5 in.
5 in.

b.
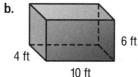
6 ft
4 ft
10 ft

EXAMPLE Use Volume to Solve a Problem

2 **FOOD** Use the information at the left. Find the approximate amount of popcorn that was contained within the popcorn box.

To find the amount of popcorn, find the volume.

Estimate 50 × 10 × 10 = 5,000

$V = \ell wh$ Volume of a rectangular prism

$V = 52.6 \times 10.1 \times 10.2$ $\ell = 52.6, w = 10.1, h = 10.2$

$V = 5,418.852$ Use a calculator.

The box contained about 5,419 cubic feet of popcorn.

Compared to the estimate, the answer is reasonable.

Skill and Concept Check

Writing Math
Exercise 1

1. **Explain** why cubic units are used to measure volume instead of linear units or square units.

2. **GEOMETRY SENSE** Visualize the three-dimensional figure shown at the right. How many of the cubes would show only 2 outside faces?

3. **OPEN ENDED** Draw a box with volume of 24 cubic units.

GUIDED PRACTICE

Find the volume of each figure. Round to the nearest tenth if necessary.

4.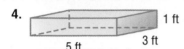
1 ft
5 ft
3 ft

5.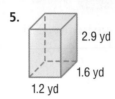
2.9 yd
1.6 yd
1.2 yd

6. **CAVES** A cave chamber is 2,300 feet long, 1,480 feet wide, and at least 230 feet high everywhere in the cave. What is the minimum volume of the cave?

Practice and Applications

Find the volume of each figure. Round to the nearest tenth if necessary.

HOMEWORK HELP

For Exercises	See Examples
7–12	1
13–14, 21	2

Extra Practice
See pages 623, 637.

7.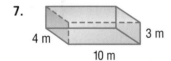
4 m
3 m
10 m

8.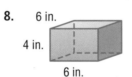
6 in.
4 in.
6 in.

9.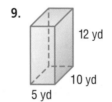
12 yd
10 yd
5 yd

10. 7 cm
3 cm
4 cm

11. 8.9 ft
1.5 ft
3.5 ft

12. 6.5 m
5.5 m
1.3 m

13. Find the volume to the nearest tenth of a rectangular prism having a length of 7.7 meters, width of 8.2 meters, and height of 9.7 meters.

14. What is the volume of a rectangular prism with a length of 10.3 feet, width of 9.9 feet, and height of 5.6 feet?

15. How many cubic feet are in 2 cubic yards?

16. How many cubic inches are in a cubic foot?

Replace each ● with <, >, or = to make a true sentence.

17. 1 ft^3 ● 1 yd^3
18. 5 m^3 ● 5 yd^3
19. 27 ft^3 ● 1 yd^3

20. **WRITE A PROBLEM** Write a problem that can be solved by finding the volume of a rectangular prism.

21. **FISH** The fish tank shown is filled to a height of 15 inches. What is the volume of water in the tank?

18 in.
22 in.
34 in.

22. **RESEARCH** Use the Internet or another source to find the dimensions and volume of the largest fish tank at an aquarium or zoo in your state or in the United States.

23. **MULTI STEP** A storage container measures 3.5 inches in length, 5.5 inches in height, and 8 inches in width. What is the volume of the container if the width is decreased by 50%?

24. **CRITICAL THINKING** If all the dimensions of a rectangular prism are doubled, does the volume double? Explain.

EXTENDING THE LESSON The volume of a cylinder is the number of cubic units needed to fill the cylinder.

To find the volume V, multiply the area of the base B by the height h. Since the base is a circle, you can replace B in $V = Bh$ with πr^2 to get $V = \pi r^2 h$.

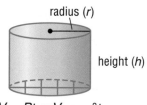
radius (r)
height (h)
$V = Bh$ or $V = \pi r^2 h$

Find the volume of each cylinder to the nearest tenth. Use 3.14 for π.

25.

5 ft
9 ft

26.
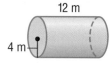
12 m
4 m

27. 6 yd

15 yd

Standardized Test Practice and Mixed Review

28. **MULTIPLE CHOICE** A rectangular prism has a volume of 288 cubic inches. Which dimensions could be the dimensions of the prism?

   Ⓐ 2 in., 4 in., 30 in.

   Ⓑ 2 in., 12 in., 12 in.

   Ⓒ 4 in., 72 in.

   Ⓓ 6 in., 8 in., 7 in.

29. **SHORT RESPONSE** Find the volume of the prism shown.

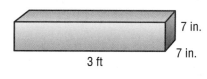
7 in.
7 in.
3 ft

Draw each figure. (Lesson 14-4)

30. cylinder

31. triangular prism

32. sphere

33. **GEOMETRY** A circle has a radius that measures 5 yards. Estimate the area of the circle. (Lesson 14-3)

GETTING READY FOR THE NEXT LESSON

PREREQUISITE SKILL Add. (Lesson 3-5)

34. $12.7 + 6.9 + 13.9$

35. $19.0 + 1.5 + 17.8$

36. $8.1 + 4.67 + 25.8$

Using a Net to Build a Cube

In this lab, you will make a two-dimensional figure called a **net** and use it to build a three-dimensional figure.

What You'll LEARN

Build a three-dimensional figure from a net and vice versa.

Materials

- cube
- scissors
- paper

ACTIVITY *Work with a partner.*

STEP 1 Place a cube on paper as shown. Trace the base of the cube, which is a square.

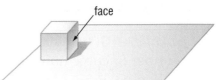
face

STEP 2 Roll the cube onto another side. Continue tracing each side to make the figure shown. This two-dimensional figure is called a net.

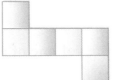

STEP 3 Cut out the net. Then build the cube.

STEP 4 Make a net like the one shown. Cut out the net and try to build a cube.

Writing Math

1. **Explain** whether both nets formed a cube. If not, describe why the net or nets did not cover the cube.

2. **Draw** three other nets that will form a cube and three other nets that will not form a cube. Describe a pattern in the nets that do form a cube.

3. **Draw** a net for a rectangular prism. Explain the difference between this net and the nets that formed a cube.

4. **Tell** what figure would be formed by each net. Explain.

a.   b.   c.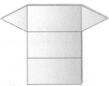

Surface Area of Rectangular Prisms

What You'll LEARN

Find the surface areas of rectangular prisms.

NEW Vocabulary

surface area

Materials
- ruler
- centimeter grid paper
- scissors
- calculator
- tape

Work with a partner.

You can use a net to explore the sum of the areas of the faces of the prism shown.

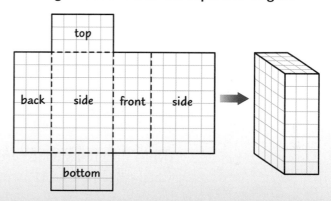

3 cm   5 cm   8 cm

STEP 1 Draw and cut out a net of the prism.

STEP 2 Fold along the dashed lines. Tape the edges.

top
back	side	front	side
bottom

1. Find the area of each face of the prism.

2. What is the sum of the areas of the faces of the prism?

3. What do you notice about the area of opposite sides of the prism? How could this simplify finding the sum of the areas?

The sum of the areas of all the faces of a prism is called the **surface area** of the prism.

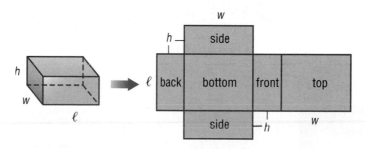

top and bottom: $(\ell \times w) + (\ell \times w) = 2\ell w$
front and back: $(\ell \times h) + (\ell \times h) = 2\ell h$
two sides: $(w \times h) + (w \times h) = 2wh$

Sum of areas of faces $= 2\ell w + 2\ell h + 2wh$

Words   The surface area S of a rectangular prism with length ℓ, width w, and height h is the sum of the areas of the faces.

Symbols $S = 2\ell w + 2\ell h + 2wh$ **Model**

EXAMPLE   **Find the Surface Area of a Rectangular Prism**

1 Find the surface area of the rectangular prism.

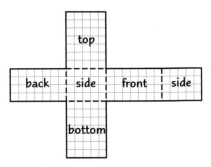

Find the area of each face.

top and bottom
$2(\ell w) = 2(7 \times 5) = 70$

front and back
$2(\ell h) = 2(7 \times 4) = 56$

two sides
$2(wh) = 2(5 \times 4) = 40$

Add to find the surface area.

The surface area is $70 + 56 + 40$ or 166 square feet.

Your Turn

a. Find the surface area of the rectangular prism.

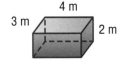

Surface area can be applied to many real-life situations.

EXAMPLE   **Use Surface Area to Solve a Problem**

2 **SPACE** An asteroid measures about 21 miles long, 8 miles wide, and 8 miles deep. Its shape resembles a rectangular prism. What is the approximate surface area of the asteroid?

Use the formula for the surface area of a rectangular prism.

$S = 2\ell w + 2\ell h + 2wh$	Surface area of a prism
$S = 2(21 \times 8) + 2(21 \times 8) + 2(8 \times 8)$	$\ell = 21$, $w = 8$, $h = 8$
$S = 2(168) + 2(168) + 2(64)$	Simplify within parentheses.
$S = 336 + 336 + 128$	Multiply.
$S = 800$	Add.

The approximate surface area of the asteroid is 800 square miles.

Skill and Concept Check

1. **Identify** each measure as *length, area, surface area,* or *volume.* Explain.

   a. the capacity of a lake

   b. the amount of land available to build a house

   c. the amount of wrapping paper needed to cover a box

2. **OPEN ENDED** Draw and label a rectangular prism that has a surface area greater than 200 square feet but less than 250 square feet.

GUIDED PRACTICE

Find the surface area of each rectangular prism.

3.
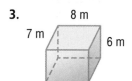
8 m
7 m
6 m

4.
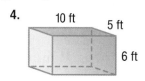
10 ft
5 ft
6 ft

5. Find the surface area of a rectangular prism that is 3.5 centimeters by 6.75 centimeters by 12 centimeters. Round to the nearest tenth.

Practice and Applications

Find the surface area of each rectangular prism. Round to the nearest tenth if necessary.

HOMEWORK HELP	
For Exercises	See Examples
6–11	1
12, 13–14	2
Extra Practice	
See pages 623, 637.	

6.

12 in.
5 in.
4 in.

7.
5 ft
3 ft
7 ft

8.
12 cm
4 cm
8 cm

9.
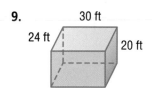
30 ft
24 ft
20 ft

10.
1.5 m
2.5 m
3.5 m

11.
2.5 mm
2.5 mm
10.5 mm

12. **AQUARIUMS** A shark petting tank is 20 feet long, 8 feet wide, and 3 feet deep. What is the surface area if the top of the tank is open?

FOOD For Exercises 13–16, use the following information.
Pretzels are to be packaged in the box shown.

4.7 in.
10.8 in.
9.9 in.

13. Estimate the surface area of the box.

14. What is the actual surface area?

15. What is the surface area if the height is increased by 100%?

16. What is the surface area if the height is decreased by 50%?

17. **WRITE A PROBLEM** Write a problem involving a rectangular prism that has a surface area of 202 square inches.

CRITICAL THINKING A cube is shown.

18. What is true about the area of the faces of a cube?

19. How could the formula $S = 2\ell w + 2\ell h + 2wh$ be simplified into a formula for the surface area of a cube?

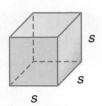

EXTENDING THE LESSON A net can also be used to show how to find the surface area of a cylinder.

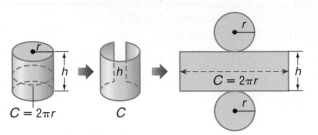

Faces	Area
top and bottom	$(\pi r^2) + (\pi r^2) = 2(\pi r^2)$
rectangular region	$(\ell \times w)$ or $2\pi rh$
Surface area $= 2(\pi r^2) + 2\pi rh$	

Find the surface area of each cylinder. Round to the nearest tenth. Use 3.14 for π.

20.

5 ft
2 ft

21.
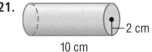
2 cm
10 cm

22.

6 in.
4 in.

Standardized Test Practice and Mixed Review

23. **MULTIPLE CHOICE** Find the surface area of the prism shown.

    Ⓐ 425 ft²   Ⓑ 440 ft²   Ⓒ 460 ft²   Ⓓ 468 ft²

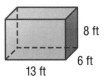
8 ft
6 ft
13 ft

24. **MULTIPLE CHOICE** Find the surface area of a cube whose sides measure 5.5 inches. Round to the neatest tenth.

    Ⓕ 225.5 in²   Ⓖ 181.5 in²   Ⓗ 125.5 in²   Ⓘ 30.3 in²

25. **GEOMETRY** Find the volume of a rectangular prism whose sides measure 5 feet, 8 feet, and $10\frac{1}{2}$ feet. (Lesson 14-5)

26. **FOOD** Draw a figure that represents a cereal box. Then identify the figure. (Lesson 14-4)

 Interdisciplinary Project

Road Trip

It's time to complete your project. Use the data you have gathered about where you are going and what you will take to prepare a Web page or poster. Be sure to include all dimensions and volume calculations with your project.

msmath1.net/webquest

Vocabulary and Concept Check

base (pp. 546, 564)	face (p. 564)	surface area (p. 575)
center (p. 565)	height (p. 546)	three-dimensional figure (p. 564)
cone (p. 565)	lateral face (p. 564)	vertex (vertices) (p. 564)
cubic units (p. 570)	prism (p. 564)	volume (p. 570)
cylinder (p. 565)	pyramid (p. 564)	
edge (p. 564)	sphere (p. 565)	

Choose the correct term to complete each sentence.

1. The flat surfaces of a three-dimensional figure are called (faces, vertices).

2. A (pyramid, cylinder) is a three-dimensional figure with one base where all other faces are triangles that meet at one point.

3. A three-dimensional figure with two circular bases is a (cone, cylinder).

4. The amount of space that a three-dimensional figure contains is called its (area, volume).

5. The total area of a three-dimensional object's faces and curved surfaces is called its (surface area, volume).

Lesson-by-Lesson Exercises and Examples

14-1 Area of Parallelograms (pp. 546–549)

Find the area of each parallelogram. Round to the nearest tenth if necessary.

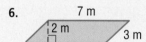

6. 7 m, 2 m, 3 m

7. $5\frac{1}{2}$ in., $7\frac{5}{8}$ in., $8\frac{3}{4}$ in.

Example 1 Find the area of the parallelogram.

$A = bh$

$A = 6 \times 5$

$A = 30 \text{ in}^2$

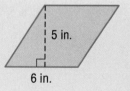

5 in., 6 in.

14-2 Area of Triangles (pp. 551–554)

Find the area of each triangle. Round to the nearest tenth if necessary.

8. 11 in., 18 in.

9. 3.4 m, 3.1 m

Example 2 Find the area of the triangle.

$A = \frac{1}{2}bh$

$A = \frac{1}{2}(75 \times 50)$

$A = 1{,}875 \text{ m}^2$

50 m, 75 m

14-3 Area of Circles (pp. 556–559)

Find the area of each circle to the nearest tenth. Use 3.14 for π.

10.
11 m

11.
14 in.

12. **RIDES** The plans for a carousel call for a circular floor with a diameter of 40 feet. Find the area of the floor.

Example 3 Find the area to the nearest tenth. Use 3.14 for π.

7 cm

$A = \pi r^2$ Area of a circle

$A \approx 3.14 \times 7^2$ Replace π with 3.14 and r with 7.

$A \approx 3.14 \times 49$ Evaluate 7^2.

$A \approx 153.9$ Multiply.

The area is about 153.9 square centimeters.

14-4 Three-Dimensional Figures (pp. 564–566)

Identify each figure.

13.

14.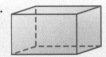

15. **SPORTS** What is the shape of a basketball?

Example 4
Identify the figure.

It has at least three rectangular lateral faces. The bases are triangles.

The figure is a triangular prism.

14-5 Volume of Rectangular Prisms (pp. 570–573)

Find the volume of each figure. Round to the nearest tenth if necessary.

16.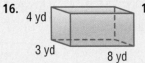
4 yd
3 yd
8 yd

17.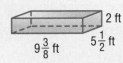
2 ft
$9\frac{3}{8}$ ft
$5\frac{1}{2}$ ft

Example 5 Find the volume of the prism.

$V = \ell wh$

$V = 8 \times 4 \times 5$

$V = 160$

The volume is 160 cubic inches.

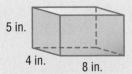
5 in.
4 in.
8 in.

14-6 Surface Area of Rectangular Prisms (pp. 575–578)

Find the surface area of each rectangular prism. Round to the nearest tenth if necessary.

18.
7 in.
6 in.
7 in.

19.
4.5 cm
6.8 cm
5.9 cm

Example 6 Find the surface area of the rectangular prism in Example 5.

top and bottom: $2(8 \times 4)$ or 64

front and back: $2(8 \times 5)$ or 80

two sides: $2(4 \times 5)$ or 40

The surface area is $64 + 80 + 40$ or 184 square feet.

Vocabulary and Concepts

1. **Write** the formula for the area of a triangle.

2. **Define** *volume*.

Skills and Applications

Find the area of each figure. Round to the nearest tenth if necessary.

3.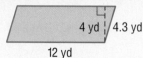
4 yd | 4.3 yd
12 yd

4.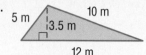
5 m   10 m
3.5 m
12 m

5.
5 in.

6. **TRAFFIC SIGN** A triangular yield sign has a base of 32 inches and a height of 30 inches. Find the area of the sign.

7. **GARDENING** A circular flowerbed has a radius of 2 meters. If you can plant 40 bulbs per square meter, how many bulbs should you buy?

Identify each figure.

8.

9.

10.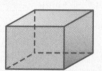

Find the volume of each figure. Round to the nearest tenth if necessary.

11.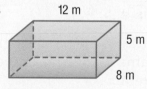
12 m
5 m
8 m

12.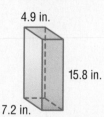
4.9 in.
15.8 in.
7.2 in.

13.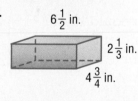
$6\frac{1}{2}$ in.
$2\frac{1}{3}$ in.
$4\frac{3}{4}$ in.

14. **POOLS** A rectangular diving pool is 20 feet by 15 feet by 8 feet. How much water is required to fill the pool?

15. Find the surface area of the prism in Exercise 11.

Standardized Test Practice

16. **MULTIPLE CHOICE** Which expression gives the surface area of a rectangular prism with length ℓ, width w, and height h?

Ⓐ $2\ell^2 + 2h^2 + 2w^2$

Ⓑ $2\ell w + 2\ell h + 2wh$

Ⓒ $2(\ell \times w \times h)$

Ⓓ $2\ell(w + h)$

PART 1   Multiple Choice

Record your answers on the answer sheet provided by your teacher or on a sheet of paper.

1. An airplane is flying at a height of 23,145.769 feet. Which of the following numbers is in the hundreds place? (Prerequisite Skill, p. 586)

    Ⓐ 1   Ⓑ 3   Ⓒ 4   Ⓓ 6

2. What is the best estimate for the total number of pounds of paper recycled by Ms. Maliqua's class? (Lesson 3-4)

Paper Recycling Drive	
Week	**Amount (lb)**
1	22.5
2	38.2
3	32.7
4	53.1

    Ⓕ 130 lb   Ⓖ 140 lb
    Ⓗ 160 lb   Ⓘ 170 lb

3. What could be the perimeter of the rectangle shown? (Lesson 4-5)

    Area = 42 m²

    Ⓐ 13 m   Ⓑ 20 m
    Ⓒ 26 m   Ⓓ 88 m

4. What is $3\frac{9}{16}$ expressed as an improper fraction? (Lesson 5-3)

    Ⓕ $\frac{48}{16}$ Ⓖ $\frac{43}{16}$
    Ⓗ $3\frac{16}{9}$ Ⓘ $\frac{57}{16}$

5. What is the value of r in the equation $6r = 30$? (Lesson 9-4)

    Ⓐ 0.2   Ⓑ 5   Ⓒ 24   Ⓓ 180

6. What is the ratio of the number of hearts to the total number of figures below? (Lesson 10-1)

    Ⓕ $\frac{2}{9}$ Ⓖ $\frac{1}{4}$ Ⓗ $\frac{1}{3}$ Ⓘ $\frac{4}{9}$

7. What is the area of a parallelogram with a base of 5 inches and a height of 3 inches? (Lesson 14-1)

    Ⓐ 8 in²   Ⓑ 15 in²
    Ⓒ 15 in.   Ⓓ 16 in²

8. What is a correct statement about the relationship between the figures shown? (Lesson 14-2)

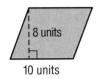 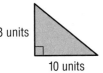

    Ⓕ The area of the parallelogram is the same as the area of the triangle squared.

    Ⓖ The area of the parallelogram is three times the area of the triangle.

    Ⓗ The area of the parallelogram is twice the area of the triangle.

    Ⓘ The area of the parallelogram is $\frac{1}{2}$ the area of the triangle.

TEST-TAKING TIP

Question 7 When answer choices include units, be sure to select an answer choice that uses the correct units.

PART 2 Short Response/Grid In

Record your answers on the answer sheet provided by your teacher or on a sheet of paper.

9. Kaley divides $3\frac{3}{5}$ pies among 9 people. How much of one pie will each person get? (Lesson 7-5)

10. Each serving of pizza is $\frac{1}{16}$ of a pizza. If $\frac{3}{4}$ of the pizza is left, how many servings are left? (Lesson 7-5)

11. What is the product of -7 and -12? (Lesson 8-4)

12. Elias bought the following items.

If the rate of sales tax that he paid was 6%, how much sales tax did he pay? (Lesson 10-7)

13. Find the probability that a randomly thrown dart will land in one of the squares labeled C. (Lesson 11-4)

14. How many inches are in 3 yards? (Lesson 12-1)

15. How many lines of symmetry does the figure shown have? (Lesson 13-5)

16. Find the area of a triangle that has a base of 12 inches and a height of 4 inches. (Lesson 14-2)

17. What is the approximate area of a circle with a radius of 10 meters? (Lesson 14-3)

18. How many faces does the rectangular prism have? (Lesson 14-4)

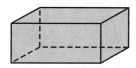

19. Write the formula that could be used to find the volume of a rectangular prism. (Use ℓ for length, w for width, h for height, and V for volume.) (Lesson 14-5)

20. What is the surface area of the rectangular prism? (Lesson 14-6)

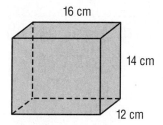

16 cm
14 cm
12 cm

PART 3 Extended Response

Record your answers on a sheet of paper. Show your work.

21. Shane built a figure using centimeter cubes. The figure stood 4 cubes high and covered a 12-centimeter by 8-centimeter area of the floor.

   a. What area of the floor did the figure cover? (Lesson 14-1)

   b. What is the volume of the figure? (Lesson 14-5)

   c. Draw Shane's structure. (Lesson 14-5)

STUDENT HANDBOOK

Skills

Reference

HOW TO...

USE THE STUDENT HANDBOOK

A Student Handbook is the additional skill and reference material found at the end of books. The Student Handbook can help answer these questions.

What If I Forget What I Learned Last Year?
Use the **Prerequisite Skills** section to refresh your memory about things you have learned in other math classes. Here's a list of the topics covered in your book.
- Place Value and Whole Numbers
- Comparing and Ordering Whole Numbers
- Adding and Subtracting Whole Numbers
- Multiplying Whole Numbers
- Dividing Whole Numbers
- Estimating with Whole Numbers

What If I Need More Practice?
You, or your teacher, may decide that working through some additional problems would be helpful. The **Extra Practice** section provides these problems for each lesson so you have ample opportunity to practice new skills.

What If I Have Trouble with Word Problems?
The **Mixed Problem Solving** portion of the book provides additional word problems that use the skills presented in each chapter. These problems give you real-life situations where the math can be applied.

What If I Forget a Vocabulary Word?
The **English-Spanish Glossary** provides a list of important, or difficult, words used throughout the textbook. It provides a definition in English and Spanish as well as the page number(s) where the word can be found.

What If I Need to Check a Homework Answer?
The answers to the odd-numbered problems are included in **Selected Answers**. Check your answers to make sure you understand how to solve all of the assigned problems.

What If I Need to Find Something Quickly?
The **Index** alphabetically lists the subjects covered throughout the entire textbook and the pages on which each subject can be found.

What If I Forget a Formula?
Inside the back cover of your math book is a list of **Formulas and Symbols** that are used in the book.

Need to Cover Your Book?
Inside the back cover are directions for a Foldable that you can use to cover your math book quickly and easily!

Prerequisite Skills

Place Value and Whole Numbers

The number system we use is based on units of 10. A number like 7,825 is a **whole number**. A digit and its **place-value** position name a number. For example, in 7,825, the digit 7 is in the thousands place, and its value is 7,000.

Place-Value Chart

1,000,000,000	100,000,000	10,000,000	1,000,000	100,000	10,000	1,000	100	10	1
one billion	hundred millions	ten millions	one million	hundred thousands	ten thousands	thousandds	hundreds	tens	ones
						7,	8	2	5

Each set of three digits is called a period. Periods are separated by a comma.

EXAMPLE Identify Place Value

1 Identify the place-value position of the digit 9 in 597,240,618.

1,000,000,000	100,000,000	10,000,000	1,000,000	100,000	10,000	1,000	100	10	1
one billion	hundred millions	ten millions	one million	hundred thousands	ten thousands	thousandds	hundreds	tens	ones
	5	9	7,	2	4	0,	6	1	8

The digit 9 is in the ten millions place.

Numbers written as 7,825 and 597,240,618 are written in **standard form**. Numbers can also be written in **word form**. When writing a number in word form, use place value. At each comma, write the name for the period.

EXAMPLES Write a Whole Number in Word Form

Write each number in word form.

2 4,567,890

Standard Form   4,567,890

Word Form   four million five hundred sixty-seven thousand eight hundred ninety

3 804,506

Standard Form   804,506

Word Form   eight hundred four thousand five hundred six

Numbers can also be written in **expanded notation** .

> ### EXAMPLE   Write a Whole Number in Expanded Notation
>
> **4** Write 28,756 in expanded notation.
>
> **Step 1** Write the product of each digit and its place value.
>
> | | |
> |---|---|
> | $20{,}000 = (2 \times 10{,}000)$ | The digit 2 is in the ten thousands place. |
> | $8{,}000 = (8 \times 1{,}000)$ | The digit 8 is in the thousands place. |
> | $700 = (7 \times 100)$ | The digit 7 is in the hundreds place. |
> | $50 = (5 \times 10)$ | The digit 5 is in the tens place. |
> | $6 = (6 \times 1)$ | The digit 6 is in the ones place. |
>
> **Step 2** Write the sum of the products.
>
> $$28{,}756 = (2 \times 10{,}000) + (8 \times 1{,}000) + (7 \times 100) + (5 \times 10) + (6 \times 1)$$

This is 28,756 written in expanded notation.

Exercises

Identify each underlined place-value position.

1. 4<u>3</u>8
2. 6,8<u>4</u>5,085
3. <u>4</u>13,467
4. 3,74<u>5</u>
5. 7<u>2</u>,567,432
6. <u>9</u>04,784,126

7. Write a number that has the digit 7 in the billions place, the digit 8 in the hundred thousands place, and the digit 4 in the tens place.

Write each number in word form.

8. 263
9. 2,013
10. 54,006
11. 47,900
12. 567,460
13. 551,002
14. 7,805,261
15. 1,125,678
16. 102,546,165
17. 582,604,072
18. 8,146,806,835
19. 67,826,657,005

Write each number in standard form.

20. forty-two
21. seven hundred fifty-one
22. three thousand four hundred twelve
23. six thousand nine hundred five
24. six hundred thousand
25. sixteen thousand fifty-two
26. seventy-six million
27. two hundred twenty-four million
28. three million four hundred thousand
29. ten billion nine hundred thousand

Write each number in expanded notation.

30. 86
31. 398
32. 620
33. 5,285
34. 4,002
35. 7,500
36. 85,430
37. 524,789
38. 8,043,967

39. **PIANOS** There are over ten million pianos in American homes, businesses, and institutions. Write ten million in expanded notation.

40. **BOOKS** The Library of Congress in Washington, D.C., has 24,616,867 books. Write this number in word form.

Comparing and Ordering Whole Numbers

When comparing the values of two whole numbers, the first number is either less than, greater than, or equal to the second number. You can use place value or a number line to compare two whole numbers.

Words	Symbol
less than	<
greater than	>
equal to	=

Method 1 Use place value.

- Line up the digits at the ones place.
- Starting at the left, compare the digits in each place-value position. In the first position where the digits differ, the number with the greater digit is the greater whole number.

Method 2 Use a number line.

- Numbers to the right are greater than numbers to the left.
- Numbers to the left are less than numbers to the right.

EXAMPLE Compare Whole Numbers

1 Replace the ● in 25,489 ● 25,589 with >, <, or = to make a true sentence.

Method 1 Use place value.

25,489   Line up the digits.
25,589   Compare.

The digits in the hundreds place are not the same. Since 4 < 5, 25,489 < 25,589.

Method 2 Use a number line.

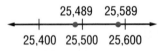

Graph and then compare the numbers.

Since 25,489 is to the left of 25,589, 25,489 < 25,589.

EXAMPLE Order Whole Numbers

2 Order 8,989, 8,957, and 8,984 from least to greatest.

8,957 is less than both 8,989 and 8,984 since 5 < 8 in the tens place. 8,984 is less than 8,989 since 4 < 9 in the ones place.

So, the order from least to greatest is 8,957, 8,984, and 8,989.

EXERCISES

Replace each ● with < or > to make a true sentence.

1. 496 ● 489
2. 5,602 ● 5,699
3. 3,455 ● 3,388
4. 9,999 ● 10,001
5. 8,993 ● 8,399
6. 13,405 ● 14,003
7. 75,495 ● 75,606
8. 400,590 ● 401,001
9. 501,222 ● 510,546
10. 245,000 ● 24,500
11. 675,656 ● 6,756,560
12. 1,000,000 ● 989,566

Order each list of whole numbers from least to greatest.

13. 678, 699, 610
14. 4,654, 4,432, 4,678
15. 42,000, 4,200, 41,898

16. **STATES** Alabama has a total area of 52,237 square miles. Arkansas has an area of 53,182 square miles. Which state is larger?

Adding and Subtracting Whole Numbers

To add or subtract whole numbers, add or subtract the digits in each place-value position. Start at the ones place.

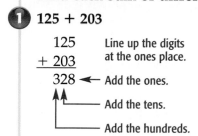 **Add or Subtract Whole Numbers**

Find each sum or difference.

1 125 + 203

```
  125       Line up the digits
+ 203       at the ones place.
  328  ◄─── Add the ones.
             Add the tens.
             Add the hundreds.
```

2 587 − 162

```
  587       Line up the digits
− 162       at the ones place.
  425  ◄─── Subtract the ones.
             Subtract the tens.
             Subtract the hundreds.
```

You may need to regroup when adding and subtracting whole numbers.

EXAMPLES **Add or Subtract Whole Numbers with Regrouping**

Find each sum or difference.

3 387 + 98

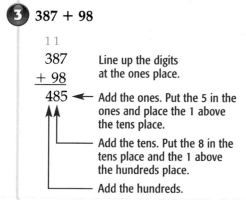

```
  1 1
  387       Line up the digits
+  98       at the ones place.
  485  ◄─── Add the ones. Put the 5 in the
             ones and place the 1 above
             the tens place.
             Add the tens. Put the 8 in the
             tens place and the 1 above
             the hundreds place.
             Add the hundreds.
```

4 612 − 59

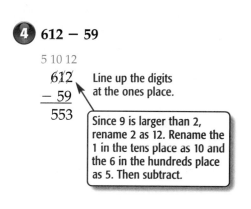

```
  5 10 12
   6̶1̶2̶      Line up the digits
−   59       at the ones place.
   553
```
> Since 9 is larger than 2, rename 2 as 12. Rename the 1 in the tens place as 10 and the 6 in the hundreds place as 5. Then subtract.

Exercises

Find each sum or difference.

1. 506 + 30	2. 315 + 583	3. 1,342 + 627	4. 5,042 + 2,143	5. 71,235 + 27,563
6. 468 − 21	7. 895 − 472	8. 1,872 − 460	9. 6,056 − 5,052	10. 74,618 − 23,311
11. 2,680 + 945	12. 5,126 + 896	13. 2,973 + 1,689	14. 1,089 + 5,239	15. 16,999 + 25,509
16. 982 − 36	17. 487 − 199	18. 6,052 − 5,456	19. 64,205 − 3,746	20. 215,000 − 12,999

21. **PLANETS** The diameter of Earth is 7,926 miles. The diameter of Mars is 4,231 miles. How much greater is the diameter of Earth than the diameter of Mars?

Multiplying Whole Numbers

To multiply a whole number by a 1-digit whole number, multiply from right to left, regrouping as necessary.

EXAMPLES **Multiply by a 1-Digit Whole Number**

Find each product.

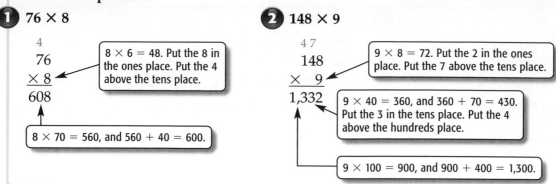

1 76 × 8

$$\begin{array}{r} 4 \\ 76 \\ \underline{\times\ 8} \\ 608 \end{array}$$

8 × 6 = 48. Put the 8 in the ones place. Put the 4 above the tens place.

8 × 70 = 560, and 560 + 40 = 600.

2 148 × 9

$$\begin{array}{r} 4\ 7 \\ 148 \\ \underline{\times\ 9} \\ 1{,}332 \end{array}$$

9 × 8 = 72. Put the 2 in the ones place. Put the 7 above the tens place.

9 × 40 = 360, and 360 + 70 = 430. Put the 3 in the tens place. Put the 4 above the hundreds place.

9 × 100 = 900, and 900 + 400 = 1,300.

When you multiply by a number with two or more digits, write individual products and then add.

EXAMPLES **Multiply by a 2-Digit Whole Number**

Find each product.

3 84 × 65

$$\begin{array}{r} 84 \\ \underline{\times\ 65} \\ 420 \\ \underline{+\ 5040} \\ 5460 \end{array}$$

Multiply. 84 × 5 = 420
Multiply. 84 × 60 = 5,040
Add. 420 + 5,040 = 5,460

4 535 × 24

$$\begin{array}{r} 535 \\ \underline{\times\ 24} \\ 2{,}140 \\ \underline{+\ 10{,}700} \\ 12{,}840 \end{array}$$

Multiply. 535 × 4 = 2,140
Multiply. 535 × 20 = 10,700
Add. 2,140 + 10,700 = 12,840

Exercises

Find each product.

1. 40
    × 5

2. 23
    × 3

3. 81
    × 6

4. 172
    × 3

5. 361
    × 7

6. 721
    × 4

7. 513
    × 3

8. 634
    × 2

9. 2,821
    × 4

10. 5,045
    × 2

11. 28
    × 10

12. 52
    × 20

13. 75
    × 19

14. 54
    × 27

15. 414
    × 22

16. 321
    × 34

17. 522
    × 32

18. 613
    × 24

19. 525
    × 150

20. 678
    × 123

21. **FOOD** The average person in the U.S. eats about 36 pounds of frozen food per year. How many pounds would an average family of four eat?

Dividing Whole Numbers

When dividing whole numbers, divide in each place-value position from left to right. Recall that in the statement $50 \div 2 = 25$, 50 is the **dividend**, 2 is the **divisor**, and 25 is the **quotient**.

EXAMPLES **Divide Whole Numbers**

Find each quotient.

1 $342 \div 9$

$$
\begin{array}{r}
38 \\
9\overline{)342} \\
-27 \\
\hline
72 \\
-72 \\
\hline
0
\end{array}
$$

Divide in each place-value position from left to right.

Since $72 - 72 = 0$, there is no remainder.

2 $6{,}493 \div 78$

$$
\begin{array}{r}
83 \text{ R}19 \\
78\overline{)6493} \\
-624 \\
\hline
253 \\
-234 \\
\hline
19
\end{array}
$$

Divide in each place-value position from left to right.

Since $253 - 234 = 19$, the remainder is 19.

Exercises

Find each quotient.

1. $3\overline{)72}$
2. $4\overline{)96}$
3. $2\overline{)84}$
4. $6\overline{)918}$

5. $8\overline{)976}$
6. $7\overline{)903}$
7. $12\overline{)60}$
8. $25\overline{)75}$

9. $15\overline{)90}$
10. $34\overline{)204}$
11. $27\overline{)135}$
12. $53\overline{)424}$

13. $24\overline{)240}$
14. $25\overline{)500}$
15. $15\overline{)600}$
16. $6\overline{)384}$

17. $34\overline{)612}$
18. $48\overline{)720}$
19. $31\overline{)1,953}$
20. $47\overline{)1,927}$

21. $19\overline{)1,045}$
22. $18\overline{)3,672}$
23. $32\overline{)9,824}$
24. $27\overline{)8,154}$

25. $8\overline{)91}$
26. $5\overline{)99}$
27. $6\overline{)80}$
28. $8\overline{)685}$

29. $4\overline{)273}$
30. $5\overline{)387}$
31. $12\overline{)75}$
32. $18\overline{)99}$

33. $27\overline{)56}$
34. $33\overline{)85}$
35. $23\overline{)97}$
36. $29\overline{)210}$

37. $62\overline{)439}$
38. $37\overline{)299}$
39. $16\overline{)134}$
40. $53\overline{)483}$

41. $50\overline{)12,575}$
42. $49\overline{)29,670}$
43. $32\overline{)38,693}$
44. $46\overline{)92,330}$

45. $26\overline{)80,311}$
46. $100\overline{)706}$
47. $100\overline{)842}$
48. $200\overline{)900}$

49. $500\overline{)705}$
50. $300\overline{)602}$
51. $400\overline{)1,632}$
52. $600\overline{)8,407}$

53. **MONEY** In 2000, there were \$3,440,000 in \$10,000 bills in circulation in the U.S. How many \$10,000 bills were there?

54. **MEASUREMENT** One acre is the same as 43,560 square feet. It is also equal to 4,840 square yards. How many square feet are there in one square yard?

Estimating with Whole Numbers

When an exact answer to a math problem is not needed, or when you want to check the reasonableness of an answer, you can use **estimation**. There are several methods of estimation. One common method is **rounding**.

To round a whole number, look at the digit to the right of the place being rounded.

- If the digit is 4 or less, the underlined digit remains the same.
- If the digit is 5 or greater, add 1 to the underlined digit.

EXAMPLES Estimate by Rounding

Estimate by rounding.

1 30,798 + 4,115 + 1,891

$$
\begin{array}{rcl}
30,798 & \rightarrow & 31,000 \\
4,115 & \rightarrow & 4,000 \\
+\ 1,891 & \rightarrow & +\ 2,000 \\
\hline
& & 37,000
\end{array}
$$

In this case, each number is rounded to the same place value.

So, the sum is about 37,000.

2 478 × 12

$$
\begin{array}{rcl}
478 & \rightarrow & 500 \\
\times\ 12 & \rightarrow & \times\ 10 \\
\hline
& & 5,000
\end{array}
$$

In this case, each number is rounded to its greatest place value.

So, the product is about 5,000.

Clustering can be used to estimate sums if all of the numbers are close to a certain number.

EXAMPLES Estimate by Clustering

Estimate by clustering.

3 97 + 102 + 99 + 104 + 101 + 98

All of the numbers are clustered around 100. There are 6 numbers. So, the sum is about 6 × 100 or 600.

4 748 + 751 + 753 + 747

All of the numbers are clustered around 750. There are 4 numbers. So, the sum is about 4 × 750 or 3,000.

Compatible numbers are two numbers that are easy to compute mentally.

EXAMPLES Estimate by Using Compatible Numbers

Estimate by using compatible numbers.

5 102 ÷ 24

$$
24\overline{)102} \quad \rightarrow \quad 25\overline{)100}^{\,4}
$$

So, the quotient is about 4.

6 71 + 19 + 28 + 83

$$
\begin{aligned}
71 + 19 + 28 + 83 &\rightarrow 70 + 20 + 30 + 80 \\
&= (70 + 30) + (20 + 80) \\
&= 100 + 100 \text{ or } 200
\end{aligned}
$$

The sum is about 200.

Another strategy that works well for some addition and subtraction problems is **front-end estimation**. In this strategy, you first add or subtract the left-most column of digits. Then, add or subtract the next column.

EXAMPLES **Use Front-End Estimation**

Estimate by using front-end estimation.

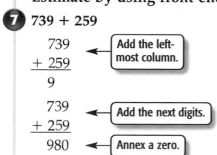

7 739 + 259

$$\begin{array}{r} 739 \\ + 259 \\ \hline 9 \end{array}$$ ← Add the left-most column.

$$\begin{array}{r} 739 \\ + 259 \\ \hline 980 \end{array}$$ ← Add the next digits. ← Annex a zero.

So, the sum is about 980.

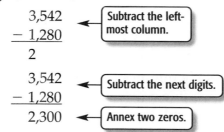

8 3,542 − 1,280

$$\begin{array}{r} 3,542 \\ - 1,280 \\ \hline 2 \end{array}$$ ← Subtract the left-most column.

$$\begin{array}{r} 3,542 \\ - 1,280 \\ \hline 2,300 \end{array}$$ ← Subtract the next digits. ← Annex two zeros.

So, the difference is about 2,300.

Exercises

Estimate by rounding.

1. 82×3
2. $435 + 219 + 121$
3. $4286 - 2089$
4. 99×12
5. $106 \div 9$
6. 192×39

Estimate by clustering.

7. $19 + 18 + 21 + 20 + 19$
8. $498 + 499 + 502 + 503$
9. $82 + 79 + 77 + 81$
10. $4 + 3 + 6 + 5 + 7 + 4 + 6 + 7$

Estimate by using compatible numbers.

11. $31 \div 4$
12. $17 + 82 + 58 + 43 + 75$
13. $122 + 79 + 232 + 68$
14. $205 \div 6$
15. $540 + 470 + 325 + 70$
16. $813 \div 9$

Estimate by using front-end estimation.

17. $852 - 312$
18. 65×4
19. 120×3
20. $895 - 246$
21. 340×2
22. $673 - 344$

Use any method to estimate.

23. $680 - 290$
24. $647 \div 9$
25. 68×11
26. $99.6 \div 18.25$
27. $43 + 62 + 22 + 77$
28. $69 + 72 + 68 + 70 + 73$
29. $390 \div 52$
30. 102×6
31. 498×19
32. $493 - 183$
33. $199 + 205 + 198$
34. $999 \div 52$
35. $1209 \div 403$
36. 2005×4
37. $490 + 514 + 520 + 483$
38. $3486 \div 5$
39. $7654 - 3254$
40. 3045×124
41. $188 + 320$
42. $12,250 \div 4008$
43. $874 - 523$

Extra Practice

Lesson 1-1
(Pages 6–9)

Use the four-step plan to solve each problem.

1. On a map of Ohio, each inch represents approximately 9 miles. Chad is planning to travel from Cleveland to Cincinnati. If the distance on the map from Cleveland to Cincinnati is about 23 inches, how far will he travel?

2. Sylvia has $102. If she made three purchases of $13, $37, and $29, how much money does she have left?

3. A cassette tape holds sixty minutes of music. If Logan has already taped eight songs that are each five minutes long on the tape, how many more minutes of music can she still tape on the cassette?

4. **PATTERNS** Complete the pattern: 1, 8, 15, 22, __?__, __?__, __?__

5. Sarah is conditioning for track. On the first day, she ran 2 laps. The second day she ran 3 laps. The third day she ran 5 laps, and on the fourth day she ran 8 laps. If this pattern continues, how many laps will she run on the seventh day?

6. The times that a new movie is showing are 9:30 A.M., 11:12 A.M., 12:54 P.M., and 2:36 P.M. What are the next three times this movie will be shown?

Lesson 1-2
(Pages 10–13)

Tell whether each number is divisible by 2, 3, 4, 5, 6, 9, or 10. Then classify each number as *even* or *odd*.

1. 48	2. 64	3. 125	4. 156
5. 216	6. 330	7. 225	8. 524
9. 1,986	10. 2,052	11. 110	12. 315
13. 405	14. 243	15. 1,233	16. 5,103
17. 8,001	18. 9,270	19. 284	20. 585
21. 1,620	22. 3,213	23. 25,365	24. 27,027

Lesson 1-3
(Pages 14–17)

Tell whether each number is *prime*, *composite*, or *neither*.

1. 20	2. 65	3. 37	4. 26
5. 54	6. 155	7. 201	8. 0
9. 49	10. 17	11. 93	12. 121
13. 29	14. 53	15. 76	16. 57

Find the prime factorization of each number.

17. 72	18. 88	19. 32
20. 86	21. 120	22. 576
23. 68	24. 240	25. 24
26. 70	27. 102	28. 121
29. 164	30. 225	31. 54

Lesson 1-4

(Pages 18–21)

Write each product using an exponent. Then find the value of the power.

1. $2 \cdot 2 \cdot 2 \cdot 2 \cdot 2$
2. $6 \cdot 6 \cdot 6$
3. $4 \cdot 4 \cdot 4 \cdot 4 \cdot 4$
4. $10 \cdot 10 \cdot 10$
5. $14 \cdot 14$
6. $3 \cdot 3 \cdot 3 \cdot 3$
7. $5 \cdot 5 \cdot 5 \cdot 5$
8. $2 \cdot 2 \cdot 2$
9. $6 \cdot 6 \cdot 6 \cdot 6 \cdot 6$
10. $8 \cdot 8 \cdot 8$
11. $7 \cdot 7 \cdot 7 \cdot 7 \cdot 7 \cdot 7$
12. $9 \cdot 9 \cdot 9$

Write each power as a product. Then find the value of the product.

13. 9^4
14. 2^3
15. 3^5
16. 4^3
17. 6^5
18. 5^4
19. 8^3
20. 12^2
21. nine cubed
22. eight to the fourth power
23. six to the fifth power
24. eleven squared

Write the prime factorization of each number using exponents.

25. 9
26. 20
27. 18
28. 63
29. 44
30. 45
31. 243
32. 175

Lesson 1-5

(Pages 24–27)

Find the value of each expression.

1. $14 - 5 + 7$
2. $12 + 10 - 5 - 6$
3. $50 - 6 + 12 + 4$
4. $12 - 2 \times 3$
5. $16 + 4 \times 5$
6. $5 + 3 \times 4 - 7$
7. $2 \times 3 + 9 \times 2$
8. $6 \times 8 + 4 \div 2$
9. $7 \times 6 - 14$
10. $8 + 12 \times 4 \div 8$
11. $13 - 6 \times 2 + 1$
12. $80 \div 10 \times 8$
13. $1 + 2 + 3 + 4$
14. $1 \times 2 \times 3 \times 4$
15. $6 + 6 \times 6$
16. $14 - 2 \times 7 + 0$
17. $156 - 6 \times 0$
18. $30 - 14 \times 2 + 8$
19. $54 \div (8 - 5)$
20. $4^2 + 3^3$
21. $(11 - 7) \times 3 - 5$
22. $25 - 9 + 4$
23. $100 \div 10 \times 2$
24. 3×4^3
25. $11 + 4 \times (12 - 7)$
26. $6^2 - 7 \times 4$
27. $12 + 5^2 - 9$

28. Find two to the fifth power divided by 8 times 2.

29. What is the value of one hundred minus seven squared times two?

Lesson 1-6

(Pages 28–31)

Evaluate each expression if $m = 2$ and $n = 4$.

1. $m + m$
2. $n - m$
3. mn
4. $2m$
5. $2n$
6. $2n + 2m$
7. $m \times 0$
8. $64 \div n$
9. $12 - m$
10. $2mn$
11. $m^2 + 3$
12. $n^2 - 5m$
13. $5n \div m$
14. $6mn$
15. $4n - 3$
16. $n \div m + 8$

Evaluate each expression if $a = 3$, $b = 4$, and $c = 12$.

17. $a + b$
18. $c - a$
19. $a + b + c$
20. $b - a$
21. $c - a \times b$
22. $a + 2 \times b$
23. $b + c \div 2$
24. ab
25. $a^2 + 3b$
26. $a^2 + c \div 6$
27. $25 + c \div b$
28. $a^2 + 2b - c$
29. $c^2 \div (ab)$
30. $c \div a + 10$
31. $2b - a$
32. $2ab$

Extra Practice

Lesson 1-7

(Pages 34–37)

Identify the solution of each equation from the list given.

1. $7 + a = 10$; 3, 13, 17

2. $14 + m = 24$; 7, 10, 34

3. $20 = 24 - n$; 2, 3, 4

4. $x + 4 = 19$; 14, 15, 16

5. $23 - p = 7$; 16, 17, 18

6. $11 = w + 6$; 3, 4, 5

7. $73 + m = 100$; 26, 27, 28

8. $44 + s = 63$; 17, 18, 19

Solve each equation mentally.

9. $b + 7 = 12$

10. $s + 10 = 23$

11. $b - 3 = 12$

12. $d + 7 = 19$

13. $23 - q = 9$

14. $21 + p = 45$

15. $17 = 23 - t$

16. $g - 13 = 5$

17. $14 - m = 6$

18. $x - 3 = 11$

19. $16 = h + 9$

20. $50 + z = 90$

Lesson 1-8

(Pages 39–41)

Find the area of each rectangle.

1.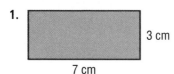
3 cm
7 cm

2. 2 in.
11 in.

3. 8 yd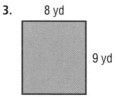
9 yd

4. 13 m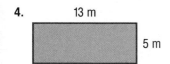
5 m

5.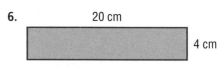
7 ft
4 ft

6. 20 cm
4 cm

Lesson 2-1

(Pages 50–53)

1. The high temperatures in Indiana cities on March 13 are listed.
52 57 48 53 52 49 48 52 51 47 51 49 57 53 48 52 52 49
Make a frequency table of the data.

2. Which scale is more appropriate for the data set 56, 85, 23, 78, 42, 63:
0 to 50 or 20 to 90? Explain your reasoning.

3. What is the best interval for the data set 132, 865, 465, 672, 318, 940, 573, 689:
10, 100, or 1,000? Explain your reasoning.

For Exercises 4 and 5, use the frequency table at the right.

4. Describe the data shown in the table.

5. Which flavor should the ice cream shop stock the most?

Ice Cream Flavors Sold in July		
Flavor	**Tally**	**Frequency**
vanilla	ЖHТ ЖHТ ЖHТ ЖHТ IIII	24
chocolate	ЖHТ ЖHТ ЖHТ III	18
strawberry	ЖHТ ЖHТ II	12
chocolate chip	ЖHТ ЖHТ ЖHТ I	16
peach	ЖHТ III	8
butter pecan	ЖHТ ЖHТ I	11

Make a bar graph for each set of data.

1. a vertical bar graph

Favorite Subject	
Subject	Frequency
Math	4
Science	6
History	2
English	8
Phys. Ed.	12

2. a horizontal bar graph

Final Grades	
Subject	Score
Math	88
Science	82
History	92
English	94

Make a line graph for each set of data.

3.

Test Scores	
Test	Score
1	62
2	75
3	81
4	83
5	78
6	92

4.

Homeroom Absences	
Day	Absences
Mon.	3
Tues.	6
Wed.	2
Thur.	1
Fri.	8

The circle graph shows the favorite subject of students at Midland Middle School.

1. What is the most popular subject?

2. Which two subjects together are preferred by half of the students?

3. Which two subjects are preferred by the least number of students?

4. How does the percent of students selecting science as their favorite subject compare to the percent of students selecting history as their favorite subject?

Students' Favorite Subjects

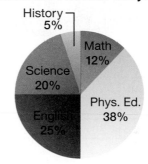

Use the graph to solve each problem.

1. Give the approximate population in 1960 for:
   a. Florida
   b. North Carolina

2. How much greater was the population of Florida than North Carolina in 1980?

3. Which state will have the greater population in 2010?

4. How much greater was the difference in population in 2000 than in 1990?

5. Make a prediction for the population of Florida in 2010.

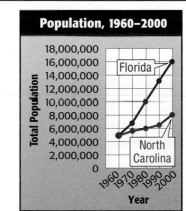

Extra Practice

Lesson 2-5

(Pages 72–75)

Make a stem-and-leaf plot for each set of data.

1. 23, 15, 39, 68, 57, 42, 51, 52, 41, 18, 29
2. 5, 14, 39, 28, 14, 6, 7, 18, 13, 28, 9, 14
3. 189, 182, 196, 184, 197, 183, 196, 194, 184
4. 71, 82, 84, 95, 76, 92, 83, 74, 81, 75, 96
5. 65, 72, 81, 68, 77, 70, 59, 63, 75, 68
6. 95, 115, 142, 127, 106, 138, 93, 118, 121, 139, 129
7. 18, 23, 35, 21, 28, 32, 17, 25, 30, 37, 25, 24
8. 3, 12, 7, 22, 19, 5, 26, 11, 16, 4, 10
9. 48, 62, 53, 41, 45, 57, 63, 50
10. 154, 138, 173, 161, 152, 166, 179, 159, 165, 139

Lesson 2-6

(Pages 76–78)

Find the mean for each set of data.

1. 1, 5, 9, 1, 2, 4, 8, 2
2. 2, 5, 8, 9, 7, 6, 6, 5
3. 1, 2, 4, 2, 2, 1, 2
4. 12, 13, 15, 12, 12, 11, 9
5. 256, 265, 247, 256
6. 957, 562, 462, 848, 721
7. 46, 54, 66, 54, 50, 66
8. 81, 82, 83, 84, 85, 86, 87
9. 7, 18, 12, 5, 22, 15, 10, 15
10. 65, 72, 83, 58, 64, 70, 78, 75, 83
11. 9, 12, 6, 15, 3, 19, 8, 10, 14, 15, 9, 12
12. 37, 29, 20, 42, 33, 30, 21, 26, 45, 37

13.

Average Depths of the Great Lakes	
Lake	Average Depth (ft)
Erie	48
Huron	195
Michigan	279
Ontario	283
Superior	500

14.

Stem	Leaf
3	8
4	5 8
5	0 5
6	0 3 4
7	2 4|5 = 45

Lesson 2-7

(Pages 80–83)

Find the mean, median, mode, and range for each set of data.

1. 16, 12, 20, 15, 12
2. 42, 38, 56, 48, 43, 43
3. 8, 3, 12, 5, 2, 9, 3
4. 85, 75, 93, 82, 73, 78
5. 25, 32, 38, 27, 35, 25, 28
6. 112, 103, 121, 104
7. 57, 63, 53, 67, 71, 67
8. 21, 25, 20, 28, 26
9. 57, 42, 86, 76, 42, 57
10. 215, 176, 194, 223, 202

11.

Students Per Class				
18	21	22	19	22
26	24	18	24	26

12.

Stem	Leaf
1	0 5
2	1 6 9
3	1 2 8
4	1 2 2 4
5	0 5 4|1 = 41

Lesson 2-8

(Pages 86–89)

The graphs shown display the same information.

1. Which graph might be misleading? Explain.

2. If Mr. Roush wishes to show that SAT scores have improved greatly since he became principal in 1999, which graph would he choose?

Graph A

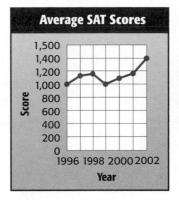

Graph B

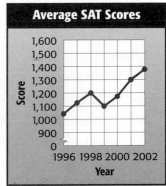

For each set of data, tell which measure (*mean*, *median*, or *mode*) might be misleading in describing the average.

3. 18, 23, 37, 94, 29

4. 76, 83, 69, 75, 80, 69

Lesson 3-1

(Pages 102–105)

Write each decimal in word form.

1. 0.8
2. 5.9
3. 1.34
4. 0.02
5. 9.21
6. 4.3
7. 13.42
8. 0.006
9. 65.083
10. 0.0072

Write each decimal in standard form and expanded form.

11. two hundredths
12. sixteen hundredths
13. four tenths
14. two and twenty-seven hundredths
15. nine and twelve hundredths
16. fifty-six and nine tenths
17. twenty-seven thousandths
18. one hundred ten-thousandths
19. two ten-thousandths
20. twenty and six hundred thousandths

Lesson 3-2

(Pages 108–110)

Use >, <, or = to compare each pair of decimals.

1. 6.0 ● 0.06
2. 1.19 ● 11.9
3. 9.4 ● 9.40
4. 1.04 ● 1.40
5. 1.2 ● 1.02
6. 8.0 ● 8.01
7. 12.13 ● 12.31
8. 0.03 ● 0.003
9. 6.15 ● 6.151
10. 0.112 ● 0.121
11. 0.556 ● 0.519
12. 8.007 ● 8.070
13. 6.005 ● 6.0050
14. 18.104 ● 18.140
15. 0.0345 ● 0.0435

Order each set of decimals from least to greatest.

16. 15.65, 51.65, 51.56, 15.56
17. 2.56, 20.56, 0.09, 25.6
18. 1.235, 1.25, 1.233, 1.23
19. 50.12, 5.012, 5.901, 50.02

Order each set of decimals from greatest to least.

20. 13.66, 13.44, 1.366, 1.633
21. 26.69, 26.09, 26.666, 26.9
22. 1.065, 1.1, 1.165, 1.056
23. 2.014, 2.010, 22.00, 22.14

Extra Practice

Lesson 3-3

(Pages 111–113)

Round each decimal to the indicated place-value position.

1. 5.64; tenths
2. 12.376; hundredths
3. 0.05362; thousandths
4. 6.17; ones
5. 15.298; tenths
6. 0.0026325; ten-thousandths
7. 758.999; hundredths
8. 4.25; ones
9. 32.6583; thousandths
10. 0.025; ones
11. 1.0049; thousandths
12. 9.25; tenths
13. 67.492; hundredths
14. 25.19; tenths
15. 26.96; tenths
16. 4.00098; hundredths
17. 34.065; hundredths
18. 0.0182; thousandths
19. 18.999; tenths
20. 74.00065; ten-thousandths

Lesson 3-4

(Pages 116–119)

Estimate using rounding.

1.	0.245	2.	2.45698	3.	0.5962	4.	17.985
	+ 0.256		− 1.26589		+ 1.2598		− 9.001

5. 0.256 + 0.6589
6. 1.2568 − 0.1569
7. 12.999 + 5.048

Estimate using front-end estimation.

8.	73.41	9.	88.42	10.	106.08
	+ 24.08		− 63.59		+ 90.83

11.	143.24	12.	64.98	13.	96.51
	− 43.65		+ 52.43		− 41.32

Estimate using clustering.

14. 4.5 + 4.95 + 5.2 + 5.49
15. 2.25 + 1.69 + 2.1 + 2.369
16. $12.15 + $11.63 + $12 + $11.89
17. 0.569 + 1.005 + 1.265 + 0.765
18. 9.85 + 10.32 + 10.18 + 9.64
19. 18.32 + 17.92 + 18.04
20. 16.3 + 15.82 + 16.01 + 15.9
21. 0.775 + 0.9 + 1.004 + 1.32

Lesson 3-5

(Pages 121–124)

Add or subtract.

1.	0.46	2.	13.7	3.	17.9	4.	19.2
	+ 0.72		+ 2.6		+ 7.41		+ 7.36

5.	0.51	6.	12.56	7.	0.21	8.	2.3
	+ 0.621		− 10.21		− 0.15		− 1.02

9.	1.025	10.	14.52	11.	2.35	12.	20
	− 0.58		− 12.6		+ 5		− 5.98

13. 15.256 + 0.236
14. 3.7 + 1.5 + 0.2
15. 0.23 + 1.2 + 0.36
16. 0.89 − 0.256
17. 25.6 − 2.3
18. 13.5 − 2.84
19. 1.265 + 1.654
20. 24.56 − 24.32
21. 0.256 − 0.255

Lesson 4-1

(Pages 135–138)

Multiply.

1. $\begin{array}{r} 0.2 \\ \times\ 6 \\ \hline \end{array}$
2. $\begin{array}{r} 0.73 \\ \times\ 5 \\ \hline \end{array}$
3. $\begin{array}{r} 0.65 \\ \times\ 3 \\ \hline \end{array}$
4. $\begin{array}{r} 9.6 \\ \times\ 4 \\ \hline \end{array}$
5. $\begin{array}{r} 12.15 \\ \times\ 6 \\ \hline \end{array}$

6. $\begin{array}{r} 0.91 \\ \times\ 8 \\ \hline \end{array}$
7. $\begin{array}{r} 0.265 \\ \times\ 7 \\ \hline \end{array}$
8. $\begin{array}{r} 2.612 \\ \times\ 4 \\ \hline \end{array}$
9. $\begin{array}{r} 0.013 \\ \times\ 5 \\ \hline \end{array}$
10. $\begin{array}{r} 0.67 \\ \times\ 2 \\ \hline \end{array}$

11. 9×0.111
12. 1.65×7
13. 9.6×3
14. 4×1.201
15. 6×7.5
16. 0.001×6
17. 5×0.0135
18. 9.2×7
19. 14.1235×4

Write each number in standard form.

20. 3×10^4
21. 2.6×10^6
22. 3.81×10^2
23. 8.4×10^3
24. 6.05×10^5
25. 7.32×10^4
26. 4.2×10^3
27. 5.06×10^7
28. 9.25×10^5

Lesson 4-2

(Pages 141–143)

Multiply.

1. 9.6×10.5
2. 3.2×0.1
3. 1.5×9.6
4. 5.42×0.21
5. 7.42×0.2
6. 0.001×0.02
7. 0.6×542
8. 6.7×5.8
9. 3.24×6.7
10. 9.8×4.62
11. 7.32×9.7
12. 0.008×0.007
13. 0.001×56
14. 4.5×0.2
15. 9.6×2.3
16. 8.3×4.9
17. 12.06×5.9
18. 0.04×8.25
19. 5.63×8.1
20. 10.35×9.1
21. 28.2×3.9
22. 2.02×1.25
23. 8.37×89.6
24. 102.1×1.221

Evaluate each expression if $x = 2.6$, $y = 0.38$, and $z = 1.02$.

25. $xy + z$
26. $y \times 9.4$
27. xyz
28. $z \times 12.34 + y$
29. $yz \times 0.8$
30. $xz - yz$

Lesson 4-3

(Pages 144–147)

Divide. Round to the nearest tenth if necessary.

1. $6\overline{)1.2}$
2. $8\overline{)23.2}$
3. $6\overline{)89.4}$
4. $15\overline{)55.5}$
5. $13\overline{)128.7}$
6. $9\overline{)2.583}$
7. $47\overline{)9.4}$
8. $26\overline{)33.8}$
9. $37.8 \div 14$
10. $5.88 \div 4$
11. $3.7 \div 5$
12. $41.4 \div 18$
13. $9.87 \div 3$
14. $8.45 \div 25$
15. $26.5 \div 4$
16. $46.25 \div 8$
17. $19.38 \div 9$
18. $8.5 \div 2$
19. $90.88 \div 14$
20. $23.1 \div 4$
21. $19.5 \div 27$
22. $26.5 \div 19$
23. $46.25 \div 25$
24. $46.25 \div 25$
25. $4.26 \div 9$
26. $18.74 \div 19$
27. $93.6 \div 8$
28. $146.24 \div 15$
29. $483.75 \div 25$
30. $268.4 \div 12$

Lesson 4-4

(Pages 152–155)

Divide.

1. $0.5\overline{)18.45}$
2. $0.08\overline{)5.2}$
3. $2.6\overline{)0.65}$
4. $1.3\overline{)12.831}$
5. $0.87\overline{)5.133}$
6. $2.54\overline{)24.13}$
7. $3.7\overline{)35.89}$
8. $26\overline{)32.5}$
9. $0.4\overline{)5.88}$
10. $3.7 \div 0.5$
11. $6.72 \div 2.4$
12. $9.87 \div 0.3$
13. $8.45 \div 2.5$
14. $90.88 \div 14.2$
15. $33.6 \div 8.4$

Divide. Round each quotient to the nearest hundredth.

16. $0.1185 \div 7.9$
17. $0.384 \div 9.6$
18. $5.364 \div 2.3$
19. $12.68 \div 3.2$
20. $43.1 \div 8.65$
21. $26.08 \div 3.3$
22. $19.436 \div 2.74$
23. $53.6 \div 12.3$
24. $0.3869 \div 5.3$
25. $38.125 \div 9.65$
26. $14.804 \div 4.3$
27. $0.849 \div 1.8$

Lesson 4-5

(Pages 158–160)

Find the perimeter of each figure.

1.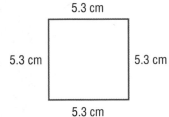
2.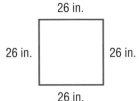
3.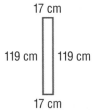

4.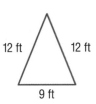
5.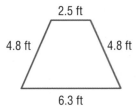
6.

Lesson 4-6

(Pages 161–164)

Find the circumference of each circle shown or described. Round to the nearest tenth.

1.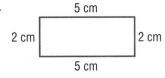
2.
3.
4.

5.
6.
7.
8.

9. $d = 5.6$ m
10. $r = 3.21$ yd
11. $r = 0.5$ in.
12. $d = 4$ m
13. $r = 16$ cm
14. $d = 9.1$ m
15. $r = 0.1$ yd
16. $d = 65.7$ m
17. $r = 1$ cm

Lesson 5-1

(Pages 177–180)

Find the GCF of each set of numbers.

1. 8 and 18
2. 6 and 9
3. 4 and 12
4. 6 and 15
5. 8 and 24
6. 5 and 20
7. 9 and 21
8. 18 and 24
9. 25 and 30
10. 36 and 54
11. 64 and 32
12. 14 and 22
13. 12 and 27
14. 17 and 51
15. 48 and 60
16. 54 and 72
17. 10 and 25
18. 12 and 28
19. 16 and 24
20. 42 and 48
21. 60 and 75
22. 24 and 40
23. 45 and 75
24. 54 and 27
25. 16, 32, 56
26. 14, 28, 42
27. 27, 45, 63
28. 12, 18, 30
29. 20, 35, 50
30. 16, 40, 72

Lesson 5-2

(Pages 182–185)

Replace each ● with a number so that the fractions are equivalent.

1. $\dfrac{12}{16} = \dfrac{●}{4}$
2. $\dfrac{7}{8} = \dfrac{●}{32}$
3. $\dfrac{3}{4} = \dfrac{75}{●}$
4. $\dfrac{8}{16} = \dfrac{●}{2}$
5. $\dfrac{6}{18} = \dfrac{1}{●}$
6. $\dfrac{27}{36} = \dfrac{3}{●}$
7. $\dfrac{1}{4} = \dfrac{16}{●}$
8. $\dfrac{9}{18} = \dfrac{●}{2}$
9. $\dfrac{9}{45} = \dfrac{●}{15}$
10. $\dfrac{2}{3} = \dfrac{18}{●}$
11. $\dfrac{24}{32} = \dfrac{●}{4}$
12. $\dfrac{48}{72} = \dfrac{6}{●}$
13. $\dfrac{5}{9} = \dfrac{●}{63}$
14. $\dfrac{7}{12} = \dfrac{35}{●}$
15. $\dfrac{21}{33} = \dfrac{7}{●}$
16. $\dfrac{25}{60} = \dfrac{●}{12}$

Write each fraction in simplest form. If the fraction is already in simplest form, write *simplest form*.

17. $\dfrac{50}{100}$
18. $\dfrac{24}{40}$
19. $\dfrac{2}{5}$
20. $\dfrac{8}{24}$
21. $\dfrac{20}{27}$
22. $\dfrac{4}{10}$
23. $\dfrac{3}{5}$
24. $\dfrac{14}{19}$
25. $\dfrac{9}{12}$
26. $\dfrac{6}{8}$
27. $\dfrac{15}{18}$
28. $\dfrac{9}{20}$
29. $\dfrac{32}{72}$
30. $\dfrac{36}{60}$
31. $\dfrac{12}{27}$

Lesson 5-3

(Pages 186–189)

Write each mixed number as an improper fraction.

1. $3\dfrac{1}{16}$
2. $2\dfrac{1}{4}$
3. $1\dfrac{3}{8}$
4. $1\dfrac{5}{12}$
5. $7\dfrac{3}{5}$
6. $6\dfrac{5}{8}$
7. $3\dfrac{1}{3}$
8. $1\dfrac{7}{9}$
9. $2\dfrac{3}{16}$
10. $1\dfrac{2}{3}$
11. $2\dfrac{3}{4}$
12. $1\dfrac{5}{9}$
13. $3\dfrac{3}{8}$
14. $4\dfrac{1}{5}$
15. $6\dfrac{2}{3}$

Write each improper fraction as a mixed number.

16. $\dfrac{33}{10}$
17. $\dfrac{103}{25}$
18. $\dfrac{22}{5}$
19. $\dfrac{13}{2}$
20. $\dfrac{29}{6}$
21. $\dfrac{101}{100}$
22. $\dfrac{21}{8}$
23. $\dfrac{19}{6}$
24. $\dfrac{23}{5}$
25. $\dfrac{99}{50}$
26. $\dfrac{39}{8}$
27. $\dfrac{13}{4}$
28. $\dfrac{26}{9}$
29. $\dfrac{18}{5}$
30. $\dfrac{11}{6}$

Find the LCM for each set of numbers.

1. 5 and 15
2. 13 and 39
3. 16 and 24
4. 18 and 20
5. 8 and 12
6. 12 and 15
7. 9 and 27
8. 5 and 6
9. 16 and 3
10. 9 and 6
11. 4 and 14
12. 14 and 49
13. 10 and 45
14. 18 and 12
15. 12 and 30
16. 5 and 7
17. 6 and 15
18. 20 and 8
19. 18 and 24
20. 11 and 33
21. 2, 3, and 6
22. 6, 2, and 22
23. 15, 12, and 8
24. 15, 24, and 30
25. 12, 18, and 3
26. 12, 35, and 10
27. 21, 14, and 6
28. 3, 6, and 9
29. 6, 10, and 15
30. 15, 75, and 25

Replace each ● with $<$, $>$ or $=$ to make a true sentence.

1. $\frac{1}{2} ● \frac{1}{3}$
2. $\frac{2}{3} ● \frac{3}{4}$
3. $\frac{5}{9} ● \frac{4}{5}$
4. $\frac{3}{6} ● \frac{6}{12}$
5. $\frac{12}{23} ● \frac{15}{19}$
6. $\frac{9}{27} ● \frac{13}{39}$
7. $\frac{7}{8} ● \frac{9}{13}$
8. $\frac{5}{9} ● \frac{7}{8}$
9. $\frac{25}{100} ● \frac{3}{8}$
10. $\frac{6}{7} ● \frac{8}{15}$
11. $\frac{5}{9} ● \frac{19}{23}$
12. $\frac{120}{567} ● \frac{1}{2}$
13. $\frac{5}{7} ● \frac{2}{3}$
14. $\frac{9}{36} ● \frac{7}{28}$
15. $\frac{2}{5} ● \frac{2}{6}$
16. $\frac{5}{9} ● \frac{12}{13}$
17. $\frac{3}{5} ● \frac{5}{8}$
18. $\frac{8}{9} ● \frac{3}{4}$
19. $\frac{11}{15} ● \frac{20}{30}$
20. $\frac{15}{24} ● \frac{6}{8}$
21. $\frac{7}{12} ● \frac{21}{36}$
22. $\frac{5}{7} ● \frac{4}{14}$
23. $\frac{3}{8} ● \frac{5}{12}$
24. $\frac{17}{25} ● \frac{68}{100}$

Order the fractions from least to greatest.

25. $\frac{1}{3}, \frac{1}{2}, \frac{2}{7}, \frac{2}{5}$
26. $\frac{3}{8}, \frac{3}{4}, \frac{1}{2}, \frac{5}{6}$
27. $\frac{5}{12}, \frac{3}{5}, \frac{3}{4}, \frac{1}{2}$
28. $\frac{7}{8}, \frac{5}{6}, \frac{8}{9}, \frac{3}{4}$
29. $\frac{7}{10}, \frac{5}{8}, \frac{3}{4}, \frac{4}{5}$
30. $\frac{5}{9}, \frac{7}{12}, \frac{3}{4}, \frac{13}{18}$

Write each decimal as a fraction or mixed number in simplest form.

1. 0.5
2. 0.8
3. 0.32
4. 0.875
5. 0.54
6. 0.38
7. 0.744
8. 0.101
9. 0.303
10. 0.486
11. 0.626
12. 0.448
13. 0.074
14. 0.008
15. 9.36
16. 10.18
17. 0.06
18. 0.75
19. 0.48
20. 0.9
21. 0.005
22. 0.4
23. 1.875
24. 5.08
25. 0.46
26. 0.128
27. 0.08
28. 6.96
29. 3.625
30. 0.006
31. 12.05
32. 24.125

Lesson 5-7

(Pages 206–209)

Write each fraction or mixed number as a decimal.

1. $\frac{3}{16}$ 2. $\frac{1}{8}$ 3. $\frac{7}{12}$ 4. $\frac{14}{25}$

5. $\frac{7}{10}$ 6. $\frac{5}{8}$ 7. $\frac{11}{15}$ 8. $\frac{8}{9}$

9. $\frac{15}{16}$ 10. $\frac{1}{12}$ 11. $\frac{7}{20}$ 12. $\frac{5}{11}$

13. $\frac{9}{10}$ 14. $\frac{11}{25}$ 15. $\frac{9}{20}$ 16. $4\frac{4}{15}$

17. $1\frac{3}{8}$ 18. $\frac{7}{16}$ 19. $2\frac{5}{12}$ 20. $6\frac{3}{5}$

21. $\frac{3}{4}$ 22. $3\frac{4}{9}$ 23. $2\frac{1}{10}$ 24. $\frac{17}{25}$

25. $5\frac{13}{20}$ 26. $\frac{11}{18}$ 27. $4\frac{11}{12}$ 28. $\frac{7}{9}$

29. $\frac{1}{3}$ 30. $\frac{37}{50}$ 31. $9\frac{5}{6}$ 32. $15\frac{3}{11}$

Lesson 6-1

(Pages 219–222)

Round each number to the nearest half.

1. $\frac{11}{12}$ 2. $\frac{5}{8}$ 3. $\frac{2}{5}$ 4. $\frac{1}{10}$ 5. $\frac{1}{6}$ 6. $\frac{2}{3}$

7. $\frac{9}{10}$ 8. $\frac{1}{8}$ 9. $\frac{4}{9}$ 10. $1\frac{1}{8}$ 11. $\frac{7}{9}$ 12. $2\frac{4}{5}$

13. $\frac{7}{9}$ 14. $7\frac{1}{10}$ 15. $10\frac{2}{3}$ 16. $\frac{1}{3}$ 17. $\frac{7}{15}$ 18. $\frac{5}{7}$

19. $\frac{7}{8}$ 20. $5\frac{1}{9}$ 21. $2\frac{5}{8}$ 22. $3\frac{1}{3}$ 23. $6\frac{7}{9}$ 24. $4\frac{2}{12}$

25. $1\frac{2}{7}$ 26. $9\frac{4}{5}$ 27. $3\frac{9}{11}$ 28. $2\frac{3}{7}$ 29. $7\frac{3}{8}$ 30. $4\frac{17}{20}$

31. $3\frac{3}{8}$ 32. $8\frac{2}{9}$ 33. $6\frac{1}{4}$ 34. $4\frac{3}{10}$ 35. $5\frac{5}{7}$ 36. $2\frac{3}{11}$

Lesson 6-2

(Pages 223–225)

Estimate.

1. $14\frac{1}{10} - 6\frac{4}{5}$ 2. $8\frac{1}{3} + 2\frac{1}{6}$ 3. $4\frac{7}{8} + 7\frac{3}{4}$ 4. $11\frac{11}{12} - 5\frac{1}{4}$

5. $3\frac{2}{5} - 1\frac{1}{4}$ 6. $4\frac{2}{5} + \frac{5}{6}$ 7. $4\frac{7}{12} - 1\frac{3}{4}$ 8. $4\frac{2}{3} + 10\frac{3}{8}$

9. $7\frac{7}{15} - 3\frac{1}{12}$ 10. $2\frac{1}{16} + 1\frac{1}{3}$ 11. $18\frac{1}{4} - 12\frac{3}{5}$ 12. $12\frac{5}{9} + 8\frac{5}{8}$

13. $8\frac{2}{3} - 5\frac{1}{2}$ 14. $3\frac{1}{8} - 2\frac{1}{5}$ 15. $9\frac{2}{7} - \frac{1}{3}$ 16. $11\frac{7}{8} - \frac{5}{6}$

17. $2\frac{2}{5} + 2\frac{1}{4}$ 18. $8\frac{1}{2} - 7\frac{4}{5}$ 19. $8\frac{3}{4} + 4\frac{2}{3}$ 20. $1\frac{1}{8} + 7\frac{1}{10}$

21. $\frac{3}{7} + \frac{8}{9}$ 22. $5\frac{11}{12} - 3\frac{1}{3}$ 23. $2\frac{1}{7} + 4\frac{3}{5}$ 24. $\frac{5}{6} - \frac{1}{4}$

25. $8\frac{2}{9} + 4\frac{7}{8}$ 26. $\frac{17}{20} + 2\frac{5}{9}$ 27. $3\frac{9}{10} - 1\frac{1}{6}$ 28. $6\frac{2}{3} + \frac{4}{9}$

Add or subtract. Write in simplest form.

1. $\frac{2}{5} + \frac{2}{5}$

2. $\frac{5}{8} + \frac{3}{8}$

3. $\frac{9}{11} - \frac{3}{11}$

4. $\frac{3}{14} + \frac{5}{14}$

5. $\frac{7}{8} - \frac{3}{8}$

6. $\frac{3}{4} - \frac{1}{4}$

7. $\frac{15}{27} - \frac{7}{27}$

8. $\frac{1}{36} + \frac{5}{36}$

9. $\frac{2}{9} - \frac{1}{9}$

10. $\frac{7}{8} - \frac{5}{8}$

11. $\frac{9}{16} - \frac{5}{16}$

12. $\frac{6}{8} + \frac{4}{8}$

13. $\frac{1}{2} + \frac{1}{2}$

14. $\frac{1}{3} - \frac{1}{3}$

15. $\frac{8}{9} + \frac{7}{9}$

16. $\frac{5}{6} - \frac{3}{6}$

17. $\frac{3}{9} + \frac{8}{9}$

18. $\frac{8}{40} + \frac{12}{40}$

19. $\frac{56}{90} - \frac{26}{90}$

20. $\frac{2}{9} + \frac{8}{9}$

21. $\frac{7}{15} - \frac{4}{15}$

22. $\frac{3}{8} + \frac{7}{8}$

23. $\frac{11}{20} - \frac{5}{20}$

24. $\frac{7}{9} - \frac{2}{9}$

25. $\frac{10}{12} + \frac{7}{12}$

26. $\frac{5}{6} + \frac{1}{6}$

27. $\frac{5}{8} + \frac{7}{8}$

28. $\frac{9}{16} + \frac{5}{16}$

Add or subtract. Write in simplest form.

1. $\frac{1}{3} + \frac{1}{2}$

2. $\frac{2}{9} + \frac{1}{3}$

3. $\frac{1}{2} + \frac{3}{4}$

4. $\frac{1}{4} + \frac{3}{12}$

5. $\frac{5}{9} - \frac{1}{3}$

6. $\frac{5}{8} - \frac{2}{5}$

7. $\frac{3}{4} - \frac{1}{2}$

8. $\frac{7}{8} - \frac{3}{16}$

9. $\frac{9}{16} + \frac{13}{24}$

10. $\frac{8}{15} + \frac{2}{3}$

11. $\frac{5}{14} + \frac{11}{28}$

12. $\frac{11}{12} + \frac{7}{8}$

13. $\frac{2}{3} - \frac{1}{6}$

14. $\frac{9}{16} - \frac{1}{2}$

15. $\frac{5}{8} - \frac{11}{20}$

16. $\frac{14}{15} - \frac{2}{9}$

17. $\frac{9}{20} + \frac{2}{15}$

18. $\frac{5}{6} + \frac{4}{5}$

19. $\frac{23}{25} - \frac{27}{50}$

20. $\frac{19}{25} - \frac{1}{2}$

21. $\frac{1}{5} + \frac{2}{3}$

22. $\frac{4}{9} - \frac{1}{6}$

23. $\frac{1}{3} + \frac{4}{5}$

24. $\frac{7}{8} - \frac{5}{12}$

25. $\frac{1}{12} + \frac{3}{10}$

26. $\frac{3}{15} + \frac{1}{25}$

27. $\frac{5}{12} - \frac{1}{6}$

28. $\frac{9}{24} - \frac{3}{8}$

Add or subtract. Write in simplest form.

1. $5\frac{1}{2} + 3\frac{1}{4}$

2. $2\frac{2}{3} + 4\frac{1}{9}$

3. $7\frac{4}{5} + 9\frac{3}{10}$

4. $9\frac{4}{7} - 3\frac{5}{14}$

5. $13\frac{1}{5} - 10$

6. $3\frac{3}{4} + 5\frac{5}{8}$

7. $3\frac{2}{5} + 7\frac{6}{15}$

8. $10\frac{2}{3} + 5\frac{6}{7}$

9. $15\frac{6}{9} - 13\frac{5}{12}$

10. $13\frac{7}{12} - 9\frac{1}{4}$

11. $5\frac{2}{3} - 3\frac{1}{2}$

12. $17\frac{2}{9} + 12\frac{1}{3}$

13. $6\frac{5}{12} + 12\frac{5}{8}$

14. $8\frac{3}{5} - 2\frac{1}{5}$

15. $23\frac{2}{3} - 4\frac{1}{2}$

16. $7\frac{1}{8} + 2\frac{5}{8}$

17. $12\frac{11}{15} - 10\frac{2}{15}$

18. $12\frac{1}{3} + 5\frac{1}{6}$

19. $2\frac{1}{4} + 3\frac{1}{8}$

20. $7\frac{9}{14} - 4\frac{3}{7}$

21. $4\frac{3}{5} - 2\frac{1}{15}$

22. $6\frac{1}{8} + 7\frac{3}{4}$

23. $5\frac{5}{6} - 2\frac{1}{2}$

24. $1\frac{2}{3} + \frac{5}{18}$

25. $2\frac{5}{9} + 1\frac{7}{12}$

26. $8\frac{7}{8} - 5\frac{1}{4}$

27. $3\frac{1}{2} - 2\frac{1}{3}$

28. $7\frac{1}{2} + 8\frac{3}{4}$

Lesson 6-6

(Pages 244–247)

Subtract. Write in simplest form.

1. $11\frac{2}{3} - 8\frac{11}{12}$
2. $3\frac{4}{7} - 1\frac{2}{3}$
3. $7\frac{1}{8} - 4\frac{1}{3}$
4. $18\frac{1}{9} - 12\frac{2}{5}$

5. $12\frac{3}{10} - 8\frac{3}{4}$
6. $43 - 5\frac{1}{5}$
7. $8\frac{1}{5} - 4\frac{1}{4}$
8. $14\frac{1}{6} - 3\frac{2}{3}$

9. $25\frac{4}{7} - 21$
10. $17\frac{3}{9} - 4\frac{3}{5}$
11. $18\frac{1}{9} - 1\frac{3}{7}$
12. $16\frac{1}{4} - 7\frac{1}{5}$

13. $18\frac{1}{5} - 6\frac{1}{4}$
14. $4 - 1\frac{2}{3}$
15. $26 - 4\frac{1}{9}$
16. $3\frac{1}{2} - 1\frac{3}{4}$

17. $4\frac{3}{8} - 2\frac{5}{6}$
18. $18\frac{1}{6} - 10\frac{3}{4}$
19. $12\frac{4}{9} - 7\frac{5}{6}$
20. $4\frac{1}{15} - 2\frac{3}{5}$

21. $7\frac{1}{14} - 4\frac{3}{7}$
22. $12\frac{3}{20} - 7\frac{7}{15}$
23. $8\frac{5}{12} - 5\frac{9}{10}$
24. $8\frac{1}{5} - 4\frac{2}{3}$

25. $7\frac{3}{16} - 2\frac{5}{6}$
26. $9\frac{2}{7} - 6\frac{5}{14}$
27. $5\frac{1}{8} - 4\frac{1}{4}$
28. $4\frac{3}{8} - 1\frac{5}{6}$

Lesson 7-1

(Pages 256–258)

Estimate each product.

1. $\frac{5}{6} \times 8$
2. $\frac{1}{3} \times 46$
3. $\frac{4}{5} \times 21$
4. $\frac{1}{9} \times 35$

5. $\frac{5}{9} \times 20$
6. $\frac{1}{8} \times 30$
7. $\frac{2}{3} \times \frac{4}{5}$
8. $\frac{1}{6} \times \frac{2}{5}$

9. $\frac{4}{9} \times \frac{3}{7}$
10. $\frac{5}{12} \times \frac{6}{11}$
11. $\frac{3}{8} \times \frac{8}{9}$
12. $\frac{3}{5} \times \frac{5}{12}$

13. $\frac{2}{5} \times \frac{5}{8}$
14. $\frac{4}{5} \times \frac{11}{12}$
15. $\frac{5}{7} \times \frac{7}{8}$
16. $\frac{1}{20} \times \frac{8}{9}$

17. $\frac{9}{11} \times \frac{14}{15}$
18. $\frac{2}{5} \times \frac{18}{19}$
19. $5\frac{3}{7} \times \frac{4}{5}$
20. $2\frac{5}{9} \times \frac{1}{8}$

21. $3\frac{9}{10} \times \frac{15}{16}$
22. $\frac{11}{12} \times 2\frac{1}{3}$
23. $3\frac{14}{15} \times \frac{3}{8}$
24. $\frac{1}{10} \times 3\frac{1}{2}$

25. $9\frac{13}{15} \times \frac{1}{2}$
26. $6\frac{7}{8} \times 2\frac{1}{5}$
27. $7\frac{1}{4} \times 4\frac{3}{4}$
28. $6\frac{2}{3} \times 5\frac{4}{5}$

Lesson 7-2

(Pages 261–264)

Multiply. Write in simplest form.

1. $\frac{1}{8} \times \frac{1}{9}$
2. $\frac{4}{7} \times 6$
3. $\frac{7}{10} \times 5$
4. $\frac{3}{8} \times 6$

5. $4 \times \frac{5}{9}$
6. $\frac{9}{10} \times \frac{3}{4}$
7. $\frac{8}{9} \times \frac{2}{3}$
8. $\frac{6}{7} \times \frac{4}{5}$

9. $\frac{7}{11} \times \frac{12}{15}$
10. $\frac{8}{13} \times \frac{2}{11}$
11. $\frac{4}{7} \times \frac{2}{9}$
12. $\frac{3}{7} \times \frac{5}{8}$

13. $\frac{5}{6} \times \frac{15}{16}$
14. $\frac{6}{14} \times \frac{12}{18}$
15. $\frac{2}{3} \times \frac{3}{13}$
16. $\frac{4}{9} \times \frac{1}{6}$

17. $\frac{3}{4} \times \frac{5}{6}$
18. $\frac{8}{11} \times \frac{11}{12}$
19. $\frac{5}{6} \times \frac{3}{5}$
20. $\frac{6}{7} \times \frac{7}{21}$

21. $\frac{8}{9} \times \frac{9}{10}$
22. $\frac{7}{9} \times \frac{5}{7}$
23. $\frac{4}{9} \times \frac{24}{25}$
24. $\frac{1}{9} \times \frac{6}{13}$

25. $\frac{5}{9} \times \frac{3}{10}$
26. $\frac{2}{3} \times \frac{7}{8}$
27. $\frac{7}{12} \times \frac{4}{9}$
28. $\frac{11}{15} \times \frac{3}{10}$

Multiply. Write in simplest form.

1. $\frac{4}{5} \times 2\frac{3}{4}$

2. $2\frac{3}{10} \times \frac{3}{5}$

3. $8\frac{5}{6} \times \frac{2}{5}$

4. $\frac{3}{4} \times 9\frac{5}{7}$

5. $6\frac{2}{3} \times 7\frac{3}{5}$

6. $7\frac{1}{5} \times 2\frac{4}{7}$

7. $8\frac{3}{4} \times 2\frac{2}{5}$

8. $4\frac{1}{3} \times 2\frac{1}{7}$

9. $4\frac{3}{5} \times 2\frac{1}{2}$

10. $5\frac{5}{6} \times 4\frac{2}{7}$

11. $6\frac{8}{9} \times 3\frac{5}{6}$

12. $2\frac{1}{9} \times 1\frac{1}{2}$

13. $4\frac{7}{15} \times 3\frac{3}{4}$

14. $5\frac{7}{9} \times 6\frac{3}{8}$

15. $1\frac{1}{4} \times 3\frac{2}{3}$

16. $2\frac{3}{5} \times 1\frac{4}{7}$

17. $4\frac{1}{5} \times 12\frac{2}{9}$

18. $3\frac{5}{8} \times 4\frac{1}{2}$

19. $6\frac{1}{2} \times 2\frac{1}{3}$

20. $3\frac{4}{5} \times 2\frac{3}{8}$

21. $5\frac{1}{4} \times 10\frac{3}{7}$

22. $1\frac{5}{9} \times 6\frac{1}{4}$

23. $2\frac{1}{6} \times 1\frac{3}{4}$

24. $3\frac{5}{7} \times 1\frac{1}{2}$

25. $2\frac{3}{5} \times 4\frac{5}{8}$

26. $6\frac{1}{8} \times 5\frac{1}{7}$

27. $2\frac{2}{3} \times 2\frac{1}{4}$

28. $2\frac{1}{2} \times 3\frac{1}{3}$

Find the reciprocal of each number.

1. $\frac{12}{13}$

2. $\frac{7}{11}$

3. 5

4. $\frac{1}{4}$

5. $\frac{7}{9}$

6. $\frac{9}{2}$

7. $\frac{1}{5}$

Divide. Write in simplest form.

8. $\frac{2}{3} \div \frac{1}{2}$

9. $\frac{3}{5} \div \frac{2}{5}$

10. $\frac{7}{10} \div \frac{3}{8}$

11. $\frac{5}{9} \div \frac{2}{3}$

12. $4 \div \frac{2}{3}$

13. $8 \div \frac{4}{5}$

14. $9 \div \frac{5}{9}$

15. $\frac{2}{7} \div 7$

16. $\frac{1}{14} \div 7$

17. $\frac{2}{13} \div \frac{5}{26}$

18. $\frac{4}{7} \div \frac{6}{7}$

19. $\frac{7}{8} \div \frac{1}{3}$

20. $15 \div \frac{3}{5}$

21. $\frac{9}{14} \div \frac{3}{4}$

22. $\frac{8}{9} \div \frac{5}{6}$

23. $\frac{4}{9} \div 36$

24. $\frac{15}{16} \div \frac{5}{8}$

25. $\frac{3}{5} \div \frac{7}{10}$

26. $\frac{5}{9} \div \frac{3}{8}$

27. $\frac{5}{6} \div \frac{3}{8}$

Divide. Write in simplest form.

1. $\frac{3}{5} \div 1\frac{2}{3}$

2. $2\frac{1}{2} \div 1\frac{1}{4}$

3. $7 \div 4\frac{9}{10}$

4. $1\frac{3}{7} \div 10$

5. $3\frac{3}{5} \div \frac{4}{5}$

6. $8\frac{2}{5} \div 4\frac{1}{2}$

7. $6\frac{1}{3} \div 2\frac{1}{2}$

8. $5\frac{1}{4} \div 2\frac{1}{3}$

9. $4\frac{1}{8} \div 3\frac{2}{3}$

10. $2\frac{5}{8} \div \frac{1}{2}$

11. $1\frac{5}{6} \div 3\frac{2}{3}$

12. $21 \div 5\frac{1}{4}$

13. $12 \div 3\frac{3}{5}$

14. $18 \div 2\frac{1}{4}$

15. $1\frac{7}{9} \div 2\frac{2}{3}$

16. $2\frac{1}{15} \div 3\frac{1}{3}$

17. $1\frac{1}{8} \div 2\frac{2}{3}$

18. $5\frac{1}{3} \div 2\frac{1}{2}$

19. $1\frac{1}{4} \div 1\frac{7}{8}$

20. $2\frac{3}{5} \div 1\frac{7}{10}$

21. $6\frac{3}{4} \div 3\frac{1}{2}$

22. $4\frac{1}{2} \div \frac{3}{8}$

23. $3\frac{1}{2} \div 1\frac{7}{9}$

24. $12\frac{1}{2} \div 5\frac{5}{6}$

25. $8\frac{1}{4} \div 2\frac{3}{4}$

26. $2\frac{3}{8} \div 5\frac{3}{7}$

27. $4\frac{5}{9} \div 5\frac{1}{3}$

28. $3\frac{1}{2} \div 5\frac{1}{4}$

Lesson 7-6

(Pages 282–284)

Describe each pattern. Then find the next two numbers in the sequence.

1. 14, 21, 28, 35, …
2. 36, 42, 48, 54, …
3. 3, 9, 27, 81, …
4. 2, 6, 10, 14, …
5. 1,600, 800, 400, 200, …
6. 192, 96, 48, 24, …
7. 15, $14\frac{1}{3}$, $13\frac{2}{3}$, 13, …
8. 11, $11\frac{1}{2}$, 12, $12\frac{1}{2}$, …
9. $\frac{1}{5}$, 2, 20, 200, …
10. 36, 6, 1, $\frac{1}{6}$, …

Find the missing number in each sequence.

11. 5, 10, $\underline{\ ?\ }$, 40, …
12. $\underline{\ ?\ }$, 193, 293, 393, …
13. 8, 32, 128, $\underline{\ ?\ }$, …
14. 11, $\underline{\ ?\ }$, 19, 23, …
15. 9, $8\frac{3}{4}$, $\underline{\ ?\ }$, $8\frac{1}{4}$, …
16. 64, $\underline{\ ?\ }$, 16, 8, …
17. $\underline{\ ?\ }$, 19, 26, 33, …
18. $\frac{1}{2}$, $1\frac{1}{2}$, $4\frac{1}{2}$, $\underline{\ ?\ }$, …
19. $\frac{1}{81}$, $\underline{\ ?\ }$, 1, 9, …
20. 12, $\underline{\ ?\ }$, 432, 2,592, …

Lesson 8-1

(Pages 294–298)

Write an integer to describe each situation.

1. a loss of 15 dollars
2. 9 degrees below zero

Draw a number line from −10 to 10. Then graph each integer on the number line.

3. −3
4. 3
5. −1
6. −8
7. 9
8. 10

Replace each ● with <, >, or = to make a true sentence.

9. −5 ● −55
10. 4 ● −66
11. −777 ● −77
12. −75 ● −75
13. −898 ● −99
14. 0 ● 44
15. 56 ● −1
16. −82 ● −9
17. −6 ● −7
18. 90 ● 101
19. 4 ● −2,000
20. −3 ● 0

Order each set of integers from least to greatest.

21. 0, 3, −21, 9, −89, 8, −65, −56
22. 70, −9, 67, −78, 0, 45, −36, −19

Write the opposite of each integer.

23. 7
24. −3
25. 11
26. −9
27. −13
28. 101

Lesson 8-2

(Pages 300–303)

Add. Use counters or a number line if needed.

1. −4 + (−7)
2. −1 + 0
3. 7 + (−13)
4. −20 + 2
5. 4 + (−6)
6. −12 + 9
7. −12 + (−10)
8. 5 + (−15)
9. 17 + 9
10. 18 + (−18)
11. −4 + (−4)
12. 0 + (−9)
13. −12 + (−9)
14. −8 + 7
15. 3 + (−6)
16. −9 + 16
17. −5 + (−3)
18. −5 + 5
19. −3 + (−3)
20. −11 + 6
21. −10 + 6
22. −5 + (−9)
23. 18 + (−20)
24. −4 + (−8)
25. 2 + (−4)
26. −3 + (−11)
27. −17 + 9

Lesson 8-3

Subtract. Use counters if necessary.

1. $7 - (-4)$	2. $-4 - (-9)$	3. $13 - (-3)$	4. $2 - (-5)$
5. $-9 - 5$	6. $-11 - (-18)$	7. $-4 - (-7)$	8. $-6 - (-6)$
9. $-6 - 6$	10. $17 - 9$	11. $-12 - (-9)$	12. $0 - (-4)$
13. $-7 - 0$	14. $-12 - (-10)$	15. $-2 - (-1)$	16. $3 - (-5)$
17. $5 - (-1)$	18. $-5 - (-6)$	19. $9 - (-1)$	20. $1 - 9$
21. $-5 - 1$	22. $-1 - 4$	23. $0 - (-7)$	24. $8 - 13$
25. $-4 - (-6)$	26. $9 - 9$	27. $-7 - (-7)$	28. $7 - 5$
29. $8 - (-5)$	30. $5 - 8$	31. $1 - 6$	32. $-8 - (-8)$

33. Find the value of $s - t$ if $s = -4$ and $t = 3$.

34. Find the value of $a - b$ if $a = 6$ and $b = -8$.

Lesson 8-4

Multiply.

1. $3 \times (-5)$	2. -5×1	3. $-8 \times (-4)$	4. $6 \times (-3)$
5. -3×2	6. $-1 \times (-4)$	7. $8 \times (-2)$	8. $-5 \times (-7)$
9. $3 \times (-9)$	10. -9×4	11. $-4 \times (-5)$	12. $5 \times (-2)$
13. $-8(3)$	14. $-9(-1)$	15. $7(-3)$	16. $2(3)$
17. $-6(0)$	18. $-5(-1)$	19. $5(-5)$	20. $-2(-3)$
21. $8(-4)$	22. $-2(4)$	23. $-4(-4)$	24. $2(9)$
25. $-2(-12)$	26. $7 \times (-4)$	27. $-5 \times (-9)$	28. -2×11
29. $4(-2)$	30. $4(-4)$	31. $-3(-11)$	32. $-3(3)$

33. Find the product of 2 and -3.

34. Evaluate qr if $q = -3$ and $r = -3$.

Lesson 8-5

Divide.

1. $12 \div (-6)$	2. $-7 \div (-1)$	3. $-4 \div 4$	4. $6 \div (-6)$
5. $0 \div (-4)$	6. $45 \div (-9)$	7. $15 \div (-5)$	8. $-6 \div 2$
9. $-28 \div (-7)$	10. $20 \div (-2)$	11. $-40 \div (-8)$	12. $12 \div (-4)$
13. $-18 \div 6$	14. $9 \div (-1)$	15. $-30 \div 6$	16. $-54 \div (-9)$
17. $28 \div (-7)$	18. $-24 \div 8$	19. $24 \div (-4)$	20. $-14 \div 7$
21. $9 \div 3$	22. $-18 \div (-6)$	23. $-9 \div (-1)$	24. $18 \div (-9)$
25. $-25 \div (-5)$	26. $15 \div (-3)$	27. $-36 \div 9$	28. $-4 \div 2$
29. $-40 \div 8$	30. $-32 \div 4$	31. $-27 \div (-9)$	32. $-8 \div 8$

33. Find the value of $b \div c$ if $b = -10$ and $c = 5$.

34. Find the value of $w \div x$ if $w = 21$ and $x = -7$.

Lesson 8-6

(Pages 320–323)

Write the ordered pair that names each point. Then identify the quadrant where each point is located.

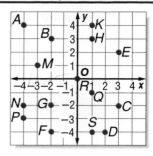

1. M
2. A
3. D
4. E
5. P
6. Q
7. B
8. C
9. F
10. G
11. N
12. R
13. K
14. H
15. S

Graph and label each point on a coordinate plane.

16. $S(4, -1)$
17. $T(-3, -2)$
18. $W(2, 1)$
19. $Y(-5, 3)$
20. $Z(-1, -3)$
21. $U(3, -3)$
22. $V(1, 2)$
23. $X(-1, 4)$

Graph on a coordinate plane.

24. $A(2, -3)$
25. $B(-4, 1)$
26. $C(-1, -5)$
27. $D(3, -2)$
28. $E(1, 3)$
29. $F(-2, -2)$
30. $G(4, -3)$
31. $H(-3, 4)$

Lesson 9-1

(Pages 333–336)

Find each product mentally. Use the Distributive Property.

1. 5×18
2. 9×27
3. 8×83
4. 7×21
5. 3×47
6. 2×10.6
7. 6×3.4
8. 5.6×3
9. 27×8

Rewrite each expression using the Distributive Property. Then evaluate.

10. $4(12 + 9)$
11. $7(8 + 3)$
12. $(6 + 13)2$
13. $(5 + 4)15$
14. $6(11) + 6(7)$
15. $3(12) + 8(12)$

Identify the property shown by each equation.

16. $17 + 12 = 12 + 17$
17. $6 \times (4 \times 9) = (6 \times 4) \times 9$
18. $(13 + 9) + 4 = 13 + (9 + 4)$
19. $8 \times 3 = 3 \times 8$
20. $4 + 6 + 12 = 6 + 4 + 12$
21. $15(8)(2) = 15(8 \times 2)$

Find each sum or product mentally.

22. $25 + 7 + 35$
23. $15 \times 9 \times 2$
24. $12 + 6 + 8$
25. $15 \times 4 \times 20$
26. $38 + 9 + 12$
27. $83 + 29 + 17$
28. $2 \times 33 \times 5$
29. $25 \times 9 \times 4$
30. $33 + 11 + 67$

Lesson 9-2

(Pages 339–342)

Solve each equation. Use models if necessary. Check your solution.

1. $x + 4 = 14$
2. $b + (-10) = 0$
3. $-2 + w = -5$
4. $k + (-3) = -5$
5. $6 = -4 + h$
6. $-7 + d = -3$
7. $9 = m + 11$
8. $f + (-9) = -19$
9. $p + 66 = 22$
10. $-34 + t = 41$
11. $-24 = e + 56$
12. $-29 + a = -54$
13. $17 + m = -33$
14. $b + (-44) = -34$
15. $w + (-39) = 55$
16. $6 + a = 13$
17. $-5 = m + 3$
18. $w + -9 = 12$
19. $8 = p + 7$
20. $-4 + c = -9$
21. $y + 11 = 8$
22. $16 = t + 5$
23. $-3 + x = 1$
24. $14 + c = 6$
25. $-9 = -12 + w$
26. $q + 6 = 4$
27. $5 + z = 13$
28. $9 = h + -5$

Solve each equation. Use models if necessary. Check your solution.

1. $y - 7 = 2$
2. $a - 10 = -22$
3. $g - 1 = 9$
4. $c - 8 = 5$
5. $z - 2 = 7$
6. $n - 1 = -87$
7. $j - 15 = -22$
8. $x - 12 = 45$
9. $y - 65 = -79$
10. $q - 16 = -31$
11. $q - 6 = 12$
12. $j - 18 = -34$
13. $k - 2 = -8$
14. $r - 76 = 41$
15. $n - 63 = -81$
16. $b - 7 = 4$
17. $-5 = g - 3$
18. $y - 2 = -6$
19. $8 = m - 3$
20. $x - 5 = 2$
21. $-6 = p - 8$
22. $h - 9 = -6$
23. $12 = w - 8$
24. $a - 6 = -1$
25. $-11 = t - 5$
26. $c - 4 = 8$
27. $18 = q - 7$
28. $r - 2 = -5$
29. $1 = z - 9$
30. $g - 10 = -4$
31. $-6 = d - 4$
32. $s - 4 = 10$

33. Find the value of c if $c - 5 = -2$.

34. Find the value of t if $4 = t - 7$.

Solve each equation. Use models if necessary.

1. $5x = 30$
2. $2w = 18$
3. $2a = 7$
4. $2d = -28$
5. $-3c = 6$
6. $11n = 77$
7. $3z = 15$
8. $9y = -63$
9. $6m = -54$
10. $5f = -75$
11. $20p = 5$
12. $4x = 16$
13. $4t = -24$
14. $7b = 21$
15. $19h = 0$
16. $22d = -66$
17. $3m = -78$
18. $8x = -2$
19. $9c = -72$
20. $5p = 35$
21. $-5k = 20$
22. $33y = 99$
23. $6z = -9$
24. $6m = -42$
25. $18 = 9x$
26. $-5p = 4$
27. $-32 = 4r$
28. $3w = 27$
29. $-12 = 16a$
30. $-4t = 6$
31. $16 = -5b$
32. $-2c = -13$

33. Solve the equation $3d = 21$.

34. What is the solution of the equation $-4x = 65$?

Solve each equation. Use models if necessary.

1. $3x + 7 = 13$
2. $2h - 5 = 7$
3. $-10 = 5x + 5$
4. $6r + 2 = 2$
5. $-2 - 3y = -11$
6. $-4y + 16 = 64$
7. $4a - 3 = 17$
8. $-8 = 3x - 5$
9. $-6 + 2m = 8$
10. $5p + 3 = 23$
11. $2 = -4x - 2$
12. $9 + 6h = 21$
13. $7y - 3 = 4$
14. $4 = 8w + 20$
15. $5 + 2g = 17$
16. $3b + 4 = 13$
17. $-5 = 4t - 13$
18. $7a - 6 = 15$
19. $8 = -3p + 5$
20. $11 + 5m = 6$
21. $-4y + 9 = 17$
22. $6 = 2w + 2$
23. $14 + 7x = 35$
24. $3c - 5 = -11$
25. $-5 = -2g + 3$
26. $7 + 3a = 25$
27. $9 = -4w - 3$
28. $-4 + 5b = 11$
29. $2x + 16 = 26$
30. $8m + 7 = -9$
31. $4 = 2t - 12$
32. $-3 + 2c = -13$

33. Three less than five times a number is seven. What is the number?

34. Twenty-four is four more than four times a number. What is the number?

Lesson 9-6

(Pages 362–365)

Copy and complete each function table.

1.

Input (n)	Output ($n - 4$)
5	■
2	■
−1	■

2.

Input (n)	Output ($3n$)
1	■
0	■
−2	■

3.

Input (n)	Output ($n + 7$)
−4	■
1	■
5	■

4.

Input (n)	Output ($-3n$)
−2	■
0	■
3	■

Find the rule for each function table.

5.

n	■
−1	4
0	5
3	8

6.

n	■
−6	−3
0	0
8	4

7.

n	■
−4	20
2	−10
6	−30

Lesson 9-7

(Pages 366–369)

Make a function table for each rule with the given input values.
Then graph the function.

1. $y = x + 1$; 2, 0, −3

2. $y = 2x$; 2, 0, −3

3. $y = x - 3$; 4, 0, −1

4. $y = \frac{x}{5}$; 10, 0, −5

5. $y = -3x$; 2, −1, −2

6. $y = 2x - 3$; 2, 0, −1

Make a function table for each graph. Then determine the function rule.

7.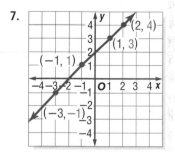

8.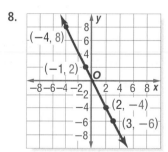

Lesson 10-1

(Pages 380–383)

Write each ratio as a fraction in simplest form.

1. 10 girls in a class of 25 students

2. 7 striped ties out of 21 ties

3. 12 golden retrievers out of 20 dogs

4. 8 red marbles in a jar of 32 marbles

5. 6 roses in a bouquet of 21 flowers

6. 21 convertibles out of 75 cars

Write each ratio as a unit rate.

7. $2 for 5 cans of tomato soup

8. $200 for 40 hours of work

9. 540 parts produced in 18 hours

10. $2.16 for one dozen cookies

11. 228 words typed in 6 minutes

12. 558 miles in 9 hours

Lesson 10-2

(Pages 386–389)

Solve each proportion. Round to the nearest hundredth if necessary.

1. $\dfrac{15}{21} = \dfrac{5}{b}$

2. $\dfrac{22}{25} = \dfrac{n}{100}$

3. $\dfrac{24}{48} = \dfrac{h}{50}$

4. $\dfrac{9}{27} = \dfrac{y}{42}$

5. $\dfrac{4}{7} = \dfrac{16}{x}$

6. $\dfrac{4}{6} = \dfrac{a}{9}$

7. $\dfrac{6}{14} = \dfrac{21}{m}$

8. $\dfrac{3}{7} = \dfrac{21}{d}$

9. $\dfrac{4}{10} = \dfrac{18}{e}$

10. $\dfrac{9}{10} = \dfrac{27}{f}$

11. $\dfrac{a}{3} = \dfrac{16}{24}$

12. $\dfrac{5}{9} = \dfrac{35}{w}$

13. $\dfrac{3}{4} = \dfrac{m}{12}$

14. $\dfrac{12}{p} = \dfrac{2}{3}$

15. $\dfrac{x}{42} = \dfrac{5}{6}$

16. $\dfrac{48}{64} = \dfrac{y}{16}$

17. $\dfrac{3}{8} = \dfrac{36}{h}$

18. $\dfrac{21}{b} = \dfrac{7}{9}$

19. $\dfrac{e}{7} = \dfrac{60}{84}$

20. $\dfrac{48}{144} = \dfrac{4}{g}$

21. $\dfrac{3}{5} = \dfrac{f}{75}$

22. $\dfrac{9}{10} = \dfrac{36}{d}$

23. $\dfrac{m}{45} = \dfrac{3}{5}$

24. $\dfrac{7}{9} = \dfrac{a}{36}$

25. $\dfrac{1}{5} = \dfrac{7}{a}$

26. $\dfrac{7}{12} = \dfrac{x}{8}$

27. $\dfrac{3}{5} = \dfrac{7}{x}$

28. $\dfrac{c}{9} = \dfrac{5}{14}$

Lesson 10-3

(Pages 391–393)

The map at the right has a scale of $\frac{1}{4}$ inch = 5 kilometers. Use a ruler to measure each map distance to the nearest $\frac{1}{4}$ inch. Then find the actual distances.

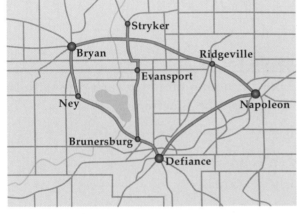

1. Bryan to Napoleon

2. Stryker to Evansport

3. Ney to Bryan

4. Defiance to Napoleon

5. Ridgeville to Bryan

6. Brunersburg to Ney

On a set of architectural drawings, the scale is 3 inches = 10 feet. Find the actual measurements.

7. length of dining room, 3 inches

8. height of living room ceiling, 2.4 inches

9. height of roof, 7.5 inches

10. width of kitchen, 3.75 inches

Lesson 10-4

(Pages 395–397)

Model each percent.

1. 35%

2. 5%

3. 75%

4. 24%

5. 68%

6. 99%

Identify each percent that is modeled.

7.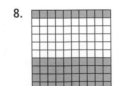

8.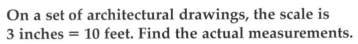

9.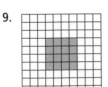

Lesson 10-5

(Pages 400–403)

Write each percent as a fraction in simplest form.

1. 13%　　　　　2. 25%　　　　　3. 8%　　　　　4. 105%

5. 60%　　　　　6. 70%　　　　　7. 80%　　　　　8. 45%

9. 20%　　　　　10. 14%　　　　　11. 75%　　　　　12. 120%

13. 5%　　　　　14. 2%　　　　　15. 450%　　　　　16. 15%

17. What percent of a dollar is a dime?

Write each fraction as a percent.

18. $\frac{77}{100}$　　　　19. $\frac{3}{4}$　　　　20. $\frac{17}{20}$　　　　21. $\frac{3}{25}$

22. $\frac{3}{10}$　　　　23. $\frac{27}{50}$　　　　24. $\frac{2}{5}$　　　　25. $\frac{3}{50}$

26. $\frac{9}{20}$　　　　27. $\frac{8}{5}$　　　　28. $\frac{1}{4}$　　　　29. $\frac{1}{5}$

30. $\frac{19}{20}$　　　　31. $\frac{7}{10}$　　　　32. $\frac{11}{25}$　　　　33. $\frac{5}{4}$

34. How is seventy-six hundredths written as a percent?

Lesson 10-6

(Pages 404–406)

Write each percent as a decimal.

1. 5%　　　　　2. 22%　　　　　3. 50%　　　　　4. 420%

5. 75%　　　　　6. 1%　　　　　7. 100%　　　　　8. 3.7%

9. 0.9%　　　　　10. 9%　　　　　11. 90%　　　　　12. 900%

13. 78%　　　　　14. 62.5%　　　　　15. 15%　　　　　16. 5.2%

17. How is thirteen percent written as a decimal?

Write each decimal as a percent.

18. 0.02　　　　　19. 0.2　　　　　20. 0.002　　　　　21. 1.02

22. 0.66　　　　　23. 0.11　　　　　24. 0.354　　　　　25. 0.31

26. 0.09　　　　　27. 5.2　　　　　28. 2.22　　　　　29. 0.008

30. 0.275　　　　　31. 0.3　　　　　32. 6.0　　　　　33. 4.12

34. Write five and twelve hundredths as a percent.

Lesson 10-7

(Pages 409–412)

Find the percent of each number.

1. 11% of 48　　　　　2. 1.9% of 50　　　　　3. 29% of 500

4. 41% of 50　　　　　5. 32% of 300　　　　　6. 411% of 50

7. 149% of 60　　　　　8. 4.1% of 50　　　　　9. 62% of 200

10. 58% of 100　　　　　11. 52% of 400　　　　　12. 68% of 30

13. 9% of 25　　　　　14. 48% of 1000　　　　　15. 98% of 725

16. 25% of 80　　　　　17. 40% of 75　　　　　18. 75% of 160

19. 10% of 250　　　　　20. 250% of 50　　　　　21. 5% of 120

22. What is 12% of 42?　　　　　23. What is 6.5% of 7?

Lesson 10-8

(Pages 415–417)

Estimate each percent.

1. 38% of 150
2. 20% of 75
3. 0.2% of 500
4. 25% of 70
5. 10% of 90
6. 16% of 30
7. 39% of 40
8. 250% of 100
9. 6% of 86
10. 12.5% of 160
11. 9% of 29
12. 3% of 46
13. $66\frac{2}{3}$% of 60
14. 89% of 47
15. 435% of 30
16. 25% of 48
17. 5% of 420
18. 55% of 134
19. 28% of 4
20. 14% of 40
21. 14% of 14
22. 90% of 140
23. 40% of 45
24. 0.5% of 200

Estimate the percent that is shaded in each figure.

25.
26.
27. ○ ○ ○ ○ ○
○ ○ ○ ○ ○
○ ○ ○ ○ ○
○ ○ ○ ○ ○
○ ○ ○ ○ ○
○ ○ ○ ○ ○

Lesson 11-1

(Pages 428–431)

A set of 30 tickets are placed in a bag. There are 6 baseball tickets, 4 hockey tickets, 4 basketball tickets, 2 football tickets, 3 symphony tickets, 2 opera tickets, 4 ballet tickets, and 5 theater tickets. One ticket is selected without looking. Find each probability. Write each answer as a fraction, a decimal rounded to the nearest hundredth, and a percent rounded to the nearest whole number.

1. P(basketball)
2. P(sports event)
3. P(opera or ballet)
4. P(soccer)
5. P(not symphony)
6. P(theater)
7. P(basketball or hockey)
8. P(not a sports event)
9. P(not opera)
10. P(baseball)
11. P(football)
12. P(not soccer)
13. P(opera)
14. P(not theater)
15. P(symphony)
16. P(soccer or football)
17. P(opera or theater)
18. P(hockey)

Lesson 11-2

(Pages 433–436)

For Exercises 1–3, draw a tree diagram to show the sample space for each situation. Tell how many outcomes are possible.

1. tossing a quarter and rolling a number cube
2. a choice of a red, blue, or green sweater with a white, black, or tan skirt
3. a choice of a chicken, ham, turkey, or bologna sandwich with coffee, milk, juice, or soda
4. How many ways can a person choose to watch two television shows from four television shows?

For Exercises 5–7, a coin is tossed three times.

5. How many outcomes are possible?
6. What is P(two heads)?
7. What is P(no heads)?

Lesson 11-3

(Pages 438–441)

Determine whether each sample is a good sample. Explain.

Type of Survey	Survey Location
1. favorite television show	department store
2. favorite movie	shopping center
3. favorite sport	soccer game
4. favorite ice cream flavor	school
5. favorite vacation	popular theme park

BASKETBALL For Exercises 6 and 7, use the following information. In basketball, Daniel made 12 of his last 18 shots.

6. Find the probability of Daniel making a shot on his next attempt.

7. Suppose Daniel takes 30 shots during his next game. About how many of the shots will he make?

Lesson 11-4

(Pages 444–447)

Find the probability that a randomly thrown dart will land in the shaded region of each dartboard.

1.

2.

3.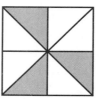

4.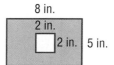
8 in.
2 in.
2 in. 5 in.

5.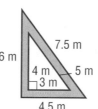
6 m   7.5 m
4 m   5 m
3 m
4.5 m

6.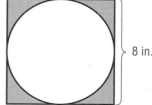
8 in.

Lesson 11-5

(Pages 450–453)

The spinner shown is spun and a card is chosen from the set of cards shown. Find the probability of each event.

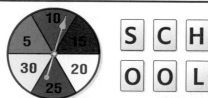

1. P(15 and C)
2. P(even and H)
3. P(odd and a vowel)
4. P(multiple of 5 and O)
5. P(composite and L)
6. P(multiple of 10 and consonant)

A bag contains 3 quarters and 5 dimes. Another bag contains 12 pennies and 8 nickels. One coin is chosen from each bag without looking. Find the probability of each event.

7. P(dime and nickel)
8. P(quarter)
9. P(quarter and penny)
10. P(not a nickel)
11. P(dime and penny)
12. P(not a dime and nickel)

Lesson 12-1

(Pages 465–468)

Complete.

1. 3 yd = __?__ in.
2. 12 ft = __?__ yd
3. 9 ft = __?__ in.
4. 48 in. = __?__ ft
5. 2 yd = __?__ in.
6. 2 mi = __?__ ft
7. $5\frac{1}{2}$ ft = __?__ in.
8. $9\frac{1}{3}$ yd = __?__ ft
9. 21,120 ft = __?__ mi
10. 100 in. = __?__ ft
11. 8 yd = __?__ in.
12. 72 in. = __?__ ft

Draw a line segment of each length.

13. $1\frac{1}{4}$ inches
14. $\frac{5}{8}$ inch
15. $1\frac{3}{4}$ inches
16. $1\frac{1}{2}$ inches
17. $3\frac{1}{8}$ inches
18. $2\frac{3}{8}$ inches
19. $1\frac{5}{8}$ inches
20. $2\frac{1}{2}$ inches
21. $3\frac{3}{4}$ inches

Find the length of each line segment to the nearest half, fourth, or eighth inch.

22. •——————•
23. •——————•
24. •——————•
25. •——————•

Lesson 12-2

(Pages 470–473)

Complete.

1. 3 gal = __?__ pt
2. 24 pt = __?__ gal
3. 20 lb = __?__ oz
4. 2 gal = __?__ fl oz
5. 20 pt = __?__ qt
6. 18 qt = __?__ pt
7. 2,000 lb = __?__ T
8. 3 T = __?__ lb
9. 6 lb = __?__ oz
10. 9 lb = __?__ oz
11. 15 qt = __?__ gal
12. 4 pt = __?__ c
13. 4 gal = __?__ qt
14. 4 qt = __?__ fl oz
15. 12 pt = __?__ c
16. 10 pt = __?__ qt
17. 24 fl oz = __?__ c
18. 1.5 pt = __?__ c
19. $\frac{1}{4}$ lb = __?__ oz
20. 5 T = __?__ lb
21. 2 lb = __?__ oz
22. 8 pt = __?__ qt
23. 48 fl oz = __?__ c
24. 6 gal = __?__ qt
25. 9 qt = __?__ c
26. 2 gal = __?__ c
27. 16 c = __?__ qt
28. 2 qt = __?__ fl oz
29. 16 pt = __?__ gal
30. 5 pt = __?__ c

Lesson 12-3

(Pages 476–479)

Write the metric unit of length that you would use to measure each of the following.

1. length of a paper clip
2. width of a classroom
3. distance from school to home
4. length of a hockey skate
5. length of a school bus
6. distance from Cleveland to Columbus
7. thickness of a calculator
8. length of a blade of grass

Measure each line segment in centimeters and millimeters.

9. •——————•
10. •——————•
11. •——————•
12. •——————•
13. •——————•
14. •——————•
15. •——•
16. •——•

Lesson 12-4

(Pages 484–487)

Write the metric unit of mass or capacity that you would use to measure each of the following. Then estimate the mass or capacity.

1. bag of sugar
2. pitcher of fruit punch
3. mass of a dime
4. amount of water in an ice cube
5. a vitamin
6. pencil
7. mass of a puppy
8. bottle of perfume
9. grain of sand
10. mass of a car
11. baseball
12. paperback book
13. juice in a small cup
14. calculator
15. large bottle of soda
16. trumpet
17. bucket of water
18. backpack with 4 books

Lesson 12-5

(Pages 490–493)

Complete.

1. 400 mm = ___?___ cm
2. 4 kg = ___?___ g
3. 660 cm = ___?___ m
4. 0.3 L = ___?___ mL
5. 30 mm = ___?___ cm
6. 84.5 g = ___?___ kg
7. ___?___ m = 54 cm
8. ___?___ L = 563 mL
9. ___?___ mg = 21 g
10. 4 L = ___?___ mL
11. 61.2 mg = ___?___ g
12. 4,497 mL = ___?___ L
13. ___?___ mm = 45 cm
14. 632 mL = ___?___ L
15. 61 g = ___?___ mg
16. ___?___ mg = 0.51 kg
17. 0.63 L = ___?___ mL
18. 18 km = ___?___ cm
19. ___?___ m = 36 cm
20. 5 kg = ___?___ g
21. 3,250 mL = ___?___ L
22. 7.3 km = ___?___ m
23. 453 g = ___?___ kg
24. 9.35 L = ___?___ mL
25. 8,500 m = ___?___ km
26. 3.6 kg = ___?___ g
27. 415 mL = ___?___ L
28. 521 cm = ___?___ m
29. 63 g = ___?___ kg
30. 2.5 L = ___?___ mL

Lesson 12-6

(Pages 494–497)

Add or subtract. Rename if necessary.

1.
   6 h 14 min
   − 2 h 8 min

2.
   5 h 35 min 25 s
   +   45 min 35 s

3.
   5 h 4 min 45 s
   − 2 h 40 min 5 s

4.
   15 h 16 min
   − 8 h 35 min 16 s

5.
   9 h 20 min 10 s
   + 1 h 39 min 55 s

6.
   2 h 40 min 20 s
   + 3 h 5 min 50 s

7.
   3 h 24 min 10 s
   − 2 h 30 min 5 s

8.
   9 h 12 min
   + 2 h 51 min 15 s

9.
   4 h 9 min 15 s
   − 4 h 3 min 20 s

Find the elapsed time.

10. 1:10 P.M. to 4:45 P.M.
11. 9:40 A.M. to 11:18 A.M.
12. 10:30 A.M. to 6:00 P.M.
13. 8:45 P.M. to 1:30 A.M.
14. 8:05 A.M. to 3:25 P.M.
15. 10:30 P.M. to 1:45 A.M.
16. 11:20 P.M. to 12:15 A.M.
17. 12:40 P.M. to 10:25 P.M.

Lesson 13-1

(Pages 506–509)

Use a protractor to find the measure of each angle. Then classify the angle as acute, obtuse, right, or straight.

1.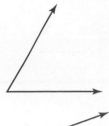

2.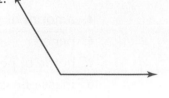

3.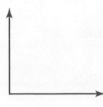

4.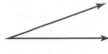

5.

6.

Classify each angle measure as acute, obtuse, right, or straight.

7. 86°

8. 101°

9. 90°

10. 180°

11. **ALGEBRA** Angles A and B are supplementary. If $m\angle A = 75°$, what is $m\angle B$?

12. **ALGEBRA** Angles C and D are complementary. If $m\angle C = 80°$, what is $m\angle D$?

Lesson 13-2

(Pages 510–512)

Estimate the measure of each angle.

1.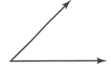

2.

3.

4.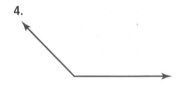

Use a protractor and a straightedge to draw angles having the following measurements.

5. 165°
6. 20°
7. 90°
8. 41°
9. 75°
10. 180°
11. 30°
12. 120°
13. 15°
14. 55°
15. 100°
16. 145°

Lesson 13-3

(Pages 515–517)

Draw each line segment or angle having the given measurement. Then use a straightedge and a compass to bisect the line segment or angle.

1. 3 in.
2. 5 cm
3. 48 mm
4. 33 mm
5. 9 in.
6. 6 cm
7. 4 in.
8. 7 cm
9. $3\frac{1}{2}$ in.
10. 65 mm
11. 110°
12. 70°
13. 25°
14. 150°
15. 90°
16. 120°
17. 30°
18. 180°
19. 142°
20. 45°

Lesson 13-4

(Pages 522–525)

Identify each polygon. Then tell if it is a regular polygon.

1.
2.
3.
4.

5.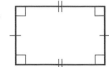
6.
7.
8.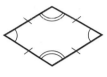

Lesson 13-5

(Pages 528–531)

Trace each figure. Then draw all lines of symmetry.

1.
2.
3.

4.
5.
6.

Tell whether each figure has rotational symmetry. Write yes **or** no.

7.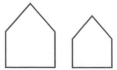
8.
9.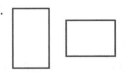

Lesson 13-6

(Pages 534–536)

Tell whether each pair of polygons is *congruent,* *similar,* **or** *neither.*

1.
2.
3.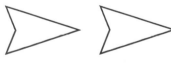

4.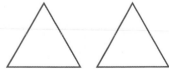
5.
6.

7.
8.
9.

Lesson 14-1

(Pages 546–549)

Find the area of each parallelogram. Round to the nearest tenth if necessary.

1.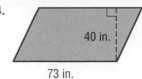
3.7 ft
7.5 ft

2.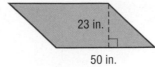
34 cm
40 cm

3.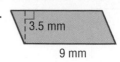
4 m
5 m

4.
40 in.
73 in.

5.
23 in.
50 in.

6.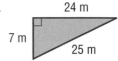
3.5 mm
9 mm

7.
5.1 m
2 m

8.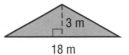
1.5 ft
11 ft

9.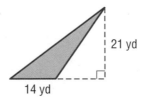
9 cm
14 cm

Lesson 14-2

(Pages 553–556)

Find the area of each triangle. Round to the nearest tenth if necessary.

1.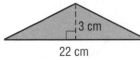
3 cm
22 cm

2.
3 m
18 m

3.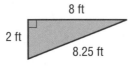
8 ft
2 ft
8.25 ft

4.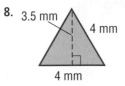
24 m
7 m
25 m

5.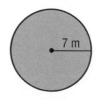
6 yd
5 yd

6.
4 in.
5 in.
3 in.

7.
21 yd
14 yd

8.
3.5 mm
4 mm
4 mm

9. base, 6 ft
   height, 3 ft

10. base, 4.2 in.
    height, 6.8 in.

11. base, 9.1 m
    height, 7.2 m

12. base, 13.2 cm
    height, 16.2 cm

Lesson 14-3

(Pages 558–561)

Find the area of each circle to the nearest tenth. Use 3.14 for π.

1.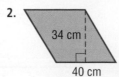
7 m

2.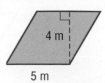
12 cm

3.
10 in.

4.
8 m

5. radius, 4 m

6. diameter, 6 in.

7. radius, 16 m

8. diameter, 11 in.

9. radius, 9 cm

10. diameter, 24 mm

Lesson 14-4

(Pages 566–568)

Identify each figure.

1.

2.

3.

4.

5.

6.

7.

8.

9.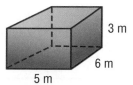

Extra Practice

Lesson 14-5

(Pages 572–575)

Find the volume of each rectangular prism.

1.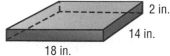
2 in.
14 in.
18 in.

2.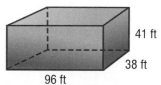
41 ft
38 ft
96 ft

3.
3 m
6 m
5 m

4.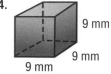
9 mm
9 mm
9 mm

5.
3 cm
3 cm
20 cm

6.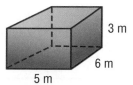
7 in.
9 in.
4 in.

7. length = 8 in.
   width = 5 in.
   height = 2 in.

8. length = 10 cm
   width = 2 cm
   height = 8 cm

9. length = 20 ft
   width = 5 ft
   height = 6 ft

Lesson 14-6

(Pages 577–580)

Find the surface area of each rectangular prism. Round to the nearest tenth if necessary.

1.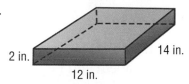
2 in.
14 in.
12 in.

2.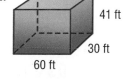
41 ft
30 ft
60 ft

3.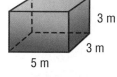
3 m
3 m
5 m

4.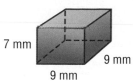
7 mm
9 mm
9 mm

5.
3 cm
3 cm
20 cm

6.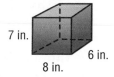
7 in.
6 in.
8 in.

7. length = 4 mm
   width = 12 mm
   height = 1.5 mm

8. length = 16 cm
   width = 20 cm
   height = 20.4 cm

9. length = 8.5 m
   width = 2.1 m
   height = 7.6 m

Mixed Problem Solving

Chapter 1 Number Patterns and Algebra

(pages 4–47)

1. **MONEY** Dave wants to buy a new computer system that is priced at $2,475. He plans to put down $300 and will pay the rest in five equal payments. Use the four-step plan to find the amount of each payment. (Lesson 1-1)

2. **BOOKS** The table below shows the number of pages of a book Elias read during the past 5 days. Use the four-step plan to find how many pages Elias will read on Saturday if he continues at this rate. (Lesson 1-1)

Pages Read by Elias	
Day	**Pages Read**
Monday	5
Tuesday	11
Wednesday	19
Thursday	29
Friday	41
Saturday	?

3. **GEOMETRY** Draw the next figure in the pattern shown below. (Lesson 1-1)

4. **SEATING** Sue is planning a holiday gathering for 117 people. She can use tables that seat 5, 6, or 9 people. If all of the tables must be the same size and all of the tables must be full, what size table should she use? (Lesson 1-2)

5. **RULERS** List the number of ways 318 rulers can be packaged for shipping to an office supply store so that each package has the same number of rulers. (Lesson 1-2)

6. **GEOGRAPHY** The United States of America has 50 states. Write 50 as a product of primes. (Lesson 1-3)

7. **ATTENDANCE** The number of students attending Blue Hills Middle School this year can be written as 5^4. How many students are enrolled at Blue Hills Middle School? (Lesson 1-4)

RETAIL For Exercises 8 and 9, use the table below. It shows the cost of several items sold at The Clothes Shack. (Lesson 1-5)

The Clothes Shack	
Item	**Price ($)**
Jeans	30
T-shirt	15
Sweatshirt	20
Shorts	10

8. Write an expression that can be used to find the total cost of 2 pairs of jeans, 3 T-shirts, and 4 pairs of shorts.

9. Find the total cost for the purchase.

10. **TRAVEL** Distance traveled can be found using the expression $r \times t$, where r represents rate and t represents time. How far will you travel if you drive for 9 hours at a rate of 65 miles per hour? (Lesson 1-6)

11. **ARCHITECTURE** The perimeter of a rectangle can be found using the expression $2\ell + 2w$, where ℓ represents length and w represents width. Find the perimeter of the front of a new building whose design is shown below. (Lesson 1-6)

    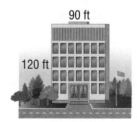

    90 ft

    120 ft

12. **AGE** The equation $13 + a = 51$ describes the sum of the ages of Elizabeth and her mother. If a is Elizabeth's mother's age, what is the age of Elizabeth's mother? (Lesson 1-7)

13. **GARDENING** Rondell has a rectangular garden that measures 18 feet long and 12 feet wide. Suppose one bag of topsoil covers 36 square feet. How many bags of topsoil does Rondell need for his garden? (Lesson 1-8)

TRAVEL For Exercises 1 and 2, use the frequency table below. (Lesson 2-1)

Speed of Cars Driving on Highway		
Speed (mph)	Tally	Frequency
51–55	I	1
56–60	IIII	4
61–65	ЖΗ II	7
66–70	ЖΗ ЖΗ II	12

1. In what speed range did most of the cars drive?

2. How many cars were driving at a speed above 65 miles per hour?

SALES For Exercises 3–5, use the table below. (Lesson 2-2)

Bookworm Book Shop	
Month	Total Sales ($)
January	9,750
February	8,200
March	7,875
April	12,300
May	10,450
June	9,900

3. Make a line graph for the data.

4. Which month had the greatest change in sales from the previous month?

5. Which month had the greatest decrease in sales from the previous month?

SPORTS For Exercises 6–8, use the circle graph below. (Lesson 2-3)

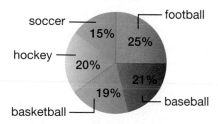

Favorite Sport

6. Which sport is the most popular?

7. Name the least popular sport.

8. Name two sports that together make up 40% of the favorite sports.

9. **SWIMMING** Use the graph below to predict how many laps Emily will swim during week 6 of training. (Lesson 2-4)

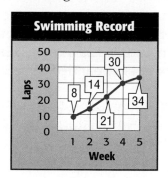

10. **JOBS** Sarah earned $64, $88, $96, $56, $48, $62, $75, $55, $50, $68, and $42. Make a stem-and-leaf plot of the data. (Lesson 2-5)

SCHOOL For Exercises 11–13, use the following information.
David's scores for five math tests are 83, 92, 88, 54, and 93. (Lesson 2-6)

11. Identify the outlier.

12. Find the mean of the data with and without the outlier.

13. Describe how the outlier affects the mean.

WEATHER For Exercises 14 and 15, use the following information.
The daily high temperatures during one week were 67°, 68°, 64°, 64°, 69°, 92°, and 66°. (Lesson 2-7)

14. Find the mean, median, and mode.

15. Which measure best describes the average temperature?

MONEY For Exercises 16 and 17, use the graph below. (Lesson 2-8)

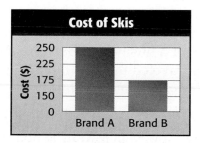

16. About how many times more does Brand A appear to cost than Brand B?

17. Explain why this graph is misleading.

Mixed Problem Solving

1. **CHEMISTRY** The weight of a particular compound is 1.0039 grams. Express this weight in words. (Lesson 3-1)

2. **PACKAGING** The length of a rectangular box of cereal measures 12.58 inches. Express this measurement in expanded form. (Lesson 3-1)

3. **TYPING** Julia finished her typing assignment in 18.109 minutes. Teresa took 18.11 minutes to complete the same assignment. Who completed the assignment faster? (Lesson 3-2)

4. **TRACK AND FIELD** The table below lists the finishing times for runners of a particular event. Order these finish times from least to greatest. (Lesson 3-2)

Finishing Times	
Runner	Time (s)
Brent	43.303
Trey	43.033
Cory	43.230
Jonas	43.3003

5. **HEIGHT** Serena and Tia are comparing their heights. Serena is 52.25 inches tall, and Tia is 52.52 inches tall. Compare Serena's height to Tia's height using <, >, or =. (Lesson 3-2)

6. **SCHOOL** The average enrollment at a middle school in a certain county is 429.86 students. Round this figure to the nearest whole number. (Lesson 3-3)

7. **GAS MILEAGE** Booker computes his gas mileage for a business trip by dividing his distance traveled in miles by the number of gallons of gasoline used during the trip. The result is 21.3742 miles per gallon. Round this gas mileage to the hundredths place. (Lesson 3-3)

8. **SCIENCE** It takes 87.97 days for the planet Mercury to revolve around the Sun. Round this number to the nearest tenth. (Lesson 3-3)

9. **MONEY** Marta plans to buy the items listed at the right. Estimate the cost of these items before tax is added. (Lesson 3-4)

THE SPORTS COVE
Baseball............6.50
Baseball glove..............37.99
Baseball cap..............13.79

10. **FUND-RAISING** During a fund-raiser at her school, Careta sold $78.35 worth of candy. Diego sold $59.94 worth of candy. Use front-end estimation to find about how much more Careta sold. (Lesson 3-4)

11. **GEOMETRY** The perimeter of a figure is the distance around it. Estimate the perimeter of the figure shown below using clustering. (Lesson 3-4)

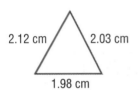

2.12 cm   2.03 cm
1.98 cm

12. **MUSIC** Use the data in the table to estimate the total number of songs Jasmine downloaded onto her digital music player. (Lesson 3-4)

Songs Downloaded	
Week	Number of Songs
1	22
2	19
3	21
4	18

13. **SHOPPING** Kyle spent $41.38 on a new pair of athletic shoes and $29.86 on a new pair of jeans. Find the total of Kyle's purchases. (Lesson 3-5)

14. **MONEY** The current balance of Tami's checking account is $237.80. Find the new balance after Tami writes a check for $29.95. (Lesson 3-5)

15. **WEATHER** During the first week of January, the National Weather Service reported that 8.07 inches of snow fell in Cleveland, Ohio. An additional 6.29 inches fell during the second week of January. Find the total snowfall for the first two weeks of January. (Lesson 3-5)

1. **BAKING** A recipe for a cake uses five 2.75-ounce packages of ground walnuts. Find the total ounces of ground walnuts needed for the recipe. (Lesson 4-1)

2. **MONEY** Nick is buying a sports magazine that costs $4.95. What will it cost him for a year if he buys one magazine every month? (Lesson 4-1)

3. **GAS MILEAGE** During a business trip, Ken's car used 18 gallons of gasoline and traveled 19.62 miles per gallon of gasoline. How far did Ken travel on his trip? (Lesson 4-2)

4. **DECORATING** Selena is considering buying new carpeting for her living room. How many square feet of carpet will Selena need? (Lesson 4-2)

16.3 ft

24.6 ft

5. **INCOME** Last week, Sakima worked 39.5 hours and earned $18.25 per hour. How much did Sakima earn before taxes? Round to the nearest cent. (Lesson 4-2)

6. **INSURANCE** Aurelia pays $414.72 per year for auto insurance. Suppose she makes 4 equal payments a year. How much does she pay every three months? (Lesson 4-3)

7. **SHIPPING** The costs for shipping the same package using three different shipping services are given in the table below. What is the average shipping cost? (Lesson 4-3)

Shipping Services Costs	
Shipping Service	Cost
A	$6.42
B	$8.57
C	$7.48

8. **SHOPPING** Logan paid $13.30 for 2.8 pounds of roast beef. Find the price per pound of the roast beef. (Lesson 4-4)

9. **GARDENING** George has a garden that covers 127.75 square feet. He wants to cover the soil of his garden with fertilizer. One bag of fertilizer covers 18.25 square feet. How many bags of fertilizer does George need? (Lesson 4-4)

10. **ARCHITECTURE** The diagram below shows the outline of a new building. Find the perimeter of the building. (Lesson 4-5)

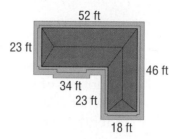

52 ft

23 ft

46 ft

34 ft

23 ft

18 ft

11. **SCHOOL** The perimeter of a desk is 60 inches. If the length of the desk measures 18 inches, what is the measure of the width? (Lesson 4-5)

12. **FENCING** Gia and Terrance are planning a vegetable garden. The dimensions of the garden are shown below. To keep out small animals, a fence will surround the garden. If the fencing costs $2.70 per yard, what will be the total cost of the fencing? (Lesson 4-5)

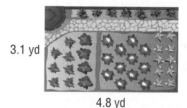

3.1 yd

4.8 yd

13. **SWIMMING** A new swimming pool is being installed at the Hillside Recreation Center. The pool is circular in shape and has a diameter of 32 feet. Find the circumference of the pool to the nearest tenth. (Lesson 4-6)

14. **RIDES** A carousel has a diameter of 36 feet. How far do passengers travel on each revolution? (Lesson 4-6)

15. **PIZZA** The Amazing Pizza Shop claims to make the largest pizza in town. The radius of their largest pizza is 15 inches. Find the circumference of the pizza. (Lesson 4-6)

Mixed Problem Solving

1. **CANDY** The table shows the amount of candy Donna purchased. She wants to place the candy in bags so that each bag has the same number of pieces of candy without mixing the different types. Find the greatest number of pieces of candy that can be put in each bag. (Lesson 5-1)

Candy Purchased	
Type	**Amount**
candy bars	48
lollipops	32
licorice sticks	24

2. **TICKETS** Mrs. Cardona collected money from her students for tickets to the school play. She recorded the amounts of money collected in the table below. What is the most a ticket could cost per student? (Lesson 5-1)

Ticket Money Collection	
Day	**Amount**
Monday	$15
Tuesday	$12
Wednesday	$18
Thursday	$21

3. **TEACHERS** Twenty-eight of the forty-two teachers on the staff at Central Middle School are female. Write this fraction in simplest form. (Lesson 5-2)

4. **WEATHER** In September, 4 out of 30 days were rainy. Express this fraction in simplest form. (Lesson 5-2)

5. **CARPENTRY** The measurement of a board being used in a carpentry project is $\frac{53}{8}$ feet. Write this improper fraction as a mixed number. (Lesson 5-3)

6. **BAKING** A bread recipe calls for $4\frac{2}{3}$ cups of flour. Write this mixed number as an improper fraction. (Lesson 5-3)

7. **BICYCLES** The front gear of a bicycle has 54 teeth, and the back gear has 18 teeth. How many complete rotations must the smaller gear make for both gears to be aligned in the original starting position? (Lesson 5-4)

8. **PATTERNS** Which three common multiples for 5 and 7 are missing from the list below? (Lesson 5-4)

35, 70, 105, ___?___, ___?___, ___?___, 245, 280, …

9. **CANDLES** Three candles have widths of $\frac{3}{8}$ inch, $\frac{5}{6}$ inch, and $\frac{2}{3}$ inch. What is the measure of the candle with the greatest width? (Lesson 5-5)

10. **HOME REPAIR** Dan has a wrench which can open to accommodate three different sizes of washers, $\frac{5}{16}$ inch, $\frac{3}{8}$ inch, and $\frac{2}{5}$ inch. Which washer is the smallest? (Lesson 5-5)

11. **TRAVEL** The road sign shows the distances from the highway exit to certain businesses. What fraction of a mile is each business from the exit? (Lesson 5-6)

Restaurant	0.65 mi
Gas Station	0.4 mi
Hotel	1.2 mi

12. **SPIDERS** The diagram shows the range in the length of a tarantula. What is one length that falls between these amounts? Write the amount as a mixed number in simplest form. (Lesson 5-6)

2.5 to 2.75 in.

13. **TRACK AND FIELD** Monifa runs a race covering $4\frac{5}{8}$ miles. Express this distance as a decimal. (Lesson 5-7)

14. **SEWING** Dana buys $3\frac{3}{4}$ yards of fabric to make a new dress. What decimal does this represent? (Lesson 5-7)

15. **COOKING** A cookie recipe calls for $2\frac{1}{3}$ cups of flour. What decimal represents this amount? (Lesson 5-7)

Mixed Problem Solving

1. **PICTURE FRAMES** A picture frame is to be wrapped in the box shown below. To the nearest half-inch, how large can the frame be? (Lesson 6-1)

2. **COOKING** A recipe for lasagna calls for $1\frac{2}{3}$ cups of cheese. Should you buy a package of cheese that contains $1\frac{1}{2}$ cups or 2 cups of cheese? (Lesson 6-1)

3. **BEVERAGES** The amount of ginger ale needed to make two types of punch is shown. About how much ginger ale is needed to make both types of punch? (Lesson 6-2)

Ginger Ale Needed	
Recipe	Ginger Ale (c)
A	$5\frac{1}{3}$
B	$\frac{3}{4}$

4. **SEWING** To make a costume, Trina needs $15\frac{3}{4}$ inches of ribbon for the skirt and $7\frac{3}{8}$ inches of ribbon for the shirt. Estimate the total amount of ribbon needed to make the costume. (Lesson 6-2)

5. **COFFEE** A coffee pot was $\frac{7}{8}$ full at the beginning of the day. By the end of the day, the pot was only $\frac{1}{8}$ full. How much of the coffee was used during the day? (Lesson 6-3)

6. **SCHOOL** Of the students in a class, $\frac{5}{16}$ handed in their report on Monday, and $\frac{7}{16}$ handed their report in on Tuesday. What fraction of the class had completed the work after two days? (Lesson 6-3)

7. **MEASUREMENT** How much more is $\frac{2}{3}$ pound than $\frac{1}{2}$ pound? (Lesson 6-4)

8. **STOCKS** The table below shows the increases in Kyle's stock. What was the total increase over these two days? (Lesson 6-4)

Kyle's Stock Increases	
Day	Increase
Monday	$\frac{9}{16}$
Tuesday	$\frac{3}{4}$

9. **CHEMISTRY** A chemist has $\frac{3}{4}$ quart of a solution. The chemist uses $\frac{1}{5}$ quart of the solution. How much of the solution remains? (Lesson 6-4)

10. **REPAIRS** A plumber repaired two sinks in an apartment. The first job took $1\frac{1}{3}$ hours, and the second job took $2\frac{3}{4}$ hours. What was the total time it took the plumber to repair both sinks? (Lesson 6-5)

11. **CARPENTRY** Daniela bought $5\frac{5}{16}$ pounds of finish nails. She used $3\frac{5}{8}$ pounds on her carpentry projects. How many pounds of nails were left over? (Lesson 6-5)

12. **BANKING** The table below shows the interest rates advertised by three different banks for a 30-year fixed rate mortgage. Find the difference between the highest and the lowest rate. (Lesson 6-6)

30-Year Fixed Mortgage Rates	
Bank	Interest Rate
A	$7\frac{1}{4}\%$
B	$6\frac{3}{5}\%$
C	$6\frac{5}{6}\%$

13. **FOOD** Wendy bought $3\frac{2}{5}$ pounds of pretzels. She shared these with her friends and together they ate $2\frac{5}{8}$ pounds of the pretzels. How much was left over? (Lesson 6-6)

Mixed Problem Solving

1. **CARPET** Seth is buying carpeting for his basement. The room's dimensions are shown below. About how much carpet will he need to buy? (Lesson 7-1)

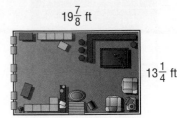

$19\frac{7}{8}$ ft

$13\frac{1}{4}$ ft

2. **SEWING** Ayana is making a quilt that will measure $4\frac{1}{3}$ feet by $2\frac{6}{7}$ feet . About how much fabric will she need to make the quilt? (Lesson 7-1)

3. **CARS** According to a survey, $\frac{2}{5}$ of people prefer tan cars. Of those people who prefer tan cars, $\frac{1}{3}$ prefer a two-door car. What fraction of the people surveyed would prefer a tan two-door car? (Lesson 7-2)

4. **SCHOOL** At Glen Middle School, $\frac{3}{8}$ of the students participate in after-school activities. How many out of 256 students can be expected to participate in after-school activities? (Lesson 7-2)

5. **TRAINS** Suppose you are building a model that is $\frac{1}{12}$ the size of the train shown. What will be the length of the model? (Lesson 7-3)

36 ft

6. **COOKING** A cookie recipe calls for $3\frac{3}{8}$ cups of sugar. Tami wants to make $1\frac{1}{2}$ times the recipe. How much sugar will Tami need? (Lesson 7-3)

7. **GEOMETRY** What is the area of a rectangle with a height of $4\frac{4}{5}$ units and a width of $2\frac{1}{8}$ units? (Lesson 7-3)

8. **CARPENTRY** The piece of wood shown below is to be cut into pieces measuring $\frac{1}{4}$ foot each. How many sections can be made? (Lesson 7-4)

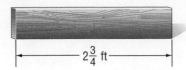

$2\frac{3}{4}$ ft

9. **SNACKS** Mika buys $\frac{3}{4}$ pound of peanuts and divides it evenly among 5 friends. Find the amount of peanuts that each friend will get. (Lesson 7-4)

10. **LANDSCAPING** The dimensions of a backyard are shown below. Decorative rocks that measure $\frac{5}{8}$ foot wide are to be placed across one length of the yard. How many rocks will be needed? (Lesson 7-5)

$11\frac{1}{4}$ ft

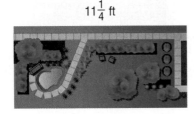

11. **BOATING** On a recent fishing trip, a boat traveled $53\frac{5}{8}$ miles in $2\frac{3}{4}$ hours. How many miles per hour did the boat average? (Lesson 7-5)

12. **SALES** An advertisement for a sale at a ski shop includes the following information.

The Ski Shop Sale	
Days Since Start of Sale	Discount on Merchandise
1	4%
2	7%
3	10%
4	13%
5	?
6	?

Find the percent discount on merchandise for days 5 and 6 of the sale. (Lesson 7-6)

1. **HIKING** Delmar hikes on a trail that has its highest point at an altitude of 5 miles above sea level. Write this number as an integer. (Lesson 8-1)

2. **BANKING** Alicia has overdrawn her checking account by $12. Write her account balance as an integer. (Lesson 8-1)

3. **GEOGRAPHY** The table below lists the record low temperatures for five states. Order these temperatures from least to greatest. (Lesson 8-1)

Record Low Temperatures	
State	**Low Temp. (°F)**
South Carolina	−19
Texas	−23
Alabama	−27
Florida	−2
Michigan	−51

Source: *The World Almanac*

4. **WEATHER** The temperature increased 14 degrees between noon and 5:00 P.M. and then decreased 6 degrees between 5:00 P.M. and 9:00 P.M. Find the temperature at 9:00 P.M. if the temperature at noon was 68°F. (Lesson 8-2)

5. **BUSINESS** The table shows the results of a company's operations over a period of three months. Find the company's overall profit or loss over this time period. (Lesson 8-2)

Month	Profit	Loss
April		$15,000
May	$19,000	
June		$12,000

6. **BICYCLING** Stacy bikes from an altitude of 2,500 feet above sea level down to a location that is 1,300 feet below sea level. How far did Stacy descend during her ride? (Lesson 8-3)

7. **TEMPERATURE** Find the difference in temperature between 41°F in the day in Anchorage, Alaska, and −27°F at night. (Lesson 8-3)

8. **FOOTBALL** The Riverdale football team lost 4 yards on each of 3 consecutive plays. Write the integer that represents the total yards lost. (Lesson 8-4)

9. **MOVIES** The attendance at a local movie theater decreases by 12 people each month. Suppose this pattern continues for 6 months. Write the integer that represents the loss in attendance at the movies. (Lesson 8-4)

10. **FLYING** An airplane coming in for a landing descended 21,000 feet over a period of 7 minutes. Write the integer that represents the average descent per minute. (Lesson 8-5)

11. **PENALTIES** A football team is penalized a total of 20 yards in 4 plays. If the team was penalized an equal number of yards on each play, write the integer that represents the yards penalized on each play. (Lesson 8-5)

MONEY For Exercises 12 and 13, use the following information.
Latisha earns $3 per hour for baby-sitting. The expression $3x$ represents the total amount earned where x is the number of hours she baby-sat. (Lesson 8-6)

12. Copy and complete the table below to find the ordered pairs (hours, total amount earned) for 1, 2, 3, and 4 hours.

x (hours)	3x	y (amount earned)	(x, y)
1			
2			
3			
4			

13. Graph the ordered pairs. Then describe the graph.

14. **COOKING** If it takes Berto 4 minutes to cook one pancake, then $4p$ represents the total time where p is the number of pancakes. Write the ordered pairs (number of pancakes, total time) for 0, 1, 2, 3, and 4 pancakes.

1. **MOVIES** Use the table shown to find the cost for 5 adults to go to the movies and each get a box of popcorn and a soft drink. (Lesson 9-1)

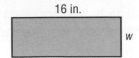

Movie Costs	
Item	Cost
ticket	$7
popcorn	$2
beverage	$2

2. **BOOKS** Julisa buys 5 hardback and 5 paperback books. Use the table shown to find the total amount she spends, not including tax. (Lesson 9-1)

Bookstore Sale	
Book Category	Cost
hardback	$12
paperback	$7

3. **CARPENTRY** A board that measures 19 meters in length is cut into two pieces. The shorter of the two pieces measures 7 meters. Write and solve an equation to find the length of the longer piece. (Lesson 9-2)

4. **BANKING** Sean withdrew $55 from his bank account. The balance of the account after the withdrawal was $123. Write and solve an equation to find the balance of the bank account before the withdrawal. (Lesson 9-3)

5. **WEATHER** As a cold front moved through town, the temperature dropped 21°F. The temperature after the cold front moved in was 36°F. Write and solve an equation to find the temperature before the cold front came through town. (Lesson 9-3)

6. **SCHOOL** The number of questions Juan answered correctly on an exam is 2 times as many as the amount Ryan answered correctly. Juan correctly answered 20 questions. Write a multiplication equation that can be used to find how many questions Ryan answered correctly. (Lesson 9-4)

7. **AGE** Kari's mother is 3 times as old as Kari. If Kari's mother is 39, how old is Kari? (Lesson 9-4)

8. **GEOMETRY** The area of the rectangle shown below is 96 square inches, and the length is 16 inches. Write and solve an equation to find the width of the rectangle. (Lesson 9-4)

16 in.

w

9. **BOWLING** Malyn and three friends went bowling. The cost for one game was $4 per person. One of Malyn's friends had her own bowling shoes, but the rest of them had to rent shoes. If they spent a total of $22 to bowl one game, how much did it cost to rent a pair of shoes? (Lesson 9-5)

10. **NUMBER SENSE** Eight more than three times a number is twenty-six. Find the number. (Lesson 9-5)

11. **GEOMETRY** The width of a rectangle is 18 inches. Find the length if the perimeter of the rectangle is 58 inches. (Lesson 9-5)

FUND-RAISING For Exercises 12 and 13, use the following information.
The school chorale is selling T-shirts and sweatshirts to raise money for a local charity. The cost of each item is shown. (Lesson 9-6)

$5
$12

12. Write a function rule that represents the total selling price of t T-shirts and s sweatshirts.

13. What would be the total amount collected if 9 T-shirts and 6 sweatshirts are sold?

JOBS For Exercises 14 and 15, use the following information.
Cory works at a fast-food restaurant. He earns $5 per hour and must pay $15 for a uniform out of his first paycheck. (Lesson 9-7)

14. Write a function that represents Cory's earnings for his first paycheck.

15. Graph the function on a coordinate plane.

1. **SNACKS** A 16-ounce bag of potato chips costs $2.08, and a 40-ounce bag costs $4.40. Which bag is less expensive per ounce? (Lesson 10-1)

2. **NATURE** Write the ratio that compares the number of leaves to the number of acorns. (Lesson 10-1)

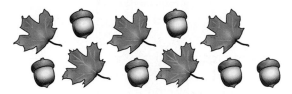

3. **SPORTS** The table shows the favorite sports for a small group of students from South Middle School. Suppose there are 400 students in the school. How many of the students can be expected to prefer football? (Lesson 10-2)

Sports Preference	
Favorite Sport	**Students**
baseball	6
basketball	5
football	9
hockey	5

4. **AIRPLANES** A model of an airplane is 12 inches long. If the model has a scale of 1 inch = 5 feet, what is the actual length of the airplane? (Lesson 10-3)

5. **HOUSES** The actual width of the lot on which a new house is going to be built is 99 feet. An architect's drawing of the lot has a scale of 2 inches = 9 feet. What is the width of the lot on the drawing? (Lesson 10-3)

6. **SHOPPING** According to a survey, 29% of people do most of their holiday shopping online. Make a model to show 29%. (Lesson 10-4)

7. **FOOD** Tamyra and her friends ate 40% of a pizza. Pablo and his friends ate 35% of a pizza. Make a model to show who ate more pizza. (Lesson 10-4)

SCHOOL For Exercises 8 and 9, use the following information and the table below. The table below shows the number of times Ben has been on time for school and the number of times he has been tardy during the past month. (Lesson 10-5)

On Time	Tardy
17	3

8. What percent of the time has Ben been tardy?

9. What is the fraction of days that Ben has been on time?

10. **BANKING** A local bank advertises a home equity loan with an interest rate of 0.1275. Express this interest rate as a percent. (Lesson 10-6)

11. **SKIING** Mateo wants to buy new skis. He finds a pair of skis that usually sell for $380. The skis are on sale for 20% off. Find the amount Mateo will save if he buys the skis during the sale. (Lesson 10-7)

12. **BASEBALL** Attendance at a university's baseball games during the 2002 season was 145% of the attendance during the 2001 season. If the total attendance during the 2001 season was 700, find the total attendance during the 2002 season. (Lesson 10-7)

13. **FOOD** About what percent of the eggs have been used? (Lesson 10-8)

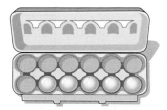

14. **ALLOWANCES** According to a survey, 58% of all teenagers receive a weekly allowance for doing household chores. Estimate how many teenagers out of 453 would receive a weekly allowance for doing household chores. (Lesson 10-8)

Mixed Problem Solving

GAMES For Exercises 1–3, use the following information.

To win a prize at a carnival, a player must choose a green beanbag from a box without looking. (Lesson 11-1)

1. What is the probability of choosing a yellow beanbag?

2. Find the probability of selecting a blue beanbag.

3. What is the probability of selecting a green beanbag?

SUNDAES For Exercises 4 and 5, use the menu shown below. (Lesson 11-2)

Ice Cream	
Vanilla	Chocolate
Topping	
Hot Fudge	Strawberry
Nuts	
Peanuts	Pecans

4. How many different ways can you choose an ice cream, a topping, and nuts?

5. Find the probability that a person will randomly choose vanilla ice cream with hot fudge and pecans.

SEASONS For Exercises 6–9, use the table that shows the results of a survey. (Lesson 11-3)

Favorite Season Survey	
Season	**Students**
Fall	4
Winter	7
Summer	9
Spring	5

6. Find the probability a student prefers summer.

7. What is the probability that a student prefers winter?

8. There are 425 students in the school where the survey was taken. Predict how many students prefer fall.

9. Suppose there are 500 students in the school. How many students can be expected to prefer spring?

GAMES For Exercises 10–12, use the following information and the dartboard shown.

To win a prize, a dart must land on red, blue, or yellow on the dartboard shown. It is equally likely that the dart will land anywhere on the dartboard. (Lesson 11-4)

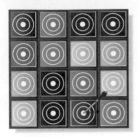

10. What is the probability of winning a prize?

11. Predict how many times a dart will land on red, blue, or yellow if it is thrown 20 times.

12. Predict how many times a dart will land on red, blue, or yellow if it is thrown 150 times.

LOLLIPOPS For Exercises 13 and 14, use the following information and the table that shows the percent of each flavor of lollipops in a bag.

One lollipop is chosen without looking and replaced. Then another lollipop is chosen. (Lesson 11-5)

Flavor of Lollipops in a Bag	
Flavor	**Percent**
strawberry	40%
grape	25%
orange	30%
lime	5%

13. Find the probability of choosing strawberry and lime.

14. What is the probability of choosing strawberry and not orange?

FIRE TRUCKS For Exercises 15 and 16, use the following information.

The probability that fire engine A is available to respond to an emergency call when needed is 94%. The probability that fire engine B is available when needed is 96%. The two fire engines operate independently of one another. (Lesson 11-5)

15. Find the probability that both fire engines are available if needed for the same emergency.

16. What is the probability that neither of the fire engines are available if needed for the same emergency?

1. **SWIMMING** A community center's swimming pool is 8 feet deep. Find the depth of the pool in yards. (Lesson 12-1)

2. **FURNITURE** The length of a desk is 48 inches. What is the measure of the length of the desk in feet? (Lesson 12-1)

3. **TYPING** Measure the length of the keyboard key shown to the nearest half, fourth, or eighth inch. (Lesson 12-1)

4. **ANIMALS** An adult male elephant can weigh up to 7 tons. How many pounds is this? (Lesson 12-2)

5. **BAKING** A recipe for a sponge cake calls for 1 pint of milk. How many fluid ounces of milk will be needed for the recipe? (Lesson 12-2)

6. **SNACKS** Based on the facts shown below, how many ounces are in each serving of potato chips? (Lesson 12-2)

Jet's Potato Chips
A 2-lb bag contains
8 servings

7. **TRAVEL** Kenji is planning to travel from Cincinnati, Ohio, to Miami, Florida. When estimating the distance, should he be accurate to the nearest centimeter, meter, or kilometer? (Lesson 12-3)

8. **ARCHITECTURE** An architect's plans for a new house shows the thickness of the doors to be used on each room. Which metric unit of length is most appropriate to use to measure the thickness of the doors? (Lesson 12-3)

9. **FOOD** Mustard comes in a 397-gram container or a 0.34-kilogram container. Which container is smaller? (Lesson 12-4)

10. **SHOPPING** A grocery store sells apple juice in a 1.24-liter bottle and in a 685-milliliter bottle. Which bottle is larger? (Lesson 12-4)

11. **BAKING** The table shows the dry ingredients needed for a chocolate chip cookie recipe. Is the total amount of the dry ingredients more or less than one kilogram? (Lesson 12-4)

Chocolate Chip Cookies	
20	grams baking powder
10	grams salt
175	grams flour
100	grams sugar
100	grams brown sugar

12. **CLEANING** A bottle of household cleaner contains 651 milliliters. How many liters does the bottle contain? (Lesson 12-5)

13. **SWIMMING** As part of a practice session for her swim team, Opal swims 24,000 centimeters of the breaststroke. Write this distance in meters. (Lesson 12-5)

14. **AUTOMOBILES** Diego is considering buying a new sport utility vehicle which is advertised as having a weight of 2,500 kilograms. What is the weight in grams? (Lesson 12-5)

15. **HOMEWORK** Jerome spends 2 hours and 20 minutes on his homework after school. If he starts working at 3:45 P.M., what time will he be done with his homework? (Lesson 12-6)

16. **RACING** A sail boat race starts at 9:35 A.M., and the first boat crosses the finish line at 11:12 A.M. Find the racing time for the winning boat. (Lesson 12-6)

17. **BASEBALL** Use the information at the right. How much longer was the Sharks baseball game than the Cardinals baseball game? (Lesson 12-6)

Length of Baseball Games	
Team	Time
Cougars	2 h 30 min
Sharks	3 h 15 min
Cardinals	2 h 58 min

Mixed Problem
Solving

1. **WINDOWS** The plans for a new home use several windows in the shape shown. Classify the angle in the window as *acute, obtuse, right,* or *straight.* (Lesson 13-1)

2. **ALGEBRA** If $m\angle A = 118°$ and $\angle A$ and $\angle B$ are supplementary angles, what is $m\angle B$? (Lesson 13-1)

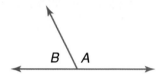

3. **ALGEBRA** Angles M and N are complementary angles. Find $m\angle N$ if $m\angle M = 62°$. (Lesson 13-1)

4. **CLOCKS** Estimate the measure of the angle made by the hour hand and the minute hand of a clock when it is 3:35 P.M. (Lesson 13-2)

5. **QUILTING** A quilt pattern uses pieces of fabric cut into triangles with an angle of 35°. Use a protractor and a straightedge to draw an angle with this measure. (Lesson 13-2)

6. **CARPENTRY** In order to build a bookcase, a 13-foot piece of lumber needs to be bisected. What will be the measure of each segment? (Lesson 13-3)

7. **BAKING** Sue is making a birthday cake for a friend. The cake will be a ladybug when complete. In order to form the wings of the ladybug, Sue must cut a 100° wedge from a round cake and then bisect the wedge to create two wings of the same size. What will be the angle measure of each of the wings? (Lesson 13-3)

8. **FLOORING** Ivan is installing a new tile floor. The tiles he is using are in the shape of a regular octagon. If the perimeter of each tile is 72 inches, how long is each side? (Lesson 13-4)

9. **TENTS** The front of a tent is shaped like an equilateral triangle. How long is each side of the tent if the perimeter is 147 inches? (Lesson 13-4)

10. **SWIMMING POOL** A swimming pool is to be built at a community center. The shape of the pool is shown below. Identify the polygon and then tell whether it appears to be a regular polygon. (Lesson 13-4)

11. **SPORTS** How many lines of symmetry does this figure of a baseball bat have? (Lesson 13-5)

12. **SCHOOL** Does a pinecone have rotational symmetry? (Lesson 13-5)

13. **MAPS** On two different-sized maps, the city park is shown using the two figures below. Tell whether the pair of figures is *congruent, similar,* or *neither.* (Lesson 13-6)

14. **GARDENING** Jessica works in two different garden areas. The first area is rectangular in shape with a length of 10 feet and a width of 6 feet. The second garden area is square in shape with sides measuring 6 feet. Tell whether these two garden plots are *congruent, similar,* or *neither.* (Lesson 13-6)

PAINTING **For Exercises 1 and 2, use the following information.**
Van is going to paint a wall in his home. The wall is in the shape of a parallelogram with a base of 8 feet and a height of 15 feet. (Lesson 14-1)

1. What is the area of the wall?

2. If one quart of paint will cover 70 square feet, how many quarts of paint does Van need to buy?

3. **SEWING** Marcie is using panels of fabric like the one shown below to make a tablecloth. Find the area of each panel. (Lesson 14-1)

30 in.

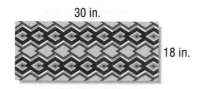

18 in.

4. **MEASUREMENT** How many square feet are in 5 square yards? (Lesson 14-1)

5. **LAND** Antoine is purchasing a plot of land on which he plans to build a house. The plot is in the shape of a triangle with a base of 96 feet and a height of 192 feet. Find the area of the plot of land. (Lesson 14-2)

6. **HISTORY** Alex is making a kite like the one shown. What will be the area of the kite? (Lesson 14-2)

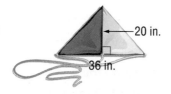

20 in.

36 in.

7. **SWIMMING POOL** A circular pool has a diameter of 22 feet. What is the area of the bottom of the pool to the nearest tenth? (Lesson 14-3)

8. **POTTERY** Benito is making a flower vase that has a circular base with a radius of $3\frac{1}{2}$ inches. Find the area of the base to the nearest tenth. (Lesson 14-3)

9. **MOUNTAINS** Suppose a mountain is circular and has a diameter of 4.2 kilometers. About how much land does the mountain cover? (Lesson 14-3)

10. **FOOD** Identify the figure that represents a box of cereal. (Lesson 14-4)

11. **BLOCKS** The figure shown below is one of the shapes found in a set of wooden building blocks. Identify the figure. (Lesson 14-4)

12. **TOY BOX** Jamila is given the toy box shown below for her birthday. What is the volume of the toy box? (Lesson 14-5)

2 ft

4 ft $2\frac{1}{2}$ ft

13. **MULTI STEP** A box has a length of 8.5 feet, a width of 5 feet, and a height of 6 feet. Find the volume of the box if the height is increased by 50%. (Lesson 14-5)

14. **PACKAGING** Find the volume of the pasta box shown below. (Lesson 14-5)

10 in.

6 in.   2 in.

15. **GIFT WRAP** Olivia is placing a gift inside a box that measures 15 centimeters by 8 centimeters by 3 centimeters. What is the surface area of the box? (Lesson 14-6)

16. **CONSTRUCTION** Find the surface area of the concrete block shown below. (Lesson 14-6)

12 in.

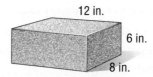

6 in.

8 in.

Mixed Problem Solving

Glossary/Glosario

Cómo usar el glosario en español:
1. Busca el término en inglés que desees encontrar.
2. El término en español, junto con la definición, se encuentran en la columna de la derecha.

English

Español

A

acute angle (p. 506) An angle with a measure greater than 0° and less than 90°.

ángulo agudo Ángulo que mide más de 0° y menos de 90°.

Addition Property of Equality (p. 345) If you add the same number to each side of an equation, the two sides remain equal.

propiedad de adición de la igualdad Si sumas el mismo número a ambos lados de una ecuación, los dos lados permanecen iguales.

Additive Identity (p. 334) The sum of any number and 0 is the number.

identidad aditiva La suma de cualquier número y 0 es el número mismo.

algebra (p. 28) A mathematical language that uses symbols, usually letters, along with numbers. The letters stand for numbers that are unknown.

álgebra Lenguaje matemático que usa símbolos, por lo general, además de números. Las letras representan números desconocidos.

algebraic expression (p. 28) A combination of variables, numbers, and at least one operation.

expresión algebraica Combinación de variables, números y, por lo menos, una operación.

angle (p. 506) Two rays with a common endpoint form an angle.

ángulo Dos rayos con un extremo común forman un ángulo.

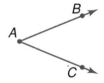

$\angle BAC$, $\angle CAB$, or $\angle A$

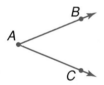

$\angle BAC$, $\angle CAB$, ó $\angle A$

area (p. 39) The number of square units needed to cover the surface enclosed by a geometric figure.

área El número de unidades cuadradas necesarias para cubrir una superficie cerrada por una figura geométrica.

Associative Property (p. 334) The way in which numbers are grouped when added or multiplied does not change the sum or product.

propiedad asociativa La manera de agrupar números al sumarlos o multiplicarlos no cambia su suma o producto.

average (p. 76) The sum of two or more quantities divided by the number of quantities; the mean.

promedio La suma de dos o más cantidades dividida entre el número de cantidades; la media.

B

bar graph (p. 56) A graph using bars to compare quantities. The height or length of each bar represents a designated number.

gráfica de barras Gráfica que usa barras para comparar cantidades. La altura o longitud de cada barra representa un número designado.

base (p. 18) In a power, the number used as a factor. In 10^3, the base is 10. That is, $10^3 = 10 \times 10 \times 10$.

base En una potencia, el número usado como factor. En 10^3, la base es 10. Es decir, $10^3 = 10 \times 10 \times 10$.

base (p. 546) Any side of a parallelogram.

base Cualquier lado de un paralelogramo.

base (p. 564) The faces on the top and bottom of a three-dimensional figure.

base Las caras superior e inferior en una figura tridimensional.

bisect (p. 515) To separate something into two congruent parts.

bisecar Separar algo en dos partes congruentes.

<hr>

C

center (p. 161) The given point from which all points on a circle or a sphere are the same distance.

centro Un punto dado del cual equidistan todos los puntos de un círculo o de una esfera.

center (p. 565) The given point from which all points on a sphere are the same distance.

centro Un punto dado del cual equidistan todos los puntos de una esfera.

centimeter (p. 476) A metric unit of length. One centimeter equals one-hundredth of a meter.

centímetro Unidad métrica de longitud. Un centímetro es igual a la centésima parte de un metro.

circle (p. 161) The set of all points in a plane that are the same distance from a given point called the center.

círculo Conjunto de todos los puntos en un plano que equidistan de un punto dado llamado centro.

circle graph (p. 62) A graph used to compare parts of a whole. The circle represents the whole and is separated into parts of the whole.

gráfica circular Tipo de gráfica estadística que se usa para comparar las partes de un todo. El círculo representa el todo y éste se separa en partes.

circumference (p. 161) The distance around a circle.

circunferencia La distancia alrededor de un círculo.

clustering (p. 117) An estimation method in which a group of numbers close in value are rounded to the same number.

agrupamiento Método de estimación en que un grupo de números cuyo valor está estrechamente relacionado, se redondean al mismo número.

coefficient (p. 350) The numerical factor of a term that contains a variable.

coeficiente El factor numérico de un término que contiene una variable.

common multiples (p. 194) Multiples that are shared by two or more numbers. For example, some common multiples of 2 and 3 are 6, 12, and 18.

múltiplos comunes Múltiplos compartidos por dos o más números. Por ejemplo, algunos múltiplos comunes de 2 y 3 son 6, 12 y 18.

Commutative Property (p. 334) The order in which numbers are added or multiplied does not change the sum or product.

propiedad conmutativa El orden en que se suman o multiplican dos números no afecta su suma o producto.

compatible numbers (p. 256) Numbers that are easy to divide mentally.

números compatibles Números que son fáciles de dividir mentalmente.

complementary (p. 507) Two angles are complementary if the sum of their measures is 90°.

ángulos complementarios Dos ángulos son complementarios si la suma de sus medidas es 90°.

complementary events (p. 429) Two events in which either one or the other must take place, but they cannot both happen at the same time. The sum of their probabilities is 1.

eventos complementarios Dos eventos tales, que uno de ellos debe ocurrir, pero ambos no pueden ocurrir simultáneamente. La suma de sus probabilidades es 1.

composite number (p. 14) A number greater than 1 with more than two factors.

número compuesto Un número entero mayor que 1 con más de dos factores.

cone (p. 565) A three-dimensional figure with curved surfaces, a circular base and one vertex.

cono Figura tridimensional con superficies curvas, una base circular y un vértice.

congruent (p. 515) Having the same measure.

congruente Que tiene la misma medida.

congruent figures (p. 534) Figures that are the same shape and size.

figuras congruentes Figuras que tienen la misma forma y tamaño.

coordinate plane (p. 320) A plane in which a horizontal number line and a vertical number line intersect at their zero points.

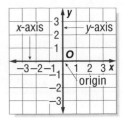

coordinate system (p. 320) See coordinate plane.

corresponding parts (p. 535) The parts of congruent figures that match.

cross products (p. 386) The products of the terms on the diagonals when two ratios are compared. If the cross products are equal, then the ratios form a proportion.

cubed (p. 18) The product in which a number is a factor three times. Two cubed is 8 because $2 \times 2 \times 2 = 8$.

cubic units (p. 570) Used to measure volume. Tells the number of cubes of a given size it will take to fill a three-dimensional figure.

cup (p. 470) A customary unit of capacity equal to 8 fluid ounces.

cylinder (p. 565) A three-dimensional figure with all curved surfaces, two circular bases and no vertices.

plano de coordenadas Plano en que una recta numérica horizontal y una recta numérica vertical se intersecan en sus puntos cero.

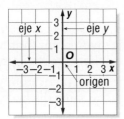

sistema de coordenadas Plano en el cual se han trazado dos rectas numéricas, una horizontal y una vertical, que se intersecan en sus puntos cero.

partes correspondientes Las partes de figuras congruentes que coinciden.

productos cruzados Productos que resultan de la comparación de los términos de las diagonales de dos razones. Si los productos son iguales, las razones forman una proporción.

al cubo El producto de un número por sí mismo, tres veces. Dos al cubo es 8 porque $2 \times 2 \times 2 = 8$.

unidades cúbicas Se usan para medir el volumen. Indican el número de cubos de cierto tamaño que se necesitan para llenar una figura tridimensional.

taza Unidad de capacidad del sistema inglés de medidas que equivale a 8 onzas líquidas.

cilindro Figura tridimensional compuesta de superficies curvas, dos bases circulares y ningún vértice.

D

data (p. 50) Information, often numerical, which is gathered for statistical purposes.

degree (p. 506) The most common unit of measure for angles.

diameter (p. 161) The distance across a circle through its center.

Distributive Property (p. 333) To multiply a sum by a number, multiply each addend of the sum by the number outside the parentheses.

divisible (p. 10) A whole number is said to be divisible by another number if the remainder is 0 when the first number is divided by the second.

datos Información, con frecuencia numérica, que se recoge con fines estadísticos.

grado La unidad más común para medir ángulos.

diámetro La distancia a través de un círculo pasando por el centro.

propiedad distributiva Para multiplicar una suma por un número, multiplica cada sumando de la suma por el número fuera del paréntesis.

divisible Se dice que un número entero es divisible entre otro si el residuo es cero cuando el primer número se divide entre el segundo.

edge (p. 564) The intersection of faces of a three-dimensional figure.

equation (p. 34) A mathematical sentence that contains an equals sign, =.

equals sign (p. 34) A symbol of equality, =.

equilateral triangle (p. 523) A triangle having all three sides congruent and all three angles congruent.

equivalent decimals (p. 109) Decimals that name the same number.

equivalent fractions (p. 182) Fractions that name the same number.

equivalent ratios (p. 381) Ratios that have the same value.

evaluate (p. 29) To find the value of an algebraic expression by replacing variables with numerals.

even (p. 10) A whole number that is divisible by 2.

event (p. 428) A specific outcome or type of outcome.

expanded form (p. 103) The sum of the products of each digit and its place value of a number.

exponent (p. 18) In a power, the number of times the base is used as a factor. In 5^3, the exponent is 3. That is, $5^3 = 5 \times 5 \times 5$.

arista La intersección de las caras de una figura tridimensional.

ecuación Un enunciado matemático que contiene el signo de igualdad, =.

signo de igualdad Símbolo que indica igualdad, =.

triángulo equilátero Triángulo cuyos tres lados y tres ángulos son congruentes.

decimales equivalentes Decimales que representan el mismo número.

fracciones equivalentes Fracciones que representan el mismo número.

razones equivalentes Dos razones que tienen el mismo valor.

evaluar Calcular el valor de una expresión sustituyendo las variables por número.

par Número entero que es divisible entre 2.

evento Resultado específico o tipo de resultado.

forma desarrollada La suma de los productos de cada dígito y el valor de posición del número.

exponente En una potencia, el número de veces que la base se usa como factor. En 5^3, el exponente es 3. Es decir, $5^3 = 5 \times 5 \times 5$.

face (p. 564) The flat surface of a three-dimensional figure.

factor (p. 14) A number that divides into a whole number with a remainder of zero.

fluid ounces (p. 470) A customary unit of capacity.

foot (p. 465) A customary unit of length equal to 12 inches.

formula (p. 39) An equation that shows a relationship among certain quantities.

frequency table (p. 50) A table for organizing a set of data that shows the number of times each item or number appears.

front-end estimation (p. 117) An estimation method in which the front digits are added or subtracted first, and then the digits in the next place value position are added or subtracted.

cara La superficie plana de una figura tridimensional.

factor Número que al dividirlo entre un número entero tiene un residuo de cero.

onzas líquidas Unidad de capacidad del sistema inglés de medidas.

pie Unidad de longitud del sistema inglés de medidas que equivale a 12 pulgadas.

fórmula Ecuación que muestra una relación entre ciertas cantidades.

tabla de frecuencias Tabla que se usa para organizar un conjunto de datos y que muestra cuántas veces aparece cada dato.

estimación frontal Método de estimación en que primero se suman o restan los dígitos del frente y a continuación se suman o restan los dígitos en el siguiente valor de posición.

function (p. 362) A relation in which each element of the input is paired with exactly one element of the output according to a specified rule.

function rule (p. 362) An expression which describes the relationship between each input and output.

function table (p. 362) A table organizing the input, rule, and output of a function.

función Relación en que cada elemento de entrada es apareado con un único elemento de salida, según una regla específica.

regla de funciones Expresión que describe la relación entre cada valor de entrada y de salida.

tabla de funciones Tabla que organiza las entradas, la regla y las salidas de una función.

gallon (p. 470) A customary unit of capacity equal to 4 quarts.

gram (p. 484) The basic unit of mass in the metric system.

graph (p. 56) A visual way to display data.

graph (p. 295) To graph an integer on a number line, draw a dot at the location on the number line that corresponds to the integer.

greatest common factor (GCF) (p. 177) The greatest of the common factors of two or more numbers. The GCF of 24 and 30 is 6.

galón Unidad de capacidad del sistema inglés de medidas que equivale a 4 cuartos de galón.

gramo Unidad fundamental de masa del sistema métrico.

gráfica Manera visual de representar datos.

graficar Para graficar un entero sobre una recta numérica, dibuja un punto en la ubicación de la recta numérica correspondiente al entero.

máximo común divisor (MCD) El mayor factor común de dos o más números. El MCD de 24 y 30 es 6.

height (p. 546) The shortest distance from the base of a parallelogram to its opposite side.

altura La distancia más corta desde la base de un paralelogramo hasta su lado opuesto.

heptagon (p. 522) A polygon having seven sides.

hexagon (p. 522) A polygon having six sides.

horizontal axis (p. 56) The axis on which the categories are shown in a bar and line graph.

hour (p. 494) A commonly used unit of time. There are 60 minutes in one hour and 24 hours in one day.

heptágono Polígono de siete lados.

hexágono Polígono de seis lados.

eje horizontal El eje sobre el cual se muestran las categorías en una gráfica de barras o gráfica lineal.

hora Unidad de tiempo de uso común. Hay 60 minutos en una hora y 24 horas en un día.

improper fraction (p. 186) A fraction that has a numerator that is greater than or equal to the denominator. The value of an improper fraction is greater than or equal to 1.

inch (p. 465) A customary unit of length. Twelve inches equal one foot.

independent events (p. 450) Two or more events in which the outcome of one event does not affect the outcome(s) of the other event(s).

integer (p. 294) The whole numbers and their opposites. …, −3, −2, −1, 0, 1, 2, 3, …

interval (p. 50) The difference between successive values on a scale.

fracción impropia Fracción cuyo numerador es mayor que o igual a su denominador. El valor de una fracción impropia es mayor que o igual a 1.

pulgada Unidad de longitud del sistema inglés de medidas. Doce pulgadas equivalen a un pie.

eventos independientes Dos o más eventos en los cuales el resultado de uno de ellos no afecta el resultado de los otros eventos.

entero Los números enteros y sus opuestos. …, −3, −2, −1, 0, 1, 2, 3, …

intervalo La diferencia entre valores sucesivos de una escala.

inverse operations (p. 339) Operations which *undo* each other. For example, addition and subtraction are inverse operations.

isosceles triangle (p. 523) A triangle in which at least two sides are congruent.

operaciones inversas Operaciones que *se anulan* mutuamente. La adición y la sustracción son operaciones inversas.

triángulo isósceles Triángulo que tiene por lo menos dos lados congruentes.

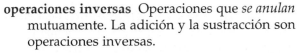

key (p. 72) A sample data point used to explain the stems and leaves.

kilogram (p. 484) A metric unit of mass. One kilogram equals one thousand grams.

kilometer (p. 476) A metric unit of length. One kilometer equals one thousand meters.

clave Punto de muestra de los datos que se usa para explicar los tallos y las hojas.

kilogramo Unidad métrica de masa. Un kilogramo equivale a mil gramos.

kilómetro Unidad métrica de longitud. Un kilómetro equivale a mil metros.

lateral face (p. 564) A side of a three-dimensional figure.

least common denominator (LCD) (p. 198) The least common multiple of the denominators of two or more fractions.

least common multiple (LCM) (p. 194) The least of the common multiples of two or more numbers. The LCM of 2 and 3 is 6.

leaves (p. 72) The units digit written to the right of the vertical line in a stem-and-leaf plot.

like fractions (p. 228) Fractions with the same denominator.

line graph (p. 56) A graph used to show how a set of data changes over a period of time.

line of symmetry (p. 528) A line that divides a figure into two halves that are reflections of each other.

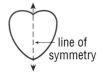
line of symmetry

cara lateral Un lado de una figura tridimensional.

mínimo común denominador (mcd) El menor múltiplo común de los denominadores de dos o más fracciones.

mínimo común múltiplo (mcm) El menor múltiplo común de dos o más números. El mcm de 2 y 3 es 6.

hojas El dígito de las unidades que se escribe a la derecha de la recta vertical en un diagrama de tallo y hojas.

fracciones semejantes Fracciones que tienen el mismo denominador.

gráfica lineal Gráfica que se usa para mostrar cómo cambian los valores durante un período de tiempo.

eje de simetría Recta que divide una figura en dos mitades que son reflexiones una de la otra.

eje de simetría

line symmetry (p. 528) A figure is said to have line symmetry when two halves of the figure match.

liter (p. 485) The basic unit of capacity in the metric system. A liter is a little more than a quart.

simetría lineal Exhiben simetría lineal las figuras que coinciden exactamente al doblarse una sobre otra.

litro Unidad básica de capacidad del sistema métrico. Un litro es un poco más de un cuarto de galón.

mean (p. 76) The sum of the numbers in a set of data divided by the number of pieces of data.

media La suma de los números en un conjunto de datos dividida entre el número total de datos.

measure of central tendency (p. 76) A number that helps describe all of the data in a data set.

median (p. 80) The middle number in a set of data when the data are arranged in numerical order. If the data has an even number, the median is the mean of the two middle numbers.

meter (p. 476) The basic unit of length in the metric system.

metric system (p. 476) A decimal system of weights and measures. The meter is the basic unit of length, the gram is the basic unit of weight, and the liter is the basic unit of capacity.

mile (p. 465) A customary unit of length equal to 5,280 feet or 1,760 yards.

milligram (p. 484) A metric unit of mass. One milligram equals one-thousandth of a gram.

milliliter (p. 485) A metric unit of capacity. One milliliter equals one-thousandth of a liter.

millimeter (p. 476) A metric unit of length. One millimeter equals one-thousandth of a meter.

minute (p. 494) A commonly used unit of time. There are 60 seconds in one minute and there are 60 minutes in one hour.

mixed number (p. 186) The sum of a whole number and a fraction. $1\frac{1}{2}$, $2\frac{3}{4}$, and $4\frac{5}{8}$ are mixed numbers.

mode (p. 80) The number(s) or item(s) that appear most often in a set of data.

multiple (p. 194) The product of the number and any whole number.

Multiplicative Identity (p. 334) The product of any number and 1 is the number.

medida de tendencia central Número que ayuda a describir todos los datos en un conjunto de datos.

mediana Número central de un conjunto de datos, una vez que los datos han sido ordenados numéricamente. Si hay un número par de datos, la mediana es el promedio de los dos datos centrales.

metro Unidad básica de longitud del sistema métrico.

sistema métrico Sistema decimal de pesos y medidas. El metro es la unidad fundamental de longitud, el gramo es la unidad fundamental de masa y el litro es la unidad fundamental de capacidad.

milla Unidad de longitud del sistema inglés que equivale a 5,280 pies ó 1,760 yardas.

miligramo Unidad métrica de masa. Un miligramo equivale a la milésima parte de un gramo.

mililitro Unidad métrica de capacidad. Un mililitro equivale a la milésima parte de un litro.

milímetro Unidad métrica de longitud. Un milímetro equivale a la milésima parte de un metro.

minuto Unidad de tiempo de uso común. Hay 60 segundos en un minuto y hay 60 minutos en una hora.

número mixto La suma de un número entero y una fracción. $1\frac{1}{2}$, $2\frac{3}{4}$ y $4\frac{5}{8}$ son números mixtos.

moda Número(s) de un conjunto de datos que aparece(n) más frecuentemente.

múltiplo El producto de un número y cualquier número entero.

identidad multiplicativa El producto de cualquier número y 1 es el número mismo.

negative integer (p. 294) An integer that is less than zero.

numerical expression (p. 24) A combination of numbers and operations.

entero negativo Entero menor que cero.

expresión numérica Una combinación de números y operaciones.

obtuse angle (p. 506) An angle that measures greater than 90° but less than 180°.

ángulo obtuso Ángulo que mide más de 90° pero menos de 180°.

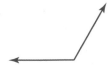

octagon (p. 522) A polygon having eight sides.

odd (p. 10) A whole number that is not divisible by 2.

octágono Polígono que tiene ocho lados.

impar Número entero que no es divisible entre 2.

opposites (p. 296) Two integers are opposites if they are represented on the number line by points that are the same distance from zero, but in opposite directions from zero. The sum of opposites is zero.

ordered pair (p. 320) A pair of numbers used to locate a point in the coordinate system. The ordered pair is written in this form: (x-coordinate, y-coordinate).

order of operations (p. 24) The rules which tell which operation to perform first when more than one operation is used.
1. Simplify the expressions inside grouping symbols, like parentheses.
2. Find the value of all powers.
3. Multiply and divide in order from left to right.
4. Add and subtract in order from left to right.

origin (p. 320) The point of intersection of the x-axis and y-axis in a coordinate system.

ounce (p. 471) A customary unit of weight. 16 ounces equals one pound.

outcomes (p. 428) Possible results of a probability event. For example, 4 is an outcome when a number cube is rolled.

outlier (p. 77) A value that is much higher or much lower than the other values in a set of data.

opuestos Dos enteros son opuestos si, en la recta numérica, están representados por puntos que equidistan de cero, pero en direcciones opuestas. La suma de opuestos es cero.

par ordenado Par de números que se utiliza para ubicar un punto en un plano de coordenadas. Se escribe de la siguiente forma: (coordenada x, coordenada y).

orden de las operaciones Reglas que establecen cuál operación debes realizar primero, cuando hay más de una operación involucrada.
1. Primero ejecuta todas las operaciones dentro de los símbolos de agrupamiento.
2. Evalúa todas las potencias.
3. Multiplica y divide en orden de izquierda a derecha.
4. Suma y resta en orden de izquierda a derecha.

origen Punto en que el eje x y el eje y se intersecan en el sistema de coordenadas.

onza Unidad de peso del sistema inglés de medidas. 16 onzas equivalen a una libra.

resultado Uno de los resultados posibles de un evento probabilístico. Por ejemplo, 4 es un resultado posible cuando se lanza un dado.

valor atípico Dato que se encuentra muy separado de los otros valores en un conjunto de datos.

P

parallelogram (p. 523) A quadrilateral that has both pairs of opposite sides congruent and parallel.

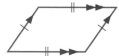

parallelogramo Cuadrilátero con ambos pares de lados opuestos paralelos y congruentes.

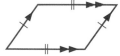

pentagon (p. 522) A polygon with five sides.

percent (p. 395) A ratio that compares a number to 100.

perimeter (p. 158) The distance around any closed geometric figure.

perpendicular (p. 515) Two line segments are perpendicular if they meet at right angles.

pint (p. 470) A customary unit of capacity equal to two cups.

polygon (p. 522) A simple closed figure in a plane formed by three or more line segments.

population (p. 438) The entire group of items or individuals from which the samples under consideration are taken.

positive integer (p. 294) An integer that is greater than zero.

pentágono Polígono que tiene cinco lados.

por ciento Razón que compara un número con 100.

perímetro La distancia alrededor de una figura geométrica cerrada.

perpendicular Dos segmentos de recta son perpendiculares si se encuentran a ángulos rectos.

pinta Unidad de capacidad del sistema inglés de medidas que equivale a dos tazas.

polígono Figura simple cerrada en un plano, formada por tres o más segmentos de recta.

población El grupo total de individuos o de artículos del cual se toman las muestras bajo estudio.

entero positivo Un entero mayor que cero.

pound (p. 471) A customary unit of weight equal to 16 ounces.

power (p. 18) A number that can be expressed using an exponent. The power 3^2 is read three to the second power, or three squared.

prime factorization (p. 15) A composite number expressed as a product of prime numbers.

prime number (p. 14) A whole number that has exactly two unique factors, 1 and the number itself.

prism (p. 564) A three-dimensional figure that has two parallel and congruent bases in the shape of polygons and at least three lateral faces shaped like rectangles. The shape of the bases tells the name of the prism.

proportion (p. 386) An equation that shows that two ratios are equivalent, $\frac{a}{b} = \frac{c}{d}$, $b \neq 0$, $d \neq 0$.

pyramid (p. 564) A solid figure that has a polygon for a base and triangles for sides. A pyramid is named for the shape of its base.

libra Unidad de peso del sistema inglés de medidas que equivale a 16 onzas.

potencias Números que se expresan usando exponentes. La potencia 3^2 se lee tres a la segunda potencia o tres al cuadrado.

factorización prima Un número compuesto expresado como el producto de números primos.

número primo Número entero que tiene exactamente dos factores, 1 y sí mismo.

prisma Figura tridimensional que tiene dos bases paralelas y congruentes en forma de polígonos y posee por lo menos tres caras laterales en forma de rectángulos. La forma de las bases indica el nombre del prisma.

proporción Enunciado que establece la igualdad de dos razones, $\frac{a}{b} = \frac{c}{d}$, $b \neq 0$, $d \neq 0$.

pirámide Figura sólida cuya base es un polígono y sus lados son triángulos. Las pirámides se nombran de acuerdo con la forma de sus bases.

Q

quadrants (p. 320) The four regions into which the two perpendicular number lines of a coordinate system separate the plane.

quadrilateral (p. 522) A polygon with four sides.

quart (p. 470) A customary unit of capacity equal to two pints.

cuadrantes Las cuatro regiones en que las dos rectas numéricas perpendiculares dividen un sistema de coordenadas.

cuadrilátero Un polígono con cuatro lados.

cuarto de galón Unidad de capacidad del sistema inglés de medidas que equivale a dos pintas.

R

radius (p. 161) The distance from the center of a circle to any point on the circle.

radius

radio Distancia desde el centro de un círculo hasta cualquier punto del mismo.

radio

random (p. 438) Outcomes occur at random if each outcome is equally likely to occur.

range (p. 82) The difference between the greatest number and the least number in a set of data.

rate (p. 381) A ratio of two measurements having different units.

aleatorio Los resultados ocurren al azar si la posibilidad de ocurrir de cada resultado es equiprobable.

rango La diferencia entre el número mayor y el número menor en un conjunto de datos.

tasa Razón que compara dos cantidades que tienen distintas unidades de medida.

ratio (p. 380) A comparison of two numbers by division. The ratio of 2 to 3 can be stated as 2 out of 3, 2 to 3, 2:3, or $\frac{2}{3}$.

reciprocals (p. 272) Any two numbers whose product is 1. Since $\frac{5}{6} \times \frac{6}{5} = 1$, $\frac{5}{6}$ and $\frac{6}{5}$ are reciprocals.

rectangle (p. 523) A quadrilateral with opposite sides congruent and parallel and all angles are right angles.

regular polygon (p. 522) A polygon having all sides congruent and all angles congruent.

repeating decimal (p. 207) A decimal whose digits repeat in groups of one or more. 0.181818… can also be written $0.\overline{18}$. The bar above the digits indicates those digits repeat.

rhombus (p. 523) A parallelogram with all sides congruent.

right angle (p. 506) An angle that measures 90°.

rotational symmetry (p. 529) A figure that can be turned or rotated less than 360° about a fixed point so that the figure looks exactly as it did before being turned is said to have rotational symmetry.

razón Comparación de dos números mediante división. La razón de 2 a 3 puede escribirse como 2 de cada 3, 2 a 3, 2:3 ó $\frac{2}{3}$.

recíproco Cualquier par de números cuyo producto es 1. Como $\frac{5}{6} \times \frac{6}{5} = 1$, $\frac{5}{6}$ y $\frac{6}{5}$ son recíprocos.

rectángulo Cuadrilátero cuyos lados opuestos son congruentes y paralelos y cuyos ángulos son todos ángulos rectos.

polígono regular Polígono cuyos lados y ángulos son todos congruentes.

decimales periódicos Decimal cuyos dígitos se repiten en grupos de uno o más. 0.181818… puede también escribirse como $0.\overline{18}$. La barra sobre los dígitos indica que esos dígitos se repiten.

rombo Paralelogramo cuyos lados son todos congruentes.

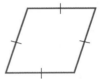

ángulo recto Ángulo que mide 90°.

simetría rotacional Se dice que una figura posee simetría rotacional si se puede girar o rotar menos de 360° en torno a un punto fijo de modo que la figura se vea exactamente como se veía antes de ser girada.

sample (p. 438) A randomly-selected group that is used to represent a whole population.

sample space (p. 433) The set of all possible outcomes in a probability experiment.

scale (p. 50) The set of all possible values of a given measurement, including the least and greatest numbers in the set, separated by the intervals used.

scale (p. 391) The ratio that compares the measurements on the drawing or model to the measurements of the real object.

muestra Grupo escogido al azar o aleatoriamente y que se usa para representar la población entera.

espacio muestral Conjunto de todos los resultados posibles de un experimento probabilístico.

escala Conjunto de todos los posibles valores de una medida dada, el cual incluye los valores máximo y mínimo del conjunto, separados mediante los intervalos que se han usado.

escala Razón que compara las medidas en un dibujo o modelo con las medidas del objeto real.

scale drawing (p. 391) A drawing that is similar but either larger or smaller than the actual object.

scale model (p. 391) A model used to represent something that is too large or too small for an actual-size model.

scalene triangle (p. 523) A triangle in which no sides are congruent.

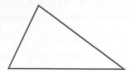

scientific notation (p. 136) A way of expressing a number as the product of a number that is at least 1 but less than 10 and a power of 10. For example, $687{,}000 = 6.87 \times 10^5$.

second (p. 494) A commonly used unit of time. There are 60 seconds in one minute.

sequence (p. 282) A list of numbers in a specific order, such as 0, 1, 2, 3, or 2, 4, 6, 8.

side (p. 506) A ray that is part of an angle.

similar figures (p. 534) Figures that have the same shape but different sizes.

simplest form (p. 183) The form of a fraction when the GCF of the numerator and the denominator is 1. The fraction $\frac{3}{4}$ is in simplest form because the GCF of 3 and 4 is 1.

solution (p. 34) The value of a variable that makes an equation true. The solution for $12 = x + 7$ is 5.

solve (p. 34) To replace a variable with a value that results in a true sentence.

sphere (p. 565) A three-dimensional figure with no faces, bases, edges, or vertices. All of its points are the same distance from a given point called the center.

center

square (p. 523) A parallelogram with all sides congruent, all angles right angles, and opposite sides parallel.

squared (p. 18) A number multiplied by itself, 4×4, or 4^2.

dibujo a escala Dibujo que es semejante, pero más grande o más pequeño que el objeto real.

modelo a escala Modelo que se usa para representar algo que es demasiado grande o demasiado pequeño como para construirlo de tamaño natural.

triángulo escaleno Triángulo sin lados congruentes.

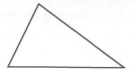

notación científica Manera de expresar números como el producto de un número que es al menos igual a 1, pero menor que 10, por una potencia de diez. Por ejemplo, $687{,}000 = 6.87 \times 10^5$.

segundo Unidad de tiempo de uso común. Hay 60 segundos en un minuto.

sucesión Lista de números en un orden específico como, por ejemplo, 0, 1, 2, 3 ó 2, 4, 6, 8.

lado Rayo que es parte de un ángulo.

figuras semejantes Figuras que tienen la misma forma, pero diferente tamaño.

forma reducida La forma de una fracción cuando el MCD del numerador y del denominador es 1. La fracción $\frac{3}{4}$ está en forma reducida porque el MCD de 3 y 4 es 1.

solución Valor de la variable de una ecuación que hace verdadera la ecuación. La solución de $12 = x + 7$ es 5.

resolver Reemplazar una variable con un valor que resulte en un enunciado verdadero.

esfera Figura tridimensional que carece de caras, bases, aristas o vértices. Todos sus puntos en el espacio equidistan de un punto dado llamado centro.

centro

cuadrado Paralelogramo con todos los lados congruentes, todos los ángulos rectos y lados opuestos paralelos.

cuadrado El producto de un número multiplicado por sí mismo, 4×4 ó 4^2.

standard form (p. 103) Numbers written without exponents.

statistics (p. 50) The study of collecting, analyzing, and presenting data.

stem-and-leaf plot (p. 72) A system used to condense a set of data where the greatest place value of the data forms the stem and the next greatest place value forms the leaves.

stems (p. 72) The greatest place value common to all the data that is written to the left of the line in a stem-and-leaf plot.

straight angle (p. 506) An angle that measures exactly 180°.

Subtraction Property of Equality (p. 340) If you subtract the same number from each side of an equation, the two sides remain equal.

supplementary (p. 507) Two angles are supplementary if the sum of their measures is 180°.

surface area (p. 575) The sum of the areas of all the surfaces (faces) of a three-dimensional figure.

survey (p. 438) A question or set of questions designed to collect data about a specific group of people.

forma estándar Números escritos sin exponentes.

estadística Estudio de la recopilación, análisis y presentación de datos.

diagrama de tallo y hojas Sistema que se usa para condensar un conjunto de datos y en el cual el mayor valor de posición de los datos forma el tallo y el siguiente valor de posición mayor forma las hojas.

tallo El mayor valor de posición común a todos los datos que se escribe a la izquierda de la línea en un diagrama de tallo y hojas.

ángulo llano Ángulo que mide exactamente 180°.

propiedad de sustracción de la igualdad Si sustraes el mismo número de ambos lados de una ecuación, los dos lados permanecen iguales.

ángulos suplementarios Dos ángulos son suplementarios si la suma de sus medidas es 180°.

área de superficie La suma de las áreas de todas las superficies (caras) de una figura tridimensional.

encuesta Pregunta o conjunto de preguntas diseñadas para recoger datos sobre un grupo específico de personas.

T

tally mark (p. 50) A counter used to record items in a group.

terminating decimal (p. 206) A decimal whose digits end. Every terminating decimal can be written as a fraction with a denominator of 10, 100, 1000, and so on.

theoretical probability (p. 428) The ratio of the number of ways an event can occur to the number of possible outcomes.

three-dimensional figure (p. 564) A figure that encloses a part of space.

ton (p. 471) A customary unit of weight equal to 2,000 pounds.

tree diagram (p. 433) A diagram used to show the total number of possible outcomes in a probability experiment.

triangle (p. 522) A polygon with three sides.

two-step equation (p. 355) An equation that has two different operations in it.

marca de conteo Marca que se usa para anotar artículos en un grupo.

decimal terminal Decimal cuyos dígitos terminan. Todo decimal terminal puede escribirse como una fracción con un denominador de 10, 100, 1000 y así sucesivamente.

probabilidad teórica La razón del número de maneras en que puede ocurrir un evento al número total de resultados posibles.

figura tridimensional Figura que encierra parte del espacio.

tonelada Unidad de peso del sistema inglés que equivale a 2,000 libras.

diagrama de árbol Diagrama que se usa para mostrar el número total de resultados posibles en un experimento probabilístico.

triángulo Polígono que posee tres lados.

ecuación de dos pasos Ecuación que contiene dos operaciones distintas.

unit rate (p. 381) A rate that has a denominator of 1.

tasa unitaria Una tasa con un denominador de 1.

variable (p. 28) A symbol, usually a letter, used to represent a number.

variable Un símbolo, por lo general, una letra, que se usa para representar un número.

Venn diagram (p. 177) A diagram which uses circles to display elements of different sets. Overlapping circles show common elements.

diagrama de Venn Diagrama que usa círculos para mostrar elementos de diferentes conjuntos. Círculos sobrepuestos indican elementos comunes.

vertex (p. 506) The common endpoint of the two rays that form an angle.

vértice El extremo común de dos rayos que forman un ángulo.

vertex (vertices) (p. 564) The point where the edges of a three-dimensional figure intersect.

vértices Punto de intersección de las aristas de una figura tridimensional.

vertical axis (p. 56) The axis on which the scale and interval are shown in a bar or line graph.

eje vertical Eje sobre el cual se muestran la escala y el intervalo en una gráfica de barras o en una gráfica lineal.

volume (p. 570) The amount of space that a three-dimensional figure contains. Volume is expressed in cubic units.

volumen Cantidad de espacio que contiene una figura tridimensional. El volumen se expresa en unidades cúbicas.

x-axis (p. 320) The horizontal line of the two perpendicular number lines in a coordinate plane.

eje x La recta horizontal de las dos rectas numéricas perpendiculares en un plano de coordenadas.

x-coordinate (p. 320) The first number of an ordered pair.

coordenada x El primer número de un par ordenado.

yard (p. 465) A customary unit of length equal to 3 feet, or 36 inches.

yarda Unidad de longitud del sistema inglés que equivale a 3 pies ó 36 pulgadas.

y-axis (p. 320) The vertical line of the two perpendicular number lines in a coordinate plane.

eje y La recta vertical de las dos rectas numéricas perpendiculares en un plano de coordenadas.

y-coordinate (p. 320) The second number of an ordered pair.

coordenada y El segundo número de un par ordenado.

Glossary/Glosario

Selected Answers

Chapter 1 Number Patterns and Algebra

Page 5   Chapter 1   Getting Started
1. ones **3.** 112 **5.** 109 **7.** 175 **9.** 36 **11.** 94 **13.** 26 **15.** 300 **17.** 630 **19.** 800 **21.** 8 **23.** 42 **25.** 52

Pages 8–9   Lesson 1-1
1. Explore: Determine what facts you know and what you need to find out; Plan: Make a plan for solving the problem and estimate the answer; Solve: Solve the problem; Examine: Check to see if your answer makes sense and compare the answer to the estimate. **3.** Sample answer: 2, 4, 6, 8, 10, …; each number increases by 2. **5.** 320 bacteria **7.** 26, 31, 36 **9.** 3,717,796 sq mi **11.** 10:22 A.M., 11:02 A.M., 11:06 A.M. **13.** 120 **15.** **17.** 21 **19.** 59

Pages 12–13   Lesson 1-2
1. Even numbers are divisible by 2, odd numbers are not. **3.** The number 56 is not divisible by 3 because the sum of the digits, 11, is not divisible by 3. **5.** 3, 5; odd **7.** 2, 3, 4, 6; even **9.** 2, 3, 4, 5, 6, 10; even **11.** 2, 3, 6; even **13.** 2, 3, 6; even **15.** 5; odd **17.** 3, 5; odd **19.** 3, 5, 9; odd **21.** 2, 3, 4, 6; even **23.** 2, 4, 5, 10; even **25.** 2, 3, 4, 6; even **27.** Sample answer: 30 **29.** Sample answer: If another state were added to the Union, there would be a total of 51 stars on the flag. 51 is only divisible by 3 and 17. The only way that the stars could be arranged in rows with the same number of stars would be 3 rows and 17 columns, or 17 rows and 3 columns. **31.** 1, 4, 7 **33.** 0, 1, 2, 3, 4, 5, 6, 7, 8, 9 **35.** 90; To find the number, multiply the different factors. Since $2 \times 3 \times 5 = 30$, you do not need to use 6 or 10 as factors because $2 \times 3 = 6$ and $2 \times 5 = 10$. You then need to multiply by 3 since 30 is not divisible by 9. The answer is then 3×30 or 90. **37.** She can package the cookies in groups of 3. **39.** 63 **41.** sometimes **43.** It is an even number. **45.** 998 **47.** 25 **49.** 12 **51.** 18

Pages 16–17   Lesson 1-3
1. 1, 2, 3, 4, 6, 12 **3.** 57; It has more than 2 factors: 1, 57, 3, 19. **5.** P **7.** P **9.** $3 \times 3 \times 3 \times 3$ **11.** 19 **13.** P **15.** C **17.** P **19.** C **21.** C **23.** P **25.** C **27.** P **29.** 2×19 **31.** $2 \times 2 \times 2 \times 3$ **33.** $2 \times 2 \times 2 \times 5$ **35.** $3 \times 3 \times 3$ **37.** 7×7 **39.** $2 \times 3 \times 7$ **41.** 17 **43.** $2 \times 3 \times 17$ **45.** 5×11 **47.** The number 2 is a prime number because it has two factors, 1 and itself. **49.** 25, 32 **51.** 3 and 5, 5 and 7, 11 and 13, 17 and 19, 29 and 31, 41 and 43, 59 and 61, 71 and 73 **53.** $3 \times 7 \times 17$ **55.** C **57.** 3, 5 **59.** 2, 5, 10 **61.** 8 **63.** 64

Pages 20–21   Lesson 1-4
1a. three to the second power or three squared **1b.** two to the first power or two **1c.** four to the fifth power **3.** 11^2, 6^3, 4^4 **5.** 2^4; 16 **7.** $2 \cdot 2 \cdot 2 \cdot 2 \cdot 2 \cdot 2$; 64 **9.** $2^2 \times 5$ **11.** 9^2; 81 **13.** 3^7; 2,187 **15.** 11^3; 1,331 **17.** 10^4 **19.** $2 \times 2 \times 2 \times 2$; 16 **21.** $5 \times 5 \times 5$; 125 **23.** $9 \times 9 \times 9$; 729 **25.** 8; 8 **27.** 7×7; 49 **29.** $4 \times 4 \times 4 \times 4 \times 4$; 1,024 **31.** $2^3 \times 7$ **33.** $2^2 \times 17$ **35.** 2×7^2 **37.** $3^3 \times 7$ **39.** 1,000,000,000 **41.** 5,832 cubic units **43.** The next value is found by dividing the previous power by 3; 1. **45.** The next value is found by dividing the previous power by 10; 10, 1. **47.** Write the number so that it has the same number of zeros as the exponent. **49.** $9 \cdot 9 \cdot 9 \cdot 9$ **51.** N **53.** P **55.** 2, 3, 6 **57.** 12 **59.** 13

Pages 26–27   Lesson 1-5
1. Sample answer: $25 \div 5 + 2 \times 6$; 17 **3.** 7 **5.** 47 **7.** 5 **9.** 29 **11.** 315 **13.** $40 **15.** 6 **17.** 13 **19.** 61 **21.** 199 **23.** 117 **25.** 35 **27.** 99 **29.** 22 **31.** 38 **33.** $7 \times 6 - 2$ **35.** $104 **37.** 60 g **39.** 27 **41.** $7 \times 7 \times 7 \times 7$; 2,401 **43.** $5 \times 5 \times 3$ **45.** $2 \times 5 \times 13$
47. **49.** 39 **51.** 60

Pages 30–31   Lesson 1-6
1.

Algebraic Expressions	Variables	Numbers	Operations
$7a - 3b$	a, b	7, 3	$\times, -$
$2w - 3x + 4y$	w, x, y	2, 3, 4	$\times, -, +$

3. $3 + 4$; It contains no variable. **5.** 5 **7.** 26 **9.** 100 **11.** 12 **13.** 7 **15.** 4 **17.** 48 **19.** 18 **21.** 51 **23.** 85 **25.** 240 **27.** 50 **29.** 12 **31.** 72 **33.** 15 **35.** 24 **37.** 29 **39.** 127 **41.** 49 **43.** 129 **45.** 1,200 **47.** 24 cm **49.** $9 \times c \div 5 + 32$ **51.** 86°F **53.** D **55.** 9 **57.** 100,000,000 mi **59.** 7 **61.** 106

Pages 36–37   Lesson 1-7
1. the value for the variable that makes the two sides of the equation equal **3.** 8 **5.** 2 **7.** 12 **9.** 15 **11.** 8 **13.** 10 **15.** 23 **17.** 7 **19.** true **21.** 9 ft **23.** 6 **25.** 8 **27.** 3 **29.** 8 **31.** 4.5 **33.** 75 **35.** 5 **37.** 75 **39.** False; this is an equation, and so both sides of the equation must equal the same value. Therefore $m + 8$ must equal 12 and m can only have one solution, 4. **41.** 3, 4, 5 **43.** 5 **45.** H **47.** 18 **49.** 4 **51.** 90 **53.** 85

Pages 40–41   Lesson 1-8
1. Multiply the length and the width. **3.** Sample answer: 3 feet by 8 feet and 2 feet by 12 feet. **5.** 120 ft² **7.** 153 m² **9.** 90 in² **11.** 512 m² **13.** 816 ft² **15.** 418 ft² **17.** Sample answer: 10×10 or 100 in² **19.** 108 in²

21.

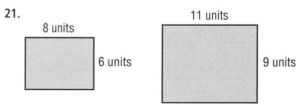

No, the area would not increase by 6 square units.
23. 2,750 ft² **25.** 4 **27.** 43

Pages 42–44   Chapter 1   Study Guide and Review
1. factor **3.** divisible **5.** prime number **7.** solution **9.** 531 votes **11.** 2, 3, 5, 6, 9, 10; even **13.** 3, 5; odd **15.** 2, 4; even **17.** 2, 4, 5, 10; even **19.** 2, 4, 5, 10; even **21.** two packages of 124 candy bars and four packages of 62 candy bars **23.** prime **25.** $3 \times 5 \times 5$ **27.** $2 \times 2 \times 2 \times 11$

29. 1^5; 1 **31.** 3^8; 6,561 **33.** 0 **35.** 116 **37.** 160 **39.** 24 **41.** 54
43. 26 **45.** 44 **47.** 52 **49.** 7 **51.** 25 **53.** 6 **55.** 8 **57.** 48 ft²

Chapter 2 Statistics and Graphs

Page 49 Chapter 2 Getting Started
1. false; order **3.** 44 **5.** 109 **7.** 81 **9.** 8 **11.** 42 **13.** 25 **15.** 13
17. 6 **19.** 51

Pages 52–53 Lesson 2-1
1. Sample answer: Find the least value and the greatest value in a data set and then choose a scale that includes these values. **3.** Sample answer: 26, 27, 28, 32, 34, 35, 37, 40, 40, 40, 45, 46

5.

Pets Owned by Various Students		
Pets	**Tally**	**Frequency**
fish	III	3
turtle	II	2
dog	⫲ IIII	9
gerbil	II	2
cat	⫲ II	7
hamster	I	1

7. 4

9. Sample answer:

Students' Monthly Trips to the Mall		
Trips	**Tally**	**Frequency**
0–2	IIII	4
3–5	⫲	5
6–8	IIII	4
9–11	III	3
12–14	II	2

11.

Favorite Type of Movie		
Type	**Tally**	**Frequency**
comedy	⫲ ⫲ I	11
romantic	II	2
action	⫲ III	8
drama	III	3
horror	IIII	4
science fiction	IIII	4

13. 100; an interval of 10 would make too many intervals and an interval of 1,000 would put all the data values in the same interval. **15.** 1980–1989 and 1990–1999
17. 2 touchdowns: 5; 2 or 4 touchdowns: 7; 2, 4, or 6 touchdowns: 10 **19.** Sample answer: 0 to 20 **21.** 322 ft²
23. 8 gallons

Pages 58–59 Lesson 2-2
1. Sample answer: Whereas a bar graph shows the frequency of each category, a line graph shows how data change over time.

3.

U.S. Endangered Species

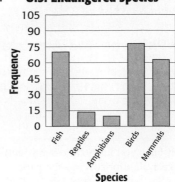

5. Sample answer: The number of endangered fish is about five times as great as the number of endangered reptiles.

7. U.S. Gulf Coast Shoreline

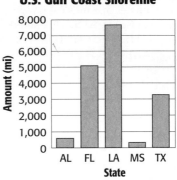

9. Alabama

11. Australia's Population

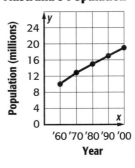

13. Sample answer: Australia's population steadily increased from 1960 to 2000. **15.** Sample answer: scale: 0–80; interval: 10
17. Sample answer: If the vertical scale is much higher than the highest value, it makes the graph flatter. Changing the interval does not affect either type of graph. **19.** G **21.** 24 ft²
23. 54 **25.** 55

Pages 64–65 Lesson 2-3
1. Sample answer: The greatest value is the largest section. The least value is the smallest section. **3.** Sample answer: Frequency table; all of the other are graphs. **5.** Almost three times as many people choose a fast-food restaurant because of quality than prices. **7.** Twice as many libraries stop asking permission at 16–17 years than at 14–15 years old. **9.** siblings and other people **11.** sunburn and yard work **13.** Sample answer: mosquitoes and sunburn
15. Choice c is least appropriate as it does not compare parts of a whole. **17.** natural gas and hydroelectric
19. Sample answer: 20 to 50

21. Distance of Daily Walks

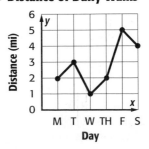

Pages 67–69 Lesson 2-4

1. Sample answer: Line graphs are often used to make a prediction because they show changes over time and they allow the viewer to see data trends, and thus, make predictions. **3.** The world population is increasing. **5.** about 8 billion **7.** Sample answer: The amount of chicken eaten is increasing. **9.** about 275 **11.** Sample answer: As the weeks increase, the time it takes her to run 1,500 meters is decreasing. **13.** At this rate, it will take between 9 and 10 weeks. **15.** about 40° **17.** about 50° warmer

19.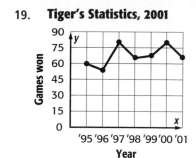

Tiger's Statistics, 2001

21. 1998 **23.** the height of the bounce will increase **25.** Sample answer: the amount of money made in baseball ticket sales over the past 5 years **27.** I **29.** Fall and Spring **31.** 25, 30, 40, 45, 50, 55, 60, 65, 75, 80

Pages 73–75 Lesson 2-5

1. The data can be easily read.

3. Sample answer:

Stem	Leaf
2	7 8 8 8 8 9 9
3	2 4 5
4	0 0 9
5	9

$3\,|\,2 = 32$

5.

Stem	Leaf
5	5
6	
7	0 0 1 5 6 9
8	0 1 2 3 5
9	0 2 3 3 4 9

$9\,|\,9 = 99$

7. 60 yr

9.

Stem	Leaf
0	1 2 7 8 9
1	3 4 6 7
2	4 5 8
3	0 5 9
4	7

$2\,|\,4 = 24$

11.

Stem	Leaf
5	1 4
6	0 1 2 2 3 5 5 7 7 7
7	0 2 2 3 8

$6\,|\,0 = 60$

13.

Stem	Leaf
5	5
6	0 5 6 7 8 8 9
7	0 1 2 3 9 9
8	2 5

$6\,|\,5 = 65°$

15.

Stem	Leaf
1	4 8
2	0 2 3 8 9
3	0 4 8
4	0 2 5 8 8 8
5	2 2 2 5 8
6	0
7	
8	
9	0 3
10	
11	2

$5\,|\,8 = 58\ mi$

17. Sample answer: What is the longest distance between checkpoints? **19.** 14 **21.** Line graph; data changes over time. **23.** Bar graph; data is sorted into categories. **25.** 300 m **27.** Sample answer: Change the stems into intervals of 10. Then count the number of pieces of data in each interval. **29.** G **31.** 1896 to 1904 **33.** 20% **35.** 16 **37.** 14

Pages 77–78 Lesson 2-6

1. Divide the sum of the data by the number of pieces of data. **3a.** 48 **3b.** 54 **5.** 44 **7.** The depth of the Arctic Ocean, 3,953 feet, is an outlier because it is much lower in value than the other depths. **9.** 14 **11.** 80 **13.** 52 **15.** 185 ft **17.** 158 ft **19.** Sample answer: 13, 13, 13, 15, 35, 15, 21 **21.** H **23.** line graph **25.** 147 **27.** 68

Pages 82–83 Lesson 2-7

1. Sample answer: To find the median, order the data from least to greatest and then find the middle number of the ordered data or the mean of the middle two numbers. To find the mode, find the number or numbers that appear most often. **3.** 18; 17; 17; 10 **5.** 8,508; 8,450; none **7.** 930 **9.** 28; 23; none; 28 **11.** 90; 89; none; 12 **13.** Sample answer: The data vary only slightly in value. **15.** $13; $12; $12 **17.** They are the same, 75. **19.** The median temperatures are the same, but the temperatures in Cleveland varied more. **21.** C **23.** 18 **25.** Matt **27.** twice as much or $2.50 more

Pages 88–89 Lesson 2-8

1. Redraw the graph so that it starts at 0. **3.** From the length of the bars, it appears that Willie Mays hit twice as many home runs as Barry Bonds. However Willie Mays hit about 660 home runs, while Barry Bonds hit 611 home runs. **5.** Brand B appears to be half the price of Brand A.

7. Sample answer:

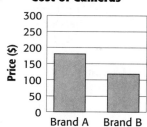

Cost of Cameras

9. The mean because the value of the mean is much higher than most of the pieces of data. **11.** Graph A; It appears that the stock has not increased or decreased rapidly. **13.** A **15.** 75°; 76°; 14° **17.** 19°

Pages 90–92 Chapter 2 Study Guide and Review

1. d **3.** a **5.** f **7.** e **9.**

Favorite Color		
Color	Tally	Frequency
red	IIII	4
blue	̷H̷I̷ II	7
green	II	2
yellow	III	3

11. banana and cinnamon **13.** 9 s

15.

Stem	Leaf
1	8 9
2	2 7 8 9
3	1 5
4	1
5	6

$5\,|\,6 = 56$

17. 27 **19.** 105 **21.** 42; 36; 46 **23.** Graph B; it looks like the company is earning more money.

Chapter 3 Adding and Subtracting Decimals

1. Plan 3. mean 5. 11 7. 12 9. 13 11. 83 13. 80 15. 27 17. 7 19. 80 21. 160

Pages 104–105   Lesson 3-1
1. Word form is the written word form of a decimal. Standard form is the usual way to write a decimal using numbers. Expanded form is a sum of the products of each digit and its place. 3. Seven and five hundredths; whereas all of the other numbers represent 0.75, this number represents 7.05. 5. eight hundredths 7. twenty-two thousandths 9. eight and six thousand two hundred eighty-four ten-thousandths 11. 0.012; $(0 \times 0.1) + (1 \times 0.01) + (2 \times 0.001)$ 13. 49.0036; $(4 \times 10) + (9 \times 1) + (0 \times 0.1) + (0 \times 0.01) + (3 \times 0.001) + (6 \times 0.0001)$ 15. four tenths 17. three and fifty-six hundredths 19. seven and seventeen hundredths 21. sixty-eight thousandths 23. seventy-eight and twenty-three thousandths 25. thirty-six ten-thousandths 27. three hundred one and nineteen ten-thousandths 29. 0.5; (5×0.1) 31. 2.05; $(2 \times 1) + (0 \times 0.1) + (5 \times 0.01)$ 33. 41.0062; $(4 \times 10) + (1 \times 1) + (0 \times 0.1) + (0 \times 0.01) + (6 \times 0.001) + (2 \times 0.0001)$ 35. 0.0083; $(0 \times 0.1) + (0 \times 0.01) + (8 \times 0.001) + (3 \times 0.0001)$ 37. fifty-two hundredths 39. 33.4 41. twenty-three dollars and seventy-nine cents 43. Sample answer: The last digit is in the second place after the decimal point. 45. Sample answer: subdivided into tenths. Decimals are used in the stock market, as in $34.50 per share, or to measure rainfall, as in 1.3 inches. 47. 0.258 49. thirty-four and fifty-six thousandths 51. 88 53. 30 55. B 57. D 59. C

Pages 109–110   Lesson 3-2
1.
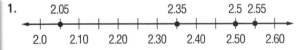

3. Carlos; 0.4 < 0.49 < 0.5 5. < 7. 0.002, 0.09, 0.19, 0.2, 0.21 9. = 11. > 13. > 15. < 17. = 19. > 21. 4.45, 4.9945, 5.545, 5.6 23. 32.32, 32.302, 32.032, 3.99 25. *The Whale, Baleen Whales, The Blue Whale* 27. B 29. 0.037 31. The mode is misleading because it is also the highest number in the set. 33. hundredths 35. thousandths

Pages 112–113   Lesson 3-3
1. To the nearest tenth, 3.47 rounds to 3.5 because 3.47 is closer to 3.5 than to 3.4.

3.4 3.41 3.42 3.43 3.44 3.45 3.46 3.47 3.48 3.49 3.50

3. 34.49; rounds to 34.5 not 34.6 5. 2 7. 0.589 9. $30 11. 7.4 13. $6 15. 2.50 17. 5.457 19. 9.5630 21. $90 23. $67 25. 6 gal 27. Sample answer: 9.95, 10.04, 9.98 29. 1,789.838 31. < 33. 32.05 35. 58 37. 62

Pages 118–119   Lesson 3-4
1. Sample answer: 1.843 is rounded to 2 and 0.328 is rounded to 0 so the answer is 2 − 0 or 2. 3. Sample answer: When you add just the whole number part, the sum is 20. The sum of the decimal part will be added on, which makes the sum greater than 20. 5. 4 − 3 = 1 7. 320

9. $3 \times 1 = 3$ 11. $50 + 16 = 66$ 13. $2 + 5 + 6 = 13$ 15. $0 + 1 = 1$ 17. $75 − $30 = $45 19. 70 21. 99 23. $340 25. 100 27. $12 29. $4 \times $3 = 12 31. $3 \times $55 = 165 33. $4 \times 100 = 400$ 35. $3 \times $20 = 60 37. Sample answer: About how much more could you expect to pay for hockey purchases than for baseball purchases? $18.00 − $15.00 = $3.00 39. C 41. 35.9 in. 43. 477 45. 465 47. 2,183

Pages 123–124   Lesson 3-5
1. Sample answer: Line up the decimal points, then add as with whole numbers. 3. Akiko is correct. Sample answer: she annexed a zero for the first number. 5. 8.7 7. 12.02 9. 4.26 11. 3.75 13. 7.9 15. 13.796 17. 14.92 19. 14.82 21. 2.23 23. 106.865 25. 95.615 27. 14.3 29. 6.7 31. 2.3 33. Sample answer: $2.55 + $3.55 = 6.1 35. Sample answer: I had $2.55 and bought a candy bar for $0.75. How much do I have left? 2.55 − 0.75 = $1.80. 37. 4.7 39. Sample answer: No, the answer depends on the trends and style of the car. 41. 0.2 − 0.1876543 = 0.0123457 43. I 45. Sample answer: 4 + 2 + 4 = 10 47. Sample answer: $11 − $6 =$5 49. 41 51. 16

Pages 127–128   Chapter 3   Study Guide and Review
1. false, less 3. false, thousandths 5. true 7. 6.5, $(6 \times 1) + (5 \times 0.1)$ 9. 53.175 11. < 13. < 15. 0.0004 17. 38 + 14 = 52 19. 7 + 4 + 4 = 15 21. 89 23. 170 25. 48 27. 4 29. 27.57 31. 116.72

Chapter 4 Multiplying and Dividing Decimals

1. round 3. $10 \times 10 = 100$ 5. $10 \times 10 \times 10 \times 10 \times 10 = 100,000$ 7. 4 9. 28 11. 38 13. 34 15. 13.7 17. 23.4

Pages 137–138   Lesson 4-1
1. Sample answer: Multiply as with whole numbers, then estimate to determine where the decimal point should be. The counting method is by counting the number of places to the right of the decimal in the factor then placing the decimal the same number of spaces counting from right to left in the product. 3. Kelly is correct. There are two places to the right of the decimal in the factor so the product should have the decimal two places from the right. 5. 4.2 7. 1.56 9. 3.6 11. 0.072 13. 374.1 15. 8.4 17. 3.9 19. 6.3 21. 14.48 23. 2.6 25. 16.2 27. 0.08 29. 0.0324 31. 25.6 in^2 33. 27.9 cm^2 35. 160.1 37. 4,000,000 39. 930,000 41. 2,170,000 43. $239.88 45. $83.88 47. 5 49. B 51. 406.593 53. Sample answer: 42 − 2 = 40 55. 1,075 57. 2,970

Pages 142–143   Lesson 4-2
1. Sample answer: $1.5 \times 4.25 = 6.375$ 3. 0.3 5. 29.87127 7. 1.092 9. $1.98 11. 4.05 13. 0.68 15. 8.352 17. 166.992 19. 0.0225 21. 19.69335 23. 5.8199 25. 0.109746 27. 9.7 29. Always; Sample answer: $0.9 \times 0.9 = 0.81$, the decimal point will move so the answer is always less than one. 31. 0.75 33. 2.7348 35. G 37. 348.8 39. 47.04 41. 7 43. 7

Pages 146–147   Lesson 4-3
1. Sample answer: Since 40 ÷ 20 = 2, the answer is about 2. 3. Less than one; you are dividing a larger number into a smaller number. 5. 13.1 7. 1.4 9. 0.6 11. $0.55 13. 0.9 15. 0.6 17. 1.2 19. 1.6 21. 10.9 23. 7.0 25. 3.58 min 27. 75 29. Bread: $0.07; Orange juice: $0.11; Cereal: $0.19 31. 24.925 33. Sample answer: 6.324 ÷ 5 35. $27.67 37. 3.68 39. 2.2512 41. 29 43. 9 45. 28.67

Pages 154–155 Lesson 4-4
1. $1.92 \div 0.51 \approx 2 \div 0.5$ or 4 **3.** $5\overline{)35}$; The quotient is 7 and the other problems have a quotient of 0.7. **5.** 12.4 **7.** 0.2 **9.** 12.2 **11.** 0.80 **13.** 45 bricks **15.** 2.3 **17.** 0.4 **19.** 450 **21.** 0.0492 **23.** 22.4 **25.** 0.605 **27.** 52,000 **29.** 0.58 **31.** 0.96 **33.** 0.03 **35.** 0.50 **37.** Sample answer: $330 \div 11 = 30$; 27.68 mpg **39.** $5.6 billion **41.** Never; by dividing by a lesser decimal, there will be more than 1 part. **43.** Sample answer: $3.813 \div 0.82 = 4.65$. **45.** 6.05 **47.** 47.04 **49.** 102.465 **51.** 8 **53.** 16

Pages 159–160 Lesson 4-5
1. Sample answer:

4 in.
3 in.

3. Luanda is correct. To find the perimeter you add all sides not multiply the sides.
5. 78 cm **7.** 264 yd **9.** 88 ft **11.** 48.6 m **13.** 24 in. **15.** 28 cm **17.** 4 **19.** 160 m **21.** 346 yd **23.** 3.02 **25.** 4.06 **27.** 391 **29.** 37.12

Pages 163–164 Lesson 4-6
1.

center

3. Alvin is correct. Jerome did not multiply the radius by 2. **5.** 65.9 ft **7.** 2.7 yd **9.** 11 in. **11.** 38.9 m **13.** 18.8 ft **15.** 131.9 mm **17.** 15.1 in. **19.** about 785 feet **21.** about 669.8 feet **23a.** 1 mi **23b.** 5 m **25.** No; since the diameter is a whole number estimate, the circumference cannot be more precise than about 37 centimeters. **27.** B **29.** 17.4 in. **31.** 82 yd **33.** 12.8

Pages 166–168 Chapter 4 Study Guide and Review
1. radius **3.** divisor **5.** perimeter **7.** 10 **9.** 8.4 **11.** 3.28 **13.** 2.40 **15.** 39.9 **17.** $119.85 **19.** 0.78 **21.** 0.204 **23.** 15.77 **25.** 0.000252 **27.** 0.34 **29.** 3.9 **31.** 0.882 **33.** $8.75 **35.** 1.8 **37.** 14.7 **39.** 65 **41.** 0.62 **43.** 26 in. **45.** 106.2 ft **47.** 33.6 in **49.** 50.2 m **51.** 391.2 ft **53.** 351.7 ft

Chapter 5 Fractions and Decimals

Page 175 Chapter 5 Getting Started
1. factors **3.** none **5.** 5 **7.** $3 \times 3 \times 7$ **9.** $2 \times 2 \times 2 \times 3 \times 5$ **11.** 5.3 **13.** 0.2 **15.** 0.4 **17.** 0.8

Pages 179–180 Lesson 5-1
1. 24: 1, 2, 3, 4, 6, 8, 12, 24; 32: 1, 2, 4, 8, 16, 32; 1, 2, 4, 8 **3.** Sample answer: 11 and 19. Their GCF is one because the numbers have no other number in common than one. **5.** 8 **7.** 10 **9.** pine seedlings: 2 rows; oak seedlings: 3 rows; maple seedlings: 5 rows **11.** 4 **13.** 5 **15.** 1 **17.** 12 **19.** 15 **21.** 17 **23.** 3 **25.** Sample answer: 45, 60, 90 **27.** 14 and 28 **29.** 12 **31.** Sample answer: 16 and 32 **33.** limes: 3 bags; oranges: 7 bags; tangerines: 5 bags **35.** D **37.** 31.4 m **39.** 38.7 m **41.** 5 **43.** 3

Pages 184–185 Lesson 5-2
1. Sample answer: $\frac{2}{8}, \frac{4}{16}, \frac{8}{32}$ **3.** 9 **5.** 5 **7.** simplest form **9.** 4 **11.** 3 **13.** 18 **15.** 6 **17.** $\frac{2}{3}$ **19.** simplest form **21.** $\frac{6}{7}$ **23.** $\frac{5}{19}$ **25.** simplest form **27.** $\frac{4}{11}$ **29.** $\frac{3}{4}$ **31.** $\frac{12}{100}$ **33.** 20

35. Sample answer: One fourth of the people surveyed prefer cats as pets. **37.** $\frac{20}{28}, \frac{10}{14}, \frac{40}{56}$ **39.** $\frac{12}{18}, \frac{28}{42}, \frac{40}{60}$
41. Sample answer: $\frac{10}{16}$;
43. 15 **45.** 42.1 m **47.** 3.2 **49.** 7.4

Pages 188–189 Lesson 5-3
1. $\frac{8}{5} = 1\frac{3}{5}$
3. If the numerator is less than the denominator, the fraction is less than 1. If the numerator is equal to the denominator, the fraction is equal to 1. If the numerator is greater than the denominator, the fraction is greater than 1.
5. $2\frac{3}{4} = \frac{11}{4}$
7. $\frac{9}{5}$ **9.** $5\frac{1}{6}$ **11.** $2\frac{5}{8}$
13.
$3\frac{1}{4} = \frac{13}{4}$
15. $2\frac{7}{8} = \frac{23}{8}$
17. $\frac{19}{3}$ **19.** $\frac{17}{5}$ **21.** $\frac{13}{8}$ **23.** $\frac{29}{4}$ **25.** $\frac{23}{6}$ **27.** $\frac{48}{7}$ **29.** $\frac{33}{5}$ **31.** $9\frac{2}{3}$ **33.** $3\frac{1}{5}$ **35.** $1\frac{1}{8}$ **37.** $1\frac{7}{9}$ **39.** $5\frac{5}{6}$ **41.** $3\frac{4}{7}$ **43.** $4\frac{7}{10}$ **45.** $11\frac{1}{9}$ **47.** $1\frac{1}{2}$ **49.** $\frac{88}{60}, \frac{76}{60}, \frac{84}{60}, \frac{69}{60}$ **51.** According to the definition of improper fraction, an improper fraction is a fraction whose numerator is greater than or equal to the denominator. So, by definition, the fraction is an improper fraction. **53.** I **55.** 3 **57.** 2 **59.** $2 \cdot 2 \cdot 3 \cdot 3$ **61.** $3 \cdot 5 \cdot 7$

Pages 196–197 Lesson 5-4
1. a. yes **b.** no **c.** no **d.** yes **3.** Ana is correct. Whereas Ana found the LCM, Kurt found the GCF. **5.** 40 **7.** 36 **9.** 10 **11.** 12 **13.** 60 **15.** 80 **17.** 40 **19.** 75 **21.** 105 **23.** 225, 270, 315 **25.** 3 rotations **27.** 60 **29.** The numerators are all multiples of 5 and the denominators are multiples of 8. **31.** C **33.** $9\frac{2}{5}$ **35.** 51 **37.** 15 **39.** D **41.** C **43.** F

Pages 200–201 Lesson 5-5
1. a. 8 **b.** 20 **c.** 18 **3.** $\frac{3}{5} - c$, $\frac{9}{10} - d$, $\frac{3}{4} - a$, $\frac{9}{20} - b$; $\frac{9}{20}, \frac{3}{5}, \frac{3}{4}, \frac{9}{10}$
5. > **7.** < **9.** $\frac{1}{4}, \frac{3}{8}, \frac{2}{3}, \frac{5}{6}$ **11.** > **13.** > **15.** = **17.** = **19.** They are equal. **21.** $\frac{3}{4}$ of a dollar **23.** $\frac{2}{9}, \frac{11}{18}, \frac{2}{3}, \frac{5}{6}$ **25.** $\frac{1}{2}, \frac{9}{16}, \frac{5}{8}, \frac{3}{4}$

27. $\frac{4}{5}, \frac{3}{8}, \frac{5}{16}, \frac{1}{4}$ 29. tobacco smoke 31. $\frac{5}{7}$ 33. B 35. 12
37. 216 39. $\frac{1}{4}$ 41. 4.6 43. 0.025

Pages 204–205 Lesson 5-6
1. Sample answer: 0.04 3. $\frac{2}{5}$ 5. $\frac{21}{40}$ 7. $1\frac{1}{4}$ 9. $\frac{7}{10}$ 11. $\frac{1}{2}$
13. $\frac{21}{100}$ 15. $\frac{41}{50}$ 17. $\frac{17}{40}$ 19. $\frac{1}{250}$ 21. $7\frac{2}{25}$ 23. $65\frac{117}{500}$ 25. $13\frac{7}{25}$
27. zoo: $\frac{4}{5}$ of a mile; camping: $\frac{1}{2}$ of a mile; hotels: $\frac{1}{5}$ of a mile
29. Sample answer: If you are staying at the camping area, how far away is the zoo? 31. $11\frac{8}{10}$ or $11\frac{4}{5}$ mi 33. Sample answer: The orange sulphur butterfly has a wingspan of 1.625 to 2.375 inches. It's wingspan is considerably longer than the western tailed-blue butterfly. 35. C 37. >
39. = 41. 300 43. 83 45. 379 students 47. 0.75 49. 0.25

Pages 208–209 Lesson 5-7
1. a. $0.\overline{5}$ b. $4.\overline{345}$ c. $13.6\overline{7}$ d. $25.15\overline{4}$ 3. Sample answer: 0.34 and $0.\overline{34}$. The digits in a terminated decimal terminate or stop. The digits in a repeating decimal repeat. 5. 0.75
7. 5.8 9. 0.0625 c 11. $0.1\overline{3}$ 13. 0.5625 15. $0.41\overline{6}$ 17. 0.09
19. 10.9 21. 4.45 23. $6.2\overline{6}$ 25. $12.\overline{5}$ 27. 0.025 29. 12.0032
31. 2.375 mi 33. $2.\overline{6}$ 35. < 37. $\frac{3}{20}, \frac{1}{6}, \frac{1}{5}, \frac{2}{9}$ 39. striate bubble
41. scotch bonnet 43. If the prime factorization of the denominator contains only 2's or 5's, the fraction is a terminating decimal. 45. I 47. $\frac{73}{100}$ 49. $11\frac{7}{50}$

Pages 210–212 Chapter 5 Study Guide and Review
1. c 3. b 5. a 7. 3 9. 14 11. 9 pieces 13. 56 15. 9 17. $\frac{3}{16}$
19. $\frac{3}{7}$ 21. $\frac{43}{8}$ 23. $\frac{65}{9}$ 25. $5\frac{1}{3}$ 27. $24\frac{1}{2}$ 29. 50 31. 36 33. 48
35. < 37. < 39. $\frac{1}{2}, \frac{5}{9}, \frac{2}{3}, \frac{3}{4}$ 41. $\frac{5}{6}$ c 43. $\frac{7}{20}$ 45. $\frac{1}{8}$ 47. $9\frac{63}{200}$
49. $\frac{6}{125}$ 51. 0.875 53. $0.41\overline{6}$ 55. 12.75 57. 0.75

Chapter 6 Adding and Subtracting Fractions

Page 217 Chapter 6 Getting Started
1. true 3. 1 + 7 = 8 5. 8 − 5 = 3 7. 10 + 10 = 20 9. $\frac{1}{6}$
11. $\frac{2}{5}$ 13. $1\frac{1}{10}$ 15. $1\frac{2}{5}$ 17. $1\frac{1}{4}$ 19. 9

Pages 221–222 Lesson 6-1
1. Sample answer: When you're buying materials for a project, you would round up because rounding down could cause you to run short. 3. 1 5. $\frac{1}{2}$ 7. 0 9. down; so it will be sure to fit 11. 1 13. 4 15. 3 17. 7 19. 2 21. $\frac{1}{2}$ 23. $\frac{1}{2}$ 25. $4\frac{1}{2}$
27. 1 29. $9\frac{1}{2}$ 31. down; so the balloon doesn't break
33. up; so the gift will fit 35. down; shelves too tall won't fit
37. 3 in. 39. 0–2: $\frac{10}{49}$; 3–5: $\frac{18}{49}$; 6–8: $\frac{1}{7}$; 9–11: $\frac{4}{49}$; 12–14: $\frac{6}{49}$;
15–17: $\frac{4}{49}$ 41. Sample answer: $4\frac{3}{8}, 4\frac{7}{16}, 4\frac{5}{8}$ 43. H 45. $0.1\overline{6}$
47. $6.08\overline{3}$ 49. 1 + 1 = 2 51. 2 + 7 + 3 = 12 53. 10 − 4 = 6

Pages 224–225 Lesson 6-2
1. Sample answer: Mark has $4\frac{3}{4}$ gallons of lemonade. If he serves $2\frac{1}{6}$ gallons at a cookout, how much will he have left?
3. $1 - \frac{1}{2} = \frac{1}{2}$ 5. 4 + 6 = 10 7. Sample answer: 3 + 3 = 6 c
9. $1 + \frac{1}{2} = 1\frac{1}{2}$ 11. $1 - \frac{1}{2} = \frac{1}{2}$ 13. 2 − 1 = 1 15. 4 + 2 = 6

17. 4 − 2 = 2 19. 6 + 3 + 3 = 12 21. 24 in. 23. Sample answer: $\frac{3}{8}$ and $\frac{7}{16}$ 25. H 27. $0.8\overline{3}$ 29. $4.\overline{3}$ 31. $1\frac{1}{5}$ 33. $1\frac{1}{3}$ 35. $1\frac{1}{2}$

Pages 230–231 Lesson 6-3
1. To add and subtract fractions with the same denominator, add or subtract the numerators. Write the result using the common denominator. 3. Della is correct. $\frac{3}{8}$ quart of pineapple juice was added to some orange juice to get $\frac{7}{8}$ quart of mixed juice. You must subtract to find the amount of orange juice you started with. 5. $\frac{1}{4}$ 7. $1\frac{1}{2}$ 9. $1\frac{1}{5}$
11. $1\frac{1}{4}$ 13. $\frac{5}{6}$ 15. $\frac{3}{5}$ 17. $1\frac{2}{3}$ 19. $\frac{5}{12}$ 21. 0 23. $1\frac{2}{7}$ 25. $\frac{1}{2}$ in.
27. $\frac{2}{3}$ 29. $\frac{7}{10}$ 31. $\frac{9}{10}$ 33. 10 35. G 37. 1 39. 2 41. $\frac{1}{2}$
43. 53.7 45. 65.8 47. 8 49. 20 51. 100

Pages 237–238 Lesson 6-4
1. Sample answer: $\frac{2}{3}$ and $\frac{4}{5}$; $\frac{10}{15}$ and $\frac{12}{15}$ 3. $\frac{8}{9}$ 5. $\frac{1}{6}$ 7. $\frac{5}{8}$
9. $1\frac{5}{12}$ 11. $\frac{9}{10}$ 13. $\frac{3}{8}$ 15. $\frac{11}{12}$ 17. $\frac{7}{20}$ 19. $\frac{1}{8}$ 21. $1\frac{1}{4}$ 23. $1\frac{5}{24}$
25. $\frac{29}{48}$ 27. $\frac{5}{16}$ in. 29. $1\frac{31}{40}$ 31. $\frac{11}{40}$ 33. $\frac{1}{24}$ 35. Fiction: $\frac{2}{5}$;
Mystery: $\frac{3}{10}$; Romance: $\frac{1}{10}$; Nonfiction: $\frac{1}{5}$ 37. sometimes;
Sample answer: For example, $\frac{1}{2}, \frac{1}{4}$, and $\frac{3}{4}$ are each less than 1. The sum of $\frac{1}{2} + \frac{1}{4}$ is $\frac{3}{4}$, which is less than 1; the sum of $\frac{1}{2} + \frac{3}{4}$ is $1\frac{1}{4}$, which is greater than 1; the sum of $\frac{1}{4} + \frac{3}{4}$ is equal to 1. 39. D 41. $\frac{1}{4}$ gal 43. Sample answer: 4 + 1 = 5
45. Sample answer: $13\frac{1}{2} - 10 = 3\frac{1}{2}$ 47. 3 49. 15

Pages 242–243 Lesson 6-5
1. Sample answer: A rectangle has a length of $3\frac{3}{8}$ feet and a width of $1\frac{1}{4}$ feet. How much longer is the length than the width? $2\frac{1}{8}$ feet 3. $4\frac{1}{2}$ 5. $8\frac{3}{10}$ 7. $15\frac{3}{20}$ 9. $8\frac{7}{12}$ 11. 8 13. $5\frac{2}{5}$
15. $7\frac{1}{4}$ 17. $6\frac{1}{6}$ 19. $7\frac{2}{5}$ 21. $3\frac{4}{9}$ 23. $13\frac{7}{8}$ 25. $4\frac{13}{60}$ 27. $5\frac{1}{5}$ s
29. $11\frac{31}{40}$ 31. $3\frac{1}{2}$ 33. $12\frac{7}{12}$ 35. $25\frac{5}{6}$ ft 37. $\frac{5}{16}$ mi 39. A
41. $\frac{2}{3}$ 43. $1\frac{1}{2}$ 45. $\frac{1}{20}$ 47. $1\frac{13}{20}$ 49. $\frac{7}{5}$ 51. $\frac{11}{8}$ 53. $\frac{25}{12}$

Pages 246–247 Lesson 6-6
1. Sample answer: Model $3\frac{1}{2}$.

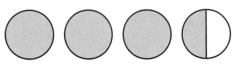

Rename $3\frac{1}{2}$ as $2\frac{12}{8}$.

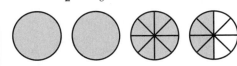

Remove $1\frac{5}{8}$.

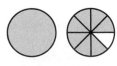

The result is $1\frac{7}{8}$.

3. a. 8 **b.** 6 **c.** 7 **d.** 5 **5.** $3\frac{5}{8} - 2\frac{3}{8}$; In the other cases, the numerator of the first fraction is smaller than the numerator of the second fraction. **7.** $1\frac{3}{4}$ **9.** $1\frac{13}{20}$ **11.** $1\frac{1}{2}$ **13.** $1\frac{1}{2}$ **15.** $6\frac{9}{10}$ **17.** $8\frac{5}{8}$ **19.** $3\frac{5}{6}$ **21.** $11\frac{19}{20}$ **23.** $\frac{3}{4}$ in. **25.** $2\frac{7}{8}$ in. **27.** $1\frac{5}{6}$ mi **29.** Sample answer: What is $3\frac{1}{4} - 2\frac{7}{8}$? **31.** $2\frac{3}{4}$ yd **33.** $3\frac{7}{12}$ **35.** $12\frac{13}{24}$ **37.** $3^2 \times 5$ **39.** 3×19

Pages 248–250   Chapter 6   Study Guide and Review
1. 5 **3.** LCD **5.** $8\frac{5}{4}$ **7.** 1 **9.** $6\frac{1}{2}$ **11.** 2 **13.** $1\frac{1}{2}$ in. **15.** $1 - 0 = 0$ **17.** $9\frac{1}{2} - 3 = 6\frac{1}{2}$ **19.** $0 + 3 = 3$ **21.** $4 - 0 = 4$ **23.** $\frac{2}{3}$ **25.** $\frac{4}{7}$ **27.** $1\frac{2}{9}$ **29.** $1\frac{1}{6}$ **31.** $\frac{25}{36}$ **33.** $1\frac{7}{12}$ **35.** $\frac{13}{24}$ **37.** 5 **39.** $2\frac{3}{10}$ **41.** $\frac{5}{9}$ **43.** $17\frac{7}{12}$ **45.** $2\frac{11}{12}$ h **47.** $2\frac{13}{24}$ **49.** $\frac{5}{6}$ **51.** $6\frac{7}{8}$ **53.** $\frac{16}{21}$ **55.** $3\frac{19}{24}$ ft

Chapter 7 Multiplying and Dividing Fractions

Page 255   Chapter 7   Getting Started
1. greatest common factor **3.** multiples **5.** 25.1 **7.** 113.1 **9.** 6 **11.** 10 **13.** 4 **15.** $\frac{13}{7}$ **17.** $\frac{25}{8}$ **19.** $\frac{1}{2}$ **21.** 0

Pages 257–258   Lesson 7-1
1. Sample answer: To estimate $\frac{2}{3}$ of 8, first find $\frac{1}{3}$ of 9, which is 3. So, $\frac{2}{3}$ of 9 is 2×3 or 6. **3.** Less than 14; sample answer: If you round 20 up to 21 when estimating $\frac{2}{3}$ of 20, your estimate will be higher than the exact answer.
For Exercises 5–23, sample answers are given.
5. $\frac{3}{4} \times 20 = 15$ **7.** $\frac{1}{2} \times 0 = 0$ **9.** $3 \times 11 = 33$ **11.** $\frac{1}{4} \times 20 = 5$ **13.** $1 \times 1 = 1$ **15.** $\frac{1}{2} \times 0 = 0$ **17.** $1 \times \frac{1}{2} = \frac{1}{2}$ **19.** $4 \times 3 = 12$ **21.** $5 \times 9 = 45$ **23.** $\frac{1}{2} \times 0 = 0$ **25.** 20 teens **27.** point N **29.** about $6 **31.** $3\frac{1}{4}$ **33.** $4\frac{17}{21}$ **35.** 3 **37.** 2 **39.** 8

Pages 263–264   Lesson 7-2
1. The overlapping shaded area is $\frac{2}{6}$ or $\frac{1}{3}$ of the whole.

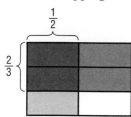

3. No; a fraction times 22 must be less than 22. **5.** $\frac{1}{16}$ **7.** $\frac{1}{4}$ **9.** 9 **11.** $\frac{5}{24}$ **13.** $\frac{2}{15}$ **15.** $1\frac{1}{2}$ **17.** $\frac{1}{6}$ **19.** $\frac{1}{6}$ **21.** $\frac{15}{32}$ **23.** $\frac{2}{9}$ **25.** $\frac{1}{24}$ **27.** $\frac{3}{16}$ **29.** $\frac{3}{10}$ **31.** $\frac{1}{5}$ **33.** 84 lb **35.** 500 mi² **37.** $\frac{1}{100}$ **39.** $\frac{4}{7}$ **41.** Sample answer: $2 \times 5 = 10$ **43.** Sample answer: $\frac{1}{2} \times 1 = \frac{1}{2}$ **45.** Debbie, Camellia, Fala, Sarah **47.** $\frac{13}{4}$ **49.** $\frac{19}{7}$ **51.** $\frac{53}{8}$

Pages 266–267   Lesson 7-3
1. Sample answer: Write the mixed numbers as improper fractions. Simplify if possible before multiplying. Then multiply the numerators and multiply the denominators. **3.** B; the product must be greater than $\frac{2}{3}$ and less than $2\frac{1}{2}$.

5. 1 **7.** $4\frac{9}{10}$ **9.** $\frac{15}{32}$ **11.** $1\frac{1}{6}$ **13.** $1\frac{1}{2}$ **15.** $1\frac{2}{3}$ **17.** 9 **19.** 22 **21.** $1\frac{1}{2}$ **23.** $2\frac{1}{3}$ **25.** $6\frac{1}{8}$ **27.** 6 **29.** $\frac{3}{4}$ **31.** Never; sample answer: A mixed number contains a whole number, and the product of two whole numbers is never a fraction less than 1. **33.** $5\frac{1}{4}$ lb **35.** $\frac{1}{9}$ **37.** $\frac{2}{7}$ **39.** $\frac{3}{32}$ **41.** $\frac{1}{12}$

Pages 274–275   Lesson 7-4
1. Sample answer: There are 6 one-thirds in 2.

3. Sample answer: $\frac{2}{5}$ and $\frac{5}{2}$ **5.** $\frac{3}{2}$ **7.** $\frac{5}{2}$ **9.** $\frac{1}{2}$ **11.** $\frac{2}{5}$ **13.** $\frac{5}{6}$ **15.** $\frac{1}{9}$ **17.** 10 **19.** $\frac{5}{2}$ **21.** $\frac{1}{8}$ **23.** $\frac{8}{3}$ **25.** $\frac{3}{4}$ **27.** 2 **29.** $1\frac{1}{8}$ **31.** 4 **33.** $1\frac{2}{3}$ **35.** $\frac{5}{16}$ **37.** $\frac{2}{7}$ **39.** $1\frac{1}{2}$ **41.** $\frac{2}{3}$ **43.** Sample answers: a) There are 3 half pies left from a bake sale. How much pie is left? $3 \times \frac{1}{2} = 1\frac{1}{2}$ pies b) There are 3 pies left from a bake sale. How many students can take home $\frac{1}{2}$ of a pie? $3 \div \frac{1}{2} = 6$ students **45. a.** $\frac{12}{11}$ **b.** $\frac{12}{11}$ **47.** I **49.** $5\frac{1}{24}$ **51.** $23\frac{1}{3}$ **53.** $\frac{5}{3}, \frac{3}{5}$ **55.** $\frac{9}{2}, \frac{2}{9}$ **57.** $\frac{34}{5}, \frac{5}{34}$

Pages 277–279   Lesson 7-5
1. Sample answer: A spool of ribbon has $12\frac{3}{4}$ yards of ribbon on it. How many pieces each $2\frac{1}{2}$ yards long can be cut from the spool? $12\frac{3}{4} \div 2\frac{1}{2} = 5\frac{1}{10}$, so 5 pieces can be cut from the spool. **3.** $1\frac{3}{4}$ **5.** $11\frac{1}{5}$ **7.** 28 slices **9.** $\frac{5}{12}$ **11.** $2\frac{2}{3}$ **13.** $7\frac{1}{2}$ **15.** 39 **17.** $1\frac{1}{2}$ **19.** $4\frac{1}{26}$ **21.** $\frac{2}{3}$ **23.** $2\frac{2}{5}$ **25.** $2\frac{2}{9}$ **27.** 5 photographs **29.** $\frac{6}{11}$ **31.** $\frac{4}{5}$ **33.** $1\frac{7}{8}$ **35.** 560 mph **37.** 22 mpg **39.** Greater than; $1\frac{2}{3}$ is less than $1\frac{3}{4}$. **41.** D **43.** $23\frac{1}{3}$ U.S. tons **45.** $\frac{3}{4}$ **47.** $8\frac{1}{3}$ **49.** $3\frac{1}{2}$ **51.** $1\frac{1}{2}$

Pages 283–284   Lesson 7-6
1. Each number is $\frac{1}{3}$ of the number before it. **3.** Drake; $1\frac{1}{2}$ is added to each number and 6 plus $1\frac{1}{2}$ is $7\frac{1}{2}$. **5.** multiply by 2; 48, 96 **7.** 29 **9.** subtract 4; 4, 0 **11.** multiply by 2; 128, 256 **13.** multiply by $\frac{1}{3}$; 2, $\frac{2}{3}$ **15.** $24\frac{1}{2}$ **17.** $\frac{1}{3}$ **19. a.** $\frac{1}{2}, \frac{1}{4}, \frac{1}{8}$, $\frac{1}{16}, \frac{1}{32}, \frac{1}{64}, \frac{1}{128}, \frac{1}{256}, \frac{1}{512}, \frac{1}{1,024}$ **b.** 1; The first 9 numbers are contained in the square, which represents 1. **21.** x^5 **23.** 5

Pages 285–286   Chapter 7   Study Guide and Review
1. false; reciprocals **3.** false; multiply **5.** true **7.** false; improper fraction **9.** Sample answer: $\frac{1}{5} \times 20 = 4$ **11.** Sample answer: $1 \times 13 = 13$ **13.** Sample answer: $5 \times 8 = 40$ **15.** Sample answer: $\frac{5}{6} \times 36 = 30$ **17.** $\frac{2}{15}$ **19.** $7\frac{1}{2}$ **21.** $26\frac{1}{2}$ **23.** $4\frac{1}{2}$ **25.** 15 **27.** $\frac{1}{6}$ **29.** $\frac{1}{16}$ **31.** $3\frac{1}{5}$ **33.** Each number is multiplied by 2; 96, 192. **35.** Each number is multiplied by $\frac{1}{5}$; 8, $1\frac{3}{5}$.

Chapter 8 Algebra: Integers

Page 293 Chapter 8 Getting Started

1. product **3.** 27 **5.** 12 **7.** 21 **9.** 8 **11.** 6 **13.** 2 **15.** 42 **17.** 45 **19.** 16 **21.** 8 **23.** 7 **25.** 8

Pages 296–298 Lesson 8-1

1. Sample answer: Graph them on a number line and then write the integers as they appear from right to left.
3. Sample answer: Negative integers are to the left of zero, positive integers are to the right of zero, and zero is neither positive nor negative. **5.** −15
6–9.

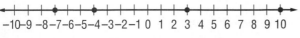

11. < **13.** −5, −4, 0, 3 **15.** −5 **17.** +5 **19.** −5 **21.** +4 or 4
23–30.

31. > **33.** > **35.** > **37.** −7, −5, 4, 6
39. −5, $\frac{1}{2}$, 0.6, $6\frac{3}{4}$, 8

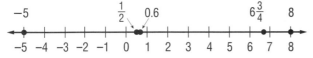

41. −4 **43.** +10 **45.** +3 **47.** −2 **49.** 0 **51.** −45° F

53.

55. Sample answer: the greatest change was in Chapter 11 since its increase was greater than any of the decreases.
57. 2 **59.** 4 **61.** A **63.** 8 **65.** $5\frac{1}{3}$ **67.** 10 **69.** 13

Pages 302–303 Lesson 8-2

1. Sample answer: −8 + (+6) = −2 **3. a.** negative **b.** zero **c.** positive **d.** negative **e.** positive **f.** negative **5.** +6 or 6
7. −10 **9.** +5 or 5 **11.** +3 or 3 **13.** +8 or 8 **15.** +11 or 11
17. −5 **19.** −6 **21.** −1 **23.** −2 **25.** +4 or 4 **27.** +1 or 1
29. −5 **31.** −10 **33.** +5 or 5 **35.** +4 or 4 **37.** −5 + (−4);
−9° F **39.** 7 yd gained **41.** −20° F **43.** Sample answer:
−9 + 2 = −7 **45.** −5
46–48.

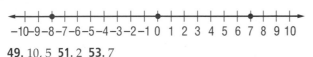

49. 10, 5 **51.** 2 **53.** 7

Pages 306–307 Lesson 8-3

1. Sample answer: −7 − (−4) **3.** Sample answer: The result is the same because 6 − 3 can be rewritten as 6 + (−3). **5.** 3
7. 0 **9.** 1 **11.** 33°F **13.** 1 **15.** 7 **17.** −4 **19.** −5 **21.** 0 **23.** −9
25. 8 **27.** −3 **29.** −20 **31.** 15°F **33.** Sample answer for 2002 MLS Regular Season: New England: 0; Columbus: 1; Chicago: 5; MetroStars: −6; D.C.: −9; Los Angeles: 11; San Jose: 10; Dallas: 1; Colorado: −5; Kansas City: −8
35. sometimes; Sample answer: −6 − (−2) = −4, −3 − (−7) = 4 **37.** never; Sample answer: 6 − (−2) = 8, 8 − (−8) = 16 **39.** 13 **41.** 0

43.

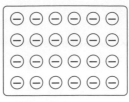

45. 56 **47.** 72

Pages 312–313 Lesson 8-4

1.

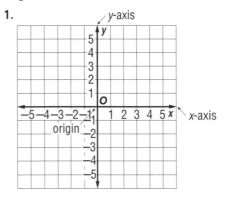

3. −8(6) is not positive. The product of a negative integer and a positive integer is negative. **5.** −56 **7.** 27 **9.** 42
11. −36 **13.** −21 **15.** −35 **17.** −30 **19.** 56 **21.** 40 **23.** 60
25. −49 **27.** −36 **29.** −120 students **31.** −162, −486; Each term is multiplied by 3. **33.** −15 cubic meters **35.** 24
37. never; Sample answer: 4 × 5 = 20 **39.** always; Sample answer: −4 × 2 = −8 **41.** the opposite of the number
43. H **45.** 2 **47.** −13 **49.** 1 **51.** −1 **53.** Sample answer: $\frac{3}{4} \times 36 = 27$ **55.** 6 **57.** 9

Pages 318–319 Lesson 8-5

1. 10 ÷ 5 = 2 **3.** Emily; the quotient of two integers with different signs is negative. **5.** −3 **7.** 5 **9.** −8 **11.** 14 **13.** −3
15. −8 **17.** 5 **19.** −9 **21.** −9 **23.** 9 **25.** −3 **27.** −5
29. 3 points **31.** −1,308,000 acres **33.** B **35.** 21 **37.** $4\frac{5}{8}$
39–42.

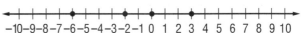

Pages 322–323 Lesson 8-6

1.

3. Sample answer: (0, 3) **5.** K **7.** (1, 4); I **9.** (−2, −5); III
10–12.

13. $(0, 0), (1, 4), (2, 8), (3, 12)$ **15.** R **17.** B **19.** T **21.** $(2, 5)$; I **23.** $(4, -3)$; IV **25.** $(-5, 4)$; II

27–35.

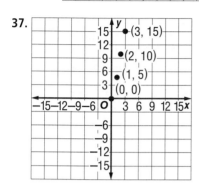

37.

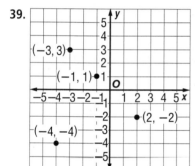

The points appear to fall in a line.

39.

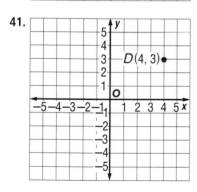

nonlinear

41.

43. -20 **45.** -24

Pages 324–326 Chapter 8 Study Guide and Review
1. b **3.** c **5.** e **7.** g
8–10.

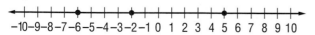

11. $<$ **13.** $<$ **15.** $-2, -1, 0, 3, 4$ **17.** -5 **19.** $+9$ or 9 **21.** -6 **23.** -3 **25.** -4 **27.** 0 **29.** $+12$ or 12 **31.** $+15$ or 15 **33.** $-3°F$ **35.** 3 **37.** 6 **39.** -4 **41.** -2 **43.** -13 **45.** 1 **47.** $15°F$ **49.** -24 **51.** 35 **53.** -32 **55.** 32 **57.** -4 **59.** 9 **61.** -3 **63.** 7 **65.** -4 **67.** $(3, 1)$; I **69.** $(-4, 3)$; II

70–73.

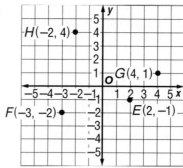

Chapter 9 Algebra: Solving Equations

Page 331 Chapter 9 Getting Started
1. true **3.** 6 **5.** 25 **7.** -2 **9.** 6 **11.** -1 **13.** -8 **15.** 9 **17.** -4 **19.** $-6, 3, 6$ **21.** $-8, -5, -4$

Pages 335–336 Lesson 9-1
1. Sample answer: Using the Distributive Property, you can think of a number in parts that may be easier to solve. **3.** The properties are true for fractions. Examples:

Commutative Property: $\frac{1}{2} + \frac{1}{4} \overset{?}{=} \frac{1}{4} + \frac{1}{2}$

$$\frac{3}{4} = \frac{3}{4} ✔$$

Associative Property: $\frac{1}{2} + \left(\frac{1}{3} + \frac{1}{4}\right) \overset{?}{=} \left(\frac{1}{2} + \frac{1}{3}\right) + \frac{1}{4}$

$$\frac{1}{2} + \left(\frac{4}{12} + \frac{3}{12}\right) \overset{?}{=} \left(\frac{3}{6} + \frac{2}{6}\right) + \frac{1}{4}$$

$$\frac{1}{2} + \frac{7}{12} \overset{?}{=} \frac{5}{6} + \frac{1}{4}$$

$$\frac{6}{12} + \frac{7}{12} \overset{?}{=} \frac{10}{12} + \frac{3}{12}$$

$$\frac{13}{12} = \frac{13}{12} ✔$$

5. 420 **7.** 16.8 **9.** $60(5) + 5(5)$; 325 **11.** Commutative ($\times$) **13.** Identity ($\times$) **15.** 48 **17.** 240 **19.** 105 **21.** 300 **23.** 216 **25.** $7(30) + 7(6)$; 252 **27.** $50(2) + 4(2)$; 108 **29.** Commutative ($+$) **31.** Associative ($+$) **33.** Identity ($+$) **35.** Commutative ($+$) **37.** 59 **39.** 700 **41.** 150 **43.** A: $50.00; B: $56.25 **45a.** 0.32 **45b.** 0.05 **45c.** 7.8 **47.** Associative ($\times$) **49.** -5 **51.** 6 **53.** -1 **55.** -2

Pages 341–342 Lesson 9-2

1.

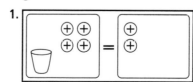

To solve the equation using models, first add two zero pairs to the right side of the mat. Then remove 4 positive counters from each side. The result is $x = -2$.
3. Negative; when you add 14 to the number, the result is less than 0. So the number must be less than 0. **5.** 5 **7.** -10 **9.** $200 + a = 450$; 250 ft **11.** 6 **13.** 3 **15.** -1 **17.** -6 **19.** -8 **21.** 6 **23.** -4 **25.** -2.1 **27.** -6.7 **29.** 2.9 **31.** $\frac{-3}{4}$ **33.** $w + 180 = 600$; 420 **35.** 50,000 lb **37.** c **39.** b **41.** 9 **43.** $30(4 + 0.5)$; 135 **45.** K **47.** $(-3, 3)$; II **49.** $(0, 2)$; none **51.** 24 **53.** -5 **55.** -6 **57.** 2

Pages 346–347 Lesson 9-3

1. Replace the value of the variable in the original equation.
3. Marcus; add 6 to undo subtracting 6. **5.** 14 **7.** 3 **9.** −5
11. $d - 35 = -75$; −40 ft **13.** 6 **15.** 4 **17.** 0 **19.** −7 **21.** 1
23. −3 **25.** 15 **27.** −7 **29.** 4.3 **31.** 3.5 **33.** $-\frac{1}{4}$ **35.** $b - 8 = 15$;
23 yd line or $b - (-8) = 15$; 7 yd line **37.** Sample answer:
How much money did you begin with if you spent
$25 and have $14 left? $x - 25 = 14$; $39 **39.** C
41. $55{,}070 + x = 56{,}000$; 930 **43.** $9\frac{1}{6}$ **45.** $3\frac{1}{5}$

47.

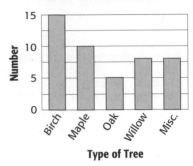

Jamil's Leaf Collection

49. 14 **51.** −3

Pages 351–352 Lesson 9-4

1.

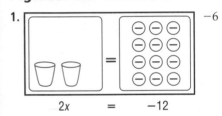

$2x = -12$

3. $3c = 4$; The solution for the other equations is 12. **5.** 3
7. −2 **9.** −4 **11.** 7 **13.** 4 **15.** 7 **17.** −6 **19.** −8 **21.** −2
23. −4 **25.** 8 **27.** 0.5 **29.** −2.5 **31.** 15 **33.** 3 **35.** 2 **37.** 30
39. −1.5 **41.** $10s = 55$; 5.5 **43.** The solution to $4x = 1{,}000$
will be greater because you divide 1,000 by a lesser number
to find the solution. **45.** 12 **47.** −20 **49.** −16 **51.** 90 **53.** F
55. 11 **57.** 0 **59.** −5 **61.** −8 **63.** 47.1 ft **65.** 2.4 m **67.** 10
69. 14

Pages 356–357 Lesson 9-5

1. addition **3.** 4 **5.** 3 **7.** 12 **9.** 3 **11.** −1 **13.** 2 **15.** −2
17. 1 **19.** −2 **21.** 2 **23.** $3 **25.** 7 **27.** G **29.** 3 **31.** −1, 2, 5
33. −1, 0, 1

Pages 364–365 Lesson 9-6

1.

Input (x)	Output (4x)
−4	−16
−2	−8
0	0
4	16

3. Olivia is correct; 5 less than a number is represented
by the expression $x - 5$. **5.** −9, 0, 18 **7.** −2x **9.** −6,−4, 4
11. $x + 2$ **13.** 2x **15.** −3x **17.** $x - 1.6$ **19.** 8 **21.** 3s + 4b

23.

Years (x)	223 million × $10 × x
1	$2,230,000,000
2	$4,460,000,000
3	$6,690,000,000

25. C **27.** $12
29. −6 **31.** 2.75

32–35.

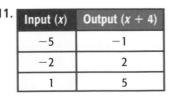

Pages 368–369 Lesson 9-7

1. A table lists selected values; the graph is a line.

3.

Input (x)	Output (x + 5)
−2	3
0	5
2	7

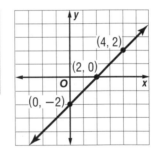

5.

Input (x)	Output (y)
−1	−2
0	0
2	4

function rule: $y = 2x$

7.

Input (x)	Output (x − 2)
0	−2
2	0
4	2

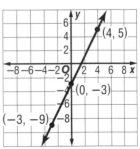

9.

Input (n)	Output (2n − 3)
−3	−9
0	−3
4	5

11.

Input (x)	Output (x + 4)
−5	−1
−2	2
1	5

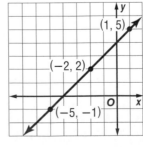

13.

Input (x)	Output ($-4x$)
2	-8
0	0
-2	8

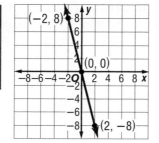

15.

Input (x)	Output (y)
2	-1
4	1
6	3

function rule: $y = x - 3$

17.

Input (x)	Output (y)
-1	-3
1	1
2	3

function rule: $y = 2x - 1$

19. B:$50w - 30$; R: $45w$ **21.** The point of intersection represents when the total earnings are the same for both people.

23.

Input (n)	Output (n^3)
-2	-8
-1	-1
0	0
1	1
2	8

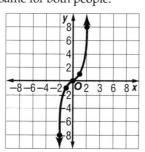

25. A **27.** $x + 2$ **29.** -4

Pages 370–372 Chapter 9 Study Guide and Review

1. Commutative **3.** function **5.** function rule **7.** $4(7) + 4(2)$; 36 **9.** $3(8 + 12)$; 60 **11.** Associative ($+$) **13.** Commutative ($+$) **15.** 3 **17.** 63 **19.** -38 **21.** 12 **23.** 6°F **25.** -3 **27.** 5 **29.** -2 **31.** -11 **33.** 7.2 **35.** 12 **37.** 3 **39.** -14 **41.** 1.5 **43.** 4 **45.** 7 **47.** 6 **49.** 1, 4, 8 **51.** $2x - 1$

53.

Input (x)	Output ($-2x$)
-3	6
-1	2
2	-4

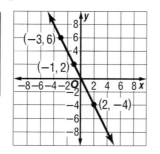

Chapter 10 Ratio, Proportion, and Percent

Page 379 Chapter 10 Getting Started

1. 100 **3.** 7 **5.** 27.72 **7.** 110

9. Sample answer:

11. Sample answer:

13. 0.375 **15.** 0.7 **17.** 90 **19.** 50

Pages 382–383 Lesson 10-1

1. $\frac{2}{5}$, 2 to 5, and 2:5 **3.** Brian is correct; Marta's rate is for 2 weeks and it needs to have a denominator of 1. **5.** $\frac{3}{4}$ **7.** $\frac{3}{7}$ **9.** $\frac{\$3}{1\ \text{case}}$ **11.** The 4-pack of batteries costs $0.90 per battery. The

8-pack costs $0.85 per battery. So, the 8-pack is less expensive per battery. **13.** $\frac{3}{5}$ **15.** $\frac{4}{11}$ **17.** $\frac{2}{9}$ **19.** $\frac{4}{9}$ **21.** $\frac{\$9}{1\ \text{ticket}}$ **23.** $\frac{\$0.12}{1\ \text{egg}}$

25. Sample answer:

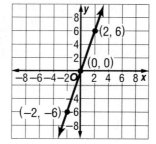

27. $\frac{19}{17}$ **29.** 15 out of 24 **31.** $\frac{4}{7}$ **33.** $x - 2$ **35.** $x + 3$ **37.** 45 **39.** 96

Pages 388–389 Lesson 10-2

1a. Yes; the cross products are equal. **1b.** No; the cross products are not equal. **1c.** Yes; the cross products are equal. **3.** $\frac{36}{44}$; All of the other ratios are equivalent to each other. **5.** 15 **7.** 18 teachers **9.** 21 **11.** 7 **13.** 3 **15.** 5 **17.** 0.3 **19.** 0.4 **21.** 0.8 **23.** $\frac{8}{20} = \frac{x}{300}$ **25.** $\frac{42}{100}$ or $\frac{21}{50}$ **27.** 210 parents **29.** 15 people **31.** 2,190 mi **33.** $2.40 per hot dog

35.

Input	Output ($3n$)
-2	-6
0	0
2	6

37. 6 **39.** 8.5

Pages 392–393 Lesson 10-3

1. Sample answer: The scale is the measurement of the drawing or model in relationship to the actual object. **3.** Greg; when writing a proportion, it is necessary to set up the proportion so that the ratios correspond. In Greg's proportion, the first ratio compares the distance on the map to the actual distance. The second ratio also compares the distance on the map to the actual distance. Jeff's proportion is incorrect because they do not correspond. The first ratio compares the distance on the map to the actual distance, and the second ratio compares the actual distance to the distance on the map. **5.** 4.5 ft **7.** 0.75 ft **9.** 30 ft **11.** 3.5 ft **13.** $1\frac{5}{16}$ inches **15.** Sample answer: It may be best to use a scale of 1:10. A scale of 1:10 means that one inch on the model is equal to 10 inches on the actual motorcycle. This would allow for the model to show more detail of the actual motorcycle. **17.** F **19.** $\frac{1}{9}$

For Exercises 21–25, sample answers are given.

21. **23.** **25.**

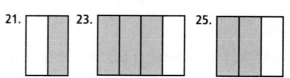

Pages 396–397 Lesson 10-4

1. Sample answer: It means you have half of a whole pizza. **3.** Yes; 43% means 43 out of 100. If Santino has 100 marbles, and gives 43% of them to Michael, he would give 43 of the marbles to Michael. Since 43 is less than 50, it is reasonable to say that Santino gave Michael fewer than 50 marbles.

5. **7.** 20% **9.** 64% **11.**

13. **15.** 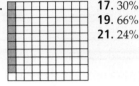 **17.** 30%
19. 66%
21. 24%

23.

25. Sample answer:

An increase of
200% means that
the photograph
is twice the size
of the original.
27. more than 2 h **29.** 6 **31.** 32 **33.** $\frac{27}{50}$ **35.** $\frac{3}{4}$

Pages 402–403   Lesson 10-5
1. Write the percent as a fraction with a denominator of
100, then simplify. **3.** Sample answer: $\frac{3}{5}, \frac{5}{8}, \frac{1}{4}$ **5.** $\frac{4}{5}$ **7.** 25%
9. 225% **11.** $\frac{7}{50}$ **13.** $\frac{1}{50}$ **15.** $1\frac{17}{20}$ **17.** 70% **19.** 125% **21.** 1%
23. 37.5% **25.** 5% **27.** $\frac{49}{50}$ **29.** 72% **31.** $\frac{19}{50}$ **33.** Sample answer:
Of the parents surveyed, $\frac{1}{25}$ feel very pressured. **35.** B

37. **39.**

41. 0.65 **43.** 0.005

Pages 405–406   Lesson 10-6
1. First write the decimal 0.34 as a fraction: $0.34 = \frac{34}{100}$.
Then write the fraction $\frac{34}{100}$ as a percent: $\frac{34}{100} = 34\%$.
3. 0.27 **5.** 0.009 **7.** 32% **9.** 12.5% **11.** 0.12 **13.** 0.06 **15.** 0.35
17. 0.003 **19.** 1.04 **21.** 40% **23.** 99% **25.** 35.5% **27.** 28.7%
29. 0.042 **31.** 0.05 **33.** = **35.** 0.0234, 20.34%, 23.4%, 2.34

37.

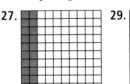

39. 0.0875 **41.** $\frac{19}{50}$ **43.** $\frac{7}{20}$ **45.** 40 **47.** 21

Pages 411–412   Lesson 10-7
1. Change 40% to a decimal and then multiply 0.40 and 65.
3. Gary; 120% does not = 12.0. **5.** 39 **7.** 0.49 **9.** 0.08 **11.** 9
13. 90 **15.** 0.5 **17.** 5.95 **19.** 135 **21.** 0.425 **23.** 206.7 **25.** 54
27. 75% **29.** 52 **31.** 27 states **33.** $69.77 **35.** 35% **37.** 70
39. $66 **41.** B **43.** 135% **45.** $\frac{7}{25}$ **47.** $\frac{17}{20}$ **49.** 10 **51.** 14

Pages 416–417   Lesson 10-8
1. Sample answer: $25\% = \frac{1}{4}$, $50\% = \frac{1}{2}$, and $75\% = \frac{3}{4}$
For Exercises 3–23, sample answers are given.
3. $\frac{2}{5} \times 50 = 20$ **5.** $\frac{3}{4} \times 32 = 24$ **7.** $\frac{1}{5} \times 100 = 20$
9. $\frac{4}{5} \times 80 = 64$ **11.** $\frac{1}{4} \times 120 = 30$ **13.** $\frac{3}{10} \times 160 = 48$
15. $\frac{7}{10} \times 200 = 140$ **17.** $\frac{2}{3} \times 300 = 200$ **19.** $\frac{1}{5} \times \$65$ or $13
21. about 25% **23.** about 50% **25.** Sample answer:
$25 - 9 = 16$ correct answers and $\frac{16}{25}$ is about $\frac{15}{25}$ or $\frac{3}{5}$.
Since $\frac{3}{5} = 60\%$, about 60% of the questions were answered
correctly. **27.** D **29.** 25.8 **31.** 5% **33.** 86.1%

Pages 418–420   Chapter 10   Study Guide and Review
1. false; division **3.** false; $2.50 **5.** false; 74% **7.** true
9. false; 34.6% **11.** $\frac{1}{4}$ **13.** $\frac{6}{7}$ **15.** $\frac{15.75 \text{ pounds}}{1 \text{ week}}$
17. $\frac{26 \text{ candy bars}}{1 \text{ package}}$ **19.** 25 **21.** 30 **23.** 30 ft **25.** 18 ft

27. 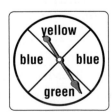 **29.**

31. 25% **33.** $\frac{9}{50}$ **35.** $1\frac{1}{5}$ **37.** 87.5% **39.** 3% **41.** 0.38 **43.** 0.66
45. 0.55 **47.** 130% **49.** 59.1% **51.** 73% **53.** 17 **55.** 10.4
57. 150 **For Exercises 59–63, sample answers are given.**
59. $\frac{3}{4} \times 20 = 15$ **61.** $\frac{1}{5} \times 100 = 20$ **63.** $\frac{2}{5} \times 240 = 96$

Chapter 11   Probability

Page 425   Chapter 11   Getting Started
1. simplest **3.** $\frac{3}{8}$ **5.** $\frac{7}{10}$ **7.** $\frac{3}{7}$ **9.** $\frac{1}{4}$ **11.** $\frac{11}{15}$ **13.** $\frac{3}{10}$ **15.** 2
17. 12 **19.** 9

Pages 430–431   Lesson 11-1
1. Sample answer: the probability of a number cube
landing on 7 **3.** Laura; she correctly listed 5 as the
favorable outcome and it is out of 6 possible outcomes.
5. $\frac{1}{5}$, 0.2, 20% **7.** $\frac{3}{5}$, 0.6, 60% **9.** $\frac{9}{10}$, 0.9, 90% **11.** $\frac{1}{8}$, 0.125, 12.5%
13. $\frac{1}{4}$, 0.25, 25% **15.** $\frac{3}{8}$, 0.375, 37.5% **17.** $\frac{1}{6}$, 0.16$\overline{6}$, 16.$\overline{6}$%
19. $\frac{1}{3}$, 0.33$\overline{3}$, 33.$\overline{3}$% **21.** $\frac{1}{2}$, 0.5, 50% **23.** $\frac{1}{2}$, 0.5, 50%

25. $\frac{5}{6}$, 0.83$\overline{3}$, 83.$\overline{3}$% **27.** Sample answer: Yes, an 85% chance
of rain means that it is likely to rain. So, you should carry
an umbrella. **29.** Since the probability of choosing a yellow
marble is 50%, choosing a yellow marble is equally likely to
happen. **31.** Since the probability of choosing a blue marble
is 0%, choosing a blue marble is impossible to happen.
33. Sample answer:

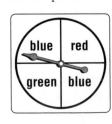

35. 4:2 or 2:1 **37.** 3:3 or 1:1 **39.** 5 **41.** 47.36 **43.** heads, tails **45.** Jan., Feb., Mar., Apr., May, June, July, Aug., Sept., Oct., Nov., Dec.

Pages 435–436 Lesson 11-2

1. the set of all possible outcomes

3.
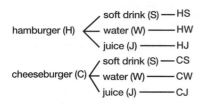
6 outcomes **5.** 12

7.
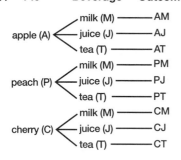
9 outcomes

9.
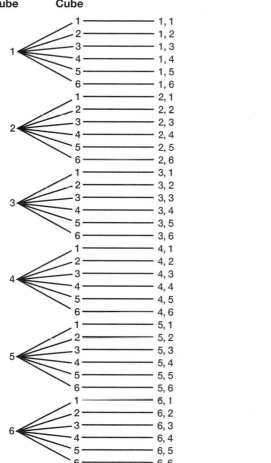
36 outcomes

11. 16 **13.** $\frac{3}{16}$, 0.1875, or 18.75% **15.** $\frac{1}{12}$, 0.08$\overline{3}$, or 8.$\overline{3}$%

17. $\frac{1}{24}$, 0.041$\overline{6}$, or 4.1$\overline{6}$% **19.** 10 ways

21.

8 outcomes

23. 24 **25.** C **27.** $\frac{2}{5}$ **29.** $\frac{3}{5}$ **31.** 2 **33.** 16

Pages 440–441 Lesson 11-3

1. Sample answer: Ask every fifth student entering the school cafeteria. **3.** Sample answer: This is a good survey because the students are selected at random, the sample is large enough to provide accurate data, and the sample is representative of the larger population. **5.** about 155 **7.** Sample answer: This is a good survey because the students are selected at random, the sample is large enough to provide accurate data, and the sample is representative of the larger population. **9.** 8 **11.** 72 **13.** 162 **15.** 15,000 **17.** C **19.** 20 ways **21.** $\frac{11}{12}$, 0.91$\overline{6}$, or 91.$\overline{6}$% **23.** $\frac{1}{4}$ **25.** $\frac{2}{5}$

Pages 446–447 Lesson 11-4

1. Sample answer:

3. $\frac{1}{2}$, 0.5, 50% **5.** about 160 **7.** $\frac{1}{4}$, 0.25, or 25% **9.** $\frac{4}{9}$, 0.$\overline{4}$, or 44.$\overline{4}$% **11.** $\frac{9}{25}$, 0.36, or 36% **13.** about 4 **15.** about 300 **17.** about 6 times **19.** C **21.** Sample answer: This would not be a good sample because people entering a Yankees' baseball game may favor a particular Yankees' baseball player. **23.** 18 **25.** 165 **27.** $\frac{2}{7}$ **29.** $\frac{8}{15}$

Pages 451–452 Lesson 11-5

1. Multiply the probability of the first event by the probability of the second event. **3.** Choosing one marble from a bag of marbles does not represent independent events as there is only one event. **5.** $\frac{1}{10}$, 0.1, or 10% **7.** $\frac{1}{4}$, 0.25, or 25% **9.** $\frac{1}{48}$, 0.021, or 2.1% **11.** $\frac{1}{24}$, 0.041, or 4.1% **13.** $\frac{5}{16}$, 0.3125, or 31.25% **15.** 0.24, 24%, or $\frac{6}{25}$ **17.** $\frac{3}{50}$, 0.06, or 6% **19.** $\frac{1}{50}$, 0.02, or 2% **21.** $\frac{3}{100}$, 0.03, or 3% **23.** $\frac{1}{10}$, 0.1, or 10% **25.** $\frac{9}{50}$, 0.18, or 18% **29.** 0.4 **31.** G **33.** about 9 **35.** $4\frac{17}{21}$ **37.** $3\frac{13}{15}$

Pages 454–456 Chapter 11 Study Guide and Review

1. f **3.** g **5.** a **7.** e **9.** $\frac{3}{4}$, 0.75, 75% **11.** $\frac{1}{2}$, 0.5, 50% **13.** $\frac{1}{2}$, 0.5, 50% **15.** $\frac{1}{3}$, 0.$\overline{3}$, 33.$\overline{3}$%

17.

| Side | Entrée | Outcome | 8 outcomes |

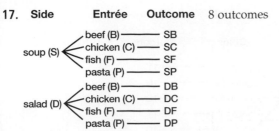

19. 12 **21.** 20 ways **23.** 147 students **25.** $\frac{5}{6}$, $0.83\overline{3}$, or $83.\overline{3}\%$
27. $\frac{21}{40}$, 0.525, or 52.5% **29.** 96 **31.** $\frac{1}{4}$, 0.25, or 25%
33. $\frac{1}{4}$, 0.25, or 25% **35.** 0.18

Chapter 12 Measurement

Page 463 Chapter 12 Getting Started
1. true **3.** 20.69 **5.** 12.96 **7.** 16.52 **9.** 14.17 **11.** 7.11 **13.** 5.14
15. 380 **17.** 67.5 **19.** 718 **21.** 0.098 **23.** 0.45 **25.** 0.073

Pages 467–468 Lesson 12-1
1. Divide 12 by 3. **3.** 12 **5.** 7,040 **7.** •————————• **9.** $\frac{7}{8}$ in.
11. Sample answer: $3\frac{3}{8}$ in., $2\frac{1}{8}$ in. **13.** 18 **15.** 15,840 **17.** $3\frac{1}{3}$

21. •————————•

23. $1\frac{3}{8}$ in. **25.** $\frac{1}{2}$ in. **27.** $1\frac{7}{8}$ in. **29.** 32 in.; $2\frac{1}{2}$ ft is the same
as 30 in., which is less than 32 in. **31.** $6\frac{1}{16}$ in. **37.** 17'
39. 7 yd 1 ft **41.** 3,600 in. **43.** $\frac{3}{5}$, 0.6, or 60% **45.** 80 **47.** 50

Pages 472–473 Lesson 12-2
1. division **5.** 6 **7.** 40 **9.** 56 **11.** 20 days **13.** 32 **15.** 8
17. $3\frac{1}{4}$ **19.** 1.5 **21.** 7 **23.** 9,000 **25.** $\frac{1}{2}$ gal **27.** 60 fl oz;
$3\frac{1}{2}$ pt = 56 fl oz, which is less than 60 fl oz. **29.** 2 fl oz
33. 1,650,000 T **35.** Sample answer: How many 1-cup
servings can be poured from a 1-gallon container of
fruit punch? **37.** 15,625 gal **39.** 32 **41.** H **43.** $\frac{1}{12}$ **45.** $\frac{1}{2}$
47. Sample answer: 1 yd

Pages 478–479 Lesson 12-3
1. Sample answer: millimeter, height of a comma;
centimeter, width of a straw; meter, distance from a
person's nose to the end of their outstretched arm;
kilometer, distance from a museum to a restaurant
3. 23 yd; It is the only measure not given in metric units.
5. kilometer **7.** centimeter **9.** millimeter **11.** meter
13. centimeter **15.** kilometer **17.** 4 cm; 40 mm **19.** 2.6 cm;
26 mm **21.** yard **23.** 3 cm; 15 mm is the same as 1.5 cm,
which is less than 3 cm. **25.** Sample answer: 14.3 cm
27. Sample answer: 7 mm **29.** meter; Sample answer:
The other measurements would be too large or too small,
respectively. **31.** 0.048 km, 4.8 m, 0.48 m, 4.8 cm, 4.8 mm
33. G **35.** 48 **37.** 4 **39.** Sample answer: shredded cheese

Pages 486–487 Lesson 12-4
1. Sample answer: a large water bottle **3.** Sample answer:
gram; 10 g **5.** Sample answer: kilogram; 4 kg **7.** Sample
answer: gram; 60 g **9.** less **11.** More; the total amount
has a mass of about 1,310 paper clips, which is a little more
than the mass of six medium apples. **13.** Sample answer:
gram; 60 g **15.** Sample answer: kilogram; 10 kg **17.** Sample

answer: liter; 8 L **19.** Sample answer: gram; 100 g
21. Sample answer: milligram; 1 mg **23.** Sample answer:
milliliter; 0.2 mL **25.** 1.7 kg; A 1.7-kilogram box of cereal has
a mass a little less than 2 times 6 or 12 apples. A 39-gram
box of cereal has a mass of about 39 paper clips. So, the
1.7-kilogram box of cereal is larger. **27.** See students' work.
The actual capacity is 355 mL. **29.** No; mass and capacity
are not the same. **31.** milliliter **33.** millimeter **35.** 25 **37.** 25

Pages 492–493 Lesson 12-5
1. Multiply by 1,000. **3.** Trina; to change from a smaller
unit to a larger unit, you need to divide. **5.** 0.205
7. 8.5 **9.** 50 **11.** 0.095 **13.** 5.2 **15.** 8,200 **17.** 400 **19.** 3.54
21. 0.0005 **23.** 82 cm **25.** 5,000 m, 700,000 m **27.** turtledove
29. 1,060 mL **31.** 1.06-L bottle **33.** $\frac{1 \text{ km}}{1,000 \text{ m}} = \frac{0.005 \text{ km}}{x \text{ m}}$; 5
35. $\frac{1 \text{ kg}}{1,000 \text{ g}} = \frac{x \text{ kg}}{263 \text{ g}}$; 0.263 **37.** 25900 **39.** 3.3 cm **41.** 8.12
43. 106.4

Pages 496–497 Lesson 12-6
1. $5\frac{1}{2}$ h **3.** 1 h 36 min 76 s; 1 h 36 min 76 s = 1 h 37 min 16 s,
but the others equal 2 h 36 min 16 s. **5.** 7 min 51 s
7. 3 h 25 min **9.** 2 h 54 min **11.** 52 min 55 s
13. 1 h 34 min 23 s **15.** 12 h 34 min **17.** 12 min 39 s
19. 8 h 9 min 8 s **21.** 3 h 44 min 48 s **23.** 3 h 40 min
25. 14 h 10 min **27.** 9:02 P.M. **29.** 12:15 P.M. **31.** 32 **33.** 40
35. 1:22 P.M. **37.** 650 **39.** liter

Pages 498–500 Lesson 12 Study Guide and Review
1. one hundredth **3.** gram **5.** 1,000 **7.** multiply **9.** longer

11. 60 **13.** 2 **15.** 5 **17.** •————————•

21. 11 **23.** 22 **25.** 16 **27.** 12 **29.** meter **31.** centimeter
33. centimeter **For Exercises 35–39, sample answers are
given. 35.** gram; 100 g **37.** milligram; 1 mg **39.** kilogram;
1 kg **41.** 355 mL **43.** 0.001 **45.** 5,020 **47.** 0.0236
49. 3,500 **51.** 6,300 mL **53.** 7 h 36 min **55.** 59 min 52 s
57. 10 h 8 min 10 s **59.** 45 min

Chapter 13 Geometry: Angles and Polygons

Page 505 Getting Started
1. square **3.** C **5.** yes **7.** 2.5 cm

Pages 508–509 Lesson 13-1
1. Sample answer:

3. 125°; obtuse **5.** 90°; right **7.** 30°; acute **9.** 90°; right
11. 130°; obtuse **13.** obtuse **15.** 53° **17.** 1–2 times:
68°; 3–6 times: 126°; Every day: 162°; Not at all: 4°
19. Sample answer: Cut into the hill and flatten the road.
21. vertical **23.** adjacent **25.** Sample answer: 40°, 140°
27. 41 h 35 min **29.** 250

31.

33.

Pages 511–512 Lesson 13-2

1. Sample answer: Draw one side of the angle and label the vertex. Then use a protractor to measure 65° and connect with a straight edge. **3.** The fourth angle does not belong as its measure is about 75°. **5.** about 30°

7. 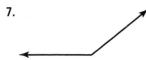 **9.** about 90° **11.** about 150°

13. 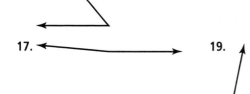 **15.**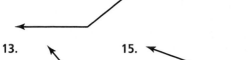

17.  **19.**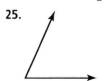

21. Sample answer:

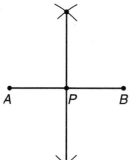 150°

23. Sample answer: A corner of a textbook is a right angle. You can use this angle as a reference to estimate the measure of an angle.

25. 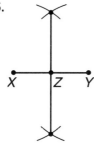 **27.** 6 hr 17 min **29.** 2 cm; 1 cm
31. 3.5 cm; 1.75 cm

Pages 516–517 Lesson 13-3

1. Two angles whose measures are equal.

3. 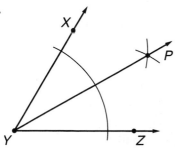 **5.**

7.

9.

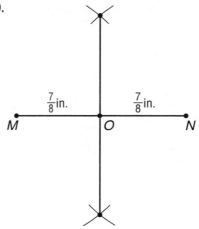

$\frac{7}{8}$ in. $\frac{7}{8}$ in.

M   O   N

11. B **13.** E **15.** $\overline{VW}$ **17.** Yes; $m\angle XVW = m\angle WVY$, $m\angle ZVY = m\angle YVU$, $m\angle YVU = m\angle UVW$ **19.** D

21.  **23.**

25. Sample answer: **27.** Sample answer:

Pages 524–525 Lesson 13-4

1a–1f are all sample answers. **1a.**

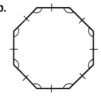

1b.  **1c.**

1d. **1e.**

1f.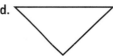

3. octagon; regular
5. square; regular
7. hexagon; regular
9. heptagon; not regular
11. octagon; not regular

13. Sample answer: **15.**

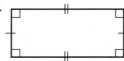 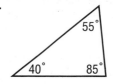

55°
40°   85°

17. Sample answer: similarities—four sides, closed figures, opposite sides parallel, opposite angles equal, all sides equal; differences—a square has four right angles.

19. Sample answer:

21. $8x = 1,080°$; $x = 135°$ **23.** sometimes **25.** always
27. 95° **29.** 115° **31.** D **33.** $\overrightarrow{KA}$

35.

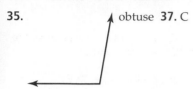

obtuse **37.** C

Pages 530–531   Lesson 13-5
1. Sample answer: Line symmetry is when you can draw a line through the figure so that each half of the figure matches. Rotational symmetry is when you are able to rotate a figure less than one complete 360° turn and the figure looks as it did in its original position.
3. Daniel; Jonas counted each line twice.

5.

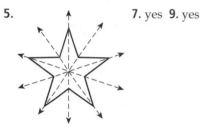

7. yes **9.** yes

11.

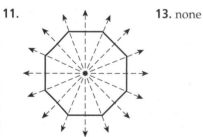

13. none

15.

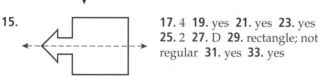

17. 4 **19.** yes **21.** yes **23.** yes
25. 2 **27.** D **29.** rectangle; not regular **31.** yes **33.** yes

Pages 535–536   Lesson 13-6
1. Similarity is when two figures have the exact same shape but not necessarily the same size. Congruency is when two figures have the exact same size and shape.
3. Sample answer:

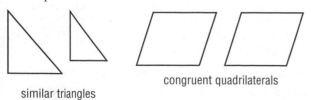

similar triangles   congruent quadrilaterals

5. similar **7.** $\overline{DF}$ **9.** similar **11.** neither **13.** similar
15. A scale model of the Statue of Liberty will have the

same shape as the actual statue, but the model will be smaller. So, they are similar figures. **17.** $\overline{AC}$ **19.** 5 ft
21. Sometimes; rectangles are similar if their corresponding sides are proportional. Two rectangles that are not proportional are not similar, as shown.

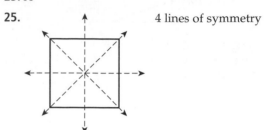

23. A

25.

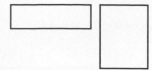

4 lines of symmetry

27.   3 lines of symmetry

Pages 538–540   Chapter 13   Study Guide and Review
1. g **3.** c **5.** f **7.** e **9.** 145°; obtuse

11.   **13.**

15. about 120° **17.**

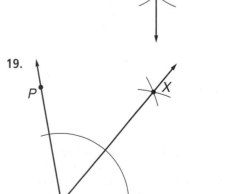

19.

21. $\overline{AB}$ or $\overline{XB}$ **23.** pentagon; regular

25.

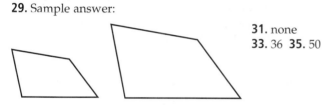

27. none **29.** no
31. yes **33.** 3 cm

Chapter 14 Geometry: Measuring Area and Volume

Page 545 Lesson 14 Getting Started
1. power **3.** circumference **5.** 1.44 **7.** 121 **9.** 100 **11.** 34
13. 35 **15.** 20 **17.** 14 **19.** 216 **21.** 160

Pages 548–549 Lesson 14-1
1. The formula for the area of a parallelogram $A = bh$ corresponds to the formula for the area of a rectangle $A = lw$ in that the base b corresponds to the length l and the height h corresponds to the width w. **3.** 18 units^2
5. 49.1 m^2 **7.** 24 units^2 **9.** 48 m^2 **11.** 31.1 ft^2 **13.** 32.4 m^2
15. Sample answer: 300 in^2 **17.** 1,296 **19.** $400 = b \times 8$ or $400 = 8b$ **21.** It is increased by four times. **23.** Sample answer: The measures of the angles are equal.
25. $\angle 1, \angle 5, \angle 7$ **27.** B
29. Sample answer:

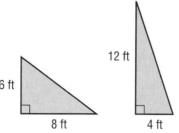

31. none
33. 36 **35.** 50

Pages 553–554 Lesson 14-2
1. Sample answer:

12 ft

6 ft

8 ft

4 ft

3. 6 units^2 **5.** 88.5 m^2 **7.** 24 units^2 **9.** 45 in^2 **11.** 39.4 m^2
13. $1\frac{3}{4}$ in^2 **15.** 25 m^2 **17.** about 19,883 ft^2 **19.** 434,700 mi^2
21. 14 in^2; 16 in. **23.** base: 37.5 units; height: 28 units
25. Yes; the larger triangle is made up of the four smaller triangles. If the area of the larger triangle is 2,100 units^2, then the area of one of the smaller triangles would be 2,100 ÷ 4 or 525 units^2. So, the answer is reasonable.
27. F **29.** similar **31.** congruent **33.** 144 **35.** 2.25

Pages 558–559 Lesson 14-3
1. Sample answer: Since π is close to 3, you can multiply the square of the radius by 3 to estimate the area of a circle. **3.** Crystal; Whitney squared the diameter instead of the radius. **5.** 11.3 yd^2 **7.** 2,640.7 mi^2 **9.** 201.0 cm^2
11. 176.6 m^2 **13.** 69.4 yd^2 **15.** 24,143.8 ft^2 **17.** 14.2 in^2
19. The area is multiplied by 4. **21.** 616 cm^2 **23.** 33.2 ft^2
25. 56 m^2

27.  **29.**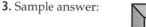

Pages 565–566 Lesson 14-4
1a. 5 **1b.** 1 **1c.** 8 **1d.** 0 **3.** square pyramid
5. 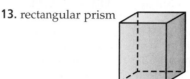 cylinder **7.** triangular pyramid **9.** sphere

11. 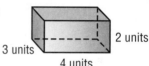 This figure is not a cone because the top of the cone is cut off

13. rectangular prism

15. Sample answer: $\overleftrightarrow{AH}$ and $\overleftrightarrow{BC}$; $\overleftrightarrow{GE}$ and $\overleftrightarrow{DC}$ **17.** F
19. 28.3 m^2 **21.** 988 ft^2 **23.** 43.1

Pages 572–573 Lesson 14-5
1. Each of the three dimensions being multiplied is expressed in a unit of measure.
3. Sample answer:

2 units

3 units

4 units

5. 5.6 yd^3 **7.** 120 m^3 **9.** 600 yd^3 **11.** 46.7 ft^3 **13.** 612.5 m^3
15. 54 **17.** < **19.** = **21.** 9,180 in^3 **23.** 77 in^3 **25.** 706.5 ft^3
27. 423.9 yd^3 **29.** 1,764 in^3

31. Sample answer:

33. about 75 yd^2
35. 38.3

Pages 577–578 Lesson 14-6
1a. volume; sample answer: Capacity is the amount of water inside the lake. **1b.** area; sample answer: the length and width of the house will determine the land area needed to build the house. **1c.** surface area; sample answer: The sum of the areas of the faces of the box tells how much paper is needed to cover the box. **3.** 292 m^2 **5.** 293.3 cm^2
7. 142 ft^2 **9.** 3,600 ft^2 **11.** 117.5 mm^2 **13.** Sample answer:
$(10 \times 11) + (10 \times 11) + (5 \times 10) + (5 \times 10) + (5 \times 11) + (5 \times 11)$ or 430 square inches **15.** $(9.9 \times 21.6) + (9.9 \times 21.6) + (4.7 \times 9.9) + (4.7 \times 9.9) + (4.7 \times 21.6) + (4.7 \times 21.6)$ or 723.78 square inches **17.** Sample answer: A tissue box measures 9 inches by 5 inches by 4 inches. How much cardboard covers the outside of the tissue box? **19.** $S = 6s^2$
21. 150.7 cm^2 **23.** C **25.** 420 ft^3

Pages 579–580 Chapter 14 Study Guide and Review
1. faces **3.** cylinder **5.** surface area **7.** $48\frac{1}{8}$ in^2 **9.** 5.3 m^2
11. 153.9 in^2 **13.** triangular pyramid **15.** sphere
17. $103\frac{1}{8}$ ft^3 **19.** 194.5 cm^2

Photo Credits

Cover (tl)Franco Vogt/Corbis, (tr)Joe Cornish/Getty Images, (b)Noel Hendrickson; **i** Noel Hendrickson; **iv v x** Aaron Haupt; **xi** Frank Lane/Parfitt/Getty Images; **xii** EyeWire/Getty Images; **xiii** AP/Wide World Photos; **xiv** Aaron Haupt; **xv** Beau Regard/Masterfile; **xvi** Joel W. Rogers/CORBIS; **xvii** Alan Thornton/Getty Images; **xviii** David Nardini/Masterfile; **xix** Aaron Haupt; **xx** Pete Saloutos/CORBIS; **xxi** Philip & Karen Smith/Getty Images; **xxii** Art Wolfe/Getty Images; **xxiii** Getty Images; **xxiv** CORBIS; **xxv** John Evans; **xxvi** (t)PhotoDisc, (b)John Evans; **xxvii** (t)PhotoDisc, (bl)PhotoDisc, (br)John Evans; **xxviii** John Evans; **2–3** Chabruken/Getty Images; **4–5** Alaska Stock; **7** AP/Wide World Photos; **11** (l)Karen Thomas/Stock Boston, (r)William Whitehurst/CORBIS; **12** Joseph Sohm; ChromoSohm Inc./CORBIS; **13** Brooke Slezak/Getty Images; **17** Frank Lane/Parfitt/GettyImages; **19** Tony Freeman/PhotoEdit; **20** Bill Frymire/Masterfile; **23** John Evans; **25** CORBIS; **27** Kreber/KS Studios; **32** John Evans; **35** Mac Donald Photography/Photo Network; **40** Chris Salvo/Getty Images; **48–49** SuperStock; **50** Rich Iwasaki/Getty Images; **51** Pierre Tremblay/Masterfile; **54** John Evans; **57** Allan Davey/Masterfile; **66** PhotoDisc; **67** John Terence Turner/Getty Images; **71** John Evans; **73** EyeWire/Getty Images; **74** Vloo Phototheque/Index Stock Imagery/PictureQuest; **76** PhotoDisc; **77** F. Stuart Westmorland/Photo Researchers; **78** David Weintraub/Stock Boston; **80** Digital Vision/Getty Images; **81** Joesph Sohm, ChromoSohm/Stock Connection/PictureQuest; **87** (l)Stephen Simpson/Getty Images, (r)Andrew J.G. Bell; Eye Ubiquitous/CORBIS; **96–97** Alexander Walter/Getty Images; **98–99** Kreber/KS Studios; **105** Matt Meadows; **108** AP/Wide World Photos; **110** Doug Martin; **111** Matt Meadows; **112** Scott Tysick/Masterfile; **113** PhotoDisc; **114** Tom Carter/Photo Edit; **115** John Evans; **117** CORBIS; **122** (l)Creasource/Series/PictureQuest, (r)AP/Wide World Photos; **124** IFA/eStock Photography/PictureQuest; **125** John Evans; **132–133** Lloyd Sutton/Masterfile; **135** Stephen Marks/Getty Images; **136** Photo Network; **138 141** Aaron Haupt; **142** Mug Shots/CORBIS; **147** Craig Hammell/CORBIS; **149 156** John Evans; **161** Photick/SuperStock; **163** Carolyn Brown/Getty Images; **172–173** Art Wolfe/Getty Images; **174–175** Tom Stack & Associates; **178** Aaron Haupt; **179** Mark Tomalty/Masterfile; **180** (t)Doug Martin, (b)Mark Steinmetz; **183** Aaron Haupt; **184** Mark Steinmetz; **187** James Urbach/SuperStock; **188** Bat Conservation International; **189** Beau Regard/Masterfile; **191** John Evans; **192** (l)John Evans, (r)Kreber/KS Studios; **195** Ken Lucas/Visuals Unlimited; **197** (t)Karen Tweedy-Holmes/CORBIS, (b)Gary W. Carter/CORBIS; **198** Matt Meadows; **202** Morrison Photography; **205** Richard Cech; **207** Dr. Darlyne A. Murawski/Getty Images; **209** (l)Fred Atwood, (c)Ronald N. Wilson, (r)Billy E. Barnes/PhotoEdit; **216–217** Kreber/KS Studios; **220** Charles Gupton/CORBIS; **221** Glencoe; **222** (l)Mark Burnett, (r)Glencoe; **223** Bethel Area Chamber of Commerce;

225 Joanna McCarthy/Getty Images; **226** (l)John Evans, (r)Laura Sifferlin; **227** Rita Maas/Getty Images; **229** Richard Stockton/Index Stock; **233** John Evans; **235** Doug Martin; **236** David Becker/Getty Images; **240** Doug Martin; **242** Jeff Gross/Getty Images; **244** Richard Laird/Getty Images; **245** Joel W. Rogers/CORBIS; **247** Mark Burnett; **254–255** Roy Ooms/Masterfile; **261** Stocktrek/CORBIS; **263** Ron Kimball/Ron Kimball Stock; **264** Maps.com/CORBIS; **266** Doug Martin; **269** John Evans; **270** Aaron Haupt; **272** PhotoDisc; **273** Alan Thornton/Getty Images; **275** CORBIS; **277** Aaron Haupt; **280** John Evans; **282** PhotoDisc; **288** United States Mint; **290–291** Andrew Wenzel/Masterfile; **292–293** Tim Davis/Getty Images; **295** (t)NASA/JPL/Malin Space Science Systems, (c)NASA/GSFC, (b)PhotoDisc; **302** Aaron Haupt; **303 307** AP/Wide World Photos; **309** John Evans; **311** (l)Tom Stack & Associates, (r)David Nardini/Masterfile; **313** (t)Reuters NewMedia Inc./CORBIS, (b)AP/Wide World Photos; **314** Laura Sifferlin; **319** Pierre Tremblay/Masterfile; **322** CORBIS; **330–331** Aaron Haupt; **333** Doug Martin; **334** Sandy Felsenthal/CORBIS; **336** Carin Krasner/Getty Images; **341** Tim Fuller; **344** Laura Sifferlin; **347** David Schultz/Getty Images; **349** John Evans; **350** Laura Sifferlin; **356** Doug Martin; **358** John Evans; **362** Joe McDonald/CORBIS; **363** Richard T. Nowitz/CORBIS; **365 366** Aaron Haupt; **376–377** DUOMO/CORBIS; **378–379** JH Pete Carmichael/Getty Images; **381** Joe McDonald/CORBIS; **387** Richard Hutchings/PhotoEdit; **388** Aaron Haupt; **392** Chad Ehlers/Getty Images; **393** Roger K. Burnard; **395** Doug Martin; **399** John Evans; **401** Pete Saloutos/CORBIS; **406** Enzo & Paolo Ragazzini/CORBIS; **413** (l)John Evans, (r)Laura Sifferlin; **415** Doug Martin; **424–425** Mike Powell/Getty Images; **429** Philip & Karen Smith/Getty Images; **439** (l)Jim Cummins/CORBIS, (r)Aaron Haupt; **440** (t)Bob Mullenix, (b)PhotoDisc; **443** John Evans; **445** Doug Martin; **448** John Evans; **452** PhotoDisc; **460–461** Martyn Goddard/CORBIS; **462–463** KS Studios; **471** Life Images; **483** John Evans; **485** Micheal Simpson/Getty Images; **486** (l)Mike Pogany, (r)Art Wolfe/Getty Images; **488** John Evans; **491** John Kelly/Getty Images; **492** PhotoDisc; **493** Mark Ransom; **494** Tim Fuller; **495** Billy Stickland/Getty Images; **504–505** (bkgd)Rob Bartee/SuperStock; **504** (l)Mark Burnett, (t)David Pollack/CORBIS, (r)Doug Martin; **508** (l)Mark Burnett, (r)CORBIS; **512** Doug Martin; **515** Matt Meadows; **517** PhotoDisc; **519 520** John Evans; **523** Geoff Butler, (bkgd)PhotoDisc; **524** (t)Mark Gibson, (b)David Pollack/CORBIS; **528** (l)Matt Meadows, (c)PhotoDisc, (r)Davies & Starr/Getty Images; **535** Getty Images; **539** Aaron Haupt; **541** Matt Meadows; **544–545** Tim Hursley/SuperStock; **547** Tony Hopewell/Getty Images; **551** Aaron Haupt; **557** CORBIS; **563** John Evans; **564** (l)Elaine Comer Shay, (r)Dominic Oldershaw; **568** John Evans; **569** Dominic Oldershaw; **571** CORBIS; **576** Denis Scott/CORBIS; **584** CORBIS.

Index

Index **671**

Index

Index

Index

Index

Index

Index

Index **679**

Symbols

Number and Operations

$+$	plus or positive
$-$	minus or negative
$a \cdot b$ $a \times b$ ab or $a(b)$	a times b
$\div$	divided by
$\pm$	plus or minus
$=$	is equal to
$\neq$	is not equal to
$>$	is greater than
$<$	is less than
$\geq$	is greater than or equal to
$\leq$	is less than or equal to
$\approx$	is approximately equal to
$\%$	percent
$a:b$	the ratio of a to b, or $\frac{a}{b}$
$0.7\overline{5}$	repeating decimal $0.75555\ldots$

Algebra and Functions

$-a$	opposite or additive inverse of a
a^n	a to the nth power
a^{-n}	$\frac{1}{a^n}$
$|x|$	absolute value of x
$\sqrt{x}$	principal (positive) square root of x
$f(n)$	function, f of n

Geometry and Measurement

$\cong$	is congruent to
$\sim$	is similar to
$^\circ$	degree(s)
$\overleftrightarrow{AB}$	line AB
$\overrightarrow{AB}$	ray AB
$\overline{AB}$	line segment AB
AB	length of $\overline{AB}$
$\llcorner$	right angle
$\perp$	is perpendicular to
$\parallel$	is parallel to
$\angle A$	angle A
$m\angle A$	measure of angle A
$\triangle ABC$	triangle ABC
(a, b)	ordered pair with x-coordinate a and y-coordinate b
O	origin
π	pi $\left(\text{approximately } 3.14 \text{ or } \frac{22}{7}\right)$

Probability and Statistics

$P(A)$	probability of event A
$n!$	n factorial
$P(n, r)$	permutation of n things taken r at a time
$C(n, r)$	combination of n things taken r at a time

Formulas

Perimeter	square	$P = 4s$
	rectangle	$P = 2\ell + 2w$ or $P = 2(\ell + w)$
Circumference	circle	$C = 2\pi r$ or $C = \pi d$
Area	square	$A = s^2$
	rectangle	$A = \ell w$
	parallelogram	$A = bh$
	triangle	$A = \frac{1}{2}bh$
	trapezoid	$A = \frac{1}{2}h(b_1 + b_2)$
	circle	$A = \pi r^2$
Surface Area	cube	$S = 6s^2$
	rectangular prism	$S = 2\ell w + 2\ell h + 2wh$
	cylinder	$S = 2\pi rh + 2\pi r^2$
Volume	cube	$V = s^3$
	prism	$V = \ell wh$ or Bh
	cylinder	$V = \pi r^2 h$ or Bh
	pyramid	$V = \frac{1}{3}Bh$
	cone	$V = \frac{1}{3}\pi r^2 h$ or $\frac{1}{3}Bh$
Pythagorean Theorem	right triangle	$a^2 + b^2 = c^2$
Temperature	Fahrenheit to Celsius	$C = \frac{5}{9}(F - 32)$
	Celsius to Fahrenheit	$F = \frac{9}{5}(C + 32)$

Measurement Conversions

Length	1 kilometer (km) = 1,000 meters (m) 1 meter = 100 centimeters (cm) 1 centimeter = 10 millimeters (mm)	1 foot (ft) = 12 inches (in.) 1 yard (yd) = 3 feet or 36 inches 1 mile (mi) = 1,760 yards or 5,280 feet
Volume and Capacity	1 liter (L) = 1,000 milliliters (mL) 1 kiloliter (kL) = 1,000 liters	1 cup (c) = 8 fluid ounces (fl oz) 1 pint (pt) = 2 cups 1 quart (qt) = 2 pints 1 gallon (gal) = 4 quarts
Weight and Mass	1 kilogram (kg) = 1,000 grams (g) 1 gram = 1,000 milligrams (mg) 1 metric ton = 1,000 kilograms	1 pound (lb) = 16 ounces (oz) 1 ton (T) = 2,000 pounds
Time	1 minute (min) = 60 seconds (s) 1 hour (h) = 60 minutes 1 day (d) = 24 hours	1 week (wk) = 7 days 1 year (yr) = 12 months (mo) or 52 weeks or 365 days 1 leap year = 366 days
Metric to Customary	1 meter $\approx$ 39.37 inches 1 kilometer $\approx$ 0.62 mile 1 centimeter $\approx$ 0.39 inch	1 kilogram $\approx$ 2.2 pounds 1 gram $\approx$ 0.035 ounce 1 liter $\approx$ 1.057 quarts